EARTH'S CLIMATE

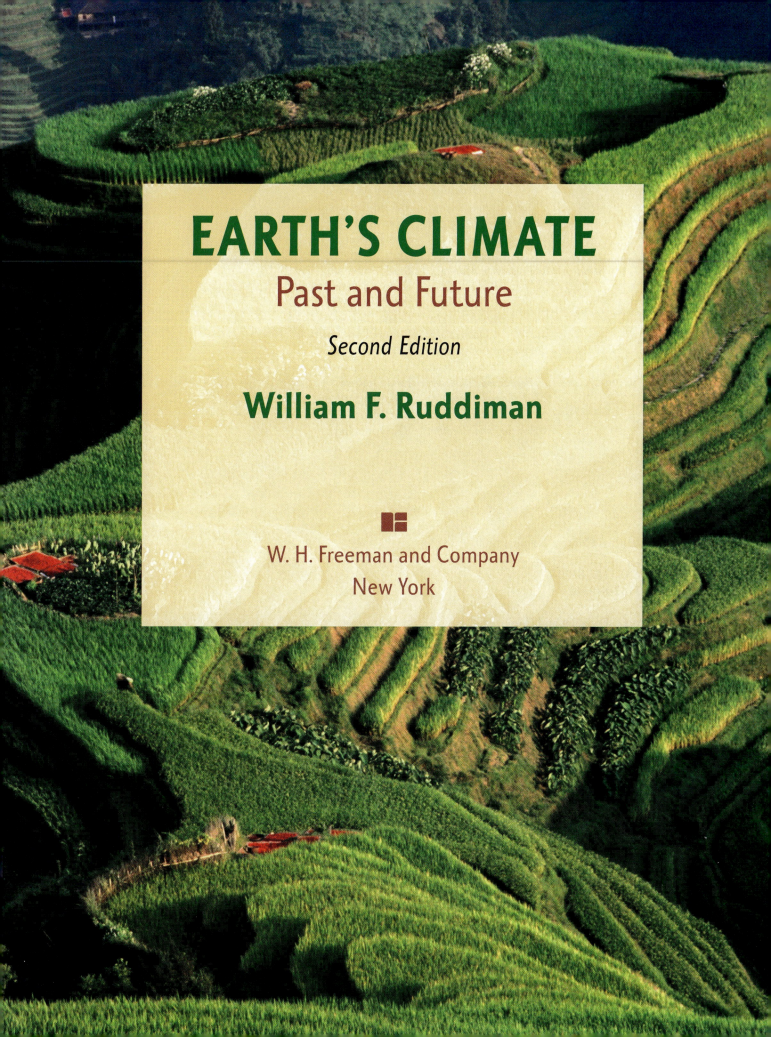

EARTH'S CLIMATE
Past and Future

Second Edition

William F. Ruddiman

W. H. Freeman and Company
New York

Publisher:	Clancy Marshall
Acquisitions Editor:	Valerie Raymond
Senior Marketing Manager:	Scott Guile
Developmental Editor:	Alysia Baker
Media Editor:	Amy Thorne
Senior Media Production Manager:	Brad Umbaugh
Project Editor:	Kerry O'Shaughnessy
Copy Editor:	Penelope Hull
Design Manager:	Diana Blume
Illustrations:	Robert Leo Smith, Jr.
Senior Illustration Coordinator:	Bill Page
Photo Editor:	Ted Szczepanski
Production Manager:	Julia DeRosa
Title Page Photograph:	© Keren Su/Lonely Planet
Composition:	Black Dot Group
Printing and Binding:	RR Donnelley

Library of Congress Control Number: 2007931323

ISBN-13: 978-0-7167-8490-6
ISBN-10: 0-7167-8490-4

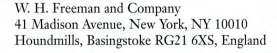

W. H. Freeman and Company
41 Madison Avenue, New York, NY 10010
Houndmills, Basingstoke RG21 6XS, England

To Ginger, for surviving another one

About the Author

William F. Ruddiman holds a 1964 undergraduate degree in geology from Williams College and a 1969 Ph.D. in marine geology from Columbia University. He worked at the U.S. Naval Oceanographic Office from 1969 until 1976 and then returned to Columbia's Lamont-Doherty Observatory as a Senior Research Associate. He was Associate Director of the Oceans and Climate Division from 1982 to 1986 and Adjunct Professor in the Department of Geology from 1982 to 1991. He moved to the University of Virginia in 1991 as a Professor in the Department of Environmental Sciences and served as department chair from 1993 to 1996. At Virginia he taught courses in climate change, physical geology, and marine geology.

Professor Ruddiman's research interests center on climate change across several time scales. His early research with Andrew McIntyre focused on orbital-scale climate change in and around the north and equatorial Atlantic Ocean. He was a member of the CLIMAP project from 1978 to 1984 and Project Director from 1980 until 1981. He was a member of the COHMAP project from 1980 until 1989 and served on the steering committee throughout that time.

His research in the late 1980s and the 1990s focused on the longer-term (tectonic-scale) physical and geochemical effects of uplift of the Tibetan Plateau and other high topography on regional and global climate. In 1997 he edited a book titled *Tectonic Uplift and Climate Change*, published by Plenum Press. His work on plateau uplift with colleagues Maureen Raymo, John Kutzbach, and Warren Prell has been featured in BBC and NOVA television documentaries.

Since 2001, his research has focused on the effects of early agriculture on greenhouse-gas concentrations. His early anthropogenic hypothesis proposes that releases of carbon dioxide and methane caused by farming drove an anomalous rise in greenhouse-gas concentrations during the last several thousand years.

Professor Ruddiman is a Fellow of the Geological Society of America and a Fellow of the American

Geophysical Union. He has participated in fifteen oceanographic cruises and was co-chief on leg 94 of the Deep-Sea Drilling Project and leg 108 of the Ocean Drilling Project.

He lives on a hillside in the Shenandoah Valley with his wife, Ginger, and an ever-changing cast of dogs and cats. His hobbies include rock-wall building, gardening, and walking the Appalachian Mountains, remnants of the continental collisions that produced the supercontinent Pangaea and the rocks now found in his walls.

BRIEF CONTENTS

Earth's Climate System Today (Chapter 2 from *Earth's Climate*, First Edition) is available at www.whfreeman.com/ruddiman2e.

CONTENTS

Earth's Climate System Today (Chapter 2 from *Earth's Climate,* First Edition) is available at www.whfreeman.com/ruddiman2e.

PREFACE

Seven years after publication of the first edition of *Earth's Climate* early in 2001, a new edition is needed because of recent evidence confirming the key role of greenhouse gases in the ongoing warming of the last century. During those seven years, climate has followed the course that climate scientists predicted would result from increased concentrations of greenhouse gases in the atmosphere. Sea ice and snow have retreated in the polar north, melting of mountain glaciers and ice caps has accelerated, and even the supposedly sluggish Greenland ice has begun to show evidence of rapid melting along its margins.

A range of new evidence also adds to the conclusion that human beings are the main cause of the recent warming: reanalysis of satellite temperature data that now show a warming comparable to that from surface stations, the realization that changes in visible and near-visible radiation from the Sun ("solar irradiance") have had a much smaller effect on global temperature during the last century than previously thought, and the discovery that enormous brown clouds emanating from southern Asia and other regions have masked (rather than amplified) part of the full effect of greenhouse-gas warming. With this new evidence, the claim that humans are playing the major role in warming the planet has gained broad acceptance in the scientific community. This fundamental shift is reflected in this new edition of the book.

Building on the First Edition

This edition of *Earth's Climate* retains several approaches used successfully in the first edition.

Multidisciplinary Scope The story of Earth's climate draws on many disciplines—geology, ecology, paleobotany, glaciology, oceanography, meteorology, biogeochemistry, climate modeling, atmospheric chemistry, and hydrology, among others. This range of disciplines is a large challenge, both to students for whom all the science in this field is new and to instructors who may specialize in one or two of the many research fields and time scales of climate change. This text eases the challenge through logical, step-by-step explanations of critical material, accompanied by attractive color graphics that summarize key points, and by including lists of follow-up resources for background material.

Following Earth's Timeline Like the first edition, the second edition explores the climatic responses of Earth's major systems (ice, water, air, vegetation, and land) as they developed through Earth's history. The structure again follows time's arrow, moving from the earliest known climate history to historical, modern, and future changes. The main reason for this structure is that this is the way Earth's climate has actually developed, so it is the most natural way to tell the story. In addition, shorter-term climate changes tend to ride on the back of longer-term changes, and the longer-term changes need to be understood first to provide context to those that are of more recent origin.

Mystery-Solving Approach Another advantage of organizing the book by time scale is that students are given a coherent, integrated view within each time scale of both the evidence of climate change and the competing hypotheses posed to explain the evidence. Because the central dynamic of science is the interconnection of data, theory, and theory testing, this integrated approach makes science come alive. The evidence of past climatic changes summarized at the start of each chapter leads to an obvious question: What caused the observed changes? The chapters then describe and evaluate the hypotheses proposed to explain the observations. Students are invited to be detectives in the problem-solving process by assessing the proposed hypotheses against a range of data and other methods, including experiments with climate models. From my own teaching experience, many of the best students are intrigued to find out that much work still remains to be done in this young field of science.

The major themes of the book remain the same as in the first edition:

- Causes (forcing) of climate change

- Natural response times of the many components of Earth's climate system

- Interactions and feedbacks among the numerous components

- Role of carbon as it moves within the climate system at each time scale

Structure The structure of *Earth's Climate*, Second Edition, is similar to its predecessor. Part I surveys the field of climate science and the approaches used to unravel Earth's climatic history. Parts II through V describe how Earth's climate has changed and are organized by time scale: tectonic-scale and earlier changes in Part II, orbital-scale changes in Part III, deglacial and millennial changes in Part IV, and historical, recent, and future changes in Part V. Progressively more recent intervals receive increasingly detailed treatment of climate changes that can be resolved in finer detail.

New to This Edition

This edition differs from the first edition in two key respects.

Streamlined Text. The major change is a shortening of 15%. Material from the first edition judged redundant or unnecessary has been removed. Now, all the chapters move from the introductory material to the final conclusions along a clearer path of logical development. At or near the end of each chapter or major section, "In summary" statements in the text are marked by tan shading to make them easier to locate. Students should find the revised chapters easier to understand.

Restructuring and Updating. The basic structure remains similar to the first edition, except for the following changes. Chapter 2 from the first edition ("Earth's Climate System Today") has been removed and posted on the W. H. Freeman and Company Web site: www.whfreeman.com/ruddiman2e. This change keeps the material readily available to instructors and students.

For many students, the sections that introduced oxygen and carbon isotopes were particularly difficult. In this edition, the full formal treatment of these two topics is moved to Appendices 1 and 2, where they do not break the "flow" of the text. Within the text, more functional definitions are given (for example, "More positive oxygen isotope ratios represent some combination of more ice and colder temperatures"). Another point of difficulty for most students was the in-depth treatment of orbital precession, which is not attempted in other texts. Despite its difficulty, this section has been retained, with the most difficult material set off in a box.

Other major structural changes have been made in the chapters in Part V. A new chapter has been added and others have been reordered so that the revised structure follows more clearly the path of time's arrow from older to younger time scales. In addition, new material covers the extensive new research on climatic changes spanning the last few centuries and millennia. This interest has largely been driven by the goal of assessing whether or not the warming of the last century has been unusual compared to natural variations during previous centuries and millennia.

Finally, this edition makes greater use of Web site resources (sites that are likely to be supported over the lifetime of this edition).

A Growing Audience for *Earth's Climate*

Over the last seven years, the first edition of *Earth's Climate* has grown to be a popular choice for upper-undergraduate courses in many earth science departments, often replacing or supplementing classical subjects like historical geology or sedimentology and stratigraphy. Adoptions of the first edition have also occurred in a wide range of other departments, such as environmental sciences, geography, ecology, botany, and oceanic and atmospheric sciences.

At the same time, many instructors have realized that this book can fill a need in courses taught at the introductory level to students who do not plan to major in science but who need to satisfy a science requirement. Because climate change is in the news every week, student interest in such courses is running at a very high level and these courses are growing. Introductory climate courses at several universities that have used the first edition of *Earth's Climate* have attracted hundreds of students, with enrollments increasing year by year. What students learn about the climate system in these courses will be put to use for the rest of their lives as they follow the news about the future of our planet.

Using a book like this at the introductory level requires lecturing and testing at an appropriately generalized (conceptual) level and deemphasizing some of the quantitative material. I think of students in introductory courses as analogous to members of an attentive jury: even without prior knowledge of a subject, they can be expected to follow the lines of argument carefully, understand the issues at hand, and draw basic conclusions from what they learn.

Instructors who are considering teaching an introductory course should find the second edition helpful. The shorter chapter lengths and improved logical flow should make this book easier for beginning students to grasp. As before, instructors can bypass the more advanced material set off in boxes titled Looking Deeper into Climate Science. The three other kinds of boxes (Climate Interactions and Feedbacks, Climate Debate, Tools of Climate Science) are part of the basic text and should not be a major problem for students in introductory courses.

Given my experience teaching this course at the introductory level, I've written a short guide for instructors, which can be found at www.whfreeman.com/ruddiman2e. And, Kristin St. John at James Madison University has written test questions to accompany this edition. Information on obtaining these questions can also be found at the Web site.

Teaching Options

Several options are available for using *Earth's Climate* as a textbook. The entire book can be the basis for a one-term course spanning all of Earth's climatic history, as outlined below. Alternatively, portions of the book can be used for shorter-length courses or parts of courses. Three such short options focused on tectonic, orbital, and recent changes are also outlined below.

Full Semester Course Full-semester courses of 14 weeks are usually taught either as three 50-minute sessions (M/W/F) or two 75-minute sessions (Tu/Th) per week. For the M/W/F option (42 sessions per term), the 19 chapters can be covered in 34 lectures, leaving 8 sessions for exams, videos and other purposes.

Course Outline (M/W/F)		No. of Lectures
Chapter 1	Framework of Climate Science	1
Chapter 2	Climate Archives, Data, and Models	2
Chapter 3	CO_2 and Long-Term Climate	1
Chapter 4	Plate Tectonics and Long-Term Climate	2
Chapter 5	Greenhouse Climate	2
Chapter 6	From Greenhouse to Icehouse: the Last 50 Million Years	2
Chapter 7	Astronomical Control of Solar Radiation	3
Chapter 8	Insolation Control of Monsoons	2
Chapter 9	Insolation Control of Ice Sheets	2
Chapter 10	Orbital-Scale Changes in Carbon Dioxide and Methane	2
Chapter 11	Orbital-Scale Interactions, Feedbacks, and Unsolved Problems	2
Chapter 12	Last Glacial Maximum	1
Chapter 13	Climate During and Since the Last Deglaciation	2
Chapter 14	Millennial Oscillations of Climate	2
Chapter 15	Humans and Preindustrial Climate	1
Chapter 16	Climate Changes During the Last 1000 Years	2
Chapter 17	Climatic Changes Since 1850	2
Chapter 18	Causes of Warming over the Last 125 Years	2
Chapter 19	Future Climatic Change	1

For the Tu/Th option (28 sessions per term), the 19 chapters can be covered in 23 lectures, leaving 5 sessions for exams, videos, and other purposes.

Course Outline (Tu/Th)		No. of Lectures
Chapter 1	Framework of Climate Science	1
Chapter 2	Climate Archives, Data, and Models	1
Chapter 3	CO_2 and Long-Term Climate	1
Chapter 4	Plate Tectonics and Long-Term Climate	1
Chapter 5	Greenhouse Climate	1
Chapter 6	From Greenhouse to Icehouse: the Last 50 Million Years	2
Chapter 7	Astronomical Control of Solar Radiation	2
Chapter 8	Insolation Control of Monsoons	1

(continued)

Course Outline (Tu/Th)		No. of Lectures *(continued)*
Chapter 9	Insolation Control of Ice Sheets	2
Chapter 10	Orbital-Scale Changes in Carbon Dioxide and Methane	2
Chapter 11	Orbital-Scale Interactions, Feedbacks, and Unsolved Problems	1
Chapter 12	Last Glacial Maximum	1
Chapter 13	Climate During and Since the Last Deglaciation	1
Chapter 14	Millennial Oscillations of Climate	1
Chapter 15	Humans and Preindustrial Climate	1
Chapter 16	Climate Changes During the Last 1000 Years	1
Chapter 17	Climatic Changes Since 1850	1
Chapter 18	Causes of Warming over the Last 125 Years	1
Chapter 19	Future Climatic Change	1

Recent Climate Some instructors prefer to focus on climatic changes during the last 150 years and those expected in the future. This material is covered in Part V of the book (Chapters 15-19). Students also need to have some understanding of the natural climatic changes that led to the pre-industrial climatic baseline from which the recent changes departed. Chapters 1 and 2 from Part 1 provide a basic introduction to reconstructing past climate changes and Chapters 12-14 from Part IV explain the origin of pre-industrial climate. Additional resources include Appendix 1, the companion web site, and the sources listed at the end of each chapter. The 10 chapters in the following outline can be covered in 10 one-hour lectures.

Recent Climate Course Outline	
Chapter 1	Framework of Climate Science
Chapter 2	Climate Archives, Data, and Models
Chapter 12	Last Glacial Maximum
Chapter 13	Climate During and Since the Last Deglaciation
Chapter 14	Millennial Oscillations of Climate
Chapter 15	Humans and Preindustrial Climate
Chapter 16	Climate Changes During the Last 1000 Years
Chapter 17	Climatic Changes Since 1850
Chapter 18	Causes of Warming Over the Last 125 Years
Chapter 19	Future Climatic Change

Orbital-Scale Climate Changes Some courses focus on orbital-scale variations in climate during the last few million years. Because the primary climatic forcing (insolation changes) and the major climatic responses (fluctuations of ice sheets and monsoons) are well known, clear cause-and-effect relationships can be explored for this interval. This material is covered in Part III (Chapters 7-11) and Part IV (Chapters 12-14). Chapters 1 and 2 in Part I provide necessary background material, and Chapter 6 in Part II covers the gradual global cooling that led to the northern hemisphere ice age cycles. In addition, Chapter 15 explores the effect of climate on long-term human evolution and the emerging role of humans in altering climate. Additional resources include Appendices 1 and 2, the companion web site, and the sources listed at the end of each chapter. The 12 chapters in this outline should take 16 one-hour lectures to cover (two lectures for chapters 6, 7, 9 and 10, and one lecture each for the others).

Orbital-Scale Course Outline

Chapter 1	Framework of Climate Science
Chapter 2	Climate Archives, Data, and Models
Chapter 6	From Greenhouse to Icehouse: the Last 50 Million Years
Chapter 7	Astronomical Control of Solar Radiation
Chapter 8	Insolation Control of Monsoons
Chapter 9	Insolation Control of Ice Sheets
Chapter 10	Orbital-Scale Changes in Carbon Dioxide and Methane
Chapter 11	Orbital-Scale Interactions, Feedbacks, and Unsolved Problems
Chapter 12	Last Glacial Maximum
Chapter 13	Climate During and Since the Last Deglaciation
Chapter 14	Millennial Oscillations of Climate
Chapter 15	Humans and Preindustrial Climate

Long-Term Climate Change   Courses that focus mainly on climate changes spanning many tens or hundreds of millions of years can make use of Chapters 3-6 in Part II, with Chapters 1 and 2 in Part I providing the necessary introductory and background material. In addition, Chapter 19 from Part V provides an intriguing perspective on the way that future climate may return to conditions that will in some ways be analogous to those that prevailed tens of millions of years ago. Supplementary materials include Appendix 1 and resources listed at the end of each chapter. The 7 chapters in the outline below should require 7 one-hour lectures.

Long-Term Course Outline

Chapter 1	Framework of Climate Science
Chapter 2	Climate Archives, Data, and Models
Chapter 3	CO_2 and Long-Term Climate
Chapter 4	Plate Tectonics and Long-Term Climate
Chapter 5	Greenhouse Climate
Chapter 6	From Greenhouse to Icehouse: the Last 50 million Years
Chapter 19	Future Climatic Change

Acknowledgments

I warmly acknowledge the efforts of Valerie Raymond, acquisitions editor at W. H. Freeman and Company. She has guided the decision to go forward with a second edition and overseen the effort, contributing many useful suggestions about the route to follow. I also thank these people for their contributions: development editor Alysia Baker, media editor Amy Thorne, senior media production manager Brad Umbaugh, project editor Kerry O'Shaughnessy, copy editor Penny Hull, design manager Diana Blume, and senior illustration coordinator Bill Page. I particularly thank Bob Smith, who created the line illustrations for both editions of *Earth's Climate*. Scores of people who have adopted the book have thanked me for the quality of his work. I also thank everyone who reviewed chapters or parts of the book:

Paul Baker
Duke University

Subir Banerjee
University of Minnesota

Jay Benner
University of Texas, Austin

William B. N. Berry
University of California, Berkeley

Julia Cole
University of Arizona

Tom Crowley
Texas A&M University

P. Thompson Davis
Bentley College

Matthew Evans
The College of William and Mary

Bart Geerts
University of Wyoming

John Gosse
University of Kansas

Andrew Ingersoll
California Institute of Technology

Emily Ito
University of Minnesota

George Jacobsen
University of Maine

David Kemp
Lakehead University

Zhuangjie Li
University of Illinois, Urbana-Champaign

Scott A. Mandia
Suffolk County Community College

Patricia Manley
Middlebury College

Isabel P. Montañez
University of California, Davis

R. Timothy Patterson
Carleton University

Maureen Raymo
Massachusetts Institute of Technology

Mike Retelle
Bates College

John Ridge
Tufts University

David Rind
Goddard Institute of Space Sciences

Lisa Sloan
University of California, Santa Cruz

Howard Spero
University of California, Davis

Alycia L. Stigall
Ohio University

Aondover Tarhule
University of Oklahoma

Lonnie Thompson
Byrd Polar Research Center

Stacey Verardo
George Mason University

Thompson Webb
Brown University

Al Werner
Mount Holyoke College

Herb Wright
University of Minnesota

EARTH'S CLIMATE

Recent fluctuations of mountain glaciers An engraving made in 1850 shows the Argentiere glacier extending far down the valley in the French Alps (top). A century later the glacier had retreated far up the mountainside (bottom). (From E. L. Ladurie, *Times of Feast, Times of Famine: A History of Climate Since the Year 1000* [New York: Doubleday, 1971] in J. Imbrie and K. P. Imbrie, *Ice Ages* [Cambridge, MA, and Springfield, NJ: Harvard University Press and Enslow Publishers, 1979].)

PART I

Framework of Climate Science

Climate change is a topic that naturally piques our curiosity. In the recent past, the hot, dry interval that produced the Dust Bowl of the 1930s drove thousands of farmers from the Great Plains west to California. A century earlier, air temperatures were cooler than now, and valley glaciers occupied positions well down the sides of mountains, compared with their locations today. Considerably further back in time, 21,000 years ago, climate was so cold that enormous ice sheets covered Canada and northern Europe, and sea level was some 120 meters (~400 feet) lower than it is today because of the amount of water stored as ice on land. Much further back, 100 million years ago, warmer conditions had eliminated ice from the face of the Earth, even at the South Pole.

These changes in the past occurred for natural reasons, some of them well understood and others still in the process of being unraveled. Climate science is a young science, and many exciting discoveries lie ahead.

Today Earth's climate is rapidly warming, and it is clear that humans are a major cause of this change. As scientists work to figure out how large the warming will be, part of the answer has come and will come from understanding the changes that occurred in the past. The chapters in Part I provide a general framework for understanding climate change by addressing several questions:

- **What are the components of Earth's climate system?**
- **How does climate change differ from day-to-day weather?**
- **What factors drive changes in Earth's climate?**
- **How do the many parts of Earth's climate system react to these driving forces and interact?**
- **How do scientists study past climates and project changes that lie in our future?**

CHAPTER 1

Overview of Climate Science

Life exists nearly everywhere on Earth because the climate is favorable. We live in the climate system: the air, land surfaces, oceans, ice, and vegetation. Climate change is also an important thread in the tapestry of Earth history, along with the evolution of life and the physical form of this planet.

But the study of climate also matters for a practical reason: it is relevant to the climatic changes we face in the near future. We have left an era when natural changes governed Earth's climate and have now entered a time when changes caused by human activity predominate.

This chapter surveys the factors that cause Earth's climate to change. It also reviews how the field of climate science came into being, how scientists study climate, and how an understanding of the history of climate change helps to inform us about changes looming in our near future.

Climate and Climate Change

Even from distant space, it is obvious that Earth is the only habitable planet in our solar system (Figure 1-1). More than 70% of its surface is a welcoming blue, the area covered by life-sustaining oceans. The remaining 30%, the land, is partly blanketed in green, darker in forested regions and lighter in regions where grass or shrubs predominate. Even the pale brown deserts and some of the white ice contain life.

Earth's favorable climate enabled life to evolve on our planet. **Climate** is a broad composite of the average condition of a region, measured by its temperature, amount of rainfall or snowfall, snow and ice cover, wind direction and strength, as well as other factors. Climate specifically applies to longer-term changes (years and longer), in contrast to the shorter fluctuations that last hours, days, or weeks and are referred to as **weather**.

Earth's climate is highly favorable to life both in an overall, planet-wide sense and at more regional scales. The surface temperature of the Earth averages a comfortable 15°C (59°F), and much of the surface ranges between 0° and 30°C (32° and 86°F) and can support life (Box 1-1).

Although we take Earth's habitability for granted, climate can change over time, and with it can change the degree to which life is possible, especially in vulnerable regions. During the several hundred years in which humans have been making scientific observations of climate, actual changes have been relatively small. Even so, climatic changes significant to human life have occurred. One striking example is the advances of valley glaciers that overran mountain farms and even some small villages in the European Alps and the mountains of Norway a few centuries ago because of a small cooling of climate. Those glaciers have since retreated to higher positions, as shown in the introduction to Part I.

Scientific studies reveal that these historical changes in climate are tiny in comparison with the much larger changes that happened earlier in Earth's history. For example, at times in the distant past ice covered much of the region that is now the Sahara Desert, and trees flourished in what are now Antarctica and Greenland.

1-1 Geologic Time

Understanding climatic changes of the past begins with a difficult challenge: coming to terms with the enormous span of time over which Earth's climatic history has developed. Human life spans are generally measured in decades. The phases of our lives, such as childhood and adolescence, come and go in a few years, and our daily lives tend to focus on needs and goals that we hope to satisfy within days or weeks.

Almost all of Earth's long history lies immensely far beyond this human perspective. Earth formed 4.55 billion years (Byr) ago (4,550,000,000 years!). Most of the earliest part of Earth's history is either a blank or is known only in a sketchy way. One reason for this gap in our knowledge is the climate system itself: the relentless

FIGURE 1-1 The habitable planet Even seen from distant space, most of Earth's surface looks inviting to life, especially its blue oceans and green forests but also its brown deserts and white ice. All these areas are prominent parts of Earth's climate system. (NASA.)

BOX 1-1 TOOLS OF CLIMATE SCIENCE

Temperature Scales

Three temperature scales are in common use in the world today. For day-to-day nonscientific purposes, most people in the United States use the Fahrenheit scale,

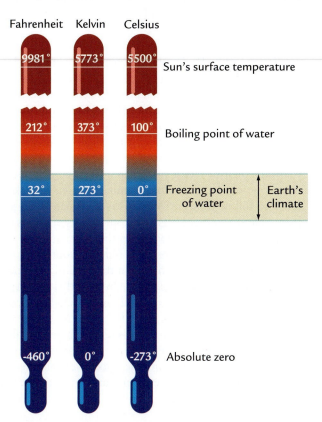

Fahrenheit   Kelvin   Celsius

9981°	5773°	5500°	Sun's surface temperature
212°	373°	100°	Boiling point of water
32°	273°	0°	Freezing point of water — Earth's climate
-460°	0°	-273°	Absolute zero

Temperature scales Scientists use the Celsius and the Kelvin temperature scales to measure climate changes. Temperatures at Earth's surface vary mainly within a small range of −50℃ to +30℃, just below and above the freezing point of water. (Adapted from W. F. Kaufman III and N. F. Comins, *Discovering the Universe,* 7th ed., © 2006 by W. H. Freeman and Company.)

developed by the German physicist Gabriel Fahrenheit. It measures temperature in degrees **Fahrenheit (°F),** with the freezing point of water at sea level set at 32°F and the boiling point at 212°F.

Most other countries in the world and most scientists, as well, routinely use the Celsius (or centigrade) scale developed by the Swedish astronomer Anders Celsius. It measures temperature in degrees **Celsius (°C),** with the scale set so that the freezing point of water is 0°C and the boiling point of water is 100°C.

These equations convert temperature values between the two scales:

$$T_C = 0.55 \, (T_F - 32) \qquad T_F = 1.8 T_C + 32$$

where T_F is the temperature in degrees Fahrenheit and T_C is the temperature in degrees Celsius.

Many scientific calculations make use of a third temperature scale developed by the British physicist Lord Kelvin (William Thomson) and known as the **Kelvin** scale. This scale is divided into units of Kelvins, rather than degrees Kelvin. The lowest point on the Kelvin scale (absolute zero, or 0K) is the coldest temperature possible, the temperature at which motions of atomic particles effectively cease. The Kelvin scale does not have negative temperatures, because no temperature colder than 0K is possible.

Temperatures above absolute zero on the Kelvin scale increase at the same rate as the Celsius scale, but with a constant offset. Absolute zero (0K) is equivalent to −273°C, and each 1K increase on the Kelvin scale above absolute zero is equivalent to a 1°C increase on the Celsius scale. As a result, 0°C is equivalent to 273K.

action of air and water on Earth's surface has eroded away many of the early deposits that might have helped us reconstruct and understand more of this history.

This book focuses mainly on the last several hundred million years of Earth's history, equivalent to less than 10% of its total age. Our focus is limited because many aspects of Earth's history are only vaguely known far back in the past, and as scientists investigate further and further back, they are forced to speculate more and more about fewer and fewer hard facts. But more information is available in the younger part of the climatic record, and our chances of measuring and understanding climate change increase.

Even the last 10% of Earth's history covers time spans beyond imagining. The climate scientists who study records spanning hundreds of thousands to hundreds of millions of years understand time only in a technical way, in effect as a means for cataloging and filing information. Geologists often refer to these unimaginably old and long intervals as "deep time," hinting at their remoteness from real understanding. Like the scientists who study climate change, you will learn in this book to catalog deep time in your own mental file, even if you cannot comprehend it in a literal sense.

The plot of time on the left in Figure 1-2 (page 6) shows that much of this book (Parts III through V) focuses on a fraction of Earth's history too small even to be shown on this simple linear scale. One way to overcome this problem is to again start with a plot of Earth's full age and then progressively expand out and magnify successively shorter intervals to show how they fit into the whole (Figure 1-2, center). The other method is to plot time on a logarithmic scale that increases by successive jumps of a factor of 10 (Figure 1-2, right). This kind of plot compresses the longer parts of the time scale and expands the shorter ones so that they all fit onto one plot.

1-2 How This Book Is Organized

Within the focus on the most recent 10% of Earth's age, this book is organized by time scale. Part II mainly covers climatic changes during the last several hundred million years, an interval during which mammals evolved from primitive to diverse forms. Part III looks at the last 3 million years, a time span when our primitive ancestors were evolving. Part IV explores changes over the last 50,000 years, an interval during which our fully human ancestors initially lived a primitive hunting-and-gathering life, then developed agriculture, and later created the first recorded human civilizations. Part V examines the last 1000 years, most of the historical era.

This progression from longer to shorter time scales is a natural one because faster changes in climate at the shorter time scales are embedded in and superimposed on the slower changes at the longer time scales (Figure 1-3). At the longest time scale, a slow warming between 300 and 100 million years (Myr) ago was followed by a gradual cooling in the last 100 million years (Figure 1-3A). This gradual cooling led to the appearance of massive northern hemisphere ice sheets that have advanced and retreated many times during the last 3 million years at cycles of several tens of thousands of years (Figure 1-3B). Superimposed on these climatic cycles were shorter oscillations that lasted a few thousand years and were largest during times when climate was colder (Figure 1-3C). The last 1000 years has been a time of relatively warm and stable climate, with much smaller oscillations (Figure 1-3D).

Each of these successive time scales reveals short oscillations embedded within longer ones, just as cycles of daily heating and nighttime cooling are embedded in the longer seasonal cycle of summer warmth and winter cold. To understand the extreme heat reached during a specific afternoon in July in the northern hemisphere, it first makes sense to consider that such an afternoon occurs in the larger context of the hottest season of the year and then to factor in the additional contribution from daytime heating. For a similar reason, it makes sense to follow time's arrow and trace climate changes from older to younger eras and from the larger cycles to the smaller ones superimposed on them.

As the book progresses from older to younger time scales, you will notice a change in the kind of information about past climate changes. In part this development reflects a change in the amount of detail that can be retrieved from climatic records, called the degree of **resolution**. Because older records tend to have less resolution, much of the focus of Part II of the book is on the longer-term average climatic states over millions of years and the way they differ from our climate today. By

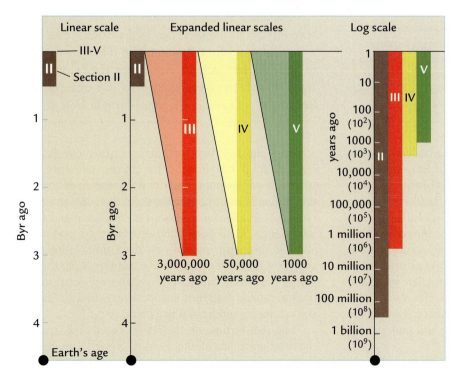

FIGURE 1-2 Earth history Earth's age is 4.55 billion years. Most of this book focuses on a very small fraction of this immense interval and can be represented only by a series of magnifications or by plotting time on a log scale that increases by factors of 10.

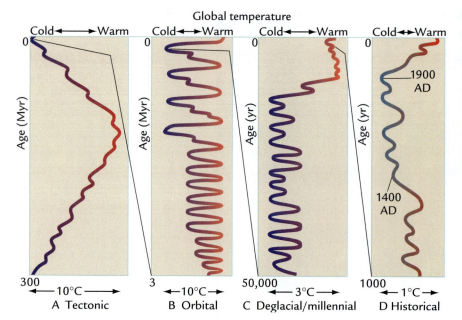

FIGURE 1-3 Time scales of climate change Changes in Earth's climate span several time scales, arrayed from longer to shorter: (A) the last 300 million years, (B) the last 3 million years, (C) the last 50,000 years, and (D) the last 1000 years. Here progressively smaller changes in climate at successively shorter time scales are magnified out from the larger changes at longer time scales.

comparison younger records tend to have progressively greater resolution, and Parts III through V look at successively shorter-term changes in climate that occur within intervals of thousands, hundreds, and finally even tens of years. We will examine the resolution issue more closely in Chapter 2.

Development of Climate Science

As scientists began to discover examples of major climatic changes earlier in Earth's history, their curiosity naturally grew about why these fluctuations had happened. The few amateur scientists and university professors who studied climate in relative isolation during the nineteenth and early twentieth centuries have now been replaced by thousands of researchers with backgrounds in geology, physics, chemistry, and biology working at universities, national laboratories, and research centers throughout the world (Figure 1-4). Today climate scientists use aircraft, ships, satellites, sophisticated new biological and chemical lab techniques, and high-powered computers, among other methods, to carry out their studies.

Studies of climate are incredibly wide-ranging. They vary according to the part of the climate system being studied, such as changes in air, water, vegetation, land surfaces, and ice. They also vary by the techniques used, including physical and chemical measurements of the properties of air, water, and ice and of life-forms fossilized in rocks; biological or botanical measurements of endless kinds of life-forms; and computer simulations to model the behavior of air, water, and vegetation.

This huge diversity of studies covers a broad array of scientific disciplines. Some studies are directed solely at improving our understanding of the modern climate system: meteorologists study the circulation of the atmosphere, oceanographers explore the circulation of the ocean, chemists investigate the composition of the ocean, atmosphere and land, glaciologists measure the behavior of ice, and ecologists analyze life-forms on land or in the water.

Other scientists study changes in climate or climate-related phenomena in Earth's recent or more distant past: *geologists* explore the broader aspects of Earth's history; *geophysicists* investigate past changes in Earth's physical configuration (continents, oceans, mountains); *geochemists* analyze past chemical changes in the ocean, air, or rocks; *paleoecologists* study past changes in vegetation and their role in the climate system; *climate modelers* evaluate possible causes of climate change; and *climate historians* comb written archives for information that will enable them to reconstruct past climates.

In recent decades, studies of Earth's climatic history have begun to cross these traditional disciplinary boundaries and merge into an interdisciplinary approach referred to as "Earth system science" or "Earth system history." Such efforts recognize that the many parts of Earth's climate system are interconnected so that investigators of climate must look at all the parts to understand the whole. This entire book is an example of this **Earth system** approach.

Similarly, this book makes no special distinction between studies of Earth's past history and investigations of the current (or very recent) climatic record. Earth's climatic history is a continuum from the distant past to the present. The book is organized by time scale because that is the way the continuum of Earth's climatic history has developed and will continue to develop in the future. Lessons learned about how the climate system has operated in the past can be applied directly to our understanding of the present and future, but the opposite is true

FIGURE 1-4 National research centers The National Center for Atmospheric Research (NCAR) in Boulder, Colorado, is one of several national laboratories and university centers at which Earth's climate is studied. (NCAR.)

as well. The broad term **climate science** refers to this vast *multidisciplinary* and *interdisciplinary* field of research and to its linkage of the past, the present, and the future.

How Scientists Study Climate Change

Climate science moves forward by an interactive mix of observation and theory. Climate scientists gather and analyze data from the kinds of climatic archives reviewed in Chapter 2, and the results of this research are written up and published. Progress in science depends on the free exchange of ideas, and climate researchers publish in order to tell the scientific community what they have discovered.

These scientists interpret their research results and occasionally come up with a new **hypothesis,** an idea proposed as an explanation for observed data. Science moves forward in part by disproving and discarding the less worthy hypotheses. Many hypotheses are eventually discarded, either because they are found to disagree with basic scientific principles or because they make predictions that subsequent observations contradict.

A hypothesis that succeeds in explaining a wide array of observations over a period of time becomes a **theory**. Scientists continue to test theories by making additional observations, developing new techniques to analyze data, and devising models to simulate the operation of the climate system. Only a few theories survive years of repeated testing. These are sometimes called "unifying theories" and are generally regarded as close approximations to "the truth," but the testing still continues.

Taken together the many expanding efforts to understand climate change have led to a scientific revolution that has accelerated through the late 1900s and early 2000s. The mystery of climate change yields its secrets slowly, and many important questions still remain to be answered, but the revolution in knowledge has been immense, as this book will show.

This revolution has reached the point where it is taking its place alongside two great earlier revolutions in knowledge of Earth history. The first was the development by Charles Darwin and others in the nineteenth century of the theory of **evolution,** which led to an understanding of the origin of the long sequence of life-forms that have appeared and disappeared during the history of this planet. The second was the synthesis during the 1960s and 1970s of the theory of **plate tectonics,** which has given us an understanding of the slow motions of continents across Earth's surface through time, as well as associated phenomena such as volcanoes, earthquakes, and mountain ranges.

Overview of the Climate System

In this section we take a first look at Earth's **climate system,** consisting of air, water, ice, land, and vegetation. At the most basic level, changes in these components through time are analyzed in terms of *cause* and *effect,* or, in the words used by climate scientists, **forcing** and **response**. The term "forcing" refers to factors that drive or cause change; the responses are the climatic changes that result.

1-4 Components of the Climate System

Figure 1-5 provides an initial impression of the vast array of factors involved in studies of Earth's climate. It shows the air, water, ice, land, and vegetation that are the major components of the climate system, as well as processes at work within the climate system, such as precipitation,

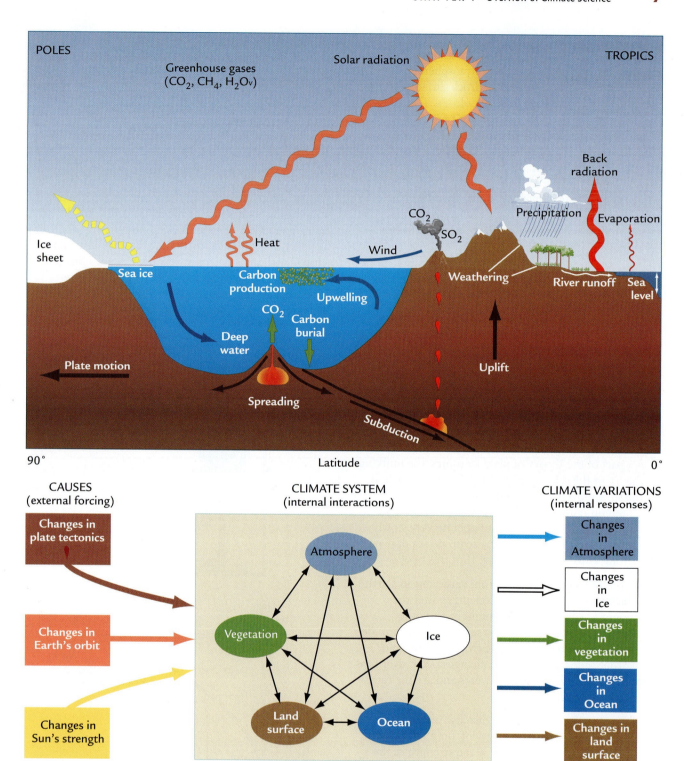

FIGURE 1-5 Earth's climate system and interactions of its components Studies of Earth's climate cover a wide range of processes, indicated at the top. Climate scientists organize and simplify this complexity, as shown at the bottom. A small number of factors drive, or "force," climate change. These factors cause interactions among the internal components of the climate system (air, water, ice, land surfaces, and vegetation). The results are the measurable variations known as climate responses.

evaporation, and winds. These processes extend from the warm tropics to the cold polar regions and from the Sun in outer space down into Earth's atmosphere, deep into its oceans, and even beneath its bedrock surface. All these processes will be explored in this book.

The complexity of the top part of Figure 1-5 is simplified in the bottom part to provide an idea of how the climate system works. The relatively small number of external factors shown on the bottom left force (or drive) changes in the climate system, and the internal components of the climate system respond by changing and interacting in many ways (bottom center). The result of all these interactions is a number of observed variations in climate that can be measured (bottom right). This complexity can be thought of as the operation of a machine: the factors that drive climate change are the input, the climate system is the machine, and the variations in climate are the output.

1-5 Climate Forcing

Three fundamental kinds of climate forcing exist in the natural world:

- *Tectonic processes* generated by Earth's internal heat affect its surface by means of processes that alter the basic geography of Earth's surface. These processes are part of the theory of plate tectonics, the unifying theory of the science of geology. Examples include the slow movement of continents, the uplift of mountain ranges, and the opening and closing of ocean basins. These processes operate very slowly over millions of years. The basic processes of plate tectonics are explained in Part II of this book.

- *Earth-orbital changes* result from variations in Earth's orbit around the Sun. These changes alter the amount of solar **radiation** (sunlight and other energy) received on Earth by season and by latitude (from the warm, low-latitude tropics to the cold, high-latitude poles). Orbital changes occur over tens to hundreds of thousands of years and are the focus of Parts III and IV.

- *Changes in the strength of the Sun* also affect the amount of solar radiation arriving on Earth. One example appears in Chapter 3: the strength of the Sun has slowly increased throughout the 4.55 Byr of Earth's existence. In addition, shorter-term variations that occur over decades or longer are part of the focus of Part V.

A fourth factor capable of influencing climate, but not in a strict sense part of the natural climate system, is the *effect of humans on climate*, referred to as **anthropogenic forcing**. This forcing is an unintended by-product of agricultural, industrial, and other human activities, and it occurs mainly by way of additions to the atmosphere of materials such as carbon dioxide (CO_2) and other **greenhouse gases**, sulfate particles, and soot. Anthropogenic effects will be covered in Part V.

1-6 Climate System Responses

The components of Earth's climate system vary widely: global mean and regional temperatures, the extent of ice of various kinds, the amounts of rainfall and snowfall, the strength and direction of the wind, the circulation of water at the ocean's surface and in its depths, and the types and amounts of vegetation. Each of these parts of the climate system responds to the factors that drive climate change with a characteristic **response time**, a measure of the time it takes to react fully to the imposed change.

Consider the example shown in Figure 1-6: a flask of water above a Bunsen burner. The Bunsen burner

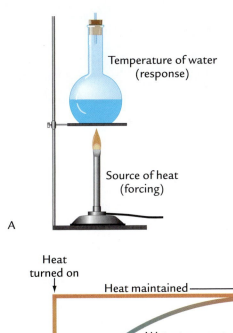

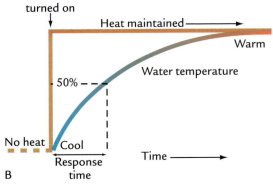

FIGURE 1-6 Response time Earth's climate system has a response time, suggested conceptually by the reaction of a beaker of water to heating by a Bunsen burner. The response time is the rate at which water in the flask warms toward an equilibrium temperature. (Adapted from J. Imbrie, "A Theoretical Framework for the Ice Ages," *Journal of the Geological Society* (London) 142 [1985]: 417–32.)

TABLE 1.1   Response Times of Various Climate System Components

Component	Response time (range)	Example
	Fast responses	
Atmosphere	Hours to weeks	Daily heating and cooling Gradual buildup of heat wave
Land surface	Hours to months	Daily heating of upper ground surface Midwinter freezing and thawing
Ocean surface	Days to months	Afternoon heating of upper few feet Warmest beach temperatures late in summer
Vegetation	Hours to decades/centuries	Sudden leaf kill by frost Slow growth of trees to maturity
Sea ice	Weeks to years	Late-winter maximum extent Historical changes near Iceland
	Slow responses	
Mountain glaciers	10–100 years	Widespread glacier retreat in 20th century
Deep ocean	100–1500 years	Time to replace world's deep water
Ice sheets	100–10,000 years	Advances/retreats of ice sheet margins Growth/decay of entire ice sheet

represents an external climate forcing (like the Sun's radiation), and the water temperature is the climatic response (such as the average temperature of Earth's surface). When the burner is lit, it begins to heat the water. The water in the flask gradually warms toward a constant temperature, and after a long interval it finally reaches and maintains an **equilibrium** value. The rate of warming (shown beneath the Bunsen burner in Figure 1-6) is rapid at first but progressively slows as time passes. It is intuitively reasonable that a response would naturally be faster when the water temperature is still far from its eventual equilibrium state and would slow as it nears equilibrium.

The rate at which the water warms toward the equilibrium temperature is its response time, defined in this case as the time it takes the water temperature to get halfway to the equilibrium value. The water temperature rises the first 50% of the way toward equilibrium during the first response time, but the same definition continues to apply later in the warming trend, as the water temperature moves from 50% $\left(\frac{1}{2}\right)$ to 75% $\left(\frac{3}{4}\right)$ to 87.5% $\left(\frac{7}{8}\right)$ to 93.75% $\left(\frac{15}{16}\right)$ of the way toward equilibrium. Each step takes one response time and moves the system half of the *remaining* way toward equilibrium. This progression can be understood in terms of the amount of the total response that remains after each step: $\frac{1}{2}, \frac{1}{4}, \frac{1}{8}, \frac{1}{16}$. This heating response has an exponential

form. Note that the *absolute amount* of change decreases through time, but the underlying response time remains exactly the same.

Each part of the climate system has its own characteristic response time (Table 1-1), ranging from hours or days up to thousands of years. The atmosphere has a very fast response time, and significant changes can occur in just hours (daily cycles of heating and cooling). The land surface reacts more slowly, but it still shows large heating and cooling changes on time scales of hours to days to weeks. Beach sand can become too hot to walk on during just a single summer afternoon, but it takes longer to chill the upper layer of soil in winter to the point where it freezes.

Liquid water has a slower response time than air or land because it can hold much more heat. The temperature response of shallow lakes or of the wind-stirred upper 100 meters of the ocean is measured in weeks to months. This slower rate is evident in the way lakes cool off seasonally but not as fast as the land does. For the deeper ocean layers that lie remote from interactions with the atmosphere, response times can range from decades to centuries or more for the deepest ocean.

Although the meter-thick layer of sea ice on polar oceans grows and melts in just months to years, thicker mountain glaciers react over longer time spans of decades to centuries. Massive (kilometers-thick) ice sheets like the one now covering the continent of Antarctica have the slowest response times in the climate system—many thousands of years, as captured in the commonly used word "glacial."

The concept also applies to vegetation, the organic part of the climate system. Unseasonable frosts can kill leaves and grass overnight, and abnormally hard freezes can do the same to the woody tissue of trees, responses measured in hours. On the other hand, seasonal spring greening of the landscape and autumn loss of leafy green material take weeks or months to complete. Pioneering vegetation that occupies newly exposed ground (for example, bare ground left behind by melting glaciers) may even take tens to hundreds of years or more to come to full development because of the slow dispersal of seeds and the time needed for them to germinate and produce mature trees.

1-7 Time Scales of Forcing versus Response

The parts of this book differ considerably in their emphasis on several factors: the forces that drive climate change, the responses of the climate system, and the interactions between forcing and response. Several hypothetical examples shown in Figure 1-7 give a sense of some basic differences:

- *The forcing is very slow in comparison with the response of the climate system.* This case is equivalent to increasing the flame of the Bunsen burner in Figure 1-6 so slowly that the water temperature has no problem keeping pace with the gradual application of more heat. If the changes in climate forcing are very slow in comparison with the response time of the climate system, the system simply passively tracks along with the forcing with no perceptible lag (Figure 1-7A).

  This case is typical of many climate changes that occur over the long tectonic time scales discussed in Part II. For example, continents can be slowly carried by plate tectonic processes toward higher or lower latitudes at rates averaging about 1 degree of latitude (100 kilometers or 60 miles) per million years. As the landmasses move toward lower latitudes, where incoming solar radiation is stronger, or toward higher latitudes, where it is weaker, temperatures over the continents react to these slow changes in solar heating with an imperceptibly tiny year-by-year response. Because the response time of air over land is short (hours to weeks; see Table 1-1), the average temperature over the continent can easily keep pace with the slow changes in average overhead solar radiation over millions of years. Shorter-term changes also occur over tectonic time scales, but they are usually harder to resolve in older records.

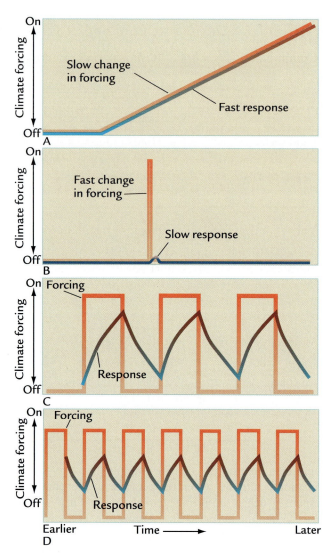

FIGURE 1-7 Rates of forcing versus response Climate responses depend on the relative rate of changes in climate forcing versus the response time of the climate system. (A) Fast response times permit the climate system to fully track slow forcing. (B) Slow response times allow little climate response to fast changes in forcing. (C, D) Roughly equal time scales of forcing and response allow varying degrees of response of the climate system to the forcing.

- *The forcing is fast in comparison with the climate system's response.* At the other extreme, the response time of the climate system may be slower than the time scale of the changes in forcing (Figure 1-7B). In this case, there is little or no response to the climate forcing. This is equivalent to turning the Bunsen burner on and off so quickly that the temperature of the water in the beaker has no time to react.

  One example of this extreme case is a total solar eclipse, which blocks Earth's only source of external heating for less than an hour. Air temperatures cool slightly during that brief interval, but then rise again. Volcanic eruptions are another example, such as the 1991 summer explosion of Mount Pinatubo in the Philippine Islands. Fine volcanic particles produced by that eruption blocked part of the Sun's radiation for several months and caused Earth's average temperature to fall by 0.5°C, but the cooling effect disappeared within a few years, because fine volcanic particles only stay in the upper layers of the atmosphere for a few years (see Table 1-1).

• *The time scales of forcing and climate response are similar.* A more interesting situation lies between the two extremes: cases in which the time scale of the climate forcing and that of the climate system's responses fall within a similar range. This situation produces a more dynamic response of the climate system, one that is typical of much of what actually happens in the real world.

Consider a different experiment with the Bunsen burner and the beaker of water. This time, the Bunsen burner (again the source of climate forcing) is abruptly turned on, left on awhile, turned off, left off awhile, turned on again, and so on (Figure 1-7C). These changes cause the water to heat up, cool off, heat up again, and so on. The water temperature responds by cycling back and forth between two different equilibrium values, one at the cold extreme with the flame off and one at the warm extreme with the flame on. But the intervals of heating and cooling do not last long enough to allow the water enough time to reach either of these equilibrium temperatures, as it did in Figure 1-6B.

The two cases shown in Figures 1-7C and 1-7D show that the frequency with which the flame is turned on and off has a direct effect on the size of the response of the water temperature. Both examples use the same equilibrium values (cold and warm) for the water temperature and the same position of the Bunsen burner relative to the beaker of water. The only difference is the length of time the flame is left on or off. If the flame is switched on and off far more rapidly than the response time of the water, the water temperature has less time to reach the equilibrium temperatures (hot or cold) and the size of the response is smaller (Figure 1-7D). But if the flame stays on or off for longer intervals, the temperature of the water has time to reach larger values nearer the full equilibrium states (Figure 1-7C).

In the real world, climate forcing rarely acts in the on-or-off way implied by the preceding examples. Instead, changes commonly occur in smooth, continuous cycles. If we again use the Bunsen burner concept, this

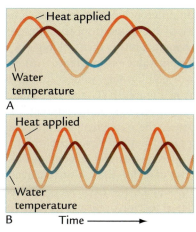

FIGURE 1-8 Cycles of forcing and response Many kinds of climate forcing vary in a cyclical way and produce cyclic climate responses. The amplitude of climate responses is related to the time allowed to attain equilibrium. (A) Climate changes are larger when the climate system has ample time to respond. (B) The same amplitude of forcing produces smaller climate changes if the climate system has less time to respond. (Adapted from J. Imbrie, "A Theoretical Framework for the Ice Ages," *Journal of the Geological Society* (London) 142 [1985]: 417–32.)

situation is analogous to keeping the burner flame (the climate forcing) on at all times but slowly and cyclically varying its intensity (Figure 1-8). The result is cycles of warming and cooling of the water that lag behind the shifts in the amount of heat applied, just as they did in Figure 1-7.

Familiar examples of this kind of forcing and response exist in daily and seasonal changes. In the northern hemisphere, the summer Sun is highest in the sky and therefore strongest at summer solstice on June 21, but the hottest air temperatures are not reached until July over the land and late August over the ocean. Similarly, the coldest winter days occur in January or February, long after the time of the weakest Sun at winter solstice on December 21. Even during a single day, the strongest solar heating occurs near noon, but the warmest temperatures are not reached until the afternoon, hours later.

Even though the smooth cycles of forcing and response in Figure 1-8 look different from the cases examined in Figure 1-7, the underlying physical response of the beaker of water (or, by extension, of the climate system) remains exactly the same. The temperature of the water in the beaker continues to react at all times with the same characteristic response time defined earlier, and the rate of response of the climate system is once again fastest when the climate system is farthest from its equilibrium value.

The main difference now is that the climate forcing (the intensity of the flame from the Bunsen burner) is constantly changing, rather than holding at a single constant equilibrium value (as in Figure 1-6) or switching between two equilibrium values in an alternating sequence (as in Figures 1-7C and D). The continuous changes in heating act as a "moving target." The climate system response (the water temperature) keeps chasing this moving target but can never catch up to it because the water temperature cannot respond quickly enough.

As was the case for the on-off changes shown in Figures 1-7C and D, the frequency with which these smooth cycles of forcing occur has a direct effect on the amplitude of the responses. This effect is apparent in the differences between the cases shown in Figures 1-8A and B. If the forcing occurs in slower (longer) cycles, it produces a larger response (larger maxima and minima) because the climate system has more time to react before the forcing turns and cycles back in the opposite direction (Figure 1-8A). In contrast, forcing that occurs in faster (shorter) cycles produces a smaller response because the climate system has less time to react before the forcing reverses direction (Figure 1-8B). These two responses differ in size even though the forcing moves back and forth between the same maximum and minimum values in both cases.

The relationships between forcing and response shown in Figure 1-8 are particularly useful for understanding the orbital-scale climatic changes explored in Parts III and IV of this book. Changes in incoming solar radiation due to changes in Earth's orbit occur over tens of thousands of years, also the response time characteristic of large ice sheets that grow and melt over the orbital time scales. This approximate match of the time scales of forcing and response sets up cyclic interactions very much like those shown in Figure 1-8.

1-8 Differing Response Rates and Climate-System Interactions

The examples shown so far summarize the response of the climate system by a single curve, as if it were capable of only a single response. But Table 1-1 showed that the system has many components with different response times. Each component responds to climatic forcing at its own tempo.

One way to grasp the impact of these differences in response is to imagine that some change is abruptly imposed on the climate system from the outside (for example, a sudden strengthening of the Sun's radiation). Each part of the climate system will respond to this sudden increase in external heating in a way analogous to the beaker of water sitting over the Bunsen burner (Figure 1-6), but in this case it reacts at a tempo dictated by its own response time (Figure 1-9). The faster-responding parts of the climate system will warm up

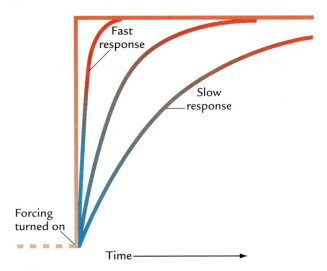

FIGURE 1-9 Variations in response times An abrupt change in climate forcing will produce climate responses ranging from slow to fast within different components of the climate system, depending on their inherent response times.

more quickly, and the slower-responding parts will do so more slowly.

We can apply this idea of differing response times to the case in which the factor causing climate change varies in smooth cycles (Figure 1-10). Here again, each part of the climate system will tend to respond at its own rate, again producing several different patterns of response. In the example shown in Figure 1-10, some fast-responding parts of the climate system respond so quickly to the climate forcing that they can track right along with it. In contrast, other slower-responding parts of the climate system lag well behind the forcing.

These differing response rates can lead to complicated interactions in the climate system. Assume that the curve in Figure 1-10 showing the initial climate forcing represents changes in the amount of the Sun's heat that

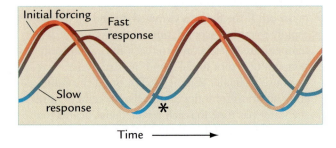

FIGURE 1-10 Variations in cycles of response If the climate forcing occurs in cycles, it will produce differing cyclic responses in the climate system, with the fast responses tracking right along with the forcing cycles while the slower responses lag well behind.

Climate Archives, Dating and Resolution

Like written chronicles of human history, climate archives hold stories of climate change for those who can read them. For the immense span of Earth's history prior to the invention of instruments in recent centuries, sediments, ice, corals, and trees are the major climatic archives.

2-1 Types of Archives

Although relatively recent climate changes can be studied in an array of archives, sediments—primarily continuous sequences of sediment deposited by water—are the major climate archive on Earth for over 99% of geologic time.

Sediments Rainfall and the runoff it produces erode rocks exposed on the continents and transport the eroded debris in streams and rivers in both physical (granular) and chemical (dissolved) forms. The sediments are eventually deposited in quieter waters where layer after layer of sediment is laid down in undisturbed succession. Most sediment is carried to the ocean either right after it is eroded or after temporary deposition on land followed by one or more cycles of additional erosion and redeposition. Sediment delivered to the seafloor may persist there for tens of millions of years until tectonic processes destroy it. The relentless action of these two processes, erosion and tectonic activity, decreases the likelihood that older sedimentary records will be preserved as time passes.

For intervals prior to the last 170 million years, all surviving sedimentary records come from the continents. Under favorable conditions, sediments may be preserved for a long time in the regions shown in Figure 2–1: continental basins that contain lakes; shallow interior seas that at times flood low-lying land; and lens-shaped piles of sediment along continental shelves (the barely submerged coasts of continents) and steeper continental slopes leading down into the deep ocean.

Sediments are useful climate archives to the extent that their deposition is uninterrupted. Major disturbances come from wave action reaching several meters below sea level and from occasional large storms that reach tens of meters deep in the water column and erode previously deposited layers. In addition, sediments deposited on steep continental slopes are vulnerable to dislodgment by disturbances such as earthquakes and tsunamis.

In the longer term, erosion tied to sea level change is a major factor that interrupts sediment deposition. Through time, the sea moves up and down along the continental margins over a total vertical range of about 200 meters. Sediments can be deposited on the upper margins when sea level is high, but these deposits are often eroded by waves and storms and carried to the deep sea when sea level subsequently falls.

All these factors ultimately determine the types of climate records preserved in sediment archives (see Figure 2–1). Sediments deposited on continental shelves when sea level is high form lens-shaped units separated by distinct surfaces where erosion has occurred. Deposition is often continuous within these sequences, with the highest rates in regions where rivers deliver sediment. Sediments deposited in interior seas on the continents during times when the ocean floods low-lying regions form continuous sequences covering wide areas.

Sediments deposited in lakes in continental basins conform to the structural framework of the depression in the bedrock. Deposition tends to be most continuous

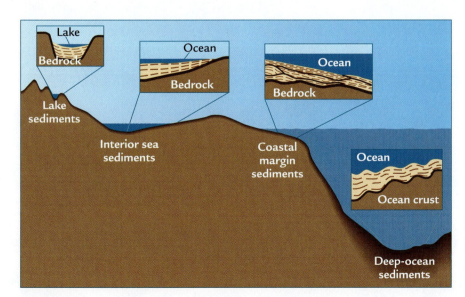

FIGURE 2-1 Sediment archives
Layered sediments are major climate archives on all time scales. The insets show typical sediment layering in sediment archives from land and sea.

CHAPTER 2

Climate Archives, Data, and Models

Climate scientists use a wide range of techniques to extract, reconstruct, and interpret the history of Earth's climate. Much of this history is recorded in four archives: sediments, ice, corals, and trees. In this chapter we first examine each of these major climate archives. Then we explore how their climate records are dated, the way Earth's climatic history is recorded in each archive, and the resolution of climate history each archive yields.

The interpretation of climate data is aided by the use of climate models to test hypotheses of climate change in a quantitative way. In this chapter we describe physical models that simulate the circulation of Earth's atmosphere and ocean and then examine the concept behind models used to track mass movements of chemical tracers through the climate system.

BOX 1-2 CLIMATE INTERACTIONS AND FEEDBACKS

Positive and Negative Feedbacks

The strength of a feedback on temperature, called the **feedback factor,** or f, is defined as

$$f = \frac{\text{temperature change with feedback}}{\text{temperature change without feedback}}$$

where "temperature change" refers to the full equilibrium response.

If f has a value of 1, no feedback exists. If the value of f is greater than 1, the net temperature change is larger than it would be without any feedback, and the climate system is characterized as having a positive feedback. If the value of f is less than 1, the temperature change is smaller than it would be in the absence of any feedback, and the climate system is characterized as having a negative feedback.

---

components of the climate system. The changes in some of these components will then further perturb climate through the action of feedbacks.

Positive feedbacks produce additional climate change beyond that triggered by the factor that initiates the change (Box 1-2). For example, a decrease in the amount of heat energy sent to Earth by the Sun would allow snow and ice to spread across high-latitude regions that had not previously been covered. Because snow and ice reflect far more sunlight (heat energy) than do bare ground or open ocean water, an increase in their extent should decrease the amount of heat taken up by Earth's surface and further cool the climate in those regions.

The positive feedback process also works in the opposite direction. If more energy from the Sun arrives and causes climate to warm, high-latitude snow and ice will retreat and allow more sunlight to be absorbed. The result will be further climatic warming. Positive feedback acts as an amplifier, regardless of the direction of change.

Negative feedbacks work in the opposite sense, by muting climate changes (see Figure 1-11). When an initial climate change is triggered, some components of Earth's climate system respond in such a way as to reduce the initial change.

Key Terms

climate (p. 4)
weather (p. 4)
Fahrenheit (p. 5)
Celsius (p. 5)
Kelvin (p. 5)
resolution (p. 6)
Earth system (p. 7)
climate science (p. 8)
hypothesis (p. 8)

theory (p. 8)
evolution (p. 8)
plate tectonics (p. 8)
climate system (p. 8)
forcing (p. 8)
response (p. 8)
radiation (p. 10)
anthropogenic forcing
  (p. 10)

greenhouse gases (p. 10)
response time (p. 10)
equilibrium (p. 11)
feedbacks (p. 15)

positive feedbacks (p. 15)
negative feedbacks
  (p. 15)
feedback factor (p. 16)

Review Questions

1. How does climate differ from weather?

2. In what ways does climate science differ from traditional sciences such as chemistry and biology?

3. How does climate forcing differ from climate response?

4. In the example in which the Bunsen burner is lit and the beaker of water at first warms quickly and then more slowly, how does the response time of the water change through time?

5. The climate system consists of many components with different response times. What is the total range of time scales over which these responses vary?

6. Do positive feedbacks always make the climate warmer?

Additional Resources

Basic Reading
Climate Change: State of Knowledge. 1997. Washington, DC: Office of Science and Technology Policy.
Understanding Climate Change. 1975. Washington, DC: U.S. National Academy of Science.

Advanced Reading
Imbrie, J. 1985. "A Theoretical Framework for the Ice Ages." *Journal of the Geological Society* 142: 417–32.

reaches a particular region over intervals of thousands of years. Also assume that the fast-response curve represents the rapid heating of landmasses at lower and middle latitudes, while the slow-response curve represents changes in the size of ice sheets lagging thousands of years later.

In this scenario, the asterisk in Figure 1-10 marks a time when large ice sheets have built up in Canada and Scandinavia (as has actually happened many times in the past, most recently 20,000 years ago). At this point in the sequence, the slow-responding ice has not yet begun to retreat, even though the heating from the Sun has begun to increase and the land far south of the ice sheets has begun to warm.

Given this situation, how do you think the air temperatures just south of the ice limits would respond? Would the air warm with the initial strengthening of the overhead Sun and heating of the land? If so, its response would track right behind the initial forcing curve in Figure 1-10.

Or would air temperatures still be under the chilling influence of the large mass of ice lying just to the north and not begin to rise until the ice starts its retreat? In this case, the ice would in a sense be acting as a semiindependent player in the climate system by exerting an influence of its own on local climate. Although the ice initially acts as a slow climate response driven by slow changes in the Sun, it then exerts its own effect on climate separate from the immediate effects of the Sun.

Both these explanations probably sound plausible, and they are. The air temperatures just south of the ice sheets will be influenced by *both* the overhead Sun and the nearby ice. The actual timing of the air-temperature response in such regions will fall somewhere in the middle, faster than the response of the ice but lagging behind the forcing from the Sun. As this example suggests, Earth's climate system is very dynamic, with numerous interactions.

The response-time concept is directly relevant to projections of climate change in the near future. Part V of this book addresses the effects of humans on climate through the buildup of greenhouse gases, primarily CO_2 produced from burning fossil fuels such as coal, oil, and natural gas. The changes in the next few centuries will be unusual in the sense that both the large climate forcing produced by humans and the warming it will cause will arrive with unusual speed. Within a few centuries, the fossil fuels that generate excess CO_2 in the atmosphere will be largely used up, CO_2 emissions will fall, and Earth's climate will begin to return toward its previous cooler state. But before that happens, Earth will face a century or more of very substantial warmth, along with many other changes.

Scientists and the public in general want to know how large the disruption caused by the several centuries of high CO_2 concentrations will be, and the answer requires an understanding of the different response

times of the major components in the climate system. Most parts of the system will begin to respond relatively quickly to the greenhouse-gas forcing, but others (those most closely tied to the ice sheets) will respond more sluggishly. A large part of the challenge facing climate scientists is to sort out these different responses and all their interactions.

1-9 Feedbacks in the Climate System

Another important kind of interaction in the climate system is the operation of **feedbacks,** processes that alter climate changes that are already underway, either by amplifying them (**positive feedbacks**) or by suppressing them (**negative feedbacks**). Figure 1-11 shows the basic way these feedbacks operate.

Assume that some external factor (again, perhaps a change in the strength of radiation from the Sun) causes Earth's climate to change. That change will consist of many different responses among the various internal

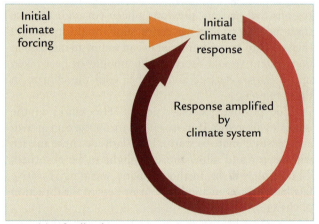

A Positive feedback

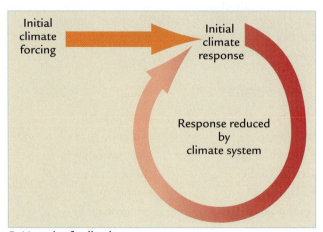

B Negative feedback

FIGURE 1-11 Climate feedbacks (A) Positive feedbacks within the climate system amplify climate changes initially caused by external factors. (B) Negative feedbacks mute or suppress the initial changes.

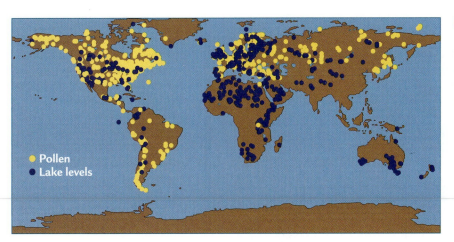

FIGURE 2-2 **Lake cores** Hundreds of cores have been taken from small lakes and analyzed for records of changes in pollen (vegetation) and lake level over the last several thousand years. (National Paleoclimate Data Center, NGDC, Boulder, CO.)

in the deeper parts of lakes. Lake sediments that fill depressions left behind by melting glaciers are especially important climate archives for the last 20,000 years (Figure 2–2).

Ice and wind are also powerful agents of sediment erosion and transport in some regions. Ice sheets that reach maximum size and then begin to retreat leave behind long curving ridges called **moraines.** Moraines contain a jumbled mix of unsorted debris carried by ice, ranging from large boulders to very fine clay. When an ice sheet readvances over moraines deposited earlier, it erodes the earlier debris and incorporates it into the newer deposits. Unraveling a climate history from this kind of record is like trying to decipher repeated episodes of writing that have been largely erased on a blackboard. By contrast, the coarse debris carried to the ocean and dropped by melting icebergs into the underlying sediments can survive in a more protected environment.

Strong winds weather rocks and form fine sediment particles in regions with dry climates. Winds form sand dunes that slowly migrate across desert areas, but continuous reworking of the sand particles complicates efforts to use dunes as climate archives. Winds also pick up smaller, silt-sized grains, lift them high in the air, and transport them far from their original sources. In regions where the winds weaken, the silt is deposited in sequences called **loess.** Loess deposits are excellent climate repositories of the last 3 million years, especially in China (Figure 2–3). Finer sediments carried off the continents by winds and deposited in ocean sediments are also useful indicators of climate.

For the portion of geologic time younger than 100 million years, climate scientists have access to an additional climate archive: sediments preserved in ocean basins. Deep-sea sediments spanning the last several million years cover almost two-thirds of Earth's

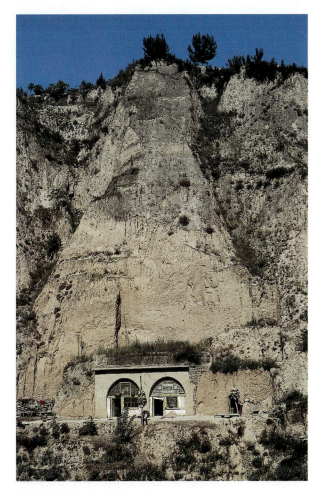

FIGURE 2-3 **Windblown loess** Strong winds have deposited thick layers of silt-sized grains in southeast China during the last 3 million years. The total thickness of these loess deposits can reach several hundred meters. In many regions people have created homes in the loess cliffs. (Courtesy of Steven Porter, University of Washington.)

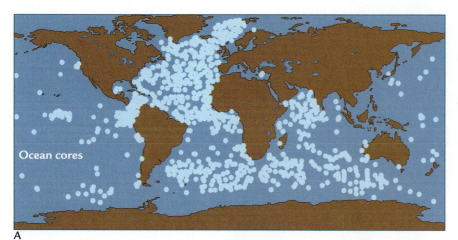

FIGURE 2-4 Ocean drilling
(A) Hundreds of ocean sediment cores
are archives of past climatic changes.
(B, C, D) The longest cores have been
retrieved by drilling operations on the
JOIDES Resolution, run by the
International Ocean Drilling Program.
(A: National Paleoclimate Data Center,
NGDC, Boulder, CO. B, C, D: Ocean
Drilling Program, Texas A&M
University.)

A

B

C

D

surface (Figure 2–4A). Older and more deeply buried sediments have been retrieved by the *JOIDES Resolution,* a ship capable of drilling into and recovering sediment sequences several kilometers thick (Figure 2–4B–D).

Because the deep ocean is generally a quiet place with relatively continuous deposition, it yields climate records of higher quality than most records from land, where water, ice, and wind actively erode deposits. Some deep-sea sediments are subject to disturbances such as dislodgment from steep slopes, physical erosion and reworking by currents on the sea floor, and chemical dissolution by corrosive water in the deeper basins. Despite these problems, many ocean basins have been sites of continuous sediment deposition for tens of millions of years. Deposition of sediments is usually much slower in the ocean than on land, but rates are higher in regions that receive influxes of sediments eroded from nearby continents, in sediments beneath productive surface waters, and in regions high above the corrosive bottom waters in the deepest ocean.

Glacial Ice At the very cold temperatures found at high latitudes and high altitudes, annual deposition of snow can pile up continuous sequences of ice that range in thickness from small mountain glaciers tens to hundreds of meters thick to much larger continent-sized ice sheets several kilometers thick (Figure 2–5). Ice core archives contain many kinds of climatic information, although it is limited geographically to the few regions where ice exists (Figure 2–6).

Ice recovered from the Antarctic ice sheet now dates back more than 700,000 years, and ice from the Greenland ice sheet dates back 120,000 years. Future ice drilling is likely to extend these records even further back in time. In contrast, most small glaciers that exist in mountain valleys (even in the tropics) record only the last 10,000 years or less of climate change. Deposition rates range from a few centimeters per year in the coldest and driest areas to meters per year in less frigid, wetter regions.

Other Climate Archives In areas of sufficient rainfall, groundwater percolating through soil and bedrock dissolves and redeposits limestone (calcite, or $CaCO_3$) in *caves.* These deposits contain records of climate over intervals that can extend back several hundred thousand years.

Trees are valuable climate archives for the interval of the last few tens, hundreds, or (in exceptional cases) thousands of years. The outer softwood layers of many kinds of trees are deposited in millimeter-thick layers that turn into hardwood. These annual layers are best developed in mid-latitude and high-latitude regions

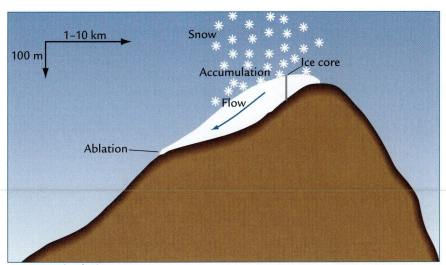

A Mountain glaciers

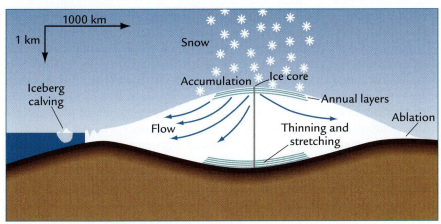

B Continental ice sheets

FIGURE 2-5 Ice archives Ice is an important archive of many climate signals. Ice cores retrieve climate records extending back (A) thousands of years in small mountain glaciers to as much as (B) hundreds of thousands of years in continent-sized ice sheets.

that experience large seasonal climate changes (see Figure 2–6).

In clear sunlit waters at tropical and subtropical latitudes, corals form annual bands of $CaCO_3$ that hold several kinds of geochemical information about climate (see Figure 2–6). Individual corals may live for time spans of years to tens or hundreds of years.

Within the last few thousand years, humans have kept **historical archives** of climate-related phenomena. Examples include the time of blooming of cherry trees in Japan, the success or failure of grape and grain harvests in Europe, and the number of days with extensive sea ice in regions such as Iceland and Hudson Bay in Canada. These records precede (and in most cases overlap) the **instrumental records** of the last 100 to 200 years. The first thermometers for measuring climate appeared in the eighteenth century, but human ingenuity has now created instruments to measure climate remotely from space (Figure 2–7).

2-2 Dating Climate Records

Climate records in older sedimentary archives are dated by a two-step process. First, scientists use the technique of **radiometric dating** to measure the decay of radioactive isotopes in rocks. (Isotopes are forms of a chemical element that have the same atomic number but differ in mass.) Dates are obtained on hard crystalline igneous rocks that once were molten and then cooled to solid form. In the second step, dates obtained from the igneous rocks provide constraints on the ages of sedimentary rocks that occur in layers between the igneous rocks and form the main archives of Earth's early climate history.

Radiometric Dating and Correlation Radiometric dating is based on the radioactive decay of a **parent isotope** to a **daughter isotope.** The parent is an unstable radioactive isotope of one element, and radioactive decay transforms it into the stable isotope of another element

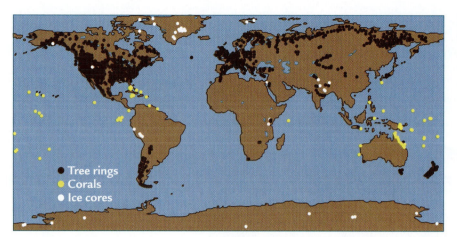

FIGURE 2-6 Ice cores, corals, and tree rings Ice cores, corals, and tree rings are archives of climate change in more recent Earth history. (National Paleoclimate Data Center, NGDC, Boulder, CO.)

(the daughter). This decay occurs at a known rate, the *decay constant*, which is a measure of the likelihood of a parent-to-daughter decay per amount of parent present per unit of time. This rate of decay in effect forms a clock with which we can measure age.

An event of some kind is required to start this clock ticking. The igneous rock that is most commonly used for dating is basalt, which cools quickly from molten outpourings of lava. The event that starts the clock ticking is the cooling of this material to the point where neither the parent nor the daughter isotope can migrate in or out of the molten mass. At this point, the rock forms a **closed system,** one in which the only changes occurring are caused by internal radioactive decay.

In the simplest example of a closed system, the decay of a parent to a daughter produces the changes shown in Figure 2–8: the parent decays away exponentially, while the daughter shows an exactly opposite (and compensating) exponential increase in abundance. The **half-life** is a convenient measure of the rate at which this process occurs: one half-life is the time needed for half the parent present to decay to the daughter. The first half-life reduces the parent to half its initial abundance, the second reduces it to half of that half (one-quarter), and so on. Notice the similarity of radioactive decay to the response time concept from Chapter 1.

Because radioactive parents have a wide range of half-lives, each is most useful over a different part of Earth's history (Table 2–1). Radioactive isotopes remain useful for at least five or six half-lives after the clock is set, but after this point too little of the parent may be left to permit reliable dating. The long, slow decay series from uranium (U) to lead (Pb) is useful for rocks that are nearly as old as Earth itself. The decay from potassium (K) to argon (Ar) is widely used for dating much of Earth's history.

Several factors can complicate radiometric dating. Unlike the simple case shown in Figure 2–8, the initial abundance of the daughter isotope is rarely zero: usually some amount was already present in the igneous rock when the decay clock was set. Other problems arise when the system does not remain fully closed to the migration of parent or daughter isotopes.

If both igneous and sedimentary rocks are present in a specific region, the igneous rocks can be used to constrain the ages of the sediment sequences. The age of each layer of sediment can be obtained from the nearby igneous rocks based on which is older or younger than the other. For example, a layer of igneous rock that spreads across the top of a layer of sediment must slightly postdate the time the sediment was deposited and so it provides a minimum age for that layer of sediment.

In actual practice, it is rare to find enough igneous rock in any one location to date sediments this way. Instead, sediment sequences are dated by a combination

FIGURE 2-7 Instrument measurements Instruments that have been used to measure climate range from the primitive thermometers of the seventeenth century to the multiple sensors flown aboard the TOPEX/Poseidon satellite. (NASA.)

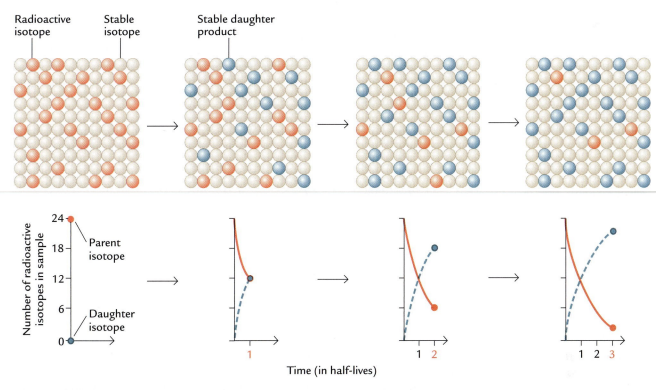

FIGURE 2-8 Radioactive decay Time is determined by measuring the gradual decay of a radioactive parent isotope to a daughter isotope. The half-life is the time needed for half the parent to decay. The relative abundances of parent and daughter isotopes follow the trends shown at the bottom. (D. Merritts et al., *Environmental Geology*, © 1997 by W. H. Freeman and Company.)

of dating and correlation using fossils or other features in the sediments. The fossil method relies on the fact that a *unique and unrepeated* sequence of organisms has appeared and disappeared through Earth's entire history and has left fossilized remains. The most useful fossils are those that are shortest-lived but geographically most widespread. If the brief existence of these species can be reliably dated in sediments in at least

a few areas using nearby igneous rocks, the ages that are obtained can be transferred to sediments in other regions that contain the same short-lived fossils but lack radiometric dating. Other physical or chemical features that show distinctively varying patterns (like ash layers) can be used in a similar way.

Radiocarbon In the younger geologic record, a different method, **radiocarbon dating**, is widely used to

TABLE 2-1	Radioactive Decay Used to Date Climate Records			
Parent isotope	**Daughter isotope**	**Half-life**	**Useful for ages:**	**Useful for dating:**
Rubidium–87 (^{87}Rb)	Strontium–87 (^{87}Sr)	47 Byr	100 Myr	Granites
Uranium–238 (^{238}U)	Lead–206 (^{206}Pb)	4.5 Byr	>100 Myr	Many rocks
Uranium–235 (^{235}U)	Lead–207 (^{207}Pb)	0.7 Byr	>100 Myr	Many rocks
Potassium–40 (^{40}K)	Argon–40 (^{40}Ar)	1.3 Byr	>100,000 years	Basalts
Thorium 230 (^{230}Th)	Radon–226*(^{226}Ra)	75,000 years	<400,000 years	Corals
Carbon–14 (^{14}C)	Nitrogen–14* (^{14}N)	5,780 years	<50,000 years	Anything that contains carbon

*Daughter in this case is a gas that has escaped and cannot be measured.

date lake sediments and other kinds of carbon-bearing archives. Neutrons that constantly stream into Earth's atmosphere from space convert ^{14}N (nitrogen gas) to ^{14}C (an unstable isotope of carbon). Vegetable and animal life forms on Earth use carbon from the atmosphere to build both their hard shells and soft tissue, and a small part of the carbon used is radioactive ^{14}C. The death of the plant or animal closes off carbon exchange with the atmosphere and starts the decay clock ticking. The ^{14}C parent decays to the ^{14}N daughter, a gas that escapes to the atmosphere. The amount of ^{14}C that has been lost when a sample is analyzed is measured by examining a stable isotope of carbon (^{12}C) that has not been removed by radioactive decay. Because half of the original amount of ^{14}C is lost by radioactive decay every 5780 years, radiocarbon dating is most useful over five or six half-lives (back to about 30,000 years ago), but in some cases it can be applied over the last 50,000 years or more (see Table 2–1).

Another technique relies on the same uranium (U) decay series used to date igneous rocks (see Table 2–1) but uses it in a different way to date corals. Ocean corals incorporate a small amount of ^{234}U and ^{238}U (but no ^{230}Th) from seawater into their shells (substituting it for calcium). When the corals die, the parent (^{238}U) slowly decays and produces ^{230}Th in the coral skeleton. In this case, however, the daughter product (^{230}Th) is not stable but radioactively decays away with a half-life of 75,000 years. Gradually the amount of ^{230}Th present in the coral moves toward a level that reflects a balance between the slow decay of the parent U and the faster loss of the daughter ^{230}Th. The clock provided by the Th/U ratio is useful for dating over the last few hundred thousand years. This technique is also used for dating stalactite and stalagmite deposits in caves.

Counting Annual Layers Some climate repositories contain annual layers that can be used to date archives by simply counting back in time year-by-year from the present. These annual layers form because of seasonal changes in the accumulation of distinctive materials.

The most visible forms of annual layering in ice (mountain glaciers and ice sheets) are alternations between darker layers that contain dust blown in from continental source regions during the dry, windy season and lighter layers marking the part of the year with little or no dust (Figure 2–9A). These dark/light couplets form annual layers that are easily visible in the upper parts of glacial ice but are gradually stretched and thinned deeper in the ice, where they cannot easily be discerned. Ages of these deeper parts of the ice are usually estimated by methods based on models of how the ice flows.

Sediments in some lakes contain annual couplets called **varves** (Figure 2–9B). These layers are particularly common in the deeper parts of lakes containing little or no life-sustaining oxygen. The lack of oxygen suppresses or eliminates bottom-dwelling organisms that would otherwise obliterate the thin annual layers by their physical activity. Varve couplets usually result from seasonal alternations between deposition of light-hued mineral-rich debris and darker sediment rich in organic material.

In regions of marked seasonal variations of climate, trees produce annual layers called **tree rings** (Figure 2–9C). These rings are alternations between thick layers of lighter wood tissue (cellulose) formed by rapid growth in spring and thin, dark layers marking cessation of growth in autumn and winter. Because most individual trees live no more than a few hundred years, the time span over which this dating technique can be used is limited, but in some areas distinctive year-to-year variations in tree ring thickness can be used to splice records from younger trees with records from older trees whose fossil trunks can still be found on the landscape.

In tropical oceans, corals record seasonal changes in the texture of the calcite ($CaCO_3$) incorporated in their skeletons (Figure 2–9D). The lighter parts of the **coral bands** are laid down in summer, during intervals of fast growth, and the darker layers are laid down during winter, when growth slows. Individual corals dated in this way rarely live more than a few decades or at most a few hundred years, but older records may be spliced into younger ones (as with tree rings).

Correlating Records with Orbital Cycles Another way to date climate records is to use the characteristic imprint of variations in Earth's solar orbit in a "tuning" exercise. Changes in Earth's orbit around the Sun alter the amount of solar radiation received by season and by latitude. The timing of these orbital variations is known very accurately from astronomical calculations (Part III), and the physical processes that link these orbital changes to climatic responses on Earth have become reasonably well understood in recent decades. The two most prominent examples are changes in the strength of low-latitude monsoons and the cyclical growth and decay of high-latitude ice sheets. Because of these relationships, climate scientists can date many of Earth's climatic responses by linking them to the well-dated external driver provided by the orbital variations. This technique provides scientists with absolute dating of Earth's climatic responses over many millions of years.

Internal Chronometers In specific instances, the techniques of counting annual layers and orbital tuning can serve a similar purpose much further back in time. Even in the absence of radiometric dates of absolute age (in years before the present), some climate archives contain internal chronometers with which climate scientists can measure *elapsed time* (duration in years).

For example, annual varves deposited in lake sediments millions of years ago still survive today in a few

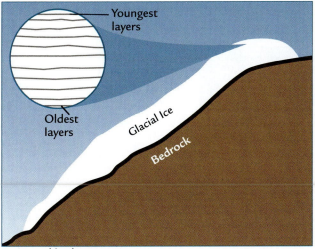

A Annual ice layers

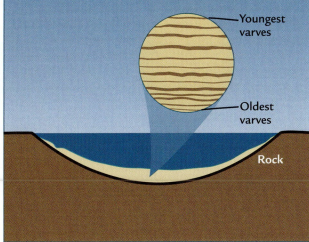

B Annual sediment varves

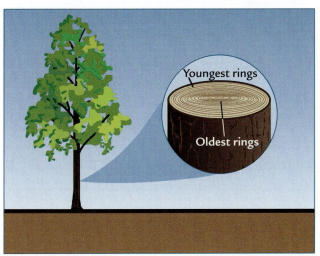

C Annual tree rings

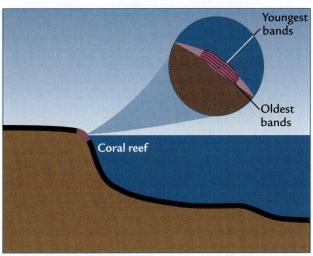

D Annual coral bands

FIGURE 2-9 Annual layering Four kinds of climate archives have annually deposited layers that can be used to date the climate records they contain: (A) ice, (B) varved lake sediments, (C) trees, and (D) corals.

protected regions. Determining the actual age of these sequences by counting varves back in time from the present is impossible because varves were not continuously deposited up until the present. However, the varves supply an internal chronometer with which to count the years that elapsed during the interval when they were deposited. This information can aid climatic interpretations.

2-3 Climatic Resolution

The extent to which the details of climatic information can be resolved depends mainly on the interplay between two factors: (1) the processes that initially disturb the climate record during and soon after deposition and (2) the rate at which the record is buried beneath additional sediments and protected from further disturbances.

Sediment Archives Most sedimentary archives used for climate studies form in *low-energy* marine environments undisturbed by turbulent waves and storms. The primary disturbance after particles settle out on the seafloor is physical stirring by deep-dwelling organisms (Figure 2–10). Organisms living on the sediment surface thoroughly mix the uppermost layers. A much smaller number of animals burrow deep into the sediments, but they do so only infrequently, and subsurface sediments are increasingly protected from most disturbances as they are buried. Eventually the sediments pass beneath the region of active mixing and become part of the permanent sedimentary record.

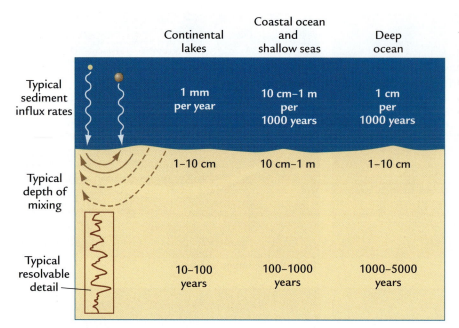

Continental lakes | Coastal ocean and shallow seas | Deep ocean

Typical sediment influx rates: 1 mm per year | 10 cm–1 m per 1000 years | 1 cm per 1000 years

Typical depth of mixing: 1–10 cm | 10 cm–1 m | 1–10 cm

Typical resolvable detail: 10–100 years | 100–1000 years | 1000–5000 years

FIGURE 2-10 Climate resolution
The degree of resolution of climate records in sediment archives is related to the rate of deposition (and burial) of sediment and to the amount of activity of organisms burrowing into the sediments.

Typical rates of sediment deposition range from as much as meters per year in coastal marine sequences and millimeters per year in lakes to millimeters per thousand years in some deep-sea sediments. Rates can vary locally around these average values by a factor of 10 because of factors such as the amount of sediment supplied locally by rivers or redistributed by currents.

The degree of disturbance by organisms that move across and burrow into the sediment surface also varies with environment. In highly productive coastal regions, large organisms burrow tens of centimeters or even meters down into the sediment. Relatively unproductive deep ocean basins have fewer and smaller bottom-dwelling organisms that typically burrow down no more than a few centimeters. Most lakes also have fewer and shallower burrowers. As a result, the resolution of sedimentary records varies with environment. Lakes usually have the best resolution and deep-ocean sediments the poorest, although locally rapid deposition can improve resolution in some ocean areas.

After particles pass through the upper layers, no further mixing occurs unless erosion reexposes the sequence back at the sediment-water interface. Increased pressure and loss of water caused by deep burial of sediments gradually compact the sediment layers and turn them into soft rock, but do not dramatically reduce the resolution they can provide.

Ice Cores Annual layers of snow are visible at the surfaces of many mountain glaciers and rapidly deposited ice sheets (see Figure 2–9A). As the snow is buried and slowly recrystallized into ice, annual layers remain resolvable to a depth that depends on their initial thickness at the time of deposition. Below this level, the layering is lost. In cores from the ice sheet on Greenland, where deposition of snow is rapid, the annual layering may remain detectable tens of thousands of years into the past. In the polar ice sheet covering eastern Antarctica, where only a small amount of snow accumulates each year, annual layering may not occur even at the ice surface.

Tree Rings and Corals At middle and high latitudes where trees produce annual layers, tree rings become a permanent record of annual climate change unless they are later disturbed by fire or by sporadic boring by insects or excavation by birds. Similarly, $CaCO_3$ bands in corals form a permanent record of seasonal to annual climate change.

The types of climate archives, the maximum time span of the records they contain, and the highest resolution achievable in each archive are summarized in Figure 2–11 in a log time scale that changes by powers of 10. Also shown at the top are the time spans covered by the major parts of this book.

Climatic Data

Climate archives contain many indicators of past climate referred to as **climate proxies.** Climate scientists use the term "proxy" (meaning "substitute") because the process of extracting climate signals from these indicators is not direct, like reading temperature from a thermometer. Instead, scientists must first determine the mechanism by which climate signals are recorded by the proxy indicators in order to decipher the climate changes. (Of course, even a typical thermometer relies on a "proxy" measurement—the height of a column of mercury calibrated to indicate temperature.)

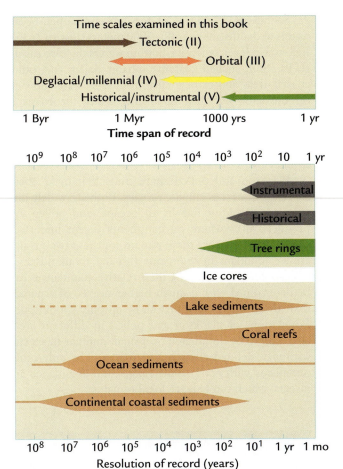

FIGURE 2-11 **Resolution of climate records** Climate
archives vary widely in the length of the records they contain
and in the degree of resolution they yield. A log scale
(changing by powers of 10) is needed to show all geologic time
in a single plot. (Adapted from J. C. Bernabo, *Proxy Data: Nature's
Record of Past Climates* [Washington, DC: National Oceanic and
Atmospheric Administration, 1978].)

The two climate proxies that are most commonly
used are (1) **biotic proxies,** which are based on changes
in composition of plant and animal groups, and (2)
geological-geochemical proxies, which are measurements of mass movements of materials through the
climate system, either as discrete (physical) particles or
in dissolved (chemical) form.

2-4 Biotic Data

Because no seafloor older than 170 million years exists,
broad-scale reconstructions of earlier oceanic environments are not possible. As a result, fossil remains from
the continents are the main climate proxy for older tectonic-scale intervals. Most of the organisms that have
ever existed on Earth are now extinct, and the further
back in time we look, the less recognizable the fossils

are. Using biotic proxies to reconstruct past climates
over longer tectonic time scales often requires a reliance
on the resemblance of past forms to their modern coun-
terparts either in general appearance or in specific features that can be measured.

Because fossil remains of plants tend to be more
numerous than those of animals in geologic records
from continents, vegetation plays a central role in the
reconstruction of ancient climates. Often the presence
of a single critical temperature-sensitive form is useful as
a climate indicator. For example, warmer climates tens
of millions of years ago are inferred from the presence of
palmlike trees at high northern latitudes (Figure 2–12).

For the younger continental record, climate scientists more commonly use the relative abundance of
climate-sensitive vegetation indicated by pollen assemblages deposited in sediments (Figure 2–13). Minute
pollen grains are produced in vast numbers by vegetation, distributed mostly by wind, and deposited in lakes,

FIGURE 2-12 **Past vegetation** For older geologic intervals,
climate on the continents can be inferred from distinctive
vegetation. The remains of trees similar to modern palms are
found in rocks from Wyoming dating to 45 million years ago.
Today frigid winters in Wyoming would kill palm trees.
(Chip Clark.)

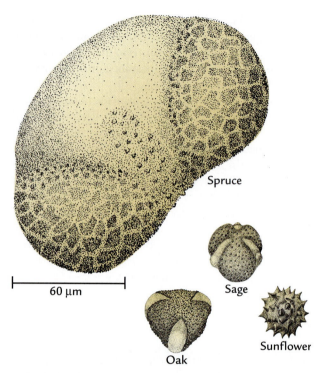

FIGURE 2-13 Pollen: a proxy indicator of climate on land
For younger intervals, climate on land can be reconstructed from changes in the relative abundance of distinctive types of pollen. For scale, small grains of sand are 60 μm or larger in diameter. (Courtesy of Alan Solomon, Environmental Protection Agency, Corvallis, OR.)

where they are preserved in oxygen-poor waters. Pollen can be identified initially by major vegetation type (trees, grass, and shrubs) and then further subdivided (spruce trees indicate cold climates; oak trees indicate warmth). Larger remains of vegetation that cannot have been carried far from their points of origin are also examined to make sure that the pollen in a lake sequence is representative of the nearby vegetation. These larger **macrofossils** include cones, seeds, and leaves.

In the oceans, four major groups of shell-forming animal and plant **plankton** are used for climate reconstructions (Figure 2–14). Two groups form shells made of calcite ($CaCO_3$). Globular sand-sized animals called **planktic foraminifera** (upper left) inhabit the upper layers of the ocean. Small spherical algae called Coccolithophoridae secrete tiny plates called **coccoliths** (lower left) in sunlit waters. Two other groups of hard-shelled plankton secrete shells of opaline silica ($SiO_2 \cdot H_2O$) and tend to thrive in productive, nutrient-rich surface waters. **Diatoms** (upper right) are silt-sized plant plankton usually shaped like either pillboxes or needles. **Radiolaria** (lower right) are sand-sized animals with ornate shells often shaped like premodern military helmets.

Sediments rich in $CaCO_3$ fossils occur in open-ocean waters at depths above 3500–4000 meters (Figure 2–15). Below that level, corrosive bottom waters dissolve calcite shells. SiO_2-shelled diatoms inhabit deltas and other coastal areas and extract silica from river water flowing off the land, but their abundance along the coasts is masked by the influx of mud eroded from the land. Radiolaria and diatoms are abundant in Antarctic and equatorial regions where highly productive waters upwell from below.

Plankton and pollen share traits that make them especially useful as climate proxies. Both are widely distributed: plankton live in all oceans, and pollen are produced everywhere on continents except under ice sheets. Also, because fossil remains of these two groups are so abundant in sediments (usually thousands in a tablespoon-sized sample), their relative abundances can be determined with a much higher degree of accuracy than those fossil types that show up only sporadically. Populations of plankton and pollen in different areas also tend to be dominated by a small number of species with well-defined climate preferences. The only other organisms with comparable ranges and abundance are insects, which rarely leave fossil remains.

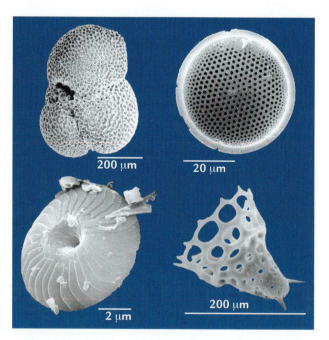

FIGURE 2-14 Plankton: a proxy indicator of climate in the ocean Four types of shelled remains of plankton are common in ocean sediments: $CaCO_3$ shells are represented by sand-sized planktic foraminifera (upper left) and small clay-sized coccoliths (lower left); SiO_2 shells include silt-sized diatoms (upper right) and sand-sized radiolaria (lower right). For scale, small grains of sand are 60 μm or larger in diameter. (Modified from W. F. Ruddiman, "Climate Studies in Ocean Cores," in *Paleoclimate Analysis and Modeling*, ed. A. D. Hecht [New York: John Wiley, 1977].)

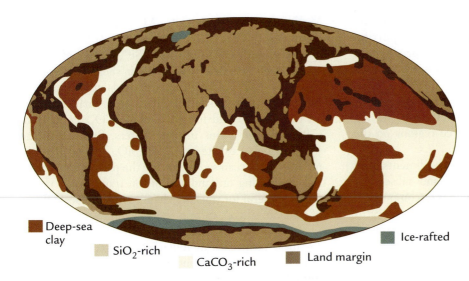

FIGURE 2-15 Distribution of ocean sediments The predominant type of sediment on the seafloor of the world ocean today varies regionally, with ice-rafted sediment in polar areas, SiO_2-rich sediment in productive areas, $CaCO_3$-rich sediment on higher rises and ridges, and windblown deep-sea silt and clay in basins far from continents. Coastal regions contain mainly debris from the land. (Modified from W. H. Berger, "Deep-Sea Sedimentation," in *The Geology of Continental Margins*, ed. C. A. Burke and C. L. Drake [New York: Springer-Verlag, 1974].)

- ■ Deep-sea clay
- ■ SiO_2-rich
- ■ $CaCO_3$-rich
- ■ Ice-rafted
- ■ Land margin

Most of the species of plankton and vegetation that live today have been present on Earth for hundreds of thousands to millions of years. The climatic preferences of these modern species can be accurately determined by comparing their present distributions to measurements of current climate. These modern climate preferences can then be used to reconstruct past climates from fossil assemblages with great accuracy in sediment archives as old as a few million years or more.

2-5 Geological and Geochemical Data

Mass movements of materials through the climate system are tied to processes of erosion, transport, and deposition, mainly by water but also by ice and wind. Most climate studies of the older parts of Earth's history rely on physical debris deposited in sedimentary archives on the continents as the main proxy for inferring past climates. For example, sediment textures can tell us about erosion and subsequent deposition of unsorted debris by ancient ice sheets in cold environments, sand dunes moving across deserts under extremely arid conditions, and deposition by water in moist environments. Although these sediment types are useful for drawing broad inferences about climate, poor dating control and the prevalence of erosion make detailed study of many older continental records difficult, and alteration of the deposits increases with the passage of time.

In contrast, ocean sediments from the last 170 million years provide relatively continuous deposition, better dating, and wide geographic coverage. As a result, the distribution of sediment types that carry distinctive information about climate can be mapped, and changes in their patterns of deposition can be quantified as **burial fluxes** (measures of the mass of sediment deposited per unit area per unit time).

Sediment is eroded from the land and deposited in ocean basins in two forms. One is debris eroded and transported as discrete particles or grains as a result of **physical weathering,** the process by which water, wind, and ice physically detach pieces of bedrock and reduce them to smaller fragments. One example shown in Figure 2–16 is coarse **ice-rafted debris** (sand and gravel) eroded by ice sheets and delivered by icebergs that melt in ocean waters. Other examples include finer **eolian sediments** (silts and clays) lifted from the continents and blown to the ocean by winds and **fluvial sediments** carried in a wide range of grain sizes by rivers to the ocean.

FIGURE 2-16 Sediment particles Deep-ocean sediments contain granular debris from land that reveals the climate of the source region. For example, sand-sized grains of quartz and other minerals rafted in from ice sheets by icebergs indicate cold climates. (Courtesy of Gerard Bond, Lamont-Doherty Earth Observatory of Columbia University.)

Geological and geochemical techniques can unravel the original sources of sediments formed by physical weathering. Microscope counts of sand-sized grains from marine sediments can distinguish different sources on the basis of distinct mineral types. In recent years, geochemical analyses of distinctive elements and isotopes have become an additional method of tracing mineral grains back to specific source regions on the continents.

The second major way of removing sediments from the land is by **chemical weathering** and subsequent transport of dissolved ions (charged ions or compounds) to the oceans in rivers (Figure 2–17). Chemical weathering occurs mainly in two ways: (1) by **dissolution,** in which carbonate rocks (such as limestone, made of $CaCO_3$) and evaporite rocks (such as rock salt, made of $NaCl$) are dissolved in water, and (2) by **hydrolysis,** in which weathering adds water to minerals derived from continental rocks made of silicates, such as basalts and granites. Both processes depend on the fact that atmospheric CO_2 and rain (H_2O) combine in soils and rock crevices to form carbonic acid (H_2CO_3), a weak acid that attacks rocks chemically. After weathering, rivers carry off many dissolved materials, including ions (Ca^{+2},

Mg^{+2}, Na^{+1}, K^{+1}, Sr^{+2}, Cd^{+2}, $A1^{+4}$, and Cl^{-1}), and ion complexes (HCO_3^-, CO_3^{-2}, and $SiO(OH)_2$).

Some of the dissolved ions (Si^{+4}, Ca^{+2}, and CO_3^{-2}) are used by plankton to form their shells (see Figures 2–14 and 2–17). A small fraction ends up in the shells of **benthic foraminifera,** sand-sized animals that live on the seafloor and form calcite ($CaCO_3$) shells from Ca^{+2} and CO_3^{-2} ions in deep waters. Because all the shells made of calcite and opal ($SiO_2 \cdot H_2O$) that are preserved in ocean sediments are the products of chemical weathering on land and of ion transport in rivers, they are useful for tracking changes in large-scale fluxes of calcium, silicon, carbon, and oxygen over time.

Because it takes a long time in the lab to analyze the chemical properties of individual samples taken from thick sedimentary sequences, many recent studies have turned to logging techniques that quickly detect and record key physical or chemical properties of the sediments. Sediment cores are moved through a detection unit that uses sound waves or other nondestructive techniques to sense the sediment properties at a rapid rate without disturbing them.

A wide range of important climatic data is also stored in the isotopes of elements in the calcite shells of planktic

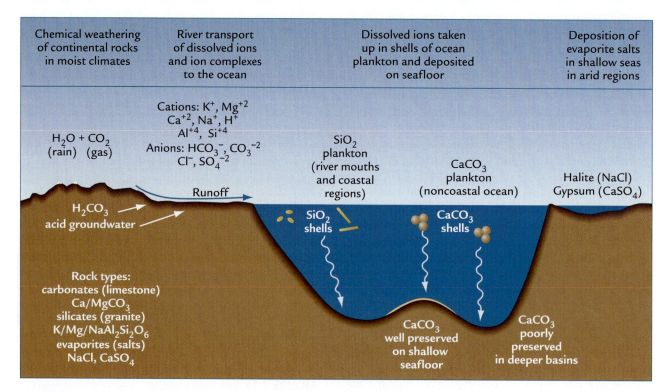

FIGURE 2-17 Chemical weathering, transport, and deposition Chemical weathering slowly attacks rocks on land and sends dissolved ions into rivers for transport to the ocean. Ocean plankton incorporate some of the dissolved ions in their shells, which fall to the seafloor and form part of the geologic record. Some dissolved ions are also deposited in shallow evaporating pools on continental margins where the climate is dry.

organisms and benthic foraminifera. Cases that will be examined in detail in later chapters include isotopes of oxygen, which record changes in the global volume of ice and in local ocean temperatures (Chapters 6 and 9); and isotopes of carbon, which trace movements of organic material among reservoirs on the continents, in the air, and in the ocean (Chapter 10).

Additional geochemical proxies gradually become available over the younger part of Earth's long climatic history. At orbital time scales, ice cores contain samples of air from past atmospheres, including concentrations of the greenhouse gases carbon dioxide (CO_2) and methane (Chapter 10). Other important proxies in ice cores include changes in the thickness of snow deposited (related to the temperature and moisture content of the air), in the amount of dust delivered by winds from various continents; and in isotopes of oxygen and hydrogen that measure air temperatures over the ice sheet.

Cave deposits contain records of groundwater derived from atmospheric precipitation. Changes in the chemical composition of this water reflect changes in the original sources of the water vapor, in the atmospheric transport path to the site of precipitation, and in the groundwater environment (Chapter 10). Sedimentary deposits in lakes record not only changes in pollen but also climatically driven fluctuations in lake levels (Chapters 12 and 13) and other chemical tracers now under active investigation.

Trees record the amount of cellulose deposited in each annual layer (determined from the width and density of tree rings) as an index of changes in precipitation during the rainy season in dry regions and changes in summer temperatures in cold regions (Chapter 16). Annual coral bands contain a wide range of chemical information, including ratios of isotopes of oxygen that record changes in temperature and precipitation (Chapter 16).

Climate Models

Scientists who extract records from Earth's climate archives inevitably discover new trends that were previously unknown. Usually, their proposed explanations for the trends are tested using climate models because models put numbers on ideas. But models also simplify some aspects of reality, and the results they provide have to be critically assessed.

In this section we examine two kinds of numerical (computer) models used by climate scientists. **Physical climate models** emphasize the physical operation of the climate system, particularly the circulation of the atmosphere and ocean but also interactions with vegetation (biology) and with atmospheric trace gases (chemistry). **Geochemical climate models** track the movement of distinctive chemical tracers through the climate system.

2-6 Physical Climate Models

Most physical models are constructed to simulate the operation of the climate system as it exists today. The modern climatic system is described on the companion Web site at www.whfreeman.com/ruddiman2e. The simulation of modern climate is called the **control case.** Models must simulate modern climate reasonably well to be trusted as a tool for exploring past climates.

Simulations of past climates occur in a three-step process (Figure 2–18). The first step is to choose the experiment to be run by specifying the input to the model. One or more aspects of the model's representation of the modern world are altered from their present form to reflect changes known to have occurred in the past. For example, the level of CO_2 in the model atmosphere might be increased or decreased, the height of its mountains raised or lowered, ice sheets removed or added, or the position of continents moved around. These features that are altered to test hypotheses of climate change are called the **boundary conditions.**

The second step is the actual operation of the model. Physical laws that drive the flow of heat energy through Earth's climate system are incorporated in the internal workings of the model. When an experiment is run, these laws come into play in a **climate simulation.**

The third step is to analyze the **climate data output** that emerges from the experiment. The data from the simulation can then be used to evaluate the hypotheses being tested. For example, does a specific change in boundary conditions cited in a hypothesis (atmospheric CO_2 level, mountain elevation, or continental position) affect climate in the way the hypothesis proposed?

Often climate data output can be tested against independent geologic data that played no part in the experimental design (Figure 2–18). For example, if a model run simulates stronger winds in a specific region for a particular interval of geologic time, scientists can sample sediment cores from that area to check whether or not larger particles of windblown dust were deposited in the locations indicated by the simulation.

Mismatches between geologic data and climate data output from physical circulation models may imply several possible problems: key boundary conditions were specified incorrectly or were omitted from the experiment; the model does not adequately simulate some part of the climate system; or the geologic data used for comparison to the model output were misinterpreted. Despite this range of possible problems, the main cause of data-model mismatches is often obvious enough to lead to useful refinements in boundary conditions, in data interpretation, or in model construction. The

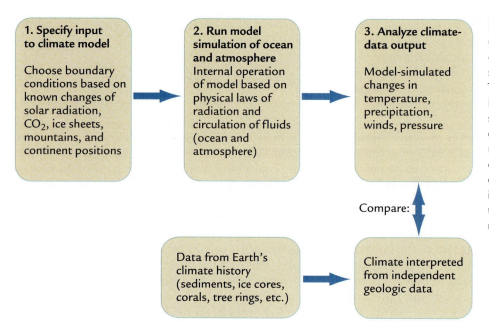

FIGURE 2-18 Data-model comparisons Models of Earth's climate are constructed to simulate present circulation. Then changes based on Earth's history (different CO_2 levels, ice sheet sizes, or mountain elevations) are inserted into the model, and simulations of past climates are run. The climate output is compared with independent geologic data to test the performance of the model.

science of reconstructing past climates moves ahead best when the strengths and limitations of both the data and the models are constantly tested against each other. This review starts with models of atmospheric circulation, then looks at ocean models, and finally briefly reviews physical models that simulate changes in ice and vegetation.

Atmospheric Models Models of Earth's atmosphere vary widely in complexity. Simpler models are less expensive to run and can simulate the evolution of climate over long intervals of time (thousands of years), but they lack or oversimplify important parts of the climate system. Complex models incorporate a more complete physical representation of the climate system, but they do so at the cost of being slower, more expensive, and able to simulate only brief snapshots of climate over a few years.

One-dimensional "column" models are the simplest kind of physical model of the atmosphere. They simulate a single vertical column of air that represents the average structure of the atmosphere of the entire planet. This air column is divided into layers that are closely spaced near Earth's surface and are more widely spaced at higher elevations. Each layer contains climatically important constituents, such as greenhouse gases and dust particles. Earth's surface is represented by a global average value that has the globally averaged properties of the water, the land, and the ice. One-dimensional (1-D) models offer a way of gaining an initial understanding of climatic effects of changes in concentrations of greenhouse gases and of airborne particles called **aerosols,** such as volcanic ash and dust.

Two-dimensional (2-D) models are a step toward a more complete portrayal of the climate system. One type of 2-D model includes an atmosphere with many vertical layers and a second dimension that represents Earth's physical properties averaged by latitude. A second dimension (even a simplified, average one) makes it possible to use these models to simulate processes that vary from pole to equator because snow and ice occur mainly at higher latitudes. Because 2-D models can simulate long intervals of time quickly and inexpensively, they are used to explore longer-term interactions among the ocean surface, sea ice, and land. They are also used in combination with models of slowly changing ice sheets (Chapter 9).

Three-dimensional **atmospheric general circulation models (A-GCMs)** provide still more complete numerical representations and simulations of the climate system. These 3-D models have the capacity to represent many key features: the spatial distribution of land, water, and ice; the elevation of mountains and ice sheets; the amount and vertical distribution of greenhouse gases in the atmosphere; and seasonal variations in solar radiation.

The boundary conditions for A-GCM experiments are specified for hundreds of model **grid boxes,** like those shown in Figure 2–19. The vertical boundaries of the grid boxes are laid out along lines of latitude and longitude at (and above) Earth's surface, and the box size shrinks near the poles because lines of longitude converge there. The horizontal boundaries of the grid boxes divide the atmosphere along lines of equal altitude above sea level. Models generally have 10 to 20 vertical layers that are more closely spaced near Earth's surface because the interactions with the land, water, and ice surfaces in the lower atmosphere are more complex than the smoother flow higher in the atmosphere.

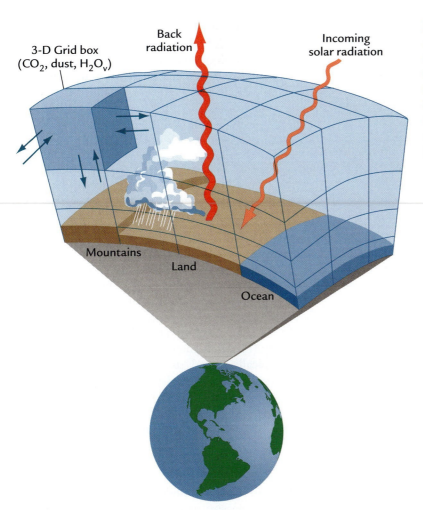

3-D Grid box
(CO_2, dust, H_2O_v)

Back radiation

Incoming solar radiation

Mountains

Land

Ocean

FIGURE 2-19 3-D GCMs General circulation models (GCMs) are full 3-D representations of Earth's surface and atmosphere, represented by individual grid boxes. Representations of Earth's surface within each grid box are entirely land, ocean, or ice. (Adapted from W. F. Ruddiman and J. E. Kutzbach, "Plateau Uplift and Climate Change," *Scientific American* 264 [1991]: 66–75.)

The operation of A-GCMs incorporates the physical laws and equations that govern the circulation of Earth's atmosphere: the fluid motion of air; conservation of mass, energy, and other properties; and gas laws covering the expansion and contraction of air. The individual grid boxes in A-GCMs interact with their immediate neighbors.

Model runs begin with the atmosphere in a state of rest. After solar heating causes air to begin to move, the model is run long enough for the atmosphere to reach a state of equilibrium (Figure 2–20). Equilibrium occurs when the long-term drift in the simulated climate data disappears. The oscillations that remain are analogous to short-term changes in weather over days and weeks.

Running climate experiments on current-generation A-GCMs requires a simulation of at least 20 years of climate. The first 15 years of the simulation are the "spin-up" interval, used to let the model attain a state of equilibrium. The last 5 years of the simulation produce the climate data that form the actual output of the model. For the control-case simulation of modern climate, the climate-data output from the GCM are compared with regional instrumental measurements of temperature, precipitation, pressure, and winds in the present climate system averaged over the last several decades (for example, Figure 2–21). Areas of major disagreement between the model output and instrumental observations often become the focus of additional improvements in the model.

As noted earlier, A-GCM experiments on past climates require scientists to specify major changes in boundary conditions on the basis of geological evidence from Earth's history. In one approach, called a **sensitivity test,** just one boundary condition is altered in relation to the present-day configuration. When the output of such an experiment is compared with the output from the modern control case, the differences in climate between the two runs isolate and reveal the unique impact caused by the change in that one boundary condition.

In contrast, a climate **reconstruction** requires changing all known boundary conditions at the same time to try to simulate the full state of the climate system at some time in the past. This approach is more demanding than a sensitivity test, because all the potentially critical

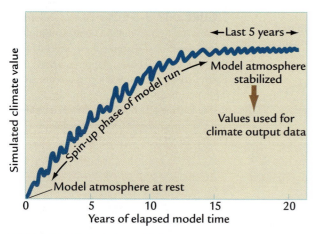

FIGURE 2-20 Model equilibrium Atmospheric GCMs require about 15 years of simulated climate change before they arrive at an equilibrium state. The final 5 years of the simulation are then averaged for use as the climate data output.

boundary conditions are rarely known well enough to specify as input to the simulation. This method is used mainly to study glacial maximum and deglacial climates of the last 20,000 years, an interval for which numerous records dated by ^{14}C methods exist.

Every 1 to 2 years the power of the world's best computers increases by a factor of 10. Over time, this increase in computing power has gradually reduced the horizontal size of the grid boxes used in GCMs. Typical grid boxes were once 8° of latitude by 10° of longitude, or as much as 1000 kilometers on a side. More recently GCM grid boxes have been reduced to 2° of latitude by 3° of longitude, or no more than 300 kilometers on a side. The result has been improved resolution of coastal outlines of continents (including narrow isthmuses) and of small seas, larger ocean islands, and large lakes. For the first time, A-GCMs can now "see" (that is, resolve) New Zealand!

The shrinking size of grid boxes has also improved the way elevation is represented in GCMs. Although low-resolution models captured the basic rounded shape of broad high plateaus and ice sheets, they smoothed the high but narrow mountain ranges such as the Andes into low-elevation blobs. Higher-resolution models increasingly distinguish these narrower features.

Increasing computer power has also allowed modelers to include more aspects of the climate system in recent A-GCMs. Features of the climate system such as soil moisture levels or vegetation types that once had to be fixed at modern values and were not allowed to interact with the model's atmosphere are now included as interactive components.

The modeling process is not a steady one-way march toward success. Initial attempts to include new

components in models are often so crude that they make the resulting climate simulations less realistic than those obtained from models that had simply held those components fixed at modern values. Only with more refined representations do the newly added components perform in a realistic way and make the resulting simulations clearly superior to the earlier versions.

Ocean GCMs Models of ocean circulation are at a slightly more primitive stage of development than atmospheric GCMs. One reason is that climate researchers know much less about the modern circulation of the oceans, especially critical processes such as the brief but intense episodes of deep-water formation at high latitudes. As a result, scientists do not have as well defined a modern target for ocean models to reproduce.

Three-dimensional ocean models (O-GCMs) are similar to A-GCMs (Figure 2–22). The lower boundary is the seafloor, broken into flat stair steps marking boundaries between individual ocean grid boxes. The upper boundary of the ocean model is the air-sea boundary. The horizontal grid boxes that subdivide the ocean typically cover 3° to 4° of latitude and longitude. The dozen or so vertical layers in the ocean are more closely spaced near the sea surface, where the flow is faster and interactions with the atmosphere are more complex, than at greater depth, where the ocean flow is slower. Typical climate-data output from O-GCM experiments includes ocean temperature, salinity, and sea-ice extent.

Like atmospheric models, most ocean GCMs are limited by the size of their grid boxes. They cannot capture the shape of very small openings, such as the modern mouth of the Mediterranean Sea at the Strait of Gibraltar. These narrow openings are important in the large-scale circulation of the ocean and critical to the success of ocean-model simulations. Most ocean models also cannot yet resolve details of flow in narrow, swift currents such as the Gulf Stream.

Models that include the full structure of the ocean are not directly coupled to atmospheric models. The problem with doing so is that air and water respond to climate changes at different rates and thus put different computational demands on each type of model. Ocean models can ignore interactions that occur on a daily cycle because these short-term changes have negligible effects on most ocean circulation. As a result, O-GCMs need to calculate changes only over time steps separated by a month or more. In contrast, daily changes are critical to models of the fast-responding atmosphere. Therefore, A-GCMs need to calculate changes in time steps separated by just a few hours, and the cost of simulating the same amount of "model time" is more expensive.

This basic incompatibility between the two kinds of models can be overcome by an approach known as *asynchronous coupling*. This procedure involves an ongoing

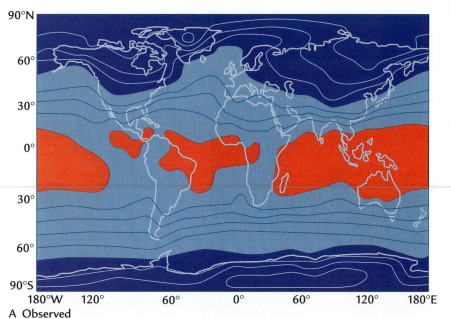

A Observed

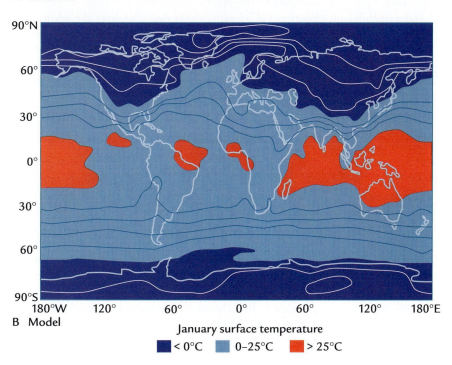

B Model

January surface temperature

■ < 0°C   ■ 0–25°C   ■ > 25°C

FIGURE 2-21 Control-case simulations GCMs are developed by testing how well they reproduce modern climate (temperature, precipitation, and winds) based on present boundary conditions (CO_2, mountains, and land-sea distribution). This case compares (A) observed January surface temperatures with (B) model-simulated values. (Adapted from J. Hansen et al., "Efficient Three-Dimensional Global Models for Climate Studies: Models I and II," *Monthly Weather Review* 111 [1983]: 609–62.)

series of runs, first using the atmosphere to drive the ocean, then the ocean to drive the atmosphere, and so on. The ocean and atmosphere exchange heat, water and water vapor, and wind-driven momentum. Going back and forth between the ocean and atmosphere models keeps the two systems from getting too far out of touch with each other. With the atmospheric model run only at selected intervals, the computer does not have to make short-term calculations of the atmospheric circulation through the entire simulation, and the overall simulation can progress much faster. In recent years, models have been developed that couple

the ocean more directly to a simplified version of the circulation of the atmosphere.

Ice Sheet Models Continent-sized ice sheets slowly grow and shrink over thousands to tens of thousands of years (see Table 1–1). A-GCMs can simulate the instantaneous effects that these high, broad, reflective masses of ice have on the rest of the climate system, including the circulation of the nearby atmosphere and ocean. The output from a GCM run spanning a few years of simulated time can also be examined to see whether an ice sheet accumulated or lost mass during the brief simulation. The answer tells modelers whether the ice

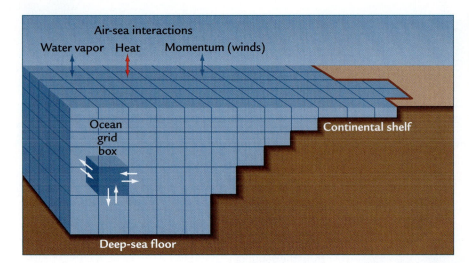

Air-sea interactions
Water vapor   Heat   Momentum (winds)

Ocean
grid
box

Continental shelf

Deep-sea floor

FIGURE 2-22 Ocean GCMs Ocean models use 3-D grid boxes that represent the shapes of ocean basins. Exchanges of water, heat, and momentum between the ocean and the atmosphere occur at the sea surface.

would have slowly melted or grown or remained at constant size under the climate conditions simulated.

A-GCMs can reproduce only short-lived snapshots of the circulation of the atmosphere, and as a result they cannot simulate the slow evolution of ice sheets over long intervals of time. To learn about this longer-term response, climate scientists create physical models of the ice sheets. One simplified type of ice-sheet model has two dimensions, one vertical and the other showing average variations with latitude but omitting any representation of longitude. These 2-D ice sheet models have been used to simulate the growth and decay of ice sheets in the northern hemisphere over tens of thousands of years in response to changes in solar radiation caused by changes in Earth's orbit. The models simulate features such as changes in ice accumulation and melting with ice elevation, flow within the ice, and depression of underlying bedrock by the weight of the ice. The 2-D ice sheet models can also be linked to 2-D atmospheric circulation models to simulate interactions among the ice sheets, atmosphere, and land surface. Some ice sheet models are three-dimensional, with the ice accumulating on a specified land surface (such as Antarctica) divided into grid boxes 50 to 100 kilometers on a side.

Vegetation Models Vegetation is an active component in the climate system, and the representation of vegetation in climate models has progressed through several stages. Early A-GCMs either ignored vegetation entirely or specified a representation of modern vegetation that did not interact with the changes in climate simulated by the model.

More recent models incorporate vegetation in an interactive way. One such modeling approach works in two steps. First, climate data derived as output from a GCM experiment (changes in temperature and precipitation) are used as input to a vegetation model that

simulates the resulting changes in vegetation. Then the simulated changes in vegetation are used as input to another GCM experiment that simulates the additional climatic feedback effects caused by the *changes* in vegetation (primarily increases or decreases in recycling of water vapor and in reflectivity of Earth's surface). Another approach embeds a vegetation submodel more directly in the main model.

2-7 Geochemical Models

Geochemical models are used to follow the movements of Earth's materials (called **geochemical tracers**) through the climate system. Unlike physical circulation models, most geochemical models do not reproduce the physical processes that govern the flow of air and water. Instead, the models trace the sources, rates of transfer, and ultimate depositional fate of two major components: sediment particles that result from physical weathering (wind, water, and ice), and dissolved ions produced by chemical weathering (dissolution or hydrolysis). Movements of tracers can be evaluated if they are not created or destroyed by radioactive decay along the way. Geochemical models can also trace exchanges of biogeochemical materials such as carbon or oxygen isotopes that cycle back and forth among the atmosphere, ocean, ice, and vegetation.

One-Way Transfer Models The most basic kind of model tracks transfers of material from its source or sources to the ultimate sites of deposition, such as debris eroded from the land and deposited in ocean sediments. If the material deposited has distinctive geochemical characteristics, it can be analyzed and its abundance quantified in terms of a flux rate—its rate of burial in that sedimentary archive (Figure 2–23). For example, scientists can quantify the rate of influx of ice-rafted debris to high-latitude polar oceans by extracting

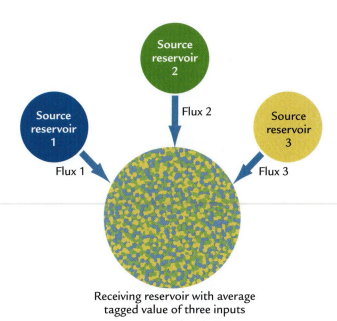

Flux 2

Source reservoir 1

Flux 1

Source reservoir 3

Flux 3

Receiving reservoir with average tagged value of three inputs

FIGURE 2-23 One-way transfers Geologists and geochemists often need to distinguish the separate contributions of several sources (usually linked to weathering of continental rocks) to a single depositional archive (such as ocean sediments).

all sediment that is sand-sized or larger and separating the mineral grains from the shells of fossil plankton. This analysis quantifies a process—changes in the production and flow of icebergs—that is directly related to climate.

The analysis can be carried a step further by counting the ice-rafted debris under a microscope to separate it into different types of grains (such as volcanic debris, quartz, and limestone). The composition of these grains can provide a general idea of source regions (for example, in the North Atlantic, volcanic debris that came from Iceland, and quartz and limestone that came from Europe or North America). Further subdivisions can be made by analyzing the grains for their isotopic composition or other distinctive chemical characteristics. This level of analysis might tell climate scientists which region within a particular continent was the source of some of the grains.

A more complicated situation arises if the material examined is fine-grained and has been derived from multiple sources. For example, fine silt and clay deposited in the North Atlantic Ocean could have been ice-rafted from North America or Europe, blown in from North Africa by dust storms, or carried in by deep currents from other sources. Although it is easy to measure the total accumulation rate of fine sediment per unit of time, it is not practical to try to separate out the individual small particles.

When subdivision of the fine material is physically impossible, chemical analysis offers an alternative, if each source of fine sediment is marked with a distinctive chemical value. One typical chemical marker is the ratio of isotopes of a single element. These different inputs combine to determine the average value of the fine-grained sediment (see Figure 2–23). The goal of this kind of analysis is to understand how the individual fluxes combine to create this average value.

Chemical Reservoirs A different modeling approach is used for geochemical tracers that are transported in dissolved form. Mass balance models divide Earth's systems into **reservoirs,** including the atmosphere, ocean, ice, vegetation, and sediments. The ocean is the most important reservoir: it receives almost all erosional products from the continents, it interacts with all of the other reservoirs, and it deposits tracers in well-preserved sedimentary archives.

The ocean reservoir is somewhat analogous to a bathtub (Figure 2–24). It gradually receives the inputs of geochemical tracers, in the same way that water slowly drips from a faucet into a large tub, and it loses geochemical-tracer outputs like water leaking slowly through a drain. The tracer also stays in the ocean for a specific amount of time, the way water does in a drippy, leaky tub.

If the flux rates of a tracer into and out of a particular reservoir (the ocean) are equal, the system is said to be at steady state: no net gain or loss of the tracer occurs in the reservoir. By analogy, if the drip from the faucet and the leak down the drain are perfectly balanced, the

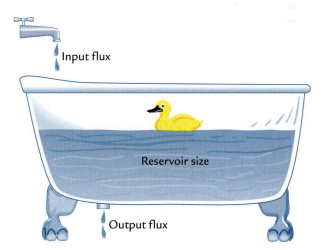

Input flux

Reservoir size

Output flux

FIGURE 2-24 Geochemical reservoirs and fluxes Geochemical reservoirs are like bathtubs with the faucet and drain both left partly open. The faucet delivers the input flux, the drain takes away the output flux, and the balance between the input and output determines the water level in the tub (reservoir). At steady state, input and output are in balance, and the water level in the tub remains constant.

water level in the tub will stay the same, even though new water continually enters and leaves the tub.

The **residence time** is the time it takes for a geochemical tracer to pass through a reservoir. In the tub analogy, the residence time is the time the average molecule of water takes to pass from the faucet to the drain. For a reservoir at steady state (a tub with an unchanging water level), the residence time is

Residence time = Reservoir size/Flux rate in (or out)

Reservoir-Exchange Models The methods discussed to this point have been based on one-way mass transfers in which geochemical tracers leave the interactive climate system by being buried in seafloor sediments and isolated out of touch with other reservoirs for millions of years. Another important exchange is the movement of a geochemical tracer back and forth between two (or more) reservoirs (Figure 2–25). In this case the tracer never comes permanently to rest in either reservoir. Instead, it is the movement between reservoirs that is of interest to climatic scientists. As before, the tracer is naturally tagged with a distinctive value, but in this case it moves back and forth between a larger reservoir (usually the ocean) and a smaller one (often ice sheets or vegetation). The history of exchanges is usually detected in the sediment record from the larger reservoir (the ocean), but the goal is to monitor changes in size of the smaller reservoirs (the volume of ice or the amount of vegetation).

One example is the transfer of water between the ocean and ice sheets on orbital time scales (discussed in Chapters 9 and 12). Exchanges of water between the relatively small reservoir stored in ice sheets on land

and the much larger reservoir left behind in the ocean can be tracked by using the fact that the isotopic composition of oxygen in the H_2O molecules in ice sheets is different from the average composition of the ocean. Measurements of the oxygen isotope composition of the ocean in shells of plankton provide a way to estimate past changes in the volume of ice stored on land.

Another useful application of reservoir-exchange analysis examines fluxes of carbon among its many reservoirs. Fluxes of carbon between the relatively small reservoir of carbon stored in land vegetation and the much larger carbon reservoir in the ocean can be tracked by using the fact that terrestrial carbon has a carbon isotope ratio distinctively different from that of marine carbon (Chapter 11). Net transfers of terrestrial carbon from land to sea can be detected by examining the average carbon isotope composition of the ocean recorded in the shells of calcite ($CaCO_3$) organisms buried in ocean sediments.

Key Terms

moraines (p. 19)
loess (p. 19)
historical archives (p. 21)
instrumental records (p. 21)
radiometric dating (p. 21)
parent isotope (p. 21)
daughter isotope (p. 21)
closed system (p. 22)
half-life (p. 22)
radiocarbon dating (p. 23)
varves (p. 24)
tree rings (p. 24)
coral bands (p. 24)
climate proxies (p. 26)
biotic proxies (p. 27)
geological-geochemical proxies (p. 27)
macrofossils (p. 28)
plankton (p. 28)
planktic foraminifera (p. 28)
coccoliths (p. 28)
diatoms (p. 28)
radiolaria (p. 28)
burial fluxes (p. 29)
physical weathering (p. 29)

ice-rafted debris (p. 29)
eolian sediments (p. 29)
fluvial sediments (p. 29)
chemical weathering (p. 30)
dissolution (p. 30)
hydrolysis (p. 30)
benthic foraminifera (p. 30)
physical climate models (p. 31)
geochemical climate models (p. 31)
control case (p. 31)
boundary conditions (p. 31)
climate simulation (p. 31)
climate data output (p. 31)
aerosols (p. 32)
atmospheric general circulation models (A-GCMs) (p. 32)
grid boxes (p. 32)
sensitivity test (p. 33)
reconstruction (p. 33)
geochemical tracers (p. 36)
reservoirs (p. 37)
residence time (p. 38)

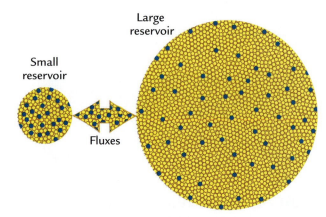

FIGURE 2-25 Reservoir exchange models Some geochemical models are designed to track reversible exchanges of important components such as water and carbon as they cycle between smaller reservoirs such as ice sheets and vegetation and the larger ocean reservoir.

Review Questions

1. Why does the importance of different climate archives change for different time scales?

2. Why are ocean sediments and ice cores important archives of climate?

3. How does the method of dating climate records vary with the type of archive?

4. How does the resolution from sedimentary archives vary with depositional environment?

5. Which two major groups of organisms are most important to climate reconstructions over the past several million years?

6. Describe how the products derived from physical and chemical weathering provide different kinds of information about the climate system.

7. Describe two ways the performance of climate models is evaluated.

8. Why aren't models of the atmosphere and ocean allowed to interact continuously?

9. Describe two features that make the ocean useful in geochemical mass balance models.

Additional Resources

Basic Reading

http://www.ngdc.noaa.gov/paleo/softlib.html. Maintained by the World Data Center for Paleoclimatology in Boulder, Colorado. Contains climate data of all kinds, as well as the locations of all sites that contain each type of data.

Advanced Reading

Bradley, R. S. 1998. *Paleoclimatology: Reconstructing Climates of the Quaternary.* International Geophysics Series, vol. 64. San Diego: Harcourt Academic Press.

Hecht, A. D., ed. 1985. *Paleoclimate Analysis and Modeling.* New York: John Wiley.

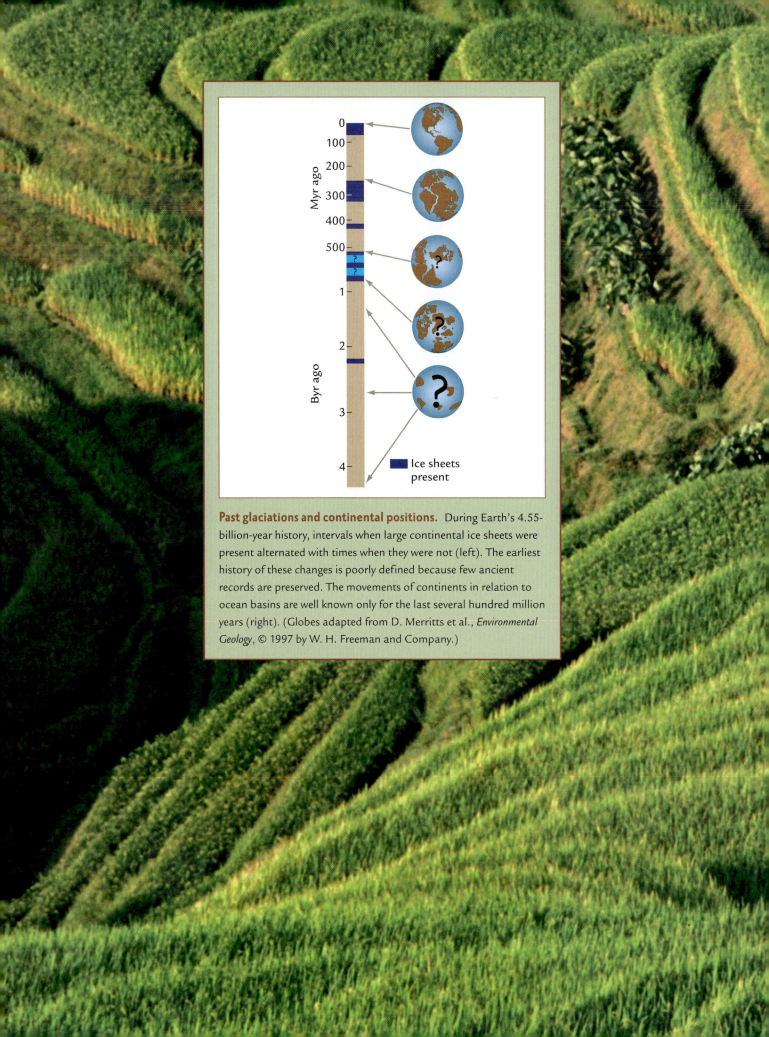

Past glaciations and continental positions. During Earth's 4.55-billion-year history, intervals when large continental ice sheets were present alternated with times when they were not (left). The earliest history of these changes is poorly defined because few ancient records are preserved. The movements of continents in relation to ocean basins are well known only for the last several hundred million years (right). (Globes adapted from D. Merritts et al., *Environmental Geology*, © 1997 by W. H. Freeman and Company.)

PART II

Tectonic-Scale Climate Change

I n this part we examine the longest time scale of climate change on Earth, the tectonic scale. Tectonic processes driven by Earth's internal heat have altered Earth's geography and altered climate for billions of years. Prior to the last several hundred million years, the record of climate change is relatively sparse, and the basic configurations of continents and oceans are poorly known. Yet we do know that Earth's climate remained relatively moderate, neither cold enough to freeze solid nor hot enough for its oceans to boil away. Earth has a built-in thermostat that allowed it to avoid those extremes.

As the movements and locations of the continents become better known, greater insights into climatic cause-and-effect relationships emerge. We know that Earth's climate has oscillated between times when ice sheets were present somewhere on Earth (such as today) and times when no ice sheets were present (about 100 Myr ago). These oscillations and their causes are the primary focus of Part II. Evidence of climate change over this long span of Earth's history comes mainly from sediments preserved on the continents and their margins.

We explore the following basic questions about Earth's tectonic-scale climate history:

- **Why has Earth remained habitable throughout its history?**
- **What explains the changes in Earth's climate over the last several hundred million years?**
- **Why was Earth ice-free even in polar regions 100 Myr ago?**
- **What are the causes and climatic effects of changes in sea level through time?**
- **How did the apocalyptic asteroid impact 65 Myr ago affect climate?**
- **What caused Earth's climate to cool over the last 55 Myr?**

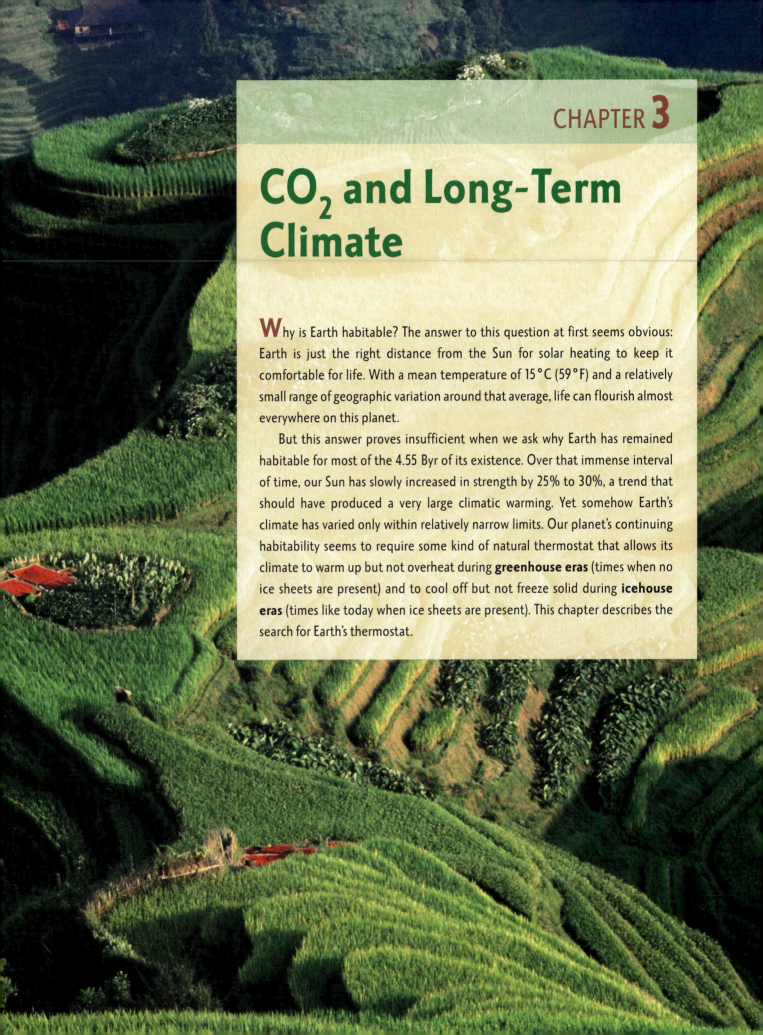

CHAPTER 3

CO$_2$ and Long-Term Climate

Why is Earth habitable? The answer to this question at first seems obvious: Earth is just the right distance from the Sun for solar heating to keep it comfortable for life. With a mean temperature of 15°C (59°F) and a relatively small range of geographic variation around that average, life can flourish almost everywhere on this planet.

But this answer proves insufficient when we ask why Earth has remained habitable for most of the 4.55 Byr of its existence. Over that immense interval of time, our Sun has slowly increased in strength by 25% to 30%, a trend that should have produced a very large climatic warming. Yet somehow Earth's climate has varied only within relatively narrow limits. Our planet's continuing habitability seems to require some kind of natural thermostat that allows its climate to warm up but not overheat during **greenhouse eras** (times when no ice sheets are present) and to cool off but not freeze solid during **icehouse eras** (times like today when ice sheets are present). This chapter describes the search for Earth's thermostat.

Greenhouse Worlds

The first clue that a factor other than distance to the Sun is involved in Earth's habitability comes from comparing it to Venus, another "terrestrial" planet with a similar overall chemical composition (Figure 3-1). Venus is a very hot planet with a mean surface temperature of 460°C, and it lies 72% as far from the Sun as Earth does.

The average amount of solar radiation sent to each planet varies inversely with the square of its distance from the Sun ($1/d^2$). Based on this relationship, Venus receives almost twice (1.93 times) as much solar radiation as Earth does:

$$\frac{Earth}{Venus} \quad \frac{(1)^2}{(0.72)^2} = \frac{1}{0.518} = 1.93$$

At first, this calculation might seem to confirm that climate depends entirely on distance from the Sun: because Venus is closer to the Sun, its surface is hotter. In fact, however, this is not the real answer, because most of the Sun's radiation never arrives at the surface. The upper atmosphere of Venus is shrouded in a thick cover of sulfuric acid clouds that reflect 80% of the incoming radiation and allow only 20% to reach the surface of the planet. In contrast, clouds on Earth reflect just 26% of the incoming radiation, allowing the other 74% to reach its surface.

This large difference in average **albedo** (the percentage of incoming radiation reflected back to space) between the atmospheres of the two planets almost exactly reverses the relative amounts of solar energy that actually reach their surfaces. Even though Venus receives almost twice as much incoming solar energy at the top of its atmosphere, its higher albedo reduces the amount that reaches its surface to just over half that received on Earth:

$$1.93 \times \frac{0.20}{0.74} = 0.52$$

With less incoming solar radiation, how can Venus be so much hotter? The answer is that Venus has an atmosphere 90 times as dense as that of Earth, and 96% of its atmosphere is composed of carbon dioxide (CO_2), a greenhouse gas that is very effective in trapping radiation. Some sunlight does penetrate the thick atmosphere and heat the surface, which causes Venus to emit radiation, just as Earth does. This kind of back radiation from its heated surface, called **longwave radiation,** is analogous to the heat emitted by a radiator. But most of the longwave back radiation never leaves the atmosphere of Venus because the CO_2 gas traps it and retains it as internal heat.

In contrast, much less of the energy radiated back from Earth's surface is trapped by water vapor, CO_2, and other greenhouse gases (companion Web site, pp. 2–3).

In summary, the main reason Venus is so hot compared to Earth is not its closer proximity to the Sun but its far greater concentrations of heat-trapping greenhouse gases.

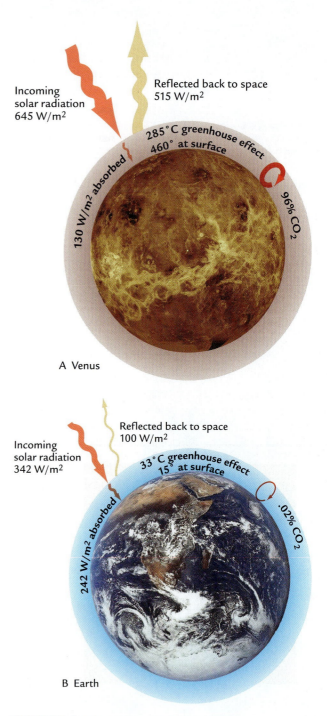

A Venus

B Earth

FIGURE 3-1 Why is Venus hot? (A) Venus receives almost twice as much solar radiation as (B) Earth, but its dense cloud cover permits less radiation to penetrate to its surface. Yet Venus is much hotter than Earth because its CO_2-enriched atmosphere creates a much stronger greenhouse effect that traps much more heat. (NASA photos.)

Because Venus and Earth both formed as rocky planets in the inner part of our solar system, they contain nearly equal amounts of carbon. Yet the two planets store their carbon in very different reservoirs. Most of Earth's carbon is tied up in its rocks, some as coal, oil, and

natural gas, while relatively little resides in the atmosphere. Combined with water vapor and other natural greenhouse gases, the net greenhouse heating of Earth's atmosphere is relatively small—about 33°C (although that difference keeps Earth from freezing solid). In complete contrast, almost all the carbon on Venus resides in its atmosphere as CO$_2$ and produces an enormous net greenhouse warming (285°C) without any significant contribution from water vapor.

This comparison shows how vital greenhouse gases can be to the climate of planets. It also highlights the fact that Earth's comfortably small greenhouse effect is an important factor in its habitability.

Faint Young Sun Paradox

By studying the evolution of stars in the universe, astronomers have recreated the history of our own Sun over the 4.55 Byr existence of our solar system. Throughout this interval, the Sun's interior has been the site of an ongoing nuclear reaction that fuses nuclei of hydrogen (H) together to form helium (He). Models developed by astronomers indicate that this process has caused our Sun to expand and gradually become brighter. These models indicate that the earliest Sun shone 25% to 30% more faintly than today, and that its luminosity, or brightness, then slowly increased to its current strength.

This insight from the field of astronomy creates an intriguing problem for climate scientists. A relatively small decrease in our Sun's present strength would cause all the water on Earth to freeze, despite the warming effect from greenhouse gases. If all our oceans and lakes were to freeze, their bright snow and ice surfaces would reflect more solar radiation and they would be difficult to melt. One-dimensional numerical climate models that simulate the mean climate of the entire planet (Chapter 2) suggest that the combination of a weak Sun and greenhouse gas levels at their present values would have kept Earth completely frozen for the first 3 billion years of its existence (Figure 3-2).

Yet evidence left in Earth's sedimentary deposits shows that Earth was not frozen for its first 3 billion years. Although the first half-billion years of Earth's existence left no record, evidence of Earth's climatic history gradually becomes more complete after that time and toward the present. Most sedimentary rocks (Chapter 2) are made up of particles that were eroded from other rocks, reworked by running water, and transported to a site of deposition. The prevalence of water-deposited sedimentary rocks throughout Earth's history is direct evidence that Earth was not completely frozen.

The first evidence of ice-deposited sediments occurs in rocks dated to about 2.3 Byr ago, but these deposits were probably the result of glaciations in polar regions similar to those on Earth today, and they are not evidence of a completely frozen planet. As summarized at

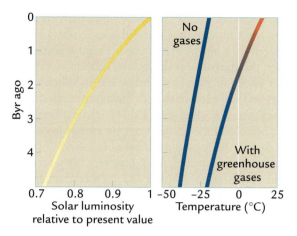

FIGURE 3-2 Faint young Sun paradox Astrophysical models of the Sun's evolution indicate that it was 25% to 30% weaker early in Earth's history (left). Climate models show that this situation would have produced a completely frozen Earth for more than half its early history if the atmosphere had had the same composition it does today (right). (Adapted from D. Merritts et al., *Environmental Geology*, ©1997 by W. H. Freeman and Company.)

the end of this chapter, a debate is currently under way as to how close Earth's climate came to a nearly frozen condition during intervals between 850 and 550 Myr ago, but for most of Earth's history the sedimentary evidence leaves no doubt that most of the water on Earth has remained unfrozen.

This conclusion is supported by the continued presence of life on Earth. Primitive life-forms date back to at least 3.5 Byr ago, and their presence on Earth is incompatible with a completely frozen planet at that time. The succession of ever more complex life-forms that have continuously occupied Earth ever since add further proof against extreme cold (or heat).

So we are confronted with a mystery: With so weak a Sun, why wasn't Earth frozen for the first two-thirds of its history? This mystery has been named the **faint young Sun paradox**.

Part of the answer to the faint young Sun paradox is obvious: something kept the early Earth warm enough to offset the Sun's weakness, but this easy answer only raises a more difficult problem. Whatever the process was that warmed the younger Earth, it must no longer be doing so today, or at least not as actively as it once did. If this same warming process had continued working at full strength right through the entire 4.55 Byr of Earth's history, it would have combined with the steadily increasing warmth from the strengthening Sun (see Figure 3-2) to overheat Earth and make it uninhabitable. Yet that has not happened: somehow Earth has stayed within a moderate temperature range throughout the entire interval when the Sun's brightness was increasing.

The solution to the faint young Sun paradox appears to require a process that works the same way a thermostat works in a house. When outside temperatures fall in winter, the **thermostat** detects the cooling and turns on a heat source that keeps the house warm. When temperatures become too hot outside in summer, the thermostat activates a cooling source that keeps the house cool. The thermostat moderates extreme swings in temperature. Such a thermostat must have been at work through Earth's history, warming its climate very early on when it would otherwise have frozen under a weak Sun and later on cutting back on the heat provided from the strengthening Sun.

One possibility is that greenhouse gases have been part of the mechanism that acts as Earth's thermostat. Our modern concentrations of greenhouse gases do not provide enough warming to have counteracted the effects of a weak early Sun, but if these gases had been more abundant earlier in Earth's history and subse-

quently decreased in abundance, that would have provided a thermostat-like control.

Our earlier comparison of Earth and Venus lends credibility to this explanation. Earth's carbon is mainly stored in its rocks, while carbon on Venus is mostly in its atmosphere. If carbon can reside in different reservoirs on different planets, why couldn't it move among reservoirs during the history of a single planet? More specifically, could the early Earth have held more carbon in its atmosphere (like Venus) and then transferred it to its rocks later in its history?

Carbon Exchanges between Rocks and the Atmosphere

To understand how carbon may have shifted among Earth's reservoirs, we need to examine the present carbon cycle (Figure 3-3A). Small amounts of carbon exist in the atmosphere, in the surface ocean, and in vegeta-

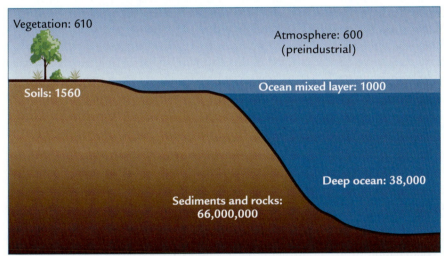

A Major carbon reservoirs (gigatons)

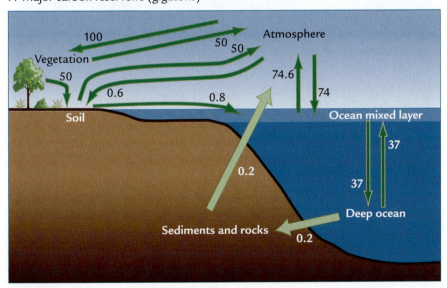

B Carbon exchange rates (gigatons/year)

FIGURE 3-3 Carbon exchanges with Earth's rocks (A) The largest reservoir of carbon on Earth lies in its rocks, not in its atmosphere, vegetation, or ocean. (B) All of Earth's reservoirs exchange carbon. Over intervals of millions of years, slow exchanges among the rock and surface reservoirs can cause large changes in atmospheric CO_2 levels. (Adapted from J. Horel and J. Geisler, *Global Environmental Change* [New York: Wiley, 1997] and from National Research Council Board on Atmospheric Sciences and Climate, *Changing Climate,* Report of the Carbon Dioxide Assessment Committee [Washington, D.C.: National Academy Press, 1993].)

tion, along with a slightly larger reservoir in soils, a much larger reservoir in the deep ocean, and an immensely larger reservoir in rocks and sediments. Carbon storage in these reservoirs is measured in billions of tons (gigatons).

The rates of carbon exchange among these reservoirs vary widely (Figure 3-3B). In general, an inverse relationship exists between the size of a reservoir and the rate at which it exchanges carbon. The smaller reservoirs (atmosphere, surface ocean, and vegetation) all exchange carbon relatively quickly, while the huge rock reservoir gains and loses carbon much more slowly. As a result of the combined effects of reservoir size and exchange rate, carbon can cycle through the smaller reservoirs at the surface within a few years but moves much more slowly through the larger and deeper reservoirs.

Because all these reservoirs exchange carbon with the atmosphere, each has the potential to alter atmospheric CO$_2$ concentrations and affect Earth's climate. The relative importance of each carbon reservoir in Earth's climate history varies according to the time scale under consideration. In this chapter, we are concerned with very gradual climate changes over tens of millions of years. Over these very long (tectonic) time scales, the effects of the slow carbon exchanges between the rocks and the surface reservoirs produce large changes in the amount of CO$_2$ in the atmosphere.

3-1 Volcanic Input of Carbon from Rocks to the Atmosphere

Carbon cycles constantly between Earth's interior and its surface. It moves from the deep rock reservoir to the surface mainly as CO$_2$ gas produced during volcanic eruptions and in the activity of hot springs (Figure 3-4).

The present rate of natural carbon input to the atmosphere from the rock reservoir is estimated at approximately 0.15 gigatons of carbon per year (see Figure 3-3B). This value is uncertain by a factor of at least 2, because volcanic explosions are irregular in time and because the amount of CO$_2$ released varies with each eruption. As we will see later, this natural rate of carbon input is roughly balanced by a similar rate of natural removal. This balance between natural input and removal rates helped to keep the size of the "natural" (preindustrial) atmospheric carbon reservoir at ~600 gigatons.

But how likely is it that this balance could have persisted over immensely long intervals of geologic time? We can evaluate this question by a simple thought experiment. Using the reservoir concept introduced in Chapter 2, we can calculate how long it would take for the atmospheric CO$_2$ level to fall to zero if all volcanic release of carbon from Earth's interior to the atmosphere abruptly ceased but carbon continued to be removed from the atmosphere at the same rate as before.

The answer, derived by dividing the preindustrial atmospheric carbon reservoir of 600 gigatons by an annual rate of carbon removal of 0.15 gigaton, is 4000 years. This number, although obviously well beyond the length of a human lifetime, is remarkably brief in the context of the several billion years of Earth's existence. It tells us that changes in volcanic input persisting over that relatively "small" span of time could have a drastic effect on the CO$_2$ content of our atmosphere.

In actuality, the atmosphere is not really this vulnerable because rapid exchanges of carbon occur continuously between the atmosphere and several other carbon reservoirs. These rapid exchanges have the effect of slowing and reducing the impact of the loss of carbon from Earth's interior.

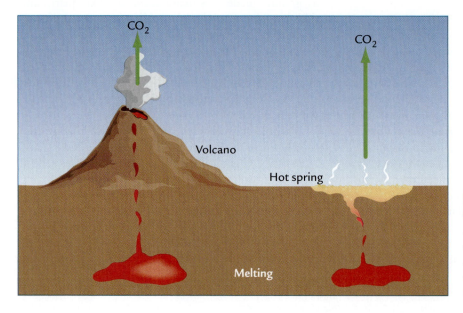

FIGURE 3-4 Input of CO$_2$ from volcanoes CO$_2$ enters Earth's atmosphere from deep in its interior through release of gases in volcanoes and at hot springs such as those found today at Yellowstone National Park in Wyoming.

In our hypothetical example of a sudden cessation of volcanic CO_2 input to the atmosphere, the actual scenario might develop more like this: As CO_2 levels in the atmosphere begin to fall, the other surface reservoirs (vegetation, surface ocean, soils) would begin to surrender some of their carbon to the atmosphere, slowing its rate of loss. For the fast-reacting reservoirs, changes in one are felt by the others almost immediately because of the rapid exchange rates.

The combined size of all the near-surface reservoirs (atmosphere, vegetation, soil, and surface ocean) is 3700 gigatons, more than six times larger than the atmospheric reservoir alone. It would take roughly 24,700 years after volcanism ceased for these reservoirs to lose all their carbon (3700 gigatons divided by 0.15 gigaton/yr).

In addition, over time spans of centuries, the large deep-ocean carbon reservoir would begin to play a role. If the surface reservoirs were all losing significant amounts of carbon, the deep ocean would feed some of its carbon to the surface ocean, from which it would be redistributed to the atmosphere and the vegetation. If we take the large deep-ocean reservoir into account, the total size of these reservoirs amounts to 41,700 gigatons. It would take 278,000 years for a total shutdown of volcanic carbon input to deplete these combined reservoirs completely (41,700 gigatons divided by 0.15 gigaton/yr).

At this point it might seem that we have shown that Earth's surface reservoirs, including the atmosphere, are actually *not* particularly vulnerable to changes in the amount of carbon coming out of (or going into) its rocks, but this conclusion would be incorrect. Even a time span as long as 278,000 years represents less than one–ten thousandth of Earth's 4.55-Byr age. Because Earth is so old, plenty of time is still available for the slow carbon exchanges with Earth's rock reservoirs to alter the amount of carbon in the surface reservoirs by large amounts.

With Earth's great antiquity taken into account, it is still amazing that over this immense span of time Earth's volcanoes have somehow managed to keep delivering just enough carbon from Earth's interior to keep the atmosphere from running out of CO_2 but not so much as to overheat the planet. This achievement requires a very delicate balance. Even more amazing is the fact that this balancing act had to be maintained as the faint young Sun was slowly increasing in strength. A simple analogy for this long-term balancing act is a tightrope walker who has to stay balanced on a narrow wire that slopes uphill over a very long distance.

We noted earlier that this balancing act requires some kind of natural thermostat to moderate Earth's temperature. Could the rate of volcanic input of CO_2 from Earth's interior have varied in such a way as to

function as the thermostat? The answer is no. The basic operating principle of a thermostat is that it first *reacts* to external changes and then *acts* to moderate their effects: a thermostat detects the chill of a cold night and sends a signal that turns on the heat.

Volcanic processes are not thought to operate in this way. The volcanic activity that has occurred on Earth throughout its history has been driven mainly by heat sources located deep in its interior and generally far removed from contact with the climate system. Climatically driven changes in temperature penetrate only the outermost few meters or tens of meters of the land (or seafloor). As a result, climate changes confined to Earth's surface have no physical way to alter deep-seated processes in Earth's interior. Without such a link, no thermostat-like changes in volcanic activity and CO_2 delivery to the surface can occur.

Earth's thermostat lies elsewhere. It must be found in a process that responds directly to the climate conditions at Earth's surface.

3-2 Removal of CO_2 from the Atmosphere by Chemical Weathering

To avoid a long-term buildup of CO_2 levels, CO_2 input to the atmosphere by volcanoes must be countered by CO_2 removal. The major long-term process of CO_2 removal is tied to chemical weathering of continental rocks (Chapter 2). Two major types of chemical weathering occur on continents: *hydrolysis* and *dissolution*.

Hydrolysis Hydrolysis is the main mechanism for removing CO_2 from the atmosphere. The three key ingredients in the process of hydrolysis are the minerals that make up typical continental rocks, water derived from rain, and CO_2 derived from the atmosphere (Figure 3-5).

Most of the continental crust consists of rocks, such as granite, made of **silicate minerals** like quartz and feldspar. Silicate minerals typically are made up of positively charged cations (Na^{+1}, K^{+1}, Fe^{+2}, Mg^{+2}, Al^{+3}, and Ca^{+2}) that are chemically bonded to negatively charged SiO_4 (silicate) structures. These silicate minerals are slowly attacked by groundwater containing carbonic acid (H_2CO_3) formed by combining atmospheric CO_2 with rainwater.

Part of the weathered rock is chemically converted to clay minerals (compounds of Si, Al, O, and H) and left as soils. Chemical weathering also produces several types of dissolved ions and ion complexes, including HCO_3^{-1}, CO_3^{-2}, H_2SiO_4, and H^{+1}. These ions are carried by rivers to the ocean, and some are incorporated in the shells of planktic organisms (see Figure 3-5).

Dozens of chemical equations describe the process of chemical weathering—in fact, there is one equation

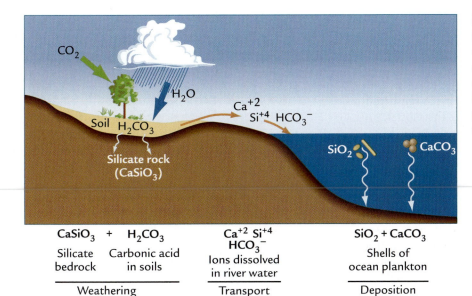

FIGURE 3-5 Chemical weathering removes atmospheric CO$_2$ Chemical weathering of silica-rich rocks on the continents removes CO$_2$ from the atmosphere, and part of the carbon is later stored in the shells of marine plankton and buried in ocean sediments.

CaSiO$_3$ + H$_2$CO$_3$	Ca^{+2} Si^{+4} HCO$_3^-$	SiO$_2$ + CaCO$_3$
Silicate Carbonic acid bedrock in soils	Ions dissolved in river water	Shells of ocean plankton
Weathering on land	Transport in rivers	Deposition in ocean

for each of the many types of silicate minerals found on continents. The part of these processes that is most important to the carbon system can be represented by these reactions:

$$H_2O + CO_2$$
Rain   From atmosphere

$$CaSiO_3 + H_2CO_3 \rightarrow CaCO_3 + SiO_2 + H_2O$$
Silicate rock   Carbonic acid   Shells of organisms
(continents)   (soil)

For simplicity, the many kinds of continental rocks and minerals are represented here by one silicate mineral, CaSiO$_3$ (wollastonite). Carbon dioxide (CO$_2$) is removed from the atmosphere, incorporated in groundwater to form carbonic acid in soils, used in the chemical weathering reaction, and deposited in the CaCO$_3$ shells of marine organisms. This reaction is a shorthand summary of the way chemical weathering removes CO$_2$ from the atmosphere and buries it in ocean sediments. This process acts slowly but persistently over long intervals of geologic time and accounts for 80% of the 0.15 gigatons of carbon buried each year in ocean sediments.

Dissolution It is important to distinguish weathering of silicates by hydrolysis from dissolution, the second type of weathering. Dissolution is the familiar process that eats away at limestone bedrock and in some areas forms limestone caves. Again, rainwater and CO$_2$ combine in soils to form carbonic acid (H$_2$CO$_3$) and attack limestone bedrock, and the dissolved ions created by dissolution again flow to the ocean in rivers. Dissolution can also be summarized by these simple reactions:

$$H_2O + CO_2$$
Rain   From atmosphere

$$CaCO_3 + H_2CO_3 \rightarrow CaCO_3 + H_2O + CO_2$$
Limestone   In soils   Shells of   Returned to
rock   organisms atmosphere

Dissolution of limestone proceeds at much faster rates than hydrolysis of silicates. Similar to hydrolysis, dissolution extracts CO$_2$ from the atmosphere to attack rock. But unlike the weathering of silicate rocks, limestone weathering causes no net removal of atmospheric CO$_2$. Within the relatively short interval of time it takes for the dissolved HCO$_3^{-1}$ and CO$_3^{-2}$ ions to reach the sea and become incorporated in the shells of organisms, all the CO$_2$ is returned to the atmosphere.

A CO$_2$ Balance In summary, slow weathering of granite and other silicate rocks on the continents by hydrolysis is the main way that CO$_2$ is pulled out of the atmosphere over very long time scales. In the context of Earth's delicate long-term balancing act, the rate of removal of carbon by chemical weathering must have very nearly balanced the rate of carbon input from volcanoes. If these rates had not been very nearly equal, the system would have been thrown off balance and caused drastic changes in CO$_2$ levels and in climate.

The existence of this delicate balance does not imply that either the (volcanic) CO$_2$ input rate or the (weathering) CO$_2$ removal rate remained absolutely constant through time. Yet the fact of Earth's long-term habitability requires that the rates of input and output must have always remained fairly closely balanced even as one or both processes changed.

How has this near-perfect balance been possible? As we noted earlier, a thermostat can provide such a balance. In our search for Earth's thermostat within its carbon system, we have ruled out volcanic input of CO_2. The only other possibility left is chemical weathering. If the rate of chemical weathering is sensitive to climate, it may be able to act as Earth's thermostat.

Climatic Factors That Control Chemical Weathering

Decades of laboratory experiments and many field studies have shown that rates of chemical weathering are influenced by three environmental factors: temperature, precipitation, and vegetation. These factors all act in a mutually reinforcing way to affect the intensity of chemical weathering.

Laboratory experiments have shown that higher temperatures cause more rapid weathering of individual silicate minerals. This trend is consistent with many temperature-dependent chemical reactions in water or other aqueous solutions. Weathering rates roughly double for each 10°C increase in temperature.

Unfortunately, it is difficult to transfer these laboratory results to studies of the real Earth. So far, experiments have examined only a few of the many silicate minerals that are common enough in Earth's crust to be important contributors to the overall rate of silicate weathering on a global scale. Natural chemical weathering rates are also difficult to determine in field studies because of the complicating effects from rapid carbonate dissolution. Because dissolution occurs many times faster than hydrolysis, the total amount of ions flowing down rivers can easily be dominated by ions derived from limestone dissolution, which does not control CO_2 levels in Earth's atmosphere, rather than from hydrolysis of silicates, which does control long-term CO_2 levels. Another problem with studying the real world is that humans have disturbed the natural chemistry of most of Earth's rivers by agricultural and industrial activities.

Still, we can apply the laboratory rule of thumb that says that silicate weathering rates double for each 10°C increase in temperature across the roughly 30°C range of mean annual temperatures found on Earth's surface (Figure 3-6A). Based on this relationship, rates of silicate weathering should increase by a factor of at least 8 ($2 \times 2 \times 2$) from the cold polar regions to the hot equatorial latitudes.

The second major control on weathering is precipitation (Figure 3-6B). Increased rainfall raises the level of groundwater held in soils, and the water combines with CO_2 to form carbonic acid and enhance the weathering process.

Temperature and precipitation are closely linked in Earth's climate system, and it is difficult to measure their

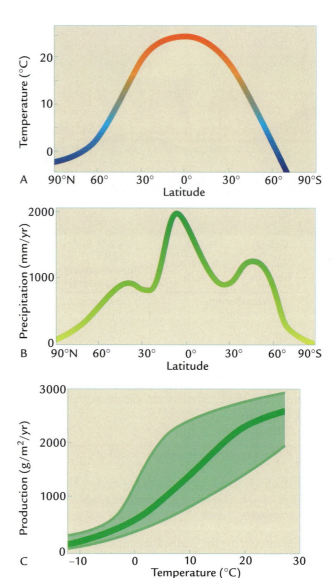

FIGURE 3-6 Climate controls on chemical weathering (A) Temperature and (B) precipitation both show a general trend from high values in warm (low) latitudes to low values in cold (high) latitudes. (C) The total amount of vegetation produced per year increases with temperature, as well as with precipitation. (A and B: Adapted from R. G. Barry and R. J. Chorley, *Atmosphere, Weather, and Climate*, 4th ed. [New York: Methuen, 1982]. C: adapted from R. L. Smith and T. M. Smith, *Elements of Ecology* [Menlo Park, CA: Addison Wesley Longman, 1998].)

separate contributions to chemical weathering. The heaviest rainfall on Earth occurs in the tropics because warm tropical air holds more moisture than cooler high-latitude air. Polar regions have much less precipitation because the atmosphere holds so little water.

This relationship breaks down to some extent at regional scales. For example, lower precipitation in some subtropical regions greatly reduces chemical weathering, even though the relatively warm temperatures in those areas would otherwise favor it. Despite these complications, temperature and precipitation generally act together. A warmer Earth is likely to be a wetter Earth, and both factors tend to act together to intensify chemical weathering.

Vegetation also enhances chemical weathering. Plants extract CO_2 from the atmosphere through the process of photosynthesis and deliver it to soils, where it combines with groundwater to form carbonic acid. Although H_2CO_3 is a weak acid, it enhances the rate of chemical breakdown of minerals. Scientists estimate that the presence of vegetation on land can increase the rate of chemical weathering by a factor of 2 to 10 over the rates typical of land that lacks vegetation.

Vegetation is closely linked to precipitation and temperature (companion Web site, pp. 47–50). Dense rain forests are found in regions with year-round rainfall, open forest or savannas in areas with a short dry season, grasslands in places with a long dry season, and deserts in areas with little or no rainfall. Each step in the direction of greater rainfall is a step toward more vegetation and more total carbon biomass stored in vegetation and soils.

In addition, the rate of production of carbon by photosynthesis across the planet is correlated with temperature (Figure 3-6C). Cold, ice-covered regions produce little plant matter, and seasonally or permanently frozen (but ice-free) polar regions produce only sparse tundra vegetation. In comparison, production of carbon in warmer mid-latitude and tropical regions is much greater.

Is Chemical Weathering Earth's Thermostat?

Now we have in hand the components of a mechanism that could act as Earth's thermostat and moderate long-term climate: the **chemical weathering thermostat.** The global rate of chemical weathering is analogous to a thermostat because it reacts to (depends on) the average state of Earth's climate and then alters that state by regulating the rate at which CO_2 is removed from the atmosphere.

Consider what would happen if Earth's climate began to warm (Figure 3-7A). Any initial climate change (for any reason) toward a warmer, moister, more heavily vegetated greenhouse Earth should enhance chemical weathering of silicate minerals, but the faster weathering in such a world should then speed up the rate of removal of CO_2 from the atmosphere. The result should be a negative feedback that removes CO_2 and moderates the size of the imposed warming.

The opposite sequence should happen if Earth's climate began to cool (Figure 3-7B). Icehouse climates are

typically cold, dry, and more sparsely vegetated, with more extensive snow and ice. An initial climate change toward a colder, drier, less vegetated Earth should reduce chemical weathering and slow the rate of removal of CO_2 from the atmosphere. Slower CO_2 removal should reduce the effect of the initial push toward climate cooling.

The action of these negative feedbacks does not mean that no climate change occurs at all. Any process that initially acts to warm Earth succeeds in doing so, but by an amount smaller than would have been the case without the negative feedback. Conversely, any process that initially acts to cool Earth succeeds in doing so, but also to a reduced degree. The existence of a climate-dependent negative feedback due to chemical weathering was proposed in 1981 by the geochemist

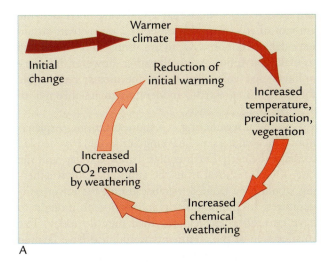

A

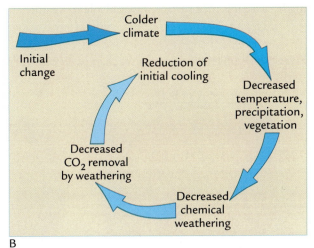

B

FIGURE 3-7 Negative feedback from chemical weathering Chemical weathering acts as a negative climate feedback by reducing the intensity of both (A) imposed climate warming and (B) imposed climate cooling.

James Walker and his colleagues Paul Hays and James Kastings.

How do we apply this concept to the mystery of the faint young Sun paradox? Recall that Earth needed a global thermostat that made it warmer early in its history to counter the weakness of the early Sun, but that later throttled back on the warming as the strengthening Sun provided greater heat.

Earth's environment very early in its history is poorly known, but it is widely thought to have included active volcanism that caused large emissions of volatile gases (including CO_2) from its interior. Many scientists believe that Earth's surface may even have been entirely molten for a few hundred million years after 4.55 Byr ago. In addition, ancient craters preserved on our moon and on other planets indicate that Earth was once under heavy bombardment by asteroids, meteors, and comets, and these collisions may have triggered greater volcanism as well. Radioactive elements deep in Earth's interior also released heat that could have increased the amount of volcanism. Increased volcanic activity would have delivered more CO_2 to the atmosphere and helped to make Earth hot. As noted earlier, however, it is very unlikely that volcanism is the thermostat responsible for maintaining Earth's moderate climate through all 4.55 Byr of its existence.

Chemical weathering is a more promising explanation. The weakness of the young Sun would have tended to make the early Earth cooler than it is today, and the rate of CO_2 removal from the atmosphere by weathering would have been slower because of the lower temperatures. In addition, early continents are thought to have covered a smaller area than they do today. The smaller area of the continents would also have favored slower CO_2 removal from the atmosphere by weathering because less rock surface was available to weather.

Slower rates of weathering would have left more CO_2 in the atmosphere over much of Earth's early history, perhaps 100 to 1000 times as much as today (Figure 3-8A). The warmth produced by this high-CO_2 atmosphere could have countered most of the cooling caused by the smaller amount of incoming solar radiation.

Then, as Earth began to receive more radiation from the brightening Sun, its surface warmed and the rate of chemical weathering gradually increased. Faster chemical weathering began to draw more CO_2 out of the atmosphere, and the resulting drop in atmospheric CO_2 levels provided a cooling effect that counteracted the gradual increase in solar warming and kept Earth's temperatures moderate (Figure 3-8B). The centerpiece of this explanation is that the slow warming of Earth by the strengthening Sun would have caused changes in weathering that moderated changes in climate.

IN SUMMARY, chemical weathering is an excellent candidate for Earth's thermostat.

If chemical weathering is Earth's thermostat, we face still another question: What happened to all that CO_2 that once resided in the atmosphere and kept Earth warm? The most likely answer is found by looking at the size of the carbon reservoirs in Figure 3-3: the carbon removed from today's atmosphere by weathering is buried in ocean sediments that eventually turn into rocks. The same process would also have been at work in the past, and over time it would have caused a slow but massive transfer of carbon from the atmosphere to rocks. If this interpretation is correct, most of Earth's early greenhouse atmosphere lies buried in its rocks instead of concentrated in the atmosphere, as on Venus.

Some scientists have suggested that greater emissions of methane (CH_4) and ammonia (NH_3) from Earth's

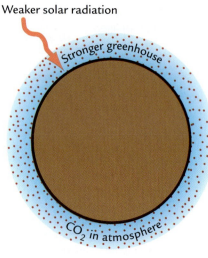

A Early Earth

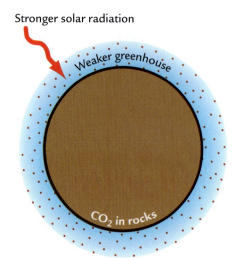

B Modern Earth

FIGURE 3-8 Earth's thermostat
A plausible explanation of the faint young Sun paradox is that (A) the weakness of the early Sun was compensated for by a stronger CO_2 greenhouse effect in the atmosphere. (B) Later, when the Sun strengthened, increased chemical weathering deposited the excess atmospheric greenhouse carbon in rocks, and the weakened greenhouse effect kept Earth's temperatures moderate. (Adapted from W. Broecker and T.-H. Peng, *Greenhouse Puzzles* [New York: Eldigio Press, 1993].)

interior warmed the early Earth. In the modern atmosphere, both of these gases tend to be broken down in the atmosphere within a few years by chemical reactions, which would seem to require large and continual additions to the early atmosphere to maintain high levels. On the other hand, the early atmosphere was devoid of oxygen, which plays a primary role in removing methane and ammonia today. These two gases could also have played a role in warming the early atmosphere.

3-3 Greenhouse Role of Water Vapor

What about the role of water vapor in these longer-term climate changes? Water vapor is by far the most important greenhouse gas in Earth's atmosphere today. It accounts for most of the 33 °C greenhouse effect that keeps our planet warm (companion Web site, pp. 2–3). Why wouldn't water vapor have played a major thermostat role in Earth's past?

The answer to this question is that water vapor and CO_2 work in fundamentally different ways in the climate system. Over tectonic time scales, CO_2 acts as a negative feedback that suppresses the size of changes in climate. In contrast, the effect of water vapor is just the reverse: it acts as a positive feedback that amplifies changes in climate.

The amount of water vapor in Earth's atmosphere varies over a wide range, from as little as 0.2% in very cold dry air to more than 3% in humid tropical air. This natural link between air temperature and water vapor produces an important positive feedback in the climate system called **water vapor feedback** (Figure 3-9).

Assume that climate warms for some reason. Because a warmer atmosphere can hold much more water vapor, the resulting increase in this greenhouse gas traps more heat. This enhanced greenhouse effect further warms Earth, amplifying the initial warming through a positive feedback loop. The same positive feedback process works in the opposite direction when climate cools: if an initial cooling reduces the amount of water vapor held in the atmosphere, an additional cooling will result from the reduced greenhouse effect. Because water vapor amplifies rather than moderates climatic changes, it cannot have acted as Earth's thermostat. The continuing presence of running water shown by geologic deposits through Earth's recorded history indicates that liquid water has been available to feed water vapor to the atmosphere and provide positive feedback to temperature changes.

Is Life the Ultimate Control on Earth's Thermostat?

We have seen that chemical weathering provides a plausible thermostat-like mechanism to moderate Earth's

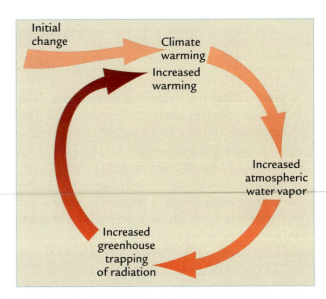

FIGURE 3-9 Water vapor feedback When climate warms, the atmosphere is able to hold more water vapor, which is the major greenhouse gas in the atmosphere. The increase in water vapor leads to further warming by means of positive feedback. This feedback works in the same way (but opposite direction) during cooling.

climate, yet we have also seen that the processes involved are not strictly physical. Biological processes participate in the carbon cycle. For this reason, some scientists infer that life itself, rather than strictly physical-chemical factors, may be the thermostat that actually regulates Earth's climate. (In addition, a substantial amount of the carbon that moves through Earth's reservoirs does so in organic form, as part of a separate and smaller subcycle (Box 3-1).

3-4 Gaia Hypothesis

The biologists James Lovelock and Lynn Margulis proposed in the 1980s that life itself has been responsible for regulating Earth's climate. They called their idea the **Gaia hypothesis,** after the ancient Greek Earth goddess. A crude analogy of how their hypothesis works is the way the fur on an animal fluffs out to create an insulated layer and keep the creature warm when the weather turns cold. The animal in effect unconsciously regulates its own environment for its own good. The Gaia hypothesis holds that life regulates climate on Earth for its own good (Figure 3-10).

Supporters of this hypothesis cite features of the chemical weathering thermostat that directly involve the action of life-forms: (1) the fact that carbon is the basis of the CO_2 cycle; (2) the action of land plants in photosynthesizing carbon dioxide and transferring it to the soil as part of the vegetation litter, thereby forming

BOX 3-1 **LOOKING DEEPER INTO CLIMATE SCIENCE**

Organic Carbon Subcycle

Nearly 20% of the carbon that cycles among Earth's carbon reservoirs today does so in organic form. Photosynthesis is critical to the organic carbon subcycle, mainly because land plants extract CO_2 from the atmosphere, and also because ocean plankton extract CO_2 from inorganic carbon dissolved in the surface ocean. Most of the organic carbon fixed and temporarily stored in land vegetation and ocean plankton is recycled and quickly returned to the ocean-atmosphere system by means of oxidation, which uses available oxygen in water or air to convert organic carbon back to inorganic form.

On land, oxidation consumes organic carbon just after the seasonal fall of leaves or die-back of green vegetation and after the death of the woody tissue of trees. In the oceans, oxidation slowly consumes organic debris sinking out of the sunlit surface layers where photosynthesis occurs.

Only a small fraction of the organic carbon formed by these processes is buried in the geologic record. Carbon from the land and carbon from the oceans contribute roughly equal amounts to this total. Burial of organic carbon is favored in water-saturated environments (marine or terrestrial) characterized by (1) low oxygen levels that minimize oxidation and (2) rapid production of organic matter that consumes the remaining oxygen and allows organic debris to escape oxidation. These conditions produce fine-grained, carbon-rich muds that eventually turn into mudstones and then into harder rocks called shales.

The carbon buried in sediments and then rocks represents a net loss of CO_2 from the interactive carbon reservoirs in the ocean, atmosphere, soil, and vegetation. Once buried, organic carbon stays in the rocks until tectonic processes return it to the surface by slow-acting processes: (1) weathering (and oxidation) of carbon-bearing rocks at Earth's surface and (2) thermal breakdown of organic carbon in rocks deep in Earth's interior, with release of liberated CO_2 through volcanoes.

Because this organic carbon subcycle carries one-fifth of the carbon moving between Earth's rocks and its surface reservoirs, it has the potential to have substantial effects on the global carbon balance and on atmospheric CO_2 over long (tectonic-scale) time intervals. Also, under conditions that cause the onset of high productivity and carbon burial in the ocean, large amounts of organic carbon can be quickly extracted from the atmosphere, causing rapid reductions of CO_2 levels and rapid climatic cooling.

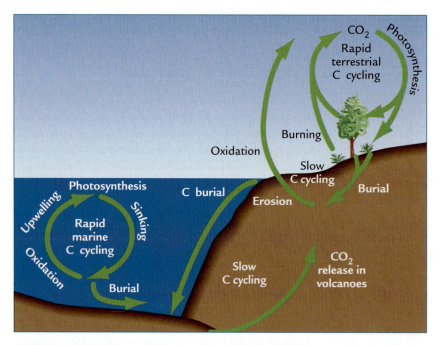

Organic carbon subcycle About 20% of the carbon that shifts between Earth's surface reservoirs (air, water, and vegetation) and its deep rock reservoirs moves in the organic carbon subcycle. Photosynthesis on land and in the surface ocean turns inorganic carbon into organic carbon, most of which is quickly returned to the atmosphere or surface ocean. A small fraction of this organic carbon is buried in continental and oceanic sediments that slowly turn into rock. This carbon is eventually returned to the atmosphere as CO_2, either by erosion of continental rocks or by melting and volcanic emissions.

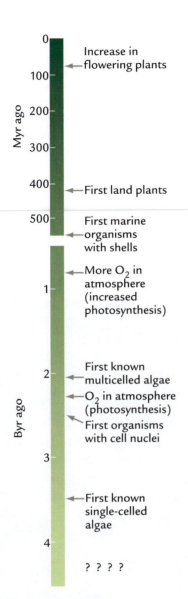

FIGURE 3-10 Gaia hypothesis Over time, life-forms gradually developed in complexity and played a progressively greater role in chemical weathering and its control of Earth's climate. In the extreme form of the Gaia hypothesis, life evolved for the purpose of regulating Earth's climate.

carbonic acid that enhances chemical weathering; and (3) the role of shell-bearing ocean plankton in extracting CO$_2$ from the ocean and storing it in their CaCO$_3$ shells. These modern biological processes are, without question, important components of the processes of chemical weathering and carbon cycling. And by extension, they contribute to the thermostat that moderates Earth's climate today.

In its more extreme form, the Gaia hypothesis states that all evolution on Earth has occurred for the greater good of the planet by producing the succession of life-forms needed to keep the planet habitable. This view is

much more controversial: it goes far beyond Darwin's concept that evolution occurs to enhance the reproductive survival of each species.

Earth's long history reveals a sequence of life-forms different from those that exist now (see Figure 3-10). No record of life exists before 3.5 Byr ago, although it is possible that primitive life-forms did exist and have simply escaped detection because the rock record is so scarce and poorly preserved. By 3.5 Byr ago, primitive single-celled marine algae capable of photosynthesis had developed (Figure 3-11A). Over the next 3 billion years, slightly more complex organisms evolved: by 2.9 Byr ago, moundlike clumps of marine algae called stromatolites that lived attached to the seafloor; by 2.5 Byr ago, organisms that contained a cell nucleus; and by 2.1 Byr ago, a variety of multicelled algae.

Most of the more complex forms of life did not arrive until late in Earth's history. Near 540 Myr ago, hard shells of many kinds of organisms abruptly appear in the fossil record. Before that time, the only fossilized records of life consisted of ghost impressions left imprinted on the surfaces of soft sediment layers. The first primitive land plants did not evolve until near 430 Myr ago (Figure 3–11B). These plants acquired the ability to survive because of stems and roots that delivered water from the ground. The first treelike plants appeared by 400 Myr ago (Figure 3–11C). Trees and grasses are important in modern chemical weathering because they acidify groundwater by adding carbon to soils as litter.

Critics of the Gaia hypothesis point out that many of the active roles played by organisms in the biosphere today are a relatively recent development in Earth's history and that the role of life in the distant past was probably negligible. In this view, early life-forms were too primitive to have had much effect on chemical weathering, and the delicate climatic balance maintained through Earth's history must have been achieved primarily by physical-chemical means (the effects of temperature and precipitation on weathering rates) rather than by biological intervention.

Critics also note that the very late appearance of shell-bearing oceanic organisms near 540 Myr ago means that life had played no obvious role in transferring the products of chemical weathering on land to the seafloor for the preceding 4 billion years. Instead, most CaCO$_3$ in the oceans was presumably deposited in warm, shallow tropical seas where concentrations of dissolved ions increased to levels that permitted chemical precipitation, apparently with little or no biological intervention. Floating planktic plants capable of photosynthesis (coccolithophorida) evolved still later, in the last 250 Myr.

Supporters of the Gaia hypothesis respond with several counterarguments. First, they claim that critics underestimate the role of primitive life-forms such as

A

B

C

FIGURE 3-11 Life-forms and weathering Over Earth's 4.55-Byr history, plants evolved toward more complicated forms capable of playing a greater role in chemical weathering. Primitive organisms similar to (A) the modern bacteria *Oscillatoria* existed by 3.5 Byr ago. The first simple land plants with roots and stems similar to those of (B) the modern plant Psilotum appeared by 430 Myr ago. Increasingly complex treelike plants similar to (C) modern tropical cycads appeared by 400 Myr ago and led to today's diversity of trees and shrubs. (A: Sinclair Stammers/Science Photo Library/Photo Researchers. B: William Ormoerod/Visuals Unlimited; C: Gerald Cubit.)

algae in the ocean and microbes on land early in Earth's history. They point to recent discoveries that modern bacteria with similarities to early primitive life-forms are now thought to play a greater role in the weathering process than has generally been recognized, and they suggest that these organisms must also have been more important than generally thought early in Earth's history, when they were the only life-forms present on land.

One indication that early life-forms were important at a global scale is the first development of an oxygen-rich atmosphere near 2.3 Byr ago, even before the first multi-celled algae (see Figure 3-10). Evidence for this important event includes the first appearance of rocks that show red staining (rusting) of iron (Fe) minerals. The appearance of oxidized iron minerals at this time coincides

roughly with the disappearance of previously widespread minerals such as FeS (pyrite, or "fool's gold"), which form only under reducing conditions (no oxygen). The only conceivable source of the oxygen that caused the widespread change to oxidized forms of iron is photosynthesis by marine organisms, implying an active global-scale role for these organisms far back in Earth's history.

Gaia supporters also point out that the general path of biological evolution matches Earth's need for progressively greater chemical weathering through time. The more primitive organisms played a much smaller role in accelerating the process of chemical weathering during a time when it was to Earth's advantage to retain CO_2 in its atmosphere to counter the weakness of the faint young Sun. Then, as the Sun strengthened and provided more heat to Earth, more advanced organisms capable of accelerating the weathering process appeared, accelerated the rates of weathering, and pulled CO$_2$ out of the atmosphere to keep the climate system in approximate balance.

> **IN SUMMARY,** the Gaia hypothesis is fascinating and is still being argued. Scientists generally agree about the "minimum" form of Gaia: the idea that living organisms have played a significant role in the history of physical-chemical processes on Earth, including chemical weathering. Still, the "maximum" claim embedded in the Gaia hypothesis—that individual life-forms regulate their own evolution for the greater benefit of all life on the planet—is not accepted by most scientists. Somewhere in between lies the answer to the role of biota in determining the presence of life on Earth.

Was There a "Thermostat Malfunction"? A Snowball Earth?

Ice sheets occur today at high latitudes, yet they coexist with hot tropics where a strong overhead Sun heats the land and the tropical oceans. With the large pole-to-equator gradient in temperature, polar ice sheets can easily coexist on a planet with tropical heat.

For a continent-sized ice sheet to have existed near the equator, temperatures in the normally hot tropics would have had to be near or below freezing through most of the year. Today's frigid polar climates would have had to have invaded the tropics to permit ice sheets to exist there.

Some climate scientists have suggested that Earth came very close to freezing totally between approximately 850 and 550 Myr ago. Sedimentary deposits from glaciers are found on several continents during this interval, evidence that ice sheets were present (Figure 3-12). These rocks contain ice-deposited mixtures of coarse boulders and cobbles along with fine silts and clays (Chapter 2). Because these ancient deposits are difficult to date and correlate accurately, scientists have inferred that as few as two or as many as four major glacial eras occurred during this long interval.

A critical question is whether these ice sheets existed near the poles or at lower latitudes. For at least one of the glacial intervals (but not some of the others), several lines of evidence suggest that the glaciated continents were in the tropics. This conclusion forms the basis of the novel idea that Earth was once nearly frozen—the **snowball Earth hypothesis.**

One obvious cause contributing to a cooler Earth was weaker solar heating from a Sun that was still 6% below its modern luminosity (see Figure 3-2). According to the thermostat concept, a cooler Earth would have reduced the rate of chemical weathering and kept CO$_2$ values higher and moderated global temperature. In this case, however, climate models suggest that CO$_2$ concentrations would have had to have been lower than today to permit ice sheets to exist in tropical latitudes. In this instance, the thermostat mechanism seems to have malfunctioned, at least for a while.

The reason for the thermostat malfunction remains unresolved. One explanation is that the continents were all clustered near the equator, where high temperatures, precipitation, and vegetation cover combined to drive unusually strong chemical weathering. Paradoxically, this tropical clustering could have reduced CO$_2$ concentrations and cooled the planet.

The debate about how cold this world was continues. Some scientists feel that Earth was frozen "hard," with sea ice extending right to the equator. Model simulations generally point toward a "softer" freeze, with sea ice reaching to middle latitudes but not into the

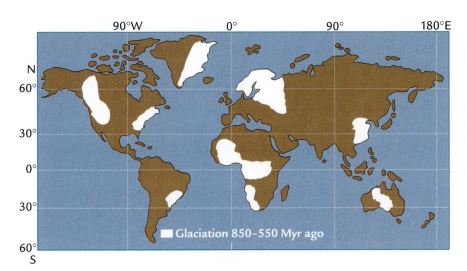

FIGURE 3-12 Snowball Earth?
Evidence of several glaciations between 850 and 550 Myr ago exists in rocks on the continents of today's Earth. If these glaciated regions were located in the tropics, Earth must have been much colder than today. (Adapted from L. A. Frakes, *Climates Through Geologic Time* [Amsterdam: Elsevier, 1979] and from J. G. Meert and R. van der Voo, "Neoproterozoic (1000–540 Myr) Glacial Intervals: No More Snowball Earth," *Earth and Planetary Science Letters* 123 [1994]: 1–13.)

tropics. A hard freeze is difficult to achieve because the large amount of solar heat stored in the ocean at low latitudes tends to keep the surface free of ice.

Key Terms

greenhouse era (p. 43)

icehouse era (p. 43)

albedo (p. 44)

longwave radiation (p. 44)

faint young Sun paradox (p. 45)

thermostat (p. 46)

silicate minerals (p. 48)

chemical weathering thermostat (p. 51)

water vapor feedback (p. 53)

Gaia hypothesis (p. 53)

snowball Earth hypothesis (p. 57)

Review Questions

1. Why is Venus so much warmer than Earth today?

2. What factors explain why Earth is habitable today?

3. What is the faint young Sun paradox?

4. What evidence suggests that Earth always had a long-term thermostat regulating its climate?

5. Why is volcanic input of CO_2 to Earth's atmosphere a poor candidate for a thermostat?

6. What climate factors affect the removal of CO_2 from the atmosphere by chemical weathering?

7. Where did the extra CO_2 from Earth's early atmosphere go?

8. What arguments support and oppose the Gaia hypothesis that life is Earth's true thermostat?

9. If Earth's surface froze solid, what would happen to CO_2 emissions from volcanoes and to CO_2 removal by chemical weathering?

Additional Resources

Basic Reading

Companion Web site at www.whfreeman.com/ruddiman2e, pp. 2–3, 30–33.

Kastings, J. F., O. B. Toon, and J. B. Pollack. 1988. "How Climate Evolved on the Terrestrial Planets." *Scientific American* (February), 90–97.

Advanced Reading

Lovelock, J. 1995. *The Ages of Gaia: A Biography of Our Living Earth.* New York: Norton.

CHAPTER 4

Plate Tectonics and Long-Term Climate

The last 550 Myr of Earth's history are far better known than the first 4 billion years. From this time forward, the locations of the continents and the shapes of the ocean basins become progressively clearer. Better-preserved sedimentary rock archives also hold more abundant evidence of past climates, including alternations between icehouse intervals (when ice sheets were present) and greenhouse intervals (times without ice on land). These fluctuations (Figure 4-1) are the focus of this chapter.

First we examine how plate tectonic processes work. Next we explore the possibility that icehouse intervals occur because plate tectonic motions cause continents to drift across cold polar regions. Then we use climate models to investigate the range of factors that controlled climate 200 Myr ago, a time when all landmasses on Earth existed as a single giant continent. These investigations reveal that changes in atmospheric CO_2 levels are needed to explain the sequence of changes from icehouse to greenhouse conditions over the last half-billion years. Finally we evaluate two hypotheses that link changes in plate tectonic processes to changes in CO_2 levels.

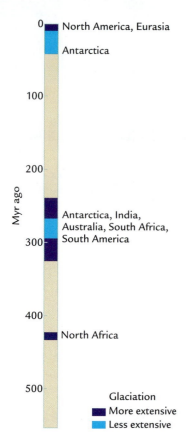

FIGURE 4-1 Icehouse intervals Three major intervals of glaciation occurred during the last 500 Myr.

Plate Tectonics

In 1914 the German meteorologist Alfred Wegener proposed that continents have slowly moved across Earth's surface for hundreds of millions of years. He based his hypothesis in part on the obvious fact that continental margins such as those of eastern South America and western Africa fit together like pieces of a jigsaw puzzle. Research in the last half of the twentieth century showed that Wegener was correct in claiming that these continents were once together and have since moved apart but that he underestimated the mobility of Earth's outer surface. In fact, Earth's entire surface is on the move.

4-1 Structure and Composition of Tectonic Plates

Wegener's assumption that continents move in relation to ocean basins had a reasonable basis. The contrast between the elevated continents and the submerged ocean basins is the most obvious division on Earth's surface. It also reflects the large difference in thickness and composition of the crustal layers that make up the continents and ocean basins (Figure 4-2).

Continental crust is 30–70 km thick, has an average composition like that of granite, and is low in

density (2.7 g/cm^3). This thick, low-density crust stands much higher than the floor of the ocean basins lying some 4000 m below sea level. **Ocean crust** is 5–10 km thick, has an average composition like that of basalt, and is higher in density (3.2 g/cm^3). Below each of these crustal layers lies the **mantle,** which is richer in heavy elements like iron (Fe) and magnesium (Mg) and has an even higher density (> 3.6 g/cm^3). The mantle extends 2890 km into the Earth's interior, almost halfway to the center of the Earth at a depth of 6370 km.

But these large differences in elevation, crustal thickness, and composition are deceptive: they are not the primary explanation for the fact that continents (and ocean basins) move. The critical reason for this mobility lies in the way different layers of rock behave.

Two rock layers characterized by very different long-term behavior exist well below Earth's surface (Figure 4-2). The outer layer, called the **lithosphere,** is 100 km thick and generally behaves the way the word "rock" suggests: as a hard, rigid substance. The lithosphere encompasses not just the crustal layers (oceanic and continental) but also the upper part of the underlying mantle.

Below the lithosphere is a layer of partly molten yet mostly solid rock called the **asthenosphere.** This layer

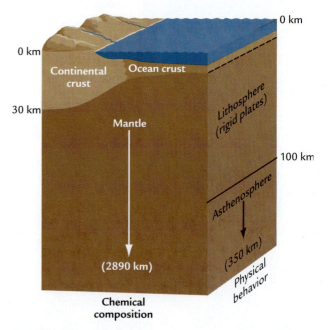

FIGURE 4-2 Earth's structure Earth's outer layers can be subdivided in two ways. The basalts of the ocean crust and the granites in continental crust differ from each other and from the underlying mantle in chemical composition. The other division is physical behavior: the lithosphere that forms the tectonic plates is a hard, rigid unit, whereas the underlying asthenosphere is softer and capable of flowing slowly.

lies entirely within the upper section of Earth's mantle at depths between 100 and 350 km. Compared to the rigid lithosphere, this deeper layer behaves like a soft, viscous fluid over long intervals of time and flows more easily. The behavior of this "softer" deeper layer allows the overlying lithosphere to move.

The lithosphere consists of a dozen **tectonic plates,** each drifting slowly across Earth's surface (Figure 4-3). These plates move at rates ranging from less than 1 up to 10 cm per year, about the same as the rate of growth of a fingernail. Over a time span of 100 Myr, 5 cm of plate motion per year adds up to 5000 km, enough to create or destroy an entire ocean basin.

Most tectonic plates consist not simply of continents or ocean basins but of combinations of the two. For example, the South American plate consists of the continent of South America and the western half of the South Atlantic Ocean, all moving as one rigid unit.

These rigid tectonic plates have three basic types of edges, or margins. Most tectonic deformation on Earth (earthquakes, faulting, and volcanoes) occurs at these plate margins (Figure 4-4).

Plates move apart at **divergent margins,** the crests of ocean ridges like the one that runs down the middle of the Atlantic Ocean (see Figure 4-3). This motion allows new ocean crust to be created, and the new crust spreads away from the ridge (see Figure 4-4 left). Plates diverging at ocean ridges carry not just the near-surface layer of ocean crust but also a much thicker layer of mantle lying underneath.

Plates come together at **convergent margins** (see Figure 4-4 left). At these locations, the ocean crust plunges deep into Earth's interior at ocean trenches in a process called **subduction.** The subducting ocean crust rides on top of a much thicker layer of upper mantle that also moves downward.

Some convergent margins occur along continent-ocean boundaries, such as the western coast of South America. In this case, narrow mountain chains such as the Andes form on the adjacent continents because of the compressive (squeezing) forces produced when the two plates move together. Subduction can also occur within the ocean, where the ocean crust of one plate plunges under another and forms volcanic ocean islands, such as those in the western Pacific. A less common but important example of converging plates is the

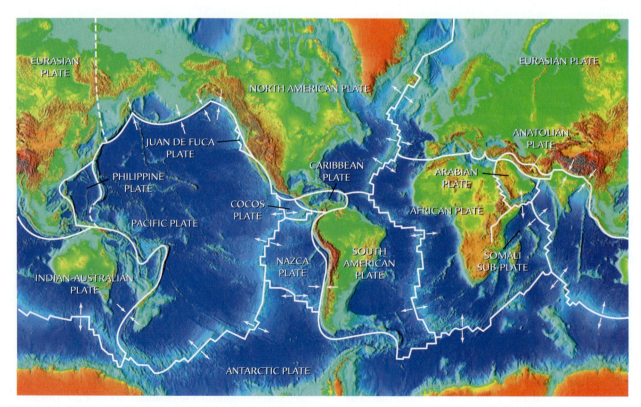

FIGURE 4-3 Tectonic plates Earth's lithosphere is divided into a dozen major tectonic plates and several smaller plates, which move as rigid units in relation to one another, as the arrows indicate. (F. Press and R. Siever, *Understanding Earth,* 2nd ed., © 1998 by W. H. Freeman and Company. False-color topography courtesy of Peter Schloss, NGDC, Boulder, Colo.)

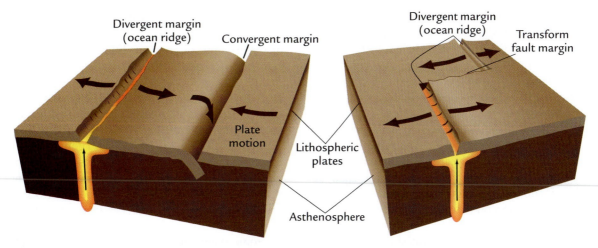

FIGURE 4-4 Plate margins Earth's tectonic plates move apart at ocean ridges (divergent margins), slide past each other at faults (transform fault margins), and push together at convergent margins. (Modified from F. Press and R. Siever, *Understanding Earth*, 2nd ed., © 1998 by W. H. Freeman and Company.)

continental collision of landmasses such as India and Asia, which can create massive high-elevation regions such as the Tibetan Plateau.

Plates also can slide past each other at **transform fault margins** (see Figure 4-4 right), moving horizontally along faults such as the San Andreas Fault in western California. Sliding of plates at transform faults involves not just the upper 30 km of continental crust but also the underlying 70 km of upper mantle.

Even though geoscientists do not yet know the balance of forces that caused past movements of plates and produced their present distributions, they can accurately measure the way these processes have changed Earth's surface during the last several hundred million years. With this knowledge, the tectonic changes can be compared with changes in climate over the same interval in order to evaluate how tectonic changes have influenced Earth's climate.

4-2 Evidence of Past Plate Motions

A broad range of evidence reveals the past effects of plate tectonics in rearranging Earth's geography. The most important evidence starts with the fact that Earth has a **magnetic field.** Molten fluids circulating in Earth's liquid iron core today create a magnetic field analogous to that of a bar magnet (Figure 4-5). Compass needles today point to magnetic north, which is located a few degrees of latitude away from the geographic North Pole, which marks Earth's axis of rotation.

This link between "magnetic north" and Earth's north polar axis of rotation is assumed to have held in the past. Some of Earth's once-molten rocks contain "fossil compasses" that record its past magnetic field.

These natural compasses were frozen into the rocks shortly after they cooled from a molten state. Today, they give scientists studying **paleomagnetism** a way to reconstruct past positions of continents and ocean basins with respect to the pole of rotation.

The best rocks to use as ancient compasses are basalts, which are rich in highly magnetic iron. Basalts form the floors of ocean basins and are also found on land in actively tectonic regions. They form from molten lavas, which cool quickly after being extruded onto Earth's surface. As the molten material cools, its iron-rich components align with Earth's magnetic field like a compass. After the lava turns into basaltic rock (when its temperature drops below 1200°C), continued cooling to temperatures near 600°C allows the "fossilized" magnetic compasses to become fixed in position in the rock. Also locked in the basalts are radioactive minerals such as potassium (K). Their slow decay (Chapter 2) can date the time when each basalt cooled and acquired its magnetic compass.

Paleomagnetism is used to reconstruct changes in the configuration of Earth's surface in two ways. (1) Back to about 500 Myr ago, paleomagnetic compasses recorded in continental basalts can be used to track movements of landmasses with respect to latitude. (2) Over the last 175 million years, paleomagnetic changes recorded in basaltic oceanic crust are used to reconstruct movements of plates and (over part of that interval) rates of spreading of the seafloor.

Paleomagnetic Determination of Past Locations of Continents Because ocean crust is constantly being destroyed at convergent plate boundaries, no crust older than 175 Myr survives. For older intervals back to about 500 Myr ago, paleomagnetism must rely on basalts found on the continents. The orientations of the magnetic

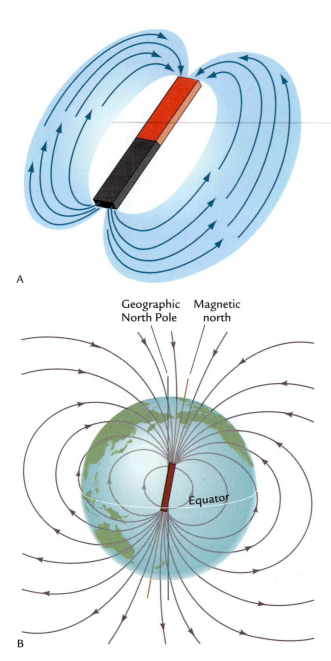

A

Geographic   Magnetic
North Pole   north

Equator

B

FIGURE 4-5 Earth's magnetic field Like (A) the magnetic field indicated by iron filings around a bar magnet, (B) Earth has a magnetic field that determines the alignment of compass needles. Basaltic rocks contain iron minerals that align with Earth's prevailing magnetic field shortly after the molten magma cools to solid rock. (B: F. Press and R. Siever, *Understanding Earth*, 2nd ed., © 1998 by W. H. Freeman and Company.)

tudes has that orientation (see Figure 4-5 bottom). In contrast, lavas that cool near the equator have internal compasses oriented closer to horizontal, nearly parallel to Earth's surface. After they form, the basaltic rocks may be carried across Earth's surface by plate tectonic processes, but their embedded magnetic compasses still record the latitude at which they formed. Rocks older than about 500 Myr are less reliable for these studies because of the increasing likelihood that their magnetic compasses have been reset to the magnetic field of a later time.

Paleomagnetic Dating of Ocean Crust Paleomagnetism is used to trace the movement of the seafloor during the last 175 Myr because of an entirely different characteristic of Earth's magnetic field: the fact that it has repeatedly reversed direction. Compasses that today point to magnetic north in the present "normal" magnetic field would have pointed to magnetic south (a position very near the South Pole) during times when the field was in a "reversed" orientation.

Past changes in the magnetic field are recorded in fossil magnetic compasses in well-dated basaltic rocks from many regions. Because these widely dispersed basaltic rocks have yielded the same sequence of reversals through time, the magnetic reversal history they record must be a worldwide phenomenon. The reversals occur at irregular intervals ranging from several million years to just a few thousand years.

Soon after this magnetic reversal sequence was established on land, marine geoscientists found stripelike magnetic patterns called **magnetic lineations** on the ocean floor (Figure 4-6). Ships surveying the ocean towed instruments that measured Earth's regional magnetic field. On the mid-ocean ridges, these magnetic lineations were found to be symmetrical around the ridge axis.

To the surprise of most scientists, the mapped pattern of highs and lows measured in the magnetic field at sea closely matched the pattern of normal and reversed intervals defined by the magnetic reversal history from sequences of basalts on land. Because of this match, scientists realized that the time framework developed on land could be transferred directly to the lineations in the ocean. Based on this link, ocean crust could be dated in any region where ships measured the magnetic lineations.

This unexpected match of magnetic patterns on land with those in the ocean proved that new (zero-age) ocean crust is being formed at the crests of ocean ridges, and that the ocean crust and underlying lithosphere then slowly spread away in both directions. As a result, the age of the ocean crust steadily increases with distance from the ridges (see Figure 4-6).

Scientists have now used this information about the age of existing ocean crust to evaluate causes of past climate changes in two ways. First, the dated magnetic lineations on the seafloor can be used to roll

compasses frozen in basalts are used to determine the past latitude of the basalt (and of the portion of continental crust in which it is embedded) in relation to the magnetic poles.

In molten lavas that cool at high latitudes, the internal magnetic compasses point in a nearly vertical direction because Earth's magnetic field at high lati-

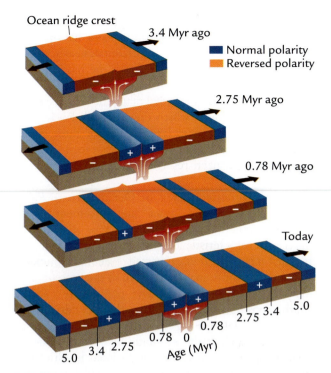

Ocean ridge crest

3.4 Myr ago

■ Normal polarity
■ Reversed polarity

2.75 Myr ago

0.78 Myr ago

Today

5.0
3.4
2.75
0.78
0
0.78
2.75 3.4 5.0
Age (Myr)

FIGURE 4-6 Magnetization of ocean crust As molten lava erupts at the seafloor, cools, and solidifies, successive bands of ocean crust form and are magnetized in the normal or reversed polarity prevailing at the time. As the plates move apart, equal amounts of magnetized crust are carried away from the ridge axis in both directions and can be used to date the seafloor. (Modified from F. Press and R. Siever, *Understanding Earth,* 2nd ed., © 1998 by W. H. Freeman and Company.)

back the recent motions of the seafloor and restore the continents and oceans to their positions during the last 175 Myr. Second, the lineations in ocean crust can be used to reconstruct the rate of **seafloor spreading.** Changes in the rate of spreading define both the rate at which new ocean crust and lithosphere are created at ocean ridges and the rate at which older ocean crust and lithosphere are subducted at ocean trenches.

IN SUMMARY, we can reconstruct the positions of continents on Earth's surface with good accuracy back to 300 Myr ago and less accurately back to 500 Myr ago or earlier. Within the last 100 Myr, we can compile spreading rates over enough of the world's ocean to attempt to estimate the global mean rate of creation and destruction of ocean crust.

Polar Position Hypothesis

An early hypothesis of long-term climate change focused on latitudinal position as a likely cause of glaciation of continents. The **polar position hypothesis** made two key predictions that can be tested over the younger part of Earth's history: (1) ice sheets should appear on continents that were located at polar or near-polar latitudes, but (2) no ice should appear if the continents were located outside polar regions. This hypothesis calls not on worldwide climate changes to explain the occurrence of icehouse intervals but simply on the movements of continents and tectonic plates across Earth's surface.

The fact that modern ice sheets occur on the polar continent of Antarctica and the near-polar landmass of Greenland makes this hypothesis seem plausible. Modern ice sheets exist at high latitudes for several reasons: cold temperatures caused by low angles of incident solar radiation, high albedos resulting from the prevalent cover of snow and sea ice, and sufficient moisture to maintain ice sheets despite melting that may occur along their lower margins (companion Web site, pp. 3-11, 27-30).

4-3 Glaciations and Continental Positions since 500 Myr Ago

We can directly test the validity of the polar position hypothesis against evidence in the geologic record. Over the last 450 Myr, seafloor spreading has slowly moved continents across Earth's surface between the warmer low-latitude climates and the colder high-latitude climates (Table 4-1). If latitudinal position alone controls climate, these movements should have produced

TABLE 4-1 Evaluation of the Polar Position Hypothesis of Glaciation

Time (Myr ago)	Ice sheets present?	Continents in polar position?	Hypothesis supported?
440	Yes	Yes	Yes
425–325	No	Yes	No
325–240	Yes	Yes	Yes
240–125	No	No	Yes
125–35	No	Yes	No
35–0	Yes	Yes	Yes

predictable changes in glaciations over intervals of tens to hundreds of millions of years.

Major continent-sized ice sheets existed on Earth during three icehouse eras: a relatively brief interval near 440 Myr ago, a long interval from 325 to 240 Myr ago, and the current icehouse era of the last 35 Myr. During most of the long intervening intervals (425-325 Myr and 240–35 Myr ago), large ice sheets do not appear to have existed.

Near 420 Myr ago, small landmasses that were later to form modern North America and Eurasia lay scattered across a wide range of latitudes (Figure 4-7A). The other land areas were combined in a southern supercontinent called **Gondwana**, equivalent to modern Africa, Arabia, Antarctica, Australia, South America, and India. This continent was located initially in the southern hemisphere on the opposite side of the globe, where the Pacific Ocean is now, but Gondwana had begun a long trip across the South Pole and then northward that would lead to a collision with the northern landmasses and formation of the giant supercontinent of **Pangaea**, meaning "All Earth" (Figure 4-7B-D).

A convenient way to represent this motion is to plot the changing position of the magnetic south pole in relation to the land (Figure 4-8). Although this presentation makes it look as if the south magnetic pole were moving southward across Gondwana, in fact the Gondwana continent was moving northward across the pole.

How well does the pattern shown in Figure 4-8 explain the intervals of glaciation and nonglaciation listed in Table 4-1? The position of the south magnetic pole 440 Myr ago agrees with the evidence of glaciation in the area of the modern Sahara Desert. The weight of the ice pressing down on the loose rubble carried in its base left striations (grooves) cut into bedrock (Figure 4-9).

At first this match seems to give us a positive confirmation of the polar position hypothesis, but on closer inspection problems emerge. One problem is that this glacial era was quite brief in terms of geologic time. Although its duration was once thought to be about 10 Myr, new evidence suggests that ice may only have been present for 1 Myr or less. This brief a glaciation is not easily explained by the slow motion of Gondwana across the South Pole (Box 4-1).

A more perplexing problem is the lack of glaciations between 425 and 325 Myr ago, even though the Gondwana continent was still continuing its slow transit across the pole (see Figure 4-8). Somehow land existed at the South Pole for almost 100 Myr without ice forming. This observation argues against the hypothesis that a polar position is the only requirement for large-scale glaciation.

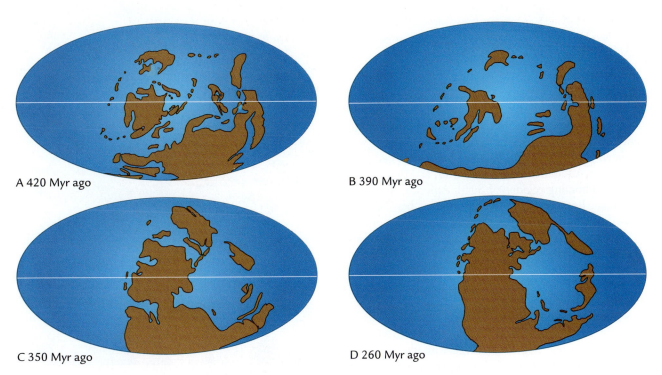

A 420 Myr ago

B 390 Myr ago

C 350 Myr ago

D 260 Myr ago

FIGURE 4-7 Moving continents (A–C) After 450 Myr ago, plate tectonic activity carried the southern continent of Gondwana across the South Pole on a path headed toward continents scattered across the northern hemisphere. (D) Subsequent collisions formed the giant continent Pangaea. (Adapted from S. Stanley, *Earth System History*, © 1999 by W. H. Freeman and Company.)

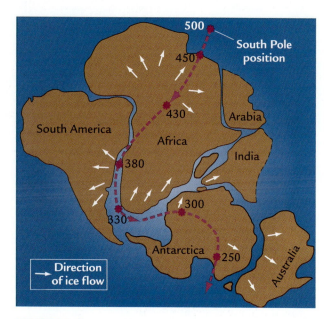

FIGURE 4-8 Gondwana glaciation and the South Pole
Changes in the position of the south magnetic pole in relation to the continent of Gondwana are largely the result of the slow movement of Gondwana. Glaciations occurred in the northern Sahara about 440 Myr ago and in southern Gondwana (South Africa, Antarctica, India, South America, and Australia) 325–240 Myr ago. The (shallow) water shown between the modern continental outlines was land during Pangaean times. (Adapted from T. J. Crowley et al., "Gondwanaland's Seasonal Cycle," *Nature* 329 [1987]: 803–7, based on P. Morel and E. Irving, "Tentative Paleocontinental Maps for the Early Phanerozoic and Proterozoic," *Journal of Geology* 86 [1978]: 535–61].)

From 325 to 240 Myr ago, Gondwana continued its slow journey across the South Pole, and a huge region centered on the south-central part of the continent was glaciated (see Figure 4-8). The ice sheets were centered on modern Antarctica and South Africa, and they spread out into adjoining regions of South America, Australia, and India. Because of the correspondence between the area of Gondwana that was glaciated and its position at or near the south magnetic pole, this long interval of glaciation is fully consistent with the polar position hypothesis.

By 240 Myr ago, Gondwana had moved northward and the glaciation had ended. The absence of ice after that time agrees with the positioning of the land away from the South Pole. By that time, the northern part of Gondwana had begun to merge with the northern continents and form the even larger supercontinent Pangaea.

After 180 Myr ago Pangaea began to break up. Its southernmost part, which included the modern continents of Antarctica, India, and Australia, moved back

over the South Pole by 125 Myr ago, yet no ice sheet developed. Antarctica remained directly over the pole but largely ice-free from 125 Myr ago until near 35 Myr ago, when ice reappeared. Here again we face the mystery encountered earlier: How could a landmass centered on a pole remain ice-free (or largely so) for almost 100 Myr?

Clearly the polar position hypothesis cannot fully explain the sequence of glaciated and unglaciated intervals over the last 500 Myr. Yet the hypothesis is successful to this extent: ice sheets developed only on landmasses that were at polar or near-polar positions, consistent with their polar locations today. This correlation (see Table 4-1) confirms that over the last 500 Myr continents had to occupy polar positions for large-scale glaciation to occur. But the same record also tells us that the presence of continents in a polar position does not guarantee that ice sheets will form.

FIGURE 4-9 Glacial striations Sediment rubble carried in the bottom layers of ice sheets about 440 Myr ago gouged striations in North African bedrock similar to those in modern ice in Alaska, shown here. (Carr Clifton.)

Brief Glaciation 440 Myr Ago

An ice sheet comparable in size to that on the continent of Antarctica today covered the North African part of the Gondwana continent near 440 Myr ago. Until recently this glaciation was thought to have lasted at least 10 Myr, and it was attributed to a combination of factors: the general cooling effect from a Sun that was 4% weaker than today, the positioning of the North African part of Gondwana directly over the South Pole, and a reduction of atmospheric CO_2 values caused by some combination of slower CO_2 input by volcanoes and faster chemical weathering. Faster weathering may have been caused by small continental collisions prior to the ones that later formed the supercontinent Pangaea, perhaps aided by the first appearance of vegetation on land and its effect in enhancing weathering (Chapter 3).

More recent dating of the geologic record suggests that this glaciation may have lasted much less than 10 Myr, possibly 1 Myr or less—a very brief episode in comparison with the 35 Myr of the present glacial era and the glaciation that lasted from 325 to 240 Myr ago. If this glaciation was indeed only a million years long, neither seafloor spreading nor chemical weathering seems likely to have changed the CO_2 concentration in the atmosphere fast enough to explain it. Volcanoes and chemical weathering rates can gradually drive CO_2 levels low enough to produce glaciation, but this episode appears to require a mechanism capable of dropping and then raising CO_2 values within 1 Myr.

One mechanism under consideration is an abrupt increase in the rate of burial of organic carbon. The organic carbon subcycle (see Box 3-1) meets several requirements for explaining a large but rapid climate cooling. Because it carries one-fifth of the total flow of carbon through the upper parts of Earth, this subcycle has the potential to alter the global carbon balance and atmospheric CO_2 levels. Also favoring this explanation is the fact that large amounts of organic carbon can be quickly buried in the sedimentary record, causing a rapid reduction of CO_2 levels.

Several kinds of changes can cause rapid burial of organic carbon: changes in wind direction that cause increased upwelling along coastal margins; an increase in the amount of organic carbon and nutrients delivered to the ocean; a change toward wetter climates on continental margins, where low relief naturally favors formation of vegetation-rich swamps; or the isolation of small ocean basins in regions of high rainfall that generates carbon-rich river runoff.

IN SUMMARY, the polar position hypothesis may be part of the story, but some other factor must also be at work, a factor that controls climate in such a way as to allow ice sheets to form over polar continents during some intervals and prohibit them from doing so during others. One likely cause is changes in concentrations of greenhouse gases.

Modeling Climate on the Supercontinent Pangaea

One fortunate aspect of studying the history of Earth's climate on tectonic time scales is the number of natural climate experiments Earth has run by greatly altering its geography. Because the locations of continents are accurately known for the past 300 Myr, climate scientists can use general circulation models (GCMs) to evaluate the impact of geographic factors on climate. Here we examine a time near 200 Myr ago when collisions of continents had formed the giant supercontinent Pangaea. Because this configuration differs considerably from the more dispersed locations of continents today,

Pangaea provides climate scientists with a very different and yet real Earth for testing the performance of climate models.

4-4 Input to the Model Simulation of Climate on Pangaea

Recall from Chapter 2 that GCM runs require the major physical aspects of a past world to be specified in advance as *boundary condition input* in order to run simulations of past climates. The most basic physical constraint is the distribution of land and sea. Pangaea remained intact from the time it formed (250 Myr ago) until it broke up after 180 Myr ago. The focus here is on this long interval of relatively stable land-sea geometry. The only tectonic change of significance during this time was a very slow northward movement of Pangaea.

At 200 Myr ago, Pangaea stretched from high northern to high southern latitudes and was almost symmetrical around the equator (Figure 4–10A). The landmasses of Gondwana (Antarctica, Australia, Africa, Arabia, South America, and India) formed its southern part. Northern Pangaea consisted of the remaining

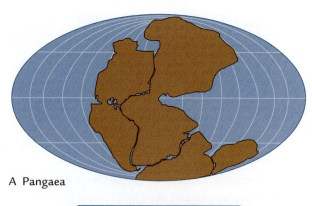

A Pangaea

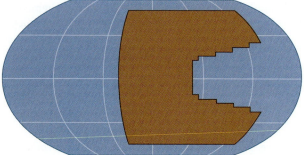

B Pangaea in model grid

FIGURE 4-10 The supercontinent Pangaea (A) Geographic reconstructions of the interval around 200 Myr ago show all the continents joined in a single landmass called Pangaea. (B) Climate modelers have simplified this configuration into an idealized continent symmetric around the equator. (A: Adapted from J. E. Kutzbach and R. G. Gallimore, "Megamonsoons of the Megacontinent," *Journal of Geophysical Research* 94 [1989]: 3341–57. B: from J. E. Kutzbach, "Idealized Pangean Climates: Sensitivity to Orbital Change," Geological Society of America Special Paper 288 [1994]: 51–55.)

landmasses, sometimes referred to as Laurasia: North America, Europe, and north-central Asia. A wedge-shaped tropical seaway indented far into Pangaea from the east, while the west coast had a smaller seaway in the north. This single landmass represented almost one-third of Earth's surface. It spanned 180° of longitude across its northern and southern limits near 70° latitude and one-quarter of Earth's circumference (90°) at the equator.

Modelers have simplified this configuration for use as input to model simulations by making the land distribution symmetrical around the equator (Figure 4–10B). This simplification requires relatively small changes in the way Pangaea is represented by grid boxes in the model. A benefit of this simplification is that each seasonal model run in each hemisphere is the exact mirror image of the same season in the other hemisphere: the seasons simply switch back and forth between hemispheres. This symmetry effectively

doubles the number of years the model simulates for each hemisphere.

A second important decision on input to the model is global sea level. Evidence from rocks on Pangaea indicates that global sea level 200 Myr ago was comparable to the level today. As a result, sea level was placed close to the structural edges of the continents, where it lies today.

A third important decision is the distribution of elevated topography on the continents, and this aspect of Pangaea is not as well known. In the simulation examined here, all land in the interior of Pangaea was represented as a low-elevation plateau at a uniform height of 1000 m, with its edges sloping gradually down to sea level along the outer margins of the continents.

Another important boundary condition that needs to be specified is the CO_2 level in the atmosphere. Unfortunately, the CO_2 concentration for 200 Myr ago is not directly known. Although long-term CO_2 levels are determined by tectonic factors, they are at the same time an integral part of the climate system. Because the choice of CO_2 concentration will have a direct impact on the climate simulated by the model, the danger of circular reasoning is present.

Fortunately, other considerations help climate modelers constrain the CO_2 level. Astronomers know that the Sun had not yet reached its present strength and was still about 1% weaker than it is today (Chapter 3). By itself, a weaker Sun should have made Pangaea significantly colder than the modern world, with snow and ice closer to the equator than today.

Yet evidence from Pangaea argues against a colder world. No ice sheets existed on Pangaea 200 Myr ago, even though its northern and southern limits lay within the Arctic and Antarctic circles (see Figure 4-10A). Today landmasses at similar latitudes are either permanently ice-covered (Greenland) or alternately ice-covered and ice-free through time (North America, Europe, and Asia). The absence of polar ice suggests that Pangaea's climate was somewhat warmer than Earth's climate is today.

Fossil evidence of vegetation on Pangaea leads to the same conclusion. Except for a few surviving types such as the ginkgo tree (Figure 4-11), Earth's vegetation has evolved to different forms since Pangaean times, and comparisons between plant types now and then have to be based on types with similar appearances rather than on actual species. Several kinds of palmlike vegetation that would have been killed by hard freezes existed on Pangaea at latitudes as high as 40°. This suggests that the equatorward limit of hard freezes on Pangaea was 40°, just above the modern limit of 30° to 40°.

The most likely reason for a warmer Pangaea is that a higher CO_2 level 200 Myr ago compensated for the weaker Sun. The model experiment examined here

FIGURE 4-11 Pangaean trees Modern gingko trees are descended from similar forms that first evolved some 200 Myr ago. (Courtesy of Mike Bowers, Blandy Farm, Boyce, VA.)

assumed a level of 1650 parts per million, almost six times the natural (preindustrial) value of 280 parts per million. As we will see, this choice not only produced temperature distributions consistent with the evidence from ice and from frost-sensitive vegetation but also simulated other climatic features that match independent evidence from the Pangaean geologic record.

With the critical boundary conditions specified, the model simulation is ready to run. After 15 years of simulated time to allow the model climate to come to a state of equilibrium, the results shown are based on the last 5 years of simulated seasonal changes.

4-5 Output from the Model Simulation of Climate on Pangaea

Because of its huge size, we might anticipate that the interior of Pangaea would have had an extremely dry continental climate, in the absence of the moderating influence of oceanic moisture. The climate model simulation confirms this expectation.

The model simulates widespread aridity at lower latitudes, especially in the Pangaean interior. Mean annual precipitation and soil moisture levels are low across large expanses of interior and western Pangaea between 40°S and 40°N (Figure 4-12). Precipitation values of 1–2 mm per day in these regions are equivalent to annual totals of 15–25 inches (35–70 cm) per year, comparable to those in semiarid grassland areas such as the western plains of the United States today (Figure 4-12A).

This pervasive aridity reflects two factors: (1) the great expanses of land at subtropical latitudes beneath the dry, downward-moving limb of the Hadley cell and (2) the large amount of land in the tropics, causing trade winds to lose most of their water vapor before reaching the continental interior (companion Web site, pp. 14–22). The ocean around Pangaea received far more rainfall than the land and more than it does today.

Geologic evidence supports the model simulation of widespread Pangaean aridity. The clearest evidence is the distribution of **evaporite** deposits, salts that precipitated out of water in lakes and in coastal margin basins with limited connections to the ocean. Evaporite salts form only in arid regions where evaporation far exceeds precipitation. More evaporite salt was deposited during the time of Pangaea than at any time in the last several hundred million years (Figure 4-13). Evaporite deposits occurred in the interior and along the tropical east coast of Pangaea, regions the model simulates as arid.

Because the moderating effects of ocean moisture failed to reach much of Pangaea's interior, the continent was left vulnerable to seasonal extremes of heating by the Sun in summer and cooling during winter. As a

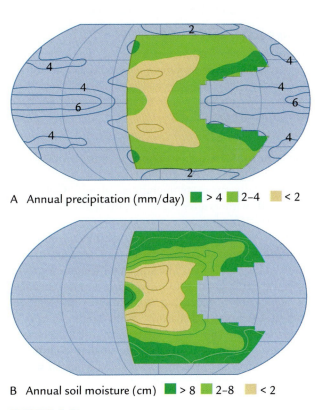

A   Annual precipitation (mm/day)   ■ > 4   ■ 2–4   ■ < 2

B   Annual soil moisture (cm)   ■ > 8   ■ 2–8   ■ < 2

FIGURE 4-12 Precipitation on Pangaea Climate models simulate patterns of (A) annual mean precipitation and (B) annual soil moisture on Pangaea. Broad areas of the tropics and subtropics were very dry. (Adapted from J. E. Kutzbach, "Idealized Pangean Climates: Sensitivity to Orbital Change," *Geological Society of America Special Paper* 288 [1994]: 41–55.)

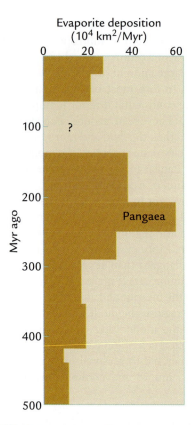

FIGURE 4-13 Pangaean evaporites The volumes of rock salt deposits (evaporites) formed on Pangaea about 200 Myr ago were larger than those formed at any other time in the last 500 Myr and indicate very dry conditions. (Adapted from W. A. Gordon, "Distribution by Latitude of Phanerozoic Evaporites," *Journal of Geology* 83 [1975]: 671–84.)

extent with the vegetation evidence. Despite the high CO_2 values used as input to the simulation, freezing still occurs farther south in the model than the evidence from past vegetation indicates.

Another characteristic of the climate of Pangaea was the strong reversal between summer and winter monsoon circulations. Monsoon circulations are driven by the different rates of response of the land and the oceans to solar heating in summer and radiative heat loss in winter (companion Web site, pp. 15–18). The large seasonal swings in land temperature and small seasonal changes in ocean temperature reflect these contrasting responses of land and ocean.

Strong solar heating over the part of Pangaea situated in the summer hemisphere caused heated air to rise over the land and a strong low-pressure cell to develop at the surface (Figure 4–15A). The rising of heated air caused a net inflow of moisture-bearing winds from the ocean, especially in the subtropics, bringing heavy rains to the subtropical east coast (Figure 4–15B).

The situation in the winter hemisphere was exactly the reverse. The weak seasonal heating from the Sun and strong heat loss by longwave back radiation caused cooling over the interior of Pangaea. The cooling caused air to sink toward the land surface, built up high pressures over the continent, and pushed cold, dry air out over the ocean. As a result, precipitation over the land was reduced.

Note that the winds on the eastern margins of Pangaea from 0° to 45° latitude reversed direction between the seasons: warm summer monsoon winds blew from the sea onto the land, but cold winter monsoon winds

result, the model simulates a huge seasonal temperature response (Figure 4-14). In some mid-latitude regions, summer daily mean temperatures of +25°C (77°F) alternated with winter daily mean temperatures of −15°C (+5°F).

The occurrence of extremely continental climates on Pangaea may help to explain the absence of ice sheets at high latitudes. The simulated winter temperatures were cold enough to provide the snowfall needed for ice sheets to grow. But hot summers on Pangaea even on the poleward margins of the landmass caused rapid melting of snow and thereby prevented glaciation. Ice sheets form more readily on smaller continents where summer temperatures are cooled by moist winds off the ocean.

The model simulation also indicates that average daily land temperatures in winter would have reached the freezing point as far equatorward as 40° latitude (see Figure 4-14), closely matching the low-latitude limit of frost-sensitive vegetation on Pangaea. But with winter nights likely to have been colder than the daily mean, the model's results actually disagree to some

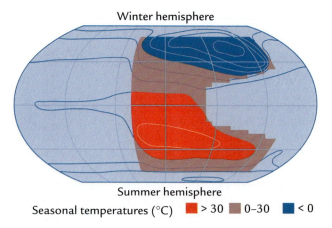

FIGURE 4-14 Temperature on Pangaea Climate model simulations show extreme seasonal temperature contrasts on Pangaea between the summer hemisphere, which was warmed by solar radiation, and the winter hemisphere, which lost heat by longwave back radiation. (Adapted from J. E. Kutzbach, "Idealized Pangean Climates: Sensitivity to Orbital Change," *Geological Society of America Special Paper* 288 [1994]: 41–55.)

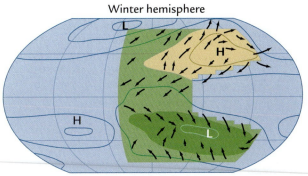

Winter hemisphere

Summer hemisphere

A Seasonal pressure and winds

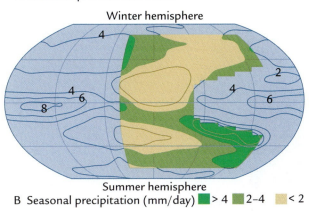

Winter hemisphere

Summer hemisphere

B Seasonal precipitation (mm/day) ■ > 4 ■ 2–4 ■ < 2

FIGURE 4-15 "Supermonsoons" on Pangaea Climate models simulate (A) very large seasonal changes in surface pressure and winds and (B) monsoonal precipitation on Pangaea. Summer heating creates a low-pressure region (L) and draws in moist oceanic winds, which drop heavy precipitation along the subtropical east coast. Winter cooling creates a high-pressure cell (H) that sends dry air out from land to sea and reduces precipitation. (Adapted from J. E. Kutzbach, "Idealized Pangean Climates: Sensitivity to Orbital Change," *Geological Society of America Special Paper* 288 [1994]: 41–55.)

blew from the land out to sea. The subtropical margins of Pangaea were places of enormous contrast in seasonal precipitation, alternating between very wet summers and dry winters.

Geologic evidence of seasonal moisture contrasts on Pangaea comes from the common occurrence of **red beds,** sandy or silty sedimentary rocks stained various shades of red by oxidation of iron minerals. Red-colored soils accumulate today in regions where the contrast in seasonal moisture is strong. The process of oxidation is analogous to rust that forms on metal tools left out in the rain. In a geologic context, the wet season provides the necessary moisture, and the rust forms during the dry season or shorter dry intervals. Red beds were more widespread on Pangaea than during other geologic intervals, a finding that is consistent with the

model simulation of highly seasonal changes in moisture between wet summer monsoons and dry winter monsoons.

Tectonic Control of CO_2 Input: BLAG Spreading Rate Hypothesis

Our examination of both the polar position hypothesis and the climate of Pangaea suggests that changes in Earth's geography alone cannot explain the climatic variations between warm greenhouse climates and cold icehouse climates during the last 450 Myr. Another likely factor in these climatic changes is variations in the CO_2 concentration of the atmosphere. In the remainder of this chapter we examine two hypotheses that attempt to explain why CO_2 has changed through time. One hypothesis emphasizes changes in CO_2 input by volcanoes; the other focuses on changes in CO_2 removal by weathering.

4-6 Control of CO_2 Input by Seafloor Spreading

A hypothesis published in 1983 proposed that climate changes during the last several hundred million years have been driven mainly by changes in the rate of CO_2 input to the atmosphere and ocean by plate tectonic processes. This hypothesis is called the BLAG hypothesis, based on the initials of its authors, the geochemists Robert **B**erner, Antonio **L**asaga, and Robert **G**arrels. We will refer to it as the **spreading rate hypothesis.**

In a world of active plate tectonic processes, carbon cycles constantly between Earth's interior and its surface (Figure 4-16). Most CO_2 is expelled to the atmosphere by volcanic activity along two kinds of locations: (1) margins of converging plates, where parts of the subducting plates melt and form molten magmas that rise to the surface in mountain belt and island arc volcanoes, delivering CO_2 and other gases from Earth's interior; and (2) margins of divergent plates (ocean ridges), where hot magma carrying CO_2 erupts directly into ocean water.

Some volcanoes also emit CO_2 at sites distant from plate boundaries where thin plumes of molten material rise from deep within the interior and reach the surface at volcanic **hot spots** (see Figure 4-16 bottom). Additional CO_2 is released to the atmosphere by the slow oxidation of old organic carbon in sedimentary rocks eroded at Earth's surface (Chapter 3).

The centerpiece of the BLAG hypothesis is the concept that changes in the rate of seafloor spreading over millions of years have controlled the rate of delivery of CO_2 to the atmosphere from the large rock reservoir of carbon, and that the resulting changes in atmospheric CO_2 concentrations have had a major impact on Earth's climate.

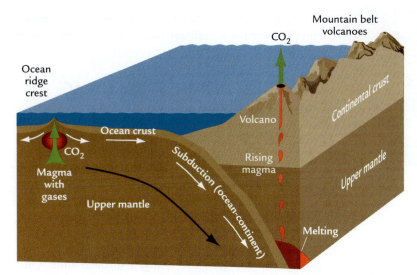

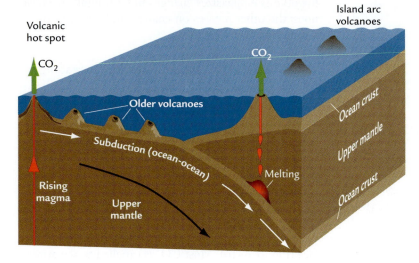

FIGURE 4-16 CO_2 input CO_2 is transferred from Earth's interior to the atmosphere-ocean system primarily at ocean ridges (top left) and subduction zones (top and bottom right). Lesser emissions of CO_2 occur when volcanoes erupt at hot spots in the middle of plates (bottom left).

Well-dated magnetic lineations show that the ocean ridges that exist today have been spreading at widely varying rates for millions of years (Figure 4-17). For example, the ridge in the South Pacific Ocean spreads as much as ten times faster than the one in the Atlantic Ocean.

The BLAG hypothesis is based on the concept that the *globally averaged* rate of seafloor spreading has changed over time. Changes in the mean rate of spreading through time should alter the transfer of CO_2 from Earth's rock reservoirs to its atmosphere at ocean ridges and subduction zone volcanoes, because these plate margins are vital participants in the process of seafloor spreading (see Figure 4-16).

Faster rates of spreading at ridge crests creates larger amounts of new ocean crust and more frequent releases of magma, which should deliver greater amounts of CO_2 to the ocean (Figure 4-18). Faster spreading also causes more rapid subduction of crust and sediment in ocean trenches and delivers larger volumes of carbon-rich sediment and rock for subsequent melting and CO_2 release through volcanoes. Conversely, slower spreading should reduce both kinds of CO_2 input to the atmosphere.

Although the BLAG hypothesis focuses on changes in spreading rates as a driver of long-term climate change, it also calls on chemical weathering for negative feedback to moderate these changes (Chapter 3). Increased volcanic emissions caused by faster seafloor spreading leads to higher atmospheric CO_2 levels and a warmer climate (see Figure 4-18 top). This initial shift toward a greenhouse climate then activates the combined effects of temperature, precipitation, and vegetation in speeding up the rate of chemical weathering and causes CO_2 to be drawn out of the atmosphere at a faster rate. The resulting CO_2 removal opposes and reduces some of the initial warming driven by faster spreading rates and higher CO_2 concentrations.

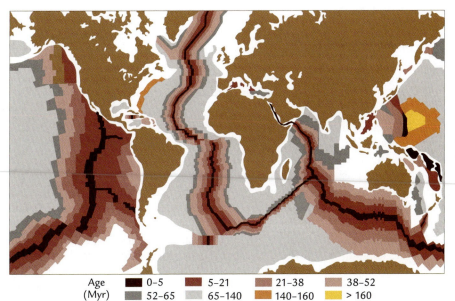

FIGURE 4-17 Age of the seafloor
Some ocean crust dates as far back as 175 Myr ago. Modern spreading rates are as much as ten times faster in the Pacific than in the Atlantic. (Modified from S. Stanley, *Earth System History*, © 1999 by W. H. Freeman and Company, after W. C. Pitman et al., Map and Chart Series MC-6 [Boulder, CO: Geological Society of America, 1974].)

Age (Myr)			
■ 0–5	■ 5–21	■ 21–38	■ 38–52
■ 52–65	■ 65–140	■ 140–160	■ > 160

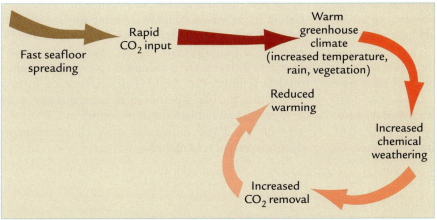

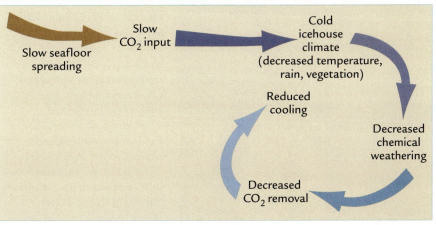

FIGURE 4-18 The spreading rate (BLAG) hypothesis This hypothesis predicts that atmospheric CO_2 concentrations and global climate are driven by the global mean rate of seafloor spreading, which controls the rate of CO_2 input at ocean ridge crests and subduction zones. The spreading rate hypothesis also invokes chemical weathering as a negative feedback that partially counters changes in atmospheric CO_2 and global climate initiated by varying rates of seafloor spreading.

Similarly, chemical weathering feedback works to offset some of the impact of cooling caused by slower volcanic input of CO_2 (see Figure 4–18 bottom). In effect, the BLAG hypothesis relies on chemical weathering to moderate any fluctuations in climate driven by changes in volcanic CO_2 input.

The BLAG hypothesis further proposes that much of the cycling of carbon between the deeper Earth and the atmosphere occurs in a closed loop (Figure 4-19). Carbon taken from the atmosphere during chemical weathering is initially stored in dissolved HCO_3 ions that are carried by rivers to the sea. As we have seen (Chapter 2), marine plankton use this dissolved carbon to form $CaCO_3$ shells, and the shells are deposited in ocean sediments when the organisms die. The movement of carbon through this part of the cycle is rapid, occurring in just a few years.

The $CaCO_3$-bearing sediments are then carried by seafloor spreading toward subduction zones at continental margins. Some sediment is scraped off at the ocean trenches, but much of it is carried downward in the subduction process (see Figure 4-19). This slow journey of carbon-bearing sediments across the ocean floor and down the trenches takes tens of millions of years.

Most of the $CaCO_3$ (and other carbon) carried down into Earth's interior by subduction melts at the hot temperatures found at great depths or is transformed in other ways. These processes eventually return CO_2 to the atmosphere through volcanoes and complete the cycle. Almost none of the subducted carbon is carried deep into the mantle. Movement of carbon through this deeper part of the cycle takes tens of millions of years.

The two chemical reactions that summarize the basic chemical changes involved at the beginning and end of this tectonic-scale carbon cycle are mirror opposites:

Chemical weathering on land

$$CaSiO_3 \quad + \quad CO_2 \quad \rightarrow \quad CaCO_3 \quad + \quad SiO_2$$
Silicate rock   Atmosphere   Plankton   Plankton

Melting and transformation in subduction zones

$$CaCO_3 \quad + \quad SiO_2 \quad \rightarrow \quad CaSiO_3 \quad + \quad CO_2$$
Ocean sediments   Silicate rock   Atmosphere

The two reactions together form a complete (closed) cycle with no net chemical change, but this cycle takes tens of millions of years. The longest part of the cycle is caused by the slow spreading and subduction of seafloor, the slow transformation of $CaCO_3$ in the lower crust and upper mantle, and the slow delivery of CO_2 to volcanoes. In contrast, changes in spreading rates can alter the rate of melting and CO_2 release to the atmosphere with little or no delay because carbon-bearing sediment is already "in the pipeline." At any interval in time, carbon-bearing sediments are in the process of being subducted into Earth's interior, and changes in the average rate of subduction will soon result in faster melting of this down-going material.

The BLAG hypothesis proposes that this cycling of carbon provides long-term stability to the climate system by moving a roughly constant amount of total carbon back and forth between the rocks and the atmosphere over long intervals of time. As a result, atmospheric CO_2 levels are constrained to vary only within moderate limits. But the long delays between carbon weathering and burial permit small imbalances to occur between the rate of burial and the return of CO_2 to the atmosphere. These imbalances drive climate changes over intervals of tens of millions of years.

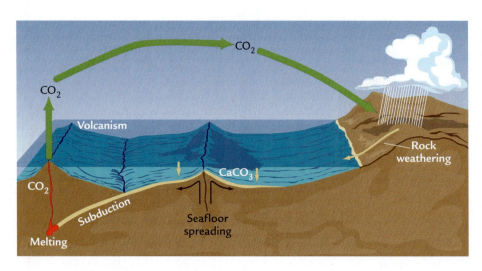

FIGURE 4-19 Carbon cycling
In the spreading rate (BLAG) hypothesis, carbon cycles continuously between rock reservoirs and the atmosphere: CO_2 is removed from the atmosphere by chemical weathering on land, deposited in the ocean, subducted, and returned to the atmosphere by volcanic activity. (Adapted from W. F. Ruddiman and J. E. Kutzbach, "Plateau uplift and climate change," *Scientific American* 264 [1991]: 66–75.)

4-7 Initial Evaluation of the BLAG Spreading Rate Hypothesis

Unfortunately, the predictions of the BLAG hypothesis cannot be directly tested over most of the geologic past because no ocean crust older than 175 Myr exists to use for calculating past spreading rates. All older crust has been subducted in ocean trenches. Half of the crust that formed 50 Myr ago has already been destroyed by rapid subduction under western South America (see Figure 4-17) and by the total disappearance of ocean crust in a former tropical seaway ("Tethys") by subduction and collision along the southern coast of Asia.

Most reconstructions suggest that the global mean spreading rate was faster 100 Myr ago than it is at present, but this issue has recently become a point of contention that will be revisited in the next two chapters. If this conclusion holds up to future scrutiny, the BLAG theory predicts that the rate of input of CO_2 to the atmosphere should have been higher 100 Myr ago than it is today (Table 4-2). This prediction agrees with geologic evidence of a warmer climate 100 Myr ago, including the absence of large polar ice sheets that now exist in both hemispheres.

Tectonic Control of CO_2 Removal: Uplift-Weathering Hypothesis

A second hypothesis that attempts to explain how plate tectonic processes control atmospheric CO_2 levels emerged from work by the marine geologist Maureen Raymo and her colleagues in the late 1980s. Parts of this concept date back to work by the geologist T. C. Chamberlain a century ago. The **uplift weathering hypothesis** proposes that chemical weathering is the active driver of climate change rather than a negative feedback that moderates climate change.

4-8 Rock Exposure and Chemical Weathering

The BLAG hypothesis emphasizes changes in CO_2 delivery to the atmosphere by seafloor spreading, and it assumes that removal of CO_2 by chemical weathering responds only to climate-related changes in temperature, precipitation, and vegetation. Although these factors do affect chemical weathering (Chapter 3), they are not the only processes that do.

The uplift weathering hypothesis starts from a different perspective. It asserts that the global mean rate of chemical weathering is heavily affected by the availability of fresh rock and mineral surfaces for the weathering process to attack, and proposes that this exposure effect can override the combined effects of the climate-related factors both locally and globally.

A simple example of the importance of rock exposure is shown in Figure 4-20. We start with a large cube of rock with six surfaces consisting of squares 1 m across, each having a surface area of 1 m². This rock cube has a total surface area of 6 m², calculated from the total areas of the six sides.

Next we slice this cube into halves along all three of its axes. This slicing creates eight smaller cubes, each 0.5 m on a side, so that each side has a surface area of 0.25 m². The total surface area of these smaller cubes is 12 m²:

$$(8 \text{ cubes}) \times (6 \text{ sides each}) \times (0.25 \text{ m}^2 \text{ of surface area per side})$$

Note that simply cutting the large cube into smaller cubes has doubled the surface area of the rock without changing its volume, at least for the imaginary laser-sharp cut assumed in this example. The fragmentation has created more exposed surface area for the weathering process to attack.

The process can be continued to finer grain sizes with similar results. Ten sequential halvings of the rock's dimensions will produce over 1 billion cubes each 1 mm on a side, about the same size as grains of sand on a beach. Together these tiny cubes have a total surface area 1000 times larger than the original 1-m³ block, yet they still retain the same total volume. Fragmentation to even smaller sizes will expose still more surface area.

With over 1000 times more surface area to act on, chemical weathering would increase by a factor of 1000 or more. This large an increase of weathering far exceeds the combined changes estimated to result from changes in temperature, precipitation, and vegetation. Clearly the climate-related factors are not the only processes to consider in evaluating chemical weathering.

4-9 Case Study: The Wind River Basin of Wyoming

Direct evidence of the importance of rock exposure in chemical weathering comes from a study of a drainage

TABLE 4-2	Evaluation of the BLAG Spreading Rate (CO_2) Input Hypothesis		
Time (Myr ago)	**Ice sheets present?**	**Spreading rate?**	**Hypothesis supported?**
100	No	Faster (?)	Yes (?)
0	Yes	Slower (?)	Yes (?)

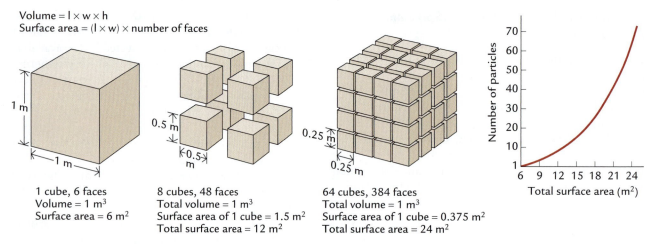

Volume = l × w × h
Surface area = (l × w) × number of faces

1 m
1 m

0.5 m
0.5 m

0.25 m
0.25 m

1 cube, 6 faces	8 cubes, 48 faces	64 cubes, 384 faces
Volume = 1 m³	Total volume = 1 m³	Total volume = 1 m³
Surface area = 6 m²	Surface area of 1 cube = 1.5 m²	Surface area of 1 cube = 0.375 m²
	Total surface area = 12 m²	Total surface area = 24 m²

Number of particles

70
60
50
40
30
20
10
1

6   9   12 15 18 21 24
Total surface area (m²)

FIGURE 4-20 **Fragmentation of rock**   Each time a cube-shaped rock is sliced into smaller cubes (with each side half as long as before), the total surface area of rock doubles, even though the volume remains the same.   (D. Merritts et al., *Environmental Geology*, © 1997 by W. H. Freeman and Company.)

basin in the Wind River Mountains of Wyoming. Because all the bedrock in this basin consists of granite, the kind of silicate rock most typical of continental crust, this watershed is reasonably representative of the average response of continental rocks to weathering.

The Wind River Mountains have been glaciated repeatedly over the last several hundred thousand years, and each glaciation has left deposits of unsorted debris (moraines) in the foothills of the valleys below. Because some of the older deposits have not been overridden by later glacial advances, undeformed moraines of various ages (from 200 to 130,000 years) can be found in the same valley.

The Wind River moraines provide an opportunity to quantify the amount of weathering of ground-up debris that is identical in composition but differs widely in age. The extent of weathering is determined by analyzing soils that have subsequently developed on the moraines. The soils gradually lose their major cations (Mg^{+2}, Na^{+1}, K^{+1}, Ca^{+2}, and others) during the chemical weathering process. The cumulative amount of chemical weathering that has occurred since each moraine was deposited can be determined by measuring the total loss of these cations. Dividing this total amount of weathering by the time elapsed since the moraine was deposited yields the *average* rate of chemical weathering over that entire interval.

The Wind River deposits show a rapid (exponential) decrease in the mean rate of weathering versus time of exposure (Figure 4-21). The younger moraines have average rates of weathering that are at least a factor of 100 faster than the older ones. The older moraines also weathered much faster during a brief interval after their deposition, but they then weathered much more slowly

later on. Why would younger glacial deposits weather so much faster?

One explanation is that freshly ground rock has more weatherable material—the kinds of fresh, unweathered silicate grains that are most vulnerable to the weathering process. These vulnerable minerals are removed through time. Once only the more resistant minerals are left, rates of weathering are slower.

Another part of the explanation relates to the effect of grain sizes on weathering (see Figure 4-20). Finer

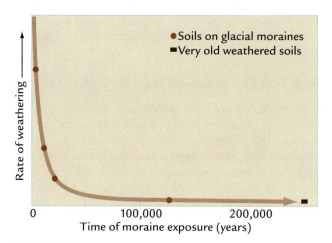

FIGURE 4-21 **Weathering and exposure time**   Glacially eroded and fragmented granite weathers quickly soon after deposition but much more slowly 100,000 years later. (Adapted from J. D. Blum, "The Effect of Late Cenozoic Glaciation and Tectonic Uplift on Silicate Weathering Rates," in *Tectonic Uplift and Climate Change*, ed. W. F. Ruddiman [New York: Plenum Press, 1997].)

grain sizes expose more surface area and cause faster weathering early in the process, but the finer sizes also disappear earlier as weathering consumes them. The coarser grain sizes that remain weather more slowly because they expose less surface area per unit of volume. Coarser fragments may also develop an outer coating or "rind" of weathering-resistant material that protects fresher material in their interiors and slows the weathering attack.

4-10 Uplift and Chemical Weathering

The uplift weathering hypothesis begins with the evidence that exposure of fragmented and unweathered rock is an important factor in the intensity of chemical weathering. It then links this evidence to the fact that exposure of freshly fragmented rock is enhanced in regions of tectonic uplift.

Several factors increase rates of exposure of fresh rock in uplifting areas. Mountains and plateaus have steep slopes both on their margins and in valleys between high peaks. Erosional processes known as **mass wasting** are unusually active on such slopes. Mass-wasting processes include rock slides and falls, flows of water-saturated debris, and a host of other processes that dislodge everything from huge slabs of rock to loose boulders, pebbles, and soil. Every event that removes overlying debris exposes fresh bedrock and unweathered material. Many high-mountain slopes consist almost entirely of fresh debris moving downslope (Figure 4-22).

Another important factor is earthquakes. Mountains and high plateaus are built by tectonic forces that push together and stack huge slivers of faulted rock at the margins of converging plates. This stacking process is accompanied by earthquakes that generate large amounts of energy, shake the ground, and dislodge debris. Even more fresh rock is exposed as a result.

A third important characteristic of steep slopes is that they are focal points for precipitation (companion Web site, p. 21). When warm air is forced up and over high terrain and cooled, water vapor condenses and precipitation occurs. High but narrow mountain belts in the tropics and mid-latitudes capture much of the moisture carried by winds. In addition, large plateaus such as the Tibetan Plateau create their own wet (monsoonal) circulations by pulling moisture in from adjacent oceans. Heavy precipitation favors chemical weathering.

Glacial ice also enhances chemical weathering in high terrain. Uplift can elevate rock surfaces to altitudes where temperatures are cold enough for mountain glaciers to form. Mountain glaciers pulverize blocks of underlying bedrock and deposit the debris in moraines at lower elevations. As we saw in the case study of the Wind River Range, glacial grinding greatly enhances rates of chemical weathering.

All these factors (steep slopes, mass wasting, earthquakes, heavy precipitation, and glaciers) are present in high mountains and plateaus. The uplift weathering hypothesis proposes that uplift accelerates chemical weathering through the combined action of these processes (Figure 4-23). Faster weathering draws more CO_2 out of the atmosphere and cools global climate toward icehouse conditions. Conversely, during times when uplift is less prevalent, chemical weathering is slower, and CO_2 stays in the atmosphere and warms the climate, producing greenhouse conditions.

The two major kinds of plate tectonic processes that cause uplift have different implications for the uplift

FIGURE 4-22 Debris on steep slopes Steep slopes of actively eroding mountains consist of highly fragmented debris periodically dislodged downslope. (Photosphere Images/Picture Quest.)

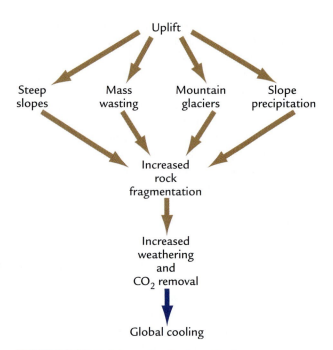

FIGURE 4-23 Uplift weathering hypothesis Active tectonic uplift produces several tectonic and climatic effects that cause strong weathering of freshly fragmented rock. This process removes CO_2 from the atmosphere and cools global climate.

weathering hypothesis. The first process, subduction of ocean crust underneath continental margins, is an integral part of plate movements and a process that is continually active in many regions on Earth. Because subduction occurs relatively steadily over time, the total amount of high mountain terrain on Earth is likely to be relatively constant through time, even though the locations and heights of individual mountain ranges vary considerably.

The second process that creates high terrain is the collision of continents, and these events are far less common. Collision between India and Asia over the last 55 Myr created the Tibetan Plateau of southern Asia, but no plateau-like feature remotely close in size to Tibet existed on Earth between 240 and 55 Myr ago because no major continental collision of this kind occurred during that interval. Earlier, between 325 and 240 Myr ago, the collisions between Gondwana and other continents

that created the supercontinent Pangaea also formed a moderate-size plateau in east-central Europe, as well as high mountain ranges in eastern North America (the Appalachians) and in northwestern Africa.

The uplift weathering hypothesis focuses mainly on plateaus created by occasional collisions of continents rather than on ever-present mountain belts. As Table 4–3 indicates, times of continental collisions that created plateaus match times of glaciations over the last 325 Myr. Like the BLAG hypothesis, the uplift weathering hypothesis is consistent with the icehouse-greenhouse-icehouse climatic sequence. But if recently discoveries prove correct, neither the uplift weathering hypothesis nor the BLAG hypothesis nor the polar position of Gondwana is a complete explanation for the short glaciation in the Sahara near 440 Myr ago (see Box 4-1).

4-11 Case Study: Weathering in the Amazon Basin

One way to evaluate the effect of uplift on chemical weathering is to examine the drainage basin of the Amazon River of South America (Figure 4-24). This basin can be divided into two major units: (1) the low-lying Amazon Basin, where easterly trade winds blowing in from the Atlantic Ocean bring frequent precipitation to the rain forests of Brazil, and (2) the high-elevation eastern slopes of the Andes Mountains, which collect most of the rest of the incoming precipitation carried by the trade winds.

Scientists have determined the regional effects of chemical weathering in this drainage basin by sampling the amount of chemically weathered ions flowing down to the Amazon River in dissolved form. They found that the upper tributaries of the Amazon emerging from the foothills of the Andes carry almost 80% of the total dissolved chemical load discharged by the river when it enters the Atlantic. Despite its vast size, the lower Amazon Basin adds only the remaining 20% of this total. Most of the chemical weathering in the Amazon drainage basin occurs in the Andes, at rates per unit area that are a factor of 40 higher than those in the lowlands.

How could this be so? This evidence seems especially at odds with what the eye actually sees in the two regions. In the lower Amazon rain forest, highly weathered clays that are the products of intense chemical

TABLE **4-3** **Evaluation of the Uplift Weathering (CO_2 Removal) Hypothesis**			
Time (Myr ago)	**Ice sheets present?**	**Continents colliding?**	**Hypothesis supported?**
325-240	Yes	Yes	Yes
240-35	No	No	Yes
35-0	Yes	Yes	Yes

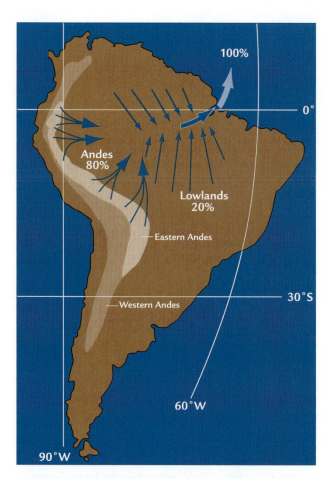

FIGURE 4-24 Weathering in the Amazon Basin Almost 80% of the chemically weathered ions that reach the Atlantic Ocean from the Amazon River come from the small area of the eastern Andes; just 20% comes from the extensive lowlands of the Amazon Basin.

heavy precipitation, and glacial erosion) combine to generate a continual supply of fresh, finely ground rock debris for weathering. Some of this weathering occurs on the steep upper slopes of exposed high terrain even in the absence of much vegetation or soil cover. Much of it occurs in basins lower in the mountains, where soils and vegetation have gained a tenuous foothold, and yet the supply of fresh unweathered rock debris from higher-elevation streams and rivers is continuous.

The absence of obvious visible chemical weathering in the Andes has two explanations. First, chemical weathering products such as clays are continually overwhelmed by the much larger supply of physically fragmented debris cascading down the steep slopes. Second, the fine clays and other products of weathering are continually removed from steep slopes and carried to the ocean by streams and rivers.

The Amazon Basin studies confirm that the rate of chemical weathering is rapid in the Andes and presumably in many of Earth's other high-elevation regions as well, even though the visible effects of chemical weathering are not apparent. These studies also show that some warm, wet, vegetated regions may be places of surprisingly slow chemical weathering.

4-12 Weathering: Both a Climate Forcing and a Feedback?

The original uplift weathering hypothesis left an important issue unresolved. It did not specify a negative feedback that would act as a thermostat and moderate the climatic effects that uplift produces. Without such a thermostat, what would stop rapid uplift from accelerating chemical weathering to the point where Earth would freeze? And why wouldn't Earth overheat during times when uplift was minimal?

One possible mechanism that could moderate the degree of uplift-induced climate change is the amount of fresh rock exposed at Earth's surface. Plate tectonic processes cause uplift across only a limited amount of Earth's surface at any one time because of the small length of plate margins involved in continental collisions and the limited areas actively involved in subduction processes on continental margins. These natural tectonic limits on the geographic extent of uplift could limit the amount of exposure of fresh rock at any one time and set a natural limit on the intensity of cooling caused by uplift.

A more plausible explanation combines the uplift weathering hypothesis with the action of the chemical weathering thermostat. Uplift of geographically limited regions (perhaps 1% of the land area) could drive climatic cooling by promoting increased chemical weathering and CO_2 removal from the atmosphere, but chemical weathering on the other 99% of the continents might well slow with the onset of colder, drier climates and the

weathering dominate. In the Andes, rock debris produced by strong physical-mechanical weathering but showing little evidence of intense chemical weathering dominates.

The answer to this mystery is deceptively simple. The lower Amazon Basin is a place where chemical weathering does indeed dominate in percentage terms, but in which the fresh minerals have long since been used up in the weathering process. The only fresh, unweathered bedrock remaining in the lowlands lies buried hundreds of meters beneath a protective cover of highly weathered clays, out of reach of intense weathering processes. These clays at and near the surface are the end products of slow bedrock weathering over many millions of years, and they have little weatherable material left. As a result, the average rate of chemical weathering in this region is extremely low.

In contrast, the physical impacts of active uplift in the Andes (steep slopes, earthquakes, mass wasting,

reduction in vegetation cover. A slowing of the rate of CO_2 removal would leave more CO_2 in the atmosphere and moderate the overall cooling. In the end, the uplift-induced weathering increase would succeed in causing a net global cooling, but it would not be nearly so large a cooling as would have occurred without the negative weathering feedback.

> **IN SUMMARY,** both the BLAG (spreading rate) hypothesis and the uplift weathering hypothesis seem to provide plausible explanations of most major icehouse-greenhouse changes of climate (see Tables 4–2 and 4–3). In Chapter 6 we will revisit both hypotheses by examining in greater detail the sequence of changes from the warm greenhouse climate of 100 Myr ago to the modern icehouse climate.

Key Terms

continental crust (p. 60)

ocean crust (p. 60)

mantle (p. 60)

lithosphere (p. 60)

asthenosphere (p. 60)

tectonic plates (p. 61)

divergent margins (p. 61)

convergent margins (p. 61)

subduction (p. 61)

continental collision (p. 62)

transform fault margins (p. 62)

magnetic field (p. 62)

paleomagnetism (p. 62)

magnetic lineations (p. 63)

seafloor spreading (p. 64)

polar position hypothesis (p. 64)

Gondwana (p. 65)

Pangaea (p. 65)

evaporite (p. 69)

red beds (p. 71)

spreading rate hypothesis (p. 71)

hot spots (p. 71)

uplift weathering hypothesis (p. 75)

mass wasting (p. 77)

Review Questions

1. Does each lithospheric plate correspond to an individual continent or ocean basin?

2. What kind of physical behavior in Earth's deeper layers allows the plates to move?

3. Explain how paleomagnetism tells us about past latitudes of continents.

4. Explain how paleomagnetism tells us about rates of spreading at ocean ridges.

5. Do glaciations always occur when continents are located in polar positions?

6. What are the major characteristics of the climate of Pangaea?

7. What is the central concept behind the BLAG (spreading rate) hypothesis?

8. What role does chemical weathering play in the BLAG hypothesis?

9. Write a chemical reaction showing how weathering removes CO_2 from the atmosphere.

10. How soon after deposition does freshly fragmented debris undergo most chemical weathering?

11. Why is chemical weathering faster in the eastern Andes than in the Amazon lowlands?

12. How could chemical weathering be both the driver and the thermostat of Earth's climate?

Additional Resources

Basic Reading

Companion Web site at www.whfreeman.com/ ruddiman2e, pp. 3–11, 14–22, 27–30.

Advanced Reading

Berner, R. A. 1999. "A New Look at the Long-Term Carbon Cycle." *GSA Today* 9: 1–6.

Blum, J. D. 1997. "The Effect of Late Cenozoic Glaciation and Tectonic Uplift on Silicate Weathering Rates and the Marine 87Sr/86Sr Record." In *Tectonic Uplift and Climate Change*, ed. W. F. Ruddiman. New York: Plenum Press.

Chamberlain, T. C. 1899. "An Attempt to Frame a Working Hypothesis of the Cause of Glacial Periods on an Atmospheric Basis." *Journal of Geology* 7: 545–84, 667–85, 751–87.

Kutzbach, J. E. 1994. "Idealized Pangaean Climates: Sensitivity to Orbital Parameters." *Geological Society of America Special Paper* 288: 41–55.

Parrish, J. T., A. M. Ziegler, and C. R. Scotese. 1982. "Rainfall Patterns and the Distribution of Coals and Evaporites in the Mesozoic and Cenozoic." *Palaeogeography, Palaeoclimate, Palaeoecology* 40: 67–101.

Raymo, M. E., W. F. Ruddiman, and P. N. Froelich. 1986. "Influence of Late Cenozoic Mountain Building on Ocean Geochemical Cycles." *Geology* 16: 649–53.

Stallard, R. F., and J. E. Edmond. 1983. "Geochemistry of the Amazon 2: The Influence of the Geology and Weathering Environments on the Dissolved Load." *Journal of Geophysical Research* 88: 9671–88.

CHAPTER 5

Greenhouse Climate

Earth's geography is much easier to reconstruct for 100 Myr ago than it is for earlier times. The positions of all the continents are known, as are the locations of the edge of the sea along continental margins and the shapes of the ocean basins. Regional and global average rates of seafloor spreading are also better constrained. In addition, climate scientists know much more about Earth's climate 100 Myr ago, including the fact that it was warm enough at the South Pole to keep ice sheets from forming. With this array of evidence, scientists can ask an important question about this interval: Was this global warmth caused by a high level of atmospheric CO_2? The answer to this question potentially holds lessons about our climatic future in a world warmed by rising levels of the same gases. Next, we explore the reasons why sea level 100 Myr ago was higher than it is today and the effects of the higher sea level on climate. Finally, we investigate the climatic and other environmental effects of a giant asteroid impact 65 Myr ago during the later stages of warm greenhouse conditions.

What Explains the Warmth 100 Myr Ago?

Around 175 Myr ago, the giant single continent of Pangaea began to break apart. By 100 Myr ago, most of today's continents had separated from one another, producing a very different-looking Earth consisting of a half-dozen smaller continents. In addition, global sea level stood at least 100 m higher than it does today, and a shallow layer of ocean water flooded continental margins and low-lying interior areas.

The geographic effect of this flooding was to fragment the existing continents into even smaller areas of land (Figure 5-1), making the geography of this greenhouse world even more unlike the single Pangaean landmass. Geologists call this interval the middle **Cretaceous,** a word meaning "abundance of chalk," because marine limestones deposited by these high seas are common around the world. This interval is important to climate scientists because geologic records contain no evidence of permanent ice anywhere on Earth, even on the parts of the Antarctic continent situated right over the South Pole. Much of this interval seems to have been a warm (greenhouse?) world.

In the northern hemisphere, evidence for unusual warmth comes from several fossil indicators, including warm-adapted vegetation such as broad-leafed evergreen trees that hold their leaves throughout the year except for a brief interval of leaf fall and regrowth. Other evidence includes warm-adapted animals such as dinosaurs, turtles, and crocodiles, all existing north of the Arctic Circle (Figure 5-2). This evidence contrasts markedly with the cold climatic conditions at such latitudes today.

The rest of Earth's surface also seems to have been warmer than today. Dinosaurs lived even on those parts of Australia and Antarctica lying south of the Antarctic Circle. At lower latitudes, coral reefs indicative of warm tropical ocean temperatures extended some 10° of latitude farther from the equator than they do today (40° versus 30°).

5-1 Model Simulations of the Cretaceous Greenhouse

To explore the reasons for the warmth 100 Myr ago, climate modeler Eric Barron and his colleagues used constraints provided by plate tectonic reconstructions of Earth's geography to run general circulation model simulations of the climate of the Cretaceous world. The surface temperature estimates from the resulting simulations were plotted in the form of zonal mean trends averaged around lines of latitude (Figure 5-3). These latitude-averaged plots represent the complex spatial pattern of temperature changes across Earth's surface in a shorthand form. The latitudes shown in this and subsequent figures are plotted so that they correct for the much larger area of Earth's surface at low latitudes than in polar regions. These plots show all latitudes in proportion to the area they actually cover.

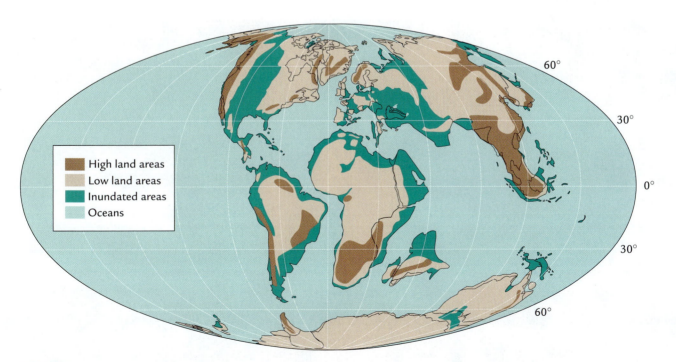

FIGURE 5-1   The world 100 Myr ago By 100 Myr ago, plate tectonic processes had broken the giant Pangaean continent into separate smaller continents that were flooded by shallow seas. (D. Merritts et al., *Environmental Geology*, 1997 by W. H. Freeman and Company.)

A

B

FIGURE 5-2　Evidence of greenhouse warmth 100 Myr ago　Vegetation and animals that appear to have been warm-adapted lived in both polar regions 100 Myr ago: (A) fossils of breadfruit trees like those that are found today in the tropics and (B) dinosaurs, many species of which lived poleward of the Arctic and Antarctic circles.　(A: Swedish Museum of Natural History, Yvonne Arremo, Stockholm. B: T. Steuberth/Institut für Geologie und Mineralogie der Universität Erlangen, Nuremberg.)

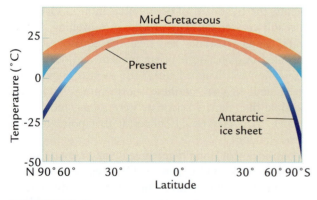

FIGURE 5-3　Cretaceous target signal　Climate scientists have used geologic data (faunal, floral, and geochemical) to compile an estimate of temperatures 100 Myr ago. Temperatures were warmer than they are today at all latitudes, especially in polar regions.　(Adapted from E. J. Barron and W. M. Washington, "Warm Cretaceous Climates: High Atmospheric CO_2 as a Plausible Mechanism," in "The Carbon Cycle and Atmospheric CO_2: Natural Variations, Archaean to Present," ed. E. T. Sundquist and W. S. Broecker, *Geophysical Monograph* 32 [Washington, DC: American Geophysical Union, 1985].)

For comparison, Earth's present temperature trend is also shown in Figure 5-3. Temperatures near the equator 100 Myr ago were a few degrees warmer than those today but as much as 20°C warmer at the North Pole and 40°C or more at the South Pole. The reason for the especially large temperature difference at the South Pole is the absence of the Antarctic ice sheet in the Cretaceous simulation. Today this ice sheet reaches elevations of 3 to 4 km, where temperatures are much colder because of the lapse-rate effect (companion Web site, p. 15). Because the top of the modern ice sheet is Earth's "surface" at the South Pole, the surface temperature there today is bitterly cold.

Geologic evidence indicates that this ice sheet did not exist during much of Cretaceous time. As noted in Chapter 2, GCMs can simulate only a few years or decades of elapsed time, and ice sheets cannot begin to grow or melt that quickly. Because of this time limitation, the presence or absence of ice sheets must be specified *in advance* as boundary condition input to GCM simulations. With geologic evidence showing that no ice sheets existed for much of mid-Cretaceous time, the

boundary condition specified for these simulations had no ice sheets. With the Antarctic ice sheet absent, Earth's surface at the South Pole is closer to sea level and the simulated polar temperatures are cold but much less frigid.

As part of this series of experiments, Barron and colleagues defined a "target signal" (an independent estimate of Cretaceous climate derived from observations) to compare against the simulations. They compiled estimates of surface temperatures 100 Myr ago from the available information—past distributions of temperature-sensitive vegetation, animals, and geochemical indicators—to use as climatic constraints on the simulation. Each target-signal estimate of past temperature includes some amount of uncertainty. The red and blue region in Figure 5-4 shows the range of possible temperature values they found.

Several experiments were then run as sensitivity tests to evaluate the significance of different factors (Chapter 2). The first experiment used the Cretaceous geography (land-sea distribution and mountain elevations) as input to the model simulation. CO_2 levels were kept at modern (preindustrial) levels. For this simulation, the tropical temperatures simulated by the model fell within the range of estimated (target) temperatures, but

temperatures poleward of 40° latitude were considerably colder than the geologic estimates (see Figure 5-4).

The second simulation retained the changes in geography but added another factor: CO_2 levels four times higher the preindustrial average. A range of indirect evidence suggests—with considerable uncertainty—that atmospheric CO_2 concentrations 100 Myr ago were somewhere between four and ten times higher than those today. Barron and his colleagues chose a CO_2 value slightly more than four times the modern (preindustrial) level. The higher CO_2 level in this second simulation further warmed the planet and removed much of the mismatch with the target signal at the poles and mid-latitudes, but this time the equatorial temperatures were hotter than the range suggested by geologic evidence.

From one perspective, the general agreement between the two model simulations and the target signal in Figure 5-4 could be judged a success; the disagreements are not large for a time so far in the past. Nevertheless, these mismatches have in subsequent years been the focus of many studies that have questioned both the data used to reconstruct the target-signal temperatures and the assumptions used in the model simulations.

5-2   What Explains the Data-Model Mismatch?

As is the case for any study of past climates, the explanation for the data-model mismatches could lie in the geologic data, in the climate model simulation, or in a combination of the two.

Problems with the Models? One possible explanation for the mismatch investigated by climate scientists has focused on shortcomings in the climate models. At the time Barron and colleagues ran their initial experiments, the treatment of ocean circulation in O-GCMs was still very crude. The process of upwelling of cool water along coastlines and near the equator was not included in global-scale ocean models, and deep-water circulation was handled with little success or not even attempted. In view of these shortcomings, it seemed possible that climate scientists were asking more from models than they could deliver at that point in their development.

An idea that initially intrigued many climate scientists was the possibility that the Cretaceous ocean operated in a fundamentally different way compared to today's ocean. Today, the surface ocean transports about half as much heat poleward as does the atmosphere. Some scientists suggested that if the ocean carried twice as much heat to the poles in the Cretaceous as it does today, the two major problems in the data-model mismatch might be resolved. The poles would be warmed by the greater influx of ocean heat, and the tropics would not be heated up as much by the CO_2 increase because of their greater loss of heat to polar regions.

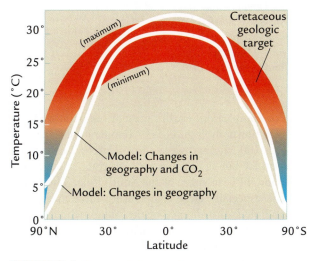

FIGURE 5-4 Data-model comparisons Climate model simulations are matched against the temperature target signal for 100 Myr ago provided by geologic data. One simulation based on the altered geography of 100 Myr ago and another that also incorporates higher CO_2 values reproduce some aspects of the target signal, but simulate a steeper pole-to-equator gradient. (Adapted from E. J. Barron and W. M. Washington, "Warm Cretaceous Climates: High Atmospheric CO_2 as a Plausible Mechanism," in "The Carbon Cycle and Atmospheric CO_2: Natural Variations, Archaean to Present, ed. E. T. Sundquist and W. S. Broecker," *Geophysical Monograph* 32 [Washington, DC: American Geophysical Union, 1985].)

This concept has been proposed in different forms by several climate scientists and can be called the **ocean heat transport hypothesis.**

A related idea was that deep water might have formed in the northern subtropics in regions of very high ocean salinity (>37%), rather than in polar regions like today. The concept was that high salinities could have made surface waters dense enough to sink into the deep ocean as **warm, saline bottom water.** A strong flow of this warm deep water from the tropics to the poles might then have contributed to the poleward heat flux needed to warm polar regions and resolve the data-model mismatch.

More recent experiments with improved ocean models have not supported the ocean heat transport hypothesis. The models do not show significant increases of poleward transport of heat by the ocean in a climate warmer than that today.

A second kind of data-model mismatch occurred in climate simulations for 100 Myr ago, and this mismatch persists in model simulations of the warm climates that continued over the next several tens of millions of years. Data from warm-adapted vegetation (early palmlike trees) and from fossil reptiles suggest that continents at high and middle latitudes had moderate climates and did not freeze in winter. In contrast, all the GCM experiments described here (those with altered geography, higher CO_2 levels, and increased ocean heat transport, alone or in combination) simulated hard freezes in winter across the interiors of the northern hemisphere continents.

As a further test on the possible role of the ocean, an experiment was run in which warm waters were specified (imposed) in the Arctic Ocean as an initial boundary condition before running the simulation. One purpose of this experiment was to find out whether a very warm Arctic ocean could explain the warm interiors of the northern continents. The simulation showed that even a warm Arctic Ocean failed to keep the interiors of the northern continents warm enough in winter to prevent freezing because the winter heat losses were too large.

Problems with the Data? A second path of investigation of the data-model mismatch focused on the possibility that the proxy data used to reconstruct past climate might have been giving an incorrect "target signal" for comparison to the models. Several studies have converged on evidence that this explanation is promising.

The tropics have been the main focus of these reassessments. Most of the data used to reconstruct Cretaceous temperatures across lower latitudes have come from geochemical analyses of ocean plankton using oxygen isotope ratios (Appendix 1). The shells of the plankton record oxygen-isotopic ratios that reveal changes in the temperature of the ocean water in which they form. It now seems likely that tropical temperatures were actually warmer by

several degrees centigrade than even the maximum target signal values shown in Figure 5-4.

After the plankton formed their shells in warm surface waters, they died and fell to the seafloor where water temperatures were 10°–15°C cooler than those at the surface. As the shells lay on the seafloor or in the uppermost sediments before being completely buried, they were bathed by cooler bottom waters. Some of these shells were partly dissolved and recrystallized. Scientists found that the more pristine the shells they examined, the warmer the temperatures their oxygen isotopic ratios indicated, whereas the shells that were more heavily altered yielded cooler temperature estimates. This trend indicated that chemical alteration on the seafloor had reset the temperatures toward those typical of the cooler bottom waters. The best-preserved shells indicated that Cretaceous temperatures were warmer by as much as 5°C than those used in the initial target signal shown in Figure 5-4.

This explanation appears to resolve much of the data-model mismatch shown in Figure 5-4. The warmer tropical temperatures match the model simulation with altered geography and higher CO_2 levels reasonably closely. The simulation is consistent with a CO_2 concentration above the one initially chosen (four times the natural modern level). Levels of CO_2 higher than this amount would also prevent the interiors of mid-latitude continents from freezing in winter.

> **IN SUMMARY,** higher CO_2 levels appear to be a major cause of the warmer climate 100 Myr ago.

5-3 Relevance of Past Greenhouse Climate to the Future

The results from studies of Cretaceous climate fit into a larger picture of the effect of CO_2 on Earth's climate. Climate scientists have run a series of GCM sensitivity tests using Earth's present geography as common boundary condition input to all simulations but allowing the level of atmospheric CO_2 to vary from as low as 100 ppm to as high as 1000 ppm. The preindustrial or natural modern value of 280 ppm lies in the lower part of this range.

This set of simulations shows that global average temperature rises with increasing CO_2 levels, but the relationship is not linear (directly proportional). Instead, Earth's temperature reacts strongly to CO_2 changes at the lower end of the range but much less so to changes at the high end of the range.

One reason for the shape of this curve is positive feedback effects from snow and sea ice. At low (<200 ppm) CO_2 values, sea ice and snow advance well past their average limits today and cover a relatively large fraction of Earth's high and middle latitudes. These bright surfaces reflect incoming solar radiation back to space, further cooling the planet (companion Web site, p. 15). With the planet covered by so much snow and ice, even small changes in CO_2 can have a relatively big impact in altering the area of Earth covered by snow and ice, especially in winter. Small CO_2 increases greatly reduce the extent of this bright reflective area, while small decreases in CO_2 greatly enlarge it. This albedo-temperature feedback makes the climate system in an icehouse world react strongly to changes in CO_2.

In contrast, much higher CO_2 values can reduce the average amount of snow and ice present at high latitudes. At CO_2 levels near 1000 ppm, little or no sea ice is present in the Arctic in summer, and only the central Arctic has ice in winter. As a result, very small surface areas of snow and ice are available to provide positive feedback to changes in CO_2. Increases in CO_2 produce warming, but little snow or ice is available to melt. As a result, the climate system in a greenhouse world is much less sensitive to changing CO_2 levels.

A second factor that accounts for the slower rise of temperature at higher CO_2 levels is **CO_2 saturation.** As CO_2 concentrations rise, the atmosphere gradually reaches the point at which further CO_2 increases have little effect in trapping additional back radiation from Earth's surface.

A third factor, but one that acts in an opposite sense, is water vapor feedback. A warm atmosphere with CO_2 values of 1000 ppm can hold much more water vapor than a cold atmosphere with values of 100 ppm (companion Web site, pp. 13–14). In this case, the feedback effect of water vapor on temperature grows stronger in warmer, higher-CO_2 conditions, in contrast to the diminished effect of albedo-temperature feedback.

These sensitivity experiments and others with different models show that large ice sheets cannot exist anywhere on Earth when CO_2 concentrations exceed 1000 ppm. At these higher CO_2 values, with global mean temperature well above 20°C, mean annual temperatures in polar regions exceed 0°C. Because ice sheets require mean annual temperatures well below freezing to persist through the summer melting season, they cannot survive in high-CO_2 greenhouse warmth.

The trend shown in Figure 5-5 is from experiments based on one GCM, but different models simulate different levels of temperature sensitivity to changes in CO_2. These variations reflect the way each model simulates the effects of such features as snow and sea ice, which provide strong positive feedback, and also of clouds, which have feedbacks that are still poorly known. The example shown here lies within the middle of the full range of model sensitivities.

In the last hundred years, the CO_2 concentration in the atmosphere has risen from 280 ppm to more than 380 ppm because of the industrial activities of humans (companion Web site, pp. 34–35). Projections for the

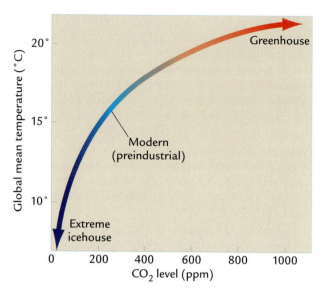

FIGURE 5-5 Effect of CO_2 on global temperature Climate model simulations of the effects of changing atmospheric CO_2 levels on global temperature show greater warmth for higher CO_2 concentrations. (Adapted from R. J. Oglesby and B. Saltzman, "Sensitivity of the Equilibrium Surface Temperature of a GCM to Changes in Atmospheric Carbon Dioxide," *Geophysical Research Letters* 17 [1990]: 1089–92.)

next two centuries indicate increases to at least 550 ppm and possibly 1000 ppm or more, nearly as high as those in the Cretaceous. We are heading back into a greenhouse world and at a very rapid rate.

Sea Level Changes and Climate

Over tectonic time scales, the average level of the world ocean has risen and fallen by 100 m or more against the margins of the continents. These rises and falls of the sea in relation to the land are called marine **transgressions** and **regressions.** Although these changes have been small in relation to the >4000-m average depth of ocean basins, they could potentially have significant effects on climate in the regions that were alternately flooded and exposed.

Many continental margins are flat, with changes of just 1 part in the vertical for every 1000 parts in the horizontal: a rise in elevation of 1 m can translate into a 1-km shift inland. As a result, vertical sea level changes of 100 m or more that occurred in the past translate into horizontal movements of the coastline measured in hundreds of kilometers. During times of high sea level, the ocean floods shallow interiors of continents and forms large seas. These long-term sea level changes also deposit and erode thick sequences of coastal sediments.

When sea level is low, the coastline tends to be situated near the break between the relatively flat continental shelf and the steeper continental slope (Figure 5-6A). At

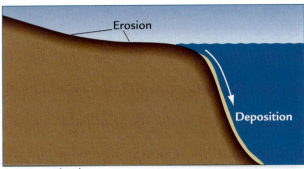

A Low sea level

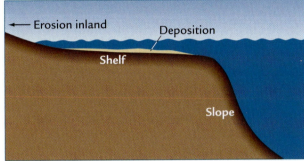

B High sea level

FIGURE 5-6 Sea level (A) When sea level is low, the coastline lies near the base of the continental shelf and sediment is deposited on the continental slope. (B) When high sea levels flood the continental margin, more sediment is trapped and deposited on the submerged shelf.

these times, erosion prevails on continental margins, and most of the eroded sediment is carried to the continental slope and dispersed down in the deeper ocean. When sea level is high, the ocean floods the low-gradient continental margin to depths of 100 m or more (Figure 5-6B). At such times, sediment is deposited on the submerged continental shelf and can survive as part of the geologic record.

Local tectonic factors that cause uplift or subsidence of the land can also affect the relative vertical position of the ocean margin against the land, even in the absence of changes in global sea level. These regional processes include mountain building, broad-scale warping of Earth's surface connected to deep-seated heating, and local depression and rebound of the land caused by the weight of ice sheets. In this chapter, we ignore these local effects and focus on changes in **eustatic sea level**—changes that are global in scale.

The most persuasive evidence for higher global sea levels in the past comes from the presence of marine sediments simultaneously deposited on coastal margins and in shallow interiors of continents at levels well above present sea level. Deposition of marine sediments on several continents at the same time indicates that the changes in sea level are global in scale, not just local features.

The best-constrained estimates of past changes in eustatic sea level are those spanning the last 100 Myr (Figure 5-7) because the geologic record of this interval is relatively complete compared to earlier times. But even for this interval, sea level estimates remain highly uncertain.

5-4 Causes of Tectonic-Scale Changes in Sea Level

In the Cretaceous world of 100 to 80 Myr ago, the coastlines and interiors of most continents were flooded by an ocean situated well above modern level (see Figure 5-1). Areas flooded included much of southern Europe (Figure 5-8) and interior regions of North America penetrated by seaways linked to the Gulf of Mexico and the Arctic Ocean. Since that time, sea level has slowly fallen to its modern position, close to the lowest level on record (except for glacial intervals with more ice on land than there is now).

Geoscientists disagree about how much higher sea level was 80 to 100 Myr ago. Estimates have ranged from 100 to 300 m above today's level. This wide range of uncertainty reflects complications that have arisen during the tens of millions of years since the Cretaceous sediments were deposited. The water-rich sediments have been compacted and lost their original thickness, the added weight of the sediments has caused the underlying rock crust to slowly subside, and ocean water has moved in over both the compacted sediments

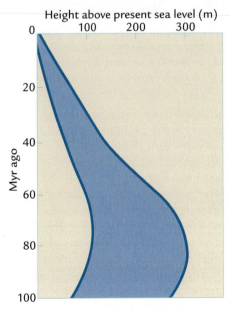

FIGURE 5-7 Sea level in the last 100 Myr Quantitative estimates of sea level change show higher sea levels 100 Myr ago. The blue area defines the large range of estimates based on different methods. (Adapted from M. Steckler, "Changes in Sea Level," in *Patterns of Change in Earth Evolution*, ed. H. H. Holland and A. F. Trendall [Berlin: Springer-Verlag, 1984].)

FIGURE 5-8 Marine limestone exposed on land Marine limestone deposits that today form the coasts of southern England and northern France are evidence of higher sea levels 100 Myr ago. (Andrew Ward/Life File/PhotoDisc.)

and the weighed-down crust and added even more weight that has caused still more subsidence. Obviously, figuring out past changes in sea level is more difficult than it might initially seem.

The evidence that sea levels in the Cretaceous (80–100 Myr ago) were higher by somewhere between 100 and 300 m than they are today has been attributed to two groups of factors: (1) tectonically driven changes in the volume of the ocean basins that altered their capacity to hold water and (2) changes in the volume of water in the ocean basins resulting from changes in climate.

Changes in the Volume of the Ocean Basins

1. *Changes in the volume of ocean ridges* Ocean ridges owe their high elevations to unusual heating from hot molten material located below the surface of the ocean crust. Heating causes the rock in these regions to expand, and expansion of the rock causes the surface of the ocean crust to rise.

The amount of thermal expansion of ocean ridges varies through time in response to changes in the rates of seafloor spreading, and the changes in the height of the ridges in turn alter the volume of ocean basins and their capacity to hold water (Figure 5-9). Ocean water is displaced up onto the continents during times when ridges spread rapidly and produce wide, high-elevation ("fat") profiles, but the sea withdraws from the continents during times when the ocean ridges spread more slowly and produce narrower, low-elevation ("thin") profiles that displace less water.

The profiles (and volumes) of ocean ridges existing in the past can be reconstructed based on a systematic relationship observed in modern ocean basins. All ocean ridges today have crests that lie at an average depth of 2500 m below the sea surface. Away from the crest, the subsurface depth profiles of these ridges follow a simple equation:

$$\text{Ridge depth} = 2500 \text{ m} + 350 \text{ (crustal age)}^{1/2}$$

$$\text{(in meters)} \quad \text{(at 0 age)} \quad \text{(in Myr)}$$

This equation describes a ridge crest that starts at an initial depth of 2500 m below the sea surface and gradually

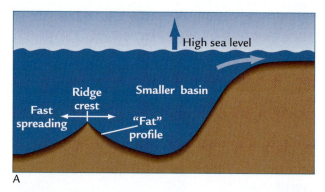

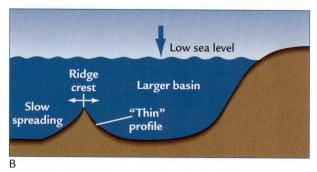

FIGURE 5-9 Spreading rates and sea level Rates of spreading at ocean ridges vary widely. (A) Fast spreading creates wider ridge profiles that reduce the volume of the ocean basins and displace more water onto the continents. (B) Slow spreading produces narrower profiles that create larger ocean basins that can hold more seawater.

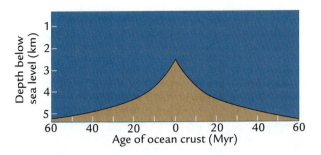

FIGURE 5-10 Subsidence of ocean ridges with time All ocean ridges show the same average profile of age (time since formation) versus depth. Heat elevates the ridge crests to a depth of 2500 m below the ocean surface, a level high above the rest of the seafloor. As the crust spreads away from the ridge crest and ages, it cools and contracts, rapidly at first and then more slowly. The crust eventually reaches a stable depth of more than 5000 m below sea level. (Adapted from J. G. Sclater et al., "The Depth of the Ocean Through the Neogene," *Geological Society of America Memoir* 163 [1985]: 1–19.)

deepens with age away from the ridge crest as the heated rock cools and contracts (Figure 5-10). The seafloor ages that were used to derive this relationship were obtained from the paleomagnetic age data examined in Chapter 4.

Ridge crests initially stand high above the rest of the seafloor because of anomalously strong heating associated with formation of new crust from molten magma. Ocean ridge elevations initially subside rapidly while moving away from the crest because of rapid heat loss, but later subsidence is more gradual as the rate of heat loss slows on the lower ridge flanks. By 60 Myr after they form, the crust and upper mantle have lost most of their excess heat, and the ridge elevations have reached a nearly stable depth of 5500 m (see Figure 5-10). Local variations in depth of a few hundred meters occur at ridge crests and down the ridge flanks as a result of small-scale tectonic irregularities, but the mean values of ocean ridge depths follow the equation remarkably well throughout the world's oceans.

Paleomagnetic evidence from today's ocean shows that different ridges spread at different rates (see Figure 4-17). Because all ridge depths are constant with age (as shown by the preceding equation), crust of a given age (and a particular depth below sea level) will have been carried much farther from the ridge crest in a given amount of time in fast-spreading areas like the South Pacific than in the slow-spreading ones like the North Atlantic. Fast spreading gives the Pacific ridge a "fatter" elevation profile than that of the Atlantic (see Figure 5-9), and the wider Pacific ridge profile displaces more water for each kilometer of its length than does the narrow Atlantic ridge.

Through time, ridge profiles may vary on a globally averaged basis. At times like the present, when globally averaged rates of seafloor spreading are relatively slow, mean ridge profiles are relatively thin, and little water is displaced onto the continents. At times when average spreading rates were faster, mean ridge profiles would have been relatively wide and more water would have been pushed up onto the continents. To use the ridge depth equation to calculate spreading rates at any time in the past requires resetting the ages of the ridge crests to zero for the time being examined and recalculating the past ages of the ridge flanks as deviations from this adjusted "zero" age.

Mean spreading rates from 100 to 80 Myr ago are generally thought to have been higher than they are today, but the amount is highly uncertain. One reason for this uncertainty is that the rate of spreading 80 Myr ago is not known for the former (now destroyed) Tethys seaway in the tropics. In addition, estimated spreading rates for the few preserved areas of ocean crust that formed 80 to 125 Myr ago have recently been revised downward. Until recently, faster spreading rates were thought to have increased global sea level by well over 200 m between 80 and 100 Myr ago. Newer estimates tend to be only half as large, and some question now exists about whether spreading rates have changed at all in the last 175 Myr.

2. Collision of continents Most plate tectonic movements do not change the net area of either the oceans or the continents: creation of new ocean crust at ocean ridge crests is balanced by destruction of ocean crust subducting into trenches, leaving the area of the ocean basins constant. However, collision of continents does alter the area of the ocean basins and also affects sea level.

Because continental crust is low in density, two colliding continents tend to float near Earth's surface rather than be pushed or pulled deep down into Earth's mantle (Figure 5-11). In the region where they collide, continental crust thickens from its normal value of 30 km to about twice that amount. This process builds a high plateau that rises well above sea level and at the same time thickens the subsurface low-density "root" of the plateau down to 60 or 70 km below Earth's surface. In the upper 15 km of Earth's crust, the thickening that creates the plateau occurs by movements along faults that cause thin slivers of crust to shear off and stack up on top of each other. Below a depth of 15 km, thickening occurs when slow flow causes rock layers to be squeezed and folded.

Because collision drives two continents together to form a plateau with a double-thick crust, this thickening must result in a net loss in the area of continental crust. To a first approximation, the area of plateau across which the crust doubles in thickness should equal the net loss of area of continental crust. This decrease in

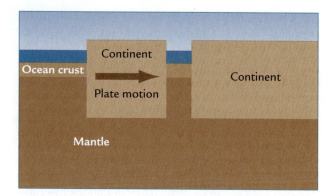
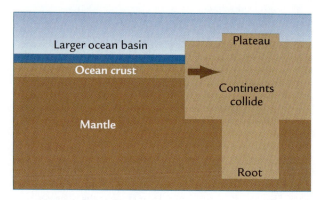

FIGURE 5-11 Continental collisions and sea level When continents collide, the continental crust doubles in thickness and creates a high plateau with a thick low-density crustal "root." This thickening reduces the original area of the continents and increases the area of the ocean basins. The increased area and volume of the ocean basins causes sea level to fall.

areal extent of the continents requires an equivalent increase in area of the ocean basins (see Figure 5-11). With a larger area of ocean to fill, the water level in the ocean should drop.

As noted in Chapter 4, continental collisions have occurred only sporadically through geologic time. The only major collision that has occurred since 100 Myr ago began when northern India first made contact with southern Asia, some 55 Myr ago. This collision, still in progress, has increased the area of the ocean by some 2 million km² over the last 55 Myr. As seawater has flowed in to fill this new area of ocean basin, global sea level has fallen an estimated 10 m below the level of 100 Myr ago, a time when no collisions were occurring.

Climatic Factors

3. *Water stored in ice sheets* Continent-sized ice sheets several kilometers in thickness and thousands of kilometers in lateral extent can extract enormous volumes of water from the ocean and store it on land (Figure 5-12). Because no permanent ice sheets existed between 100 and 80 Myr ago, little or no water was stored on the land as ice. Today the Antarctic ice sheet holds the equivalent in seawater of 66 m of global sea level, and the ice sheet on Greenland contains another 6 m. The Antarctic ice sheet has come into existence and grown to its present size within the last 35 Myr, and the Greenland ice sheet has done the same within the last few million years. Together these ice sheets have extracted a volume of ocean water equivalent to 72 m of global sea level.

4. *Thermal contraction of seawater* Ocean water has the capacity to expand and contract with temperature changes. The **thermal expansion coefficient** of water (the fractional change in its volume per degree of change in temperature) averages about 1 part in 7000 for each 1°C of temperature change. Because of this thermal behavior, even a constant amount of seawater would have lost volume during the cooling of the last 80 to 100 Myr. The temperature of low-latitude surface

waters has cooled by about 5°C over that interval, while the high-latitude surface ocean and the deep ocean have both cooled by 10°–15°C. The contraction of seawater caused by this cooling has reduced global sea level by an estimated 7 m.

The quantitative effects of several of the above factors on global sea level have to be adjusted for further complications (Box 5-1). One problem is the fact that water moving into (or out of) the ocean basins represents

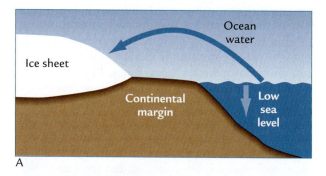

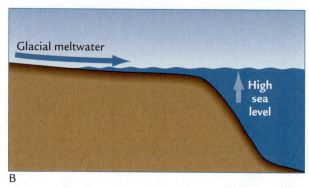

FIGURE 5-12 Ice sheets and sea level Ice sheets covering large parts of continents hold volumes of water equivalent to tens of meters of global ocean level. Sea level (A) falls when ice sheets are present on the land and (B) rises when they melt.

BOX 5-1 LOOKING DEEPER INTO CLIMATE SCIENCE

Calculating Changes in Sea Level

Water transferred between the continents and oceans represents weight added to (or removed from) the bedrock underlying the ocean basins. When water is added to the ocean, the underlying bedrock sags under the load. Similarly, the bedrock rebounds if part of the water load is removed.

This bedrock response reduces the change in sea level that would otherwise occur. For example, adding meltwater to the oceans raises sea level, but the depression of ocean bedrock under the load of the added water cancels about 30% of the sea level change. This 30% reduction is a direct result of the difference in density between water ($1 \ g/cm^3$) and bedrock ($3.3 \ g/cm^3$): $1 \div 3.3 = 0.3$.

The second complication is that the margins of the oceans have widely varying profiles (changes in elevation versus changes in distance) from region to region. Some areas have simple low-gradient profiles, but gradients in other regions are steepened by mountain building or other regional factors. As a result, changes in the total amount of water in the ocean or in the volume of water the ocean can hold need to be translated into actual net changes in the level of the global ocean produced by the complications contained in all the locally varying profiles. All regional profiles are summed into a single **hypsometric curve**, a graph that displays the proportions of Earth's surface that lie at various altitudes above and depths below sea level. Such a graph is easily constructed for today's Earth, but the exercise becomes more speculative for past intervals of plate tectonic configurations.

---

a large weight added to (or removed from) the underlying ocean crust, which sags (or rebounds) accordingly. This response of the ocean crust decreases the net magnitude of the change in sea level. The other complication has to do with translating a change in the volume of ocean water or in the volume of the ocean basins into actual movement of sea level against the complex shapes of the world's continental margins.

Taken together, these factors can explain why sea levels were some 120 to 220 m higher than the modern value 80 to 100 Myr ago (Table 5-1). Within a large range of uncertainty, these values match the estimated amount of flooding of Cretaceous continents reasonably well. The effect of seafloor spreading remains by far the largest source of uncertainty.

TABLE 5-1 Factors Contributing to Higher Sea Levels 100 to 80 Million Years Ago

Cause of sea level change	Estimated change (meters)
Decrease in ocean ridge volume	+50 to +150
Collision of India and Asia	+10
Water stored in ice sheets	+50
Thermal contraction of seawater	+10
All factors	+120 to +220

5-5 Effect of Changes in Sea Level on Climate

Climate scientists have at times cited sea level as a potential factor in long-term climate changes, although in varying and often even contradictory ways. The most likely effect of sea level changes on climate is linked to the very different thermal responses of land and water (companion Web site, pp. 9–11). The shallow (10–200 m) layer of ocean that overlaps the continental margins and invades the interior seaways has the large heat capacity typical of water, in contrast to the small heat capacity typical of land. As a result, flooding of the land tends to moderate continental extremes of climate and produce milder winters and cooler summers. Withdrawal of the sea should have the opposite effect. For large changes in global sea level, the synchronous invasion and withdrawal of the sea on many continents should result in simultaneous fluctuations between harsh continental and mild maritime climates around the world.

On continental margins flooded by rising sea level, the maritime climates of the coastal regions may simply shift landward, displacing formerly continental climates with more maritime conditions. Given the low (1:1000) gradients typical of some continental margins, such changes in climate may affect large regions. In addition, the invasion of seaways into low-lying interior regions of the continents (for example, Figure 5-1) should produce dramatically milder maritime climates. Prior to flooding, these interior regions would have had more arid and harsh climates because of their distance from the ocean.

Decades ago climate scientists thought that sea level might be a factor or even the critical control in the long-term succession of glacial (icehouse) versus nonglacial (greenhouse) climates. In this view, high sea levels caused warm climates by moderating the harsh winters, and low sea levels caused cold climates by permitting the very cold winters typical of continental conditions. Although the timing of the high sea level and largely ice-free climate 100 Myr ago compared with the low sea level and glacial climate today fits this explanation, this view is no longer viable.

The major criticism of this idea centers on the fact that summer-season ablation is a powerful factor in determining the extent of snow and ice (companion Web site, pp. 9-11). The problem is that low sea levels and withdrawal of the ocean from continental interiors lead to more extreme continental climates, including very hot summers. No matter how cold winters become in a continental climate, hot summers should easily melt any snow that accumulated and thereby oppose glaciation. Conversely, high sea levels should cause cooler, more maritime summers that favor the persistence of snow and ice through the summer ablation season at very high latitudes.

The record of the last 100 Myr supports this criticism. The high sea levels of 100 Myr ago were not accompanied by glaciation, and the low sea levels of today are. As a result, the hypothesis that sea level is the major control of long-term glaciation finds little or no support today. Glaciation is now seen as a *cause* of low sea level (because of storage of ocean water in ice sheets) rather than a *result*.

Asteroid Impact (65 Myr Ago)

The greatest catastrophes known to have affected Earth are the rare but massive impacts of large extraterrestrial asteroids and comets. An inverse relationship exists between the sizes of these objects and the frequency with which they hit Earth. The largest bodies (more than 10 km in diameter) arrive only every 50 to 100 Myr but result in much greater environmental effects than the smaller, more frequent impacts.

The impact event 65 Myr ago coincided with a global-scale extinction of some 70% of the species and 40% of the genera living at the time, including all the dinosaurs and all but one of twenty-five species of planktic foraminifera. The geologic evidence for this impact includes the worldwide distribution of a thin layer of sediment enriched in iridium (Ir), an element that is rare on Earth but 10,000 times more abundant in some kinds of meteorites (Figure 5-13A). This element was deposited in a thin layer that was later mixed by burrowing animals.

Other supporting evidence for an impact event includes small grains of quartz with distinctive textures called "shock lamellae" that are formed by the shock wave

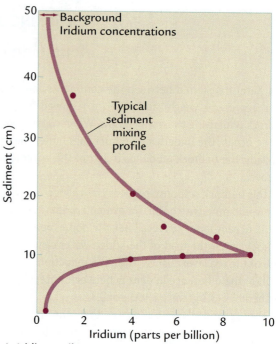

A Iridium spike

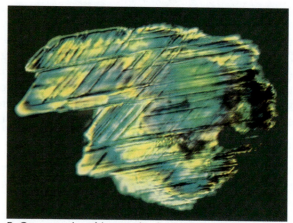

B Quartz grain subject to shock waves

FIGURE 5-13 Evidence of an asteroid impact (A) Ocean sediments containing a layer enriched in the element iridium are evidence of a large asteroid impact 65 Myr ago. (B) Sediments deposited in Montana 65 Myr ago contain grains of quartz crisscrossed by multiple lineations produced by high-pressure shock waves from an asteroid impact. (A: adapted from W. Alvarez et al., "Extraterrestrial Cause for the Cretaceous-Tertiary Extinction," *Science* 280 [1095–1108]. B: Glenn Izett, Williamsburg, VA.)

of sudden pressures much larger than those found on Earth, even in highly explosive volcanoes (Figure 5-13B). The best candidate for the site of the impact 65 Myr ago is a crater in eastern Mexico on the Yucatán Peninsula, between the Caribbean Sea and the Gulf of Mexico (Figure 5-14).

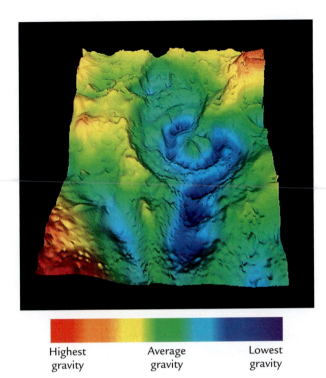

FIGURE 5-14 **65-Myr-old impact crater?** Mexico's Yucatán Peninsula has a circular area more than 200 km in diameter that is a good candidate for the site of the asteroid impact 65 Myr ago. The pattern shown is a result of measurements of Earth's gravity that can detect low-density pulverized rock (in blue) and higher-density rock (in green and yellow). (B. Sharpton, Lunar and Planetary Institute.)

Climate scientists have concluded that asteroid impacts could have affected Earth on time scales ranging in duration from instantaneous to as long as a few hundred or even a few thousand years (Figure 5-15). The instantaneous effects were caused when the asteroid blasted a hole through Earth's atmosphere. The speed of the incoming asteroid, 20 km/s, created a shock wave that moved outward, flattening objects for hundreds of miles around the impact site and heating Earth's atmosphere. Seismic waves sent through Earth's interior are thought to have been equivalent to those caused by an earthquake that would have measured 11 on the Richter scale, 100 to 1000 times stronger than the strongest earthquakes in recorded human history.

Some of the water and rock in the vicinity of the impact were instantly vaporized by the heat of the impact and blasted back out into space through the hole created by the incoming asteroid. The rest of the hot debris remained in Earth's atmosphere, heating it still further. The combined heating caused large-scale (possibly global) wildfires that ignited much of the above-ground vegetation and sent a thick layer of soot into the atmosphere.

Over the slightly longer term of days to years, the dust and soot that remained within the atmosphere spread around the planet, blocking most incoming solar radiation (see Figure 5-15). The debris injected into the lower atmosphere (the troposphere) would probably have been removed over a period of days to at most weeks because rainfall clears debris from that level of the atmosphere. In contrast, it would have taken months or years for the dust and soot injected into the stratosphere to settle out. The only means of removing debris at those altitudes is the slow pull of gravity on small particles. As a result, stratospheric particles (particularly dark soot) would have blocked significant amounts of sunlight for a year or more and cooled Earth's climate. Another likely effect over the course of a few years would have been the partial acidification of the oceans due to the creation of nitric acid (a component of acid rain) from atmospheric nitrogen, oxygen, and water vapor by the heat of the impact.

On the longer term of decades to centuries, the initial injection of carbon biomass into the atmosphere by burning should have produced higher CO_2 levels (see Figure 5-15). The plants that recovered from the firestorm and began to grow would have pulled some of the excess CO_2 out of the atmosphere, but full recovery and the development of new forms of vegetation to replace the ones that became extinct took longer. The warming induced by higher CO_2 may have lasted for centuries or longer.

By any standard used, this impact event was an enormous short-term environmental catastrophe, but what

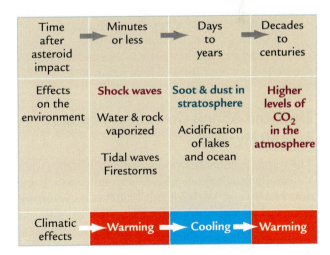

FIGURE 5-15 **Climatic and environmental effects of asteroid impacts** The asteroid impact 65 Myr ago is thought to have had major effects on Earth's environment, including the extinction of over two-thirds of the species then alive. The likely climatic effects vary with the amount of elapsed time after the initial impact and appear to have been restricted to a few centuries.

were the long-range effects on Earth's climate? Surprisingly, little long-term climatic effect is evident. Earth's climate was in a warm greenhouse state at the time of the impact, and there it remained afterward.

Several problems make it difficult to determine the climatic effects of such a brief impact event. One is that rapid changes in sedimentary records are blurred and smeared by burrowing animals and by bottom currents (Chapter 2). Distinctive impact-related features such as the iridium layer can still be detected despite this blurring (see Figure 5-13A) because they contrast clearly with the material in which they are deposited. But for most kinds of sedimentary archives, the signal from a single year (or even a decade or century) of climate that is warmer or cooler than normal will be mixed into and combined with the signals left by "normal" years and blurred beyond recognition.

A second problem with detecting the climatic effect of the impacts of even huge asteroids is that climate change was already occurring for other reasons before the impact. Records from marine sediments show a large (3°–4°C) warming and then cooling during the 500,000 years before the impact, but little or no additional change after it. Some land vegetation records suggest a long-term warming for hundreds of thousands of years after the impact, but it has not been demonstrated that this warming was related to the impact event. The increase in CO_2 levels caused by the immediate effects of the event seems unlikely to have persisted for that long.

> **IN SUMMARY,** impact events such as the one at the Cretaceous-Tertiary boundary have clearly had apocalypse-like effects on the environment, including mass extinctions of organisms that transformed life on Earth. Despite this environmental apocalypse, the background state of the climate system 65 Myr ago seems to have been changed little or not at all.

Large and Abrupt Greenhouse Episode near 50 Myr Ago

The warm greenhouse world was still in existence when a relatively brief episode of even warmer climate began near 55 Myr ago. Within about 10,000 years, ocean and terrestrial climate warmed by 5°C at low latitudes and 9°C at high latitudes. The excess warmth persisted at or near full strength for about 70,000 years and then slowly faded away over the next 100,000 years. Compared to the slow scale of tectonic changes, this thermal maximum was a brief event.

Evidence for the warming comes from the spread of plants and mammals into high latitudes and also from the decreasing values of the oxygen isotope "paleothermometers" explained in Appendix 1 (Figure 5-16). The

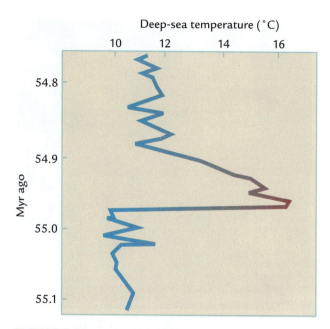

FIGURE 5-16 Unusual warmth 55 Myr ago A pulse of unusual warmth that developed near 55 Myr ago and persisted for tens of thousands of years warmed the deep ocean by several degrees Celsius.

changes associated with this relatively abrupt warming included a major acidification of ocean waters that caused widespread dissolution of $CaCO_3$ sediment on the seafloor and the extinction of nearly half of the species of benthic foraminifera living on the seafloor.

Also observed at this time is a large shift toward more negative carbon isotopic values in marine plankton. Such a change requires a major release of carbon enriched in the ^{13}C isotope (see Appendix 2). Early attention focused on methane, which exists in the atmosphere as a familiar greenhouse gas (Chapter 2) but is also present in solid form frozen (**methane clathrates**) into a slushy mixture not far below the ocean floor. If the subsurface ocean were to warm significantly, this ^{13}C-rich slush could be converted to a gas and released to the ocean and then to the atmosphere.

The large shift in $\delta^{13}C$ values showed, however, that the available methane sources 55 Myr ago were inadequate to account for the amount of carbon in the atmosphere and that immense additional releases of CO_2 were also required. The source of this huge amount of extra carbon is not entirely clear, but one obvious candidate is the deep ocean, which is currently a very large carbon reservoir. Also unclear at this point is the initial trigger for the carbon and methane releases. In any case, the addition of large amounts of CO_2 and methane considerably warmed climate for several tens of thousands of years. The recovery from this thermal perturbation took about 100,000 years. The most plausible mechanism for removing the extra carbon from the atmosphere is chemical weathering and burial of the excess carbon in ocean sediments.

This episode is potentially important to our future because it is at least a partial analog to the changes humans are now producing by our industrial releases of carbon into the climate system. The thermal maximum 55 Myr ago is estimated to have required the addition of several tens of trillions of carbon to the atmosphere. In comparison, during the last two centuries since the start of the Industrial Revolution, humans have added several hundred billion tons of carbon to the atmosphere. According to conservative extrapolations of recent trends, we will add several thousand more billions of tons of carbon to the atmosphere in the next 200 to 300 years. We face a smaller future warming than that 55 Myr ago, but it will arrive within a few hundred years rather than over 10,000 years.

Key Terms

Cretaceous (p. 82)

ocean heat transport hypothesis (p. 85)

warm, saline bottom water (p. 85)

CO_2 saturation (p. 86)

transgressions (p. 86)

regressions (p. 86)

eustatic sea level (p. 87)

thermal expansion coefficient (p. 90)

hypsometric curve (p. 91)

methane clathrate (p. 94)

Review Questions

1. What evidence shows that the world was warmer 100 Myr ago than today?

2. Why do higher CO_2 levels make sense as an explanation for this greater warmth?

3. How well do model simulations capture the distribution of temperatures 100 Myr ago?

4. What are the possible causes of mismatches between the models and geologic observations?

5. Why do some climate scientists believe that increased ocean heat transport is required to explain polar warmth 100 Myr ago?

6. Which regions of the continents were flooded by high seas 100 Myr ago?

7. What were the major factors that explain higher sea level 100 Myr ago?

8. How did higher sea levels affect global climate?

9. Do sea level changes explain past glaciations?

10. Did asteroid impacts have long-lasting effects on climate?

Additional Resources

Basic Reading

Companion Web site at www.whfreeman.com/ ruddiman2e, pp. 9–11, 13–15, 32, 34–35.

Alvarez, L. W., W. Alvarez, F. Asaro, and H. V. Michel. 1980. "Extraterrestrial Cause for the Cretaceous-Tertiary Extinction." *Science* 208: 1095–1108.

Advanced Reading

Barron, E. J., S. L. Thompson, and S. H. Schneider. 1981. "An Ice-Free Cretaceous? Results from a Model Simulation." *Science* 212: 501–8.

Miller, K., et al. 2005. "The Phanerozoic Record of Global Sea-Level Change." *Science* 312: 1293–98.

Oglesby, R. J., and B. Saltzman. 1992. "Equilibrium Climate Statistics of a General Circulation Model as a Function of Atmospheric Carbon Dioxide." Part I: "Geographic Distribution of Primary Variables." *Journal of Climate* 5: 66–92.

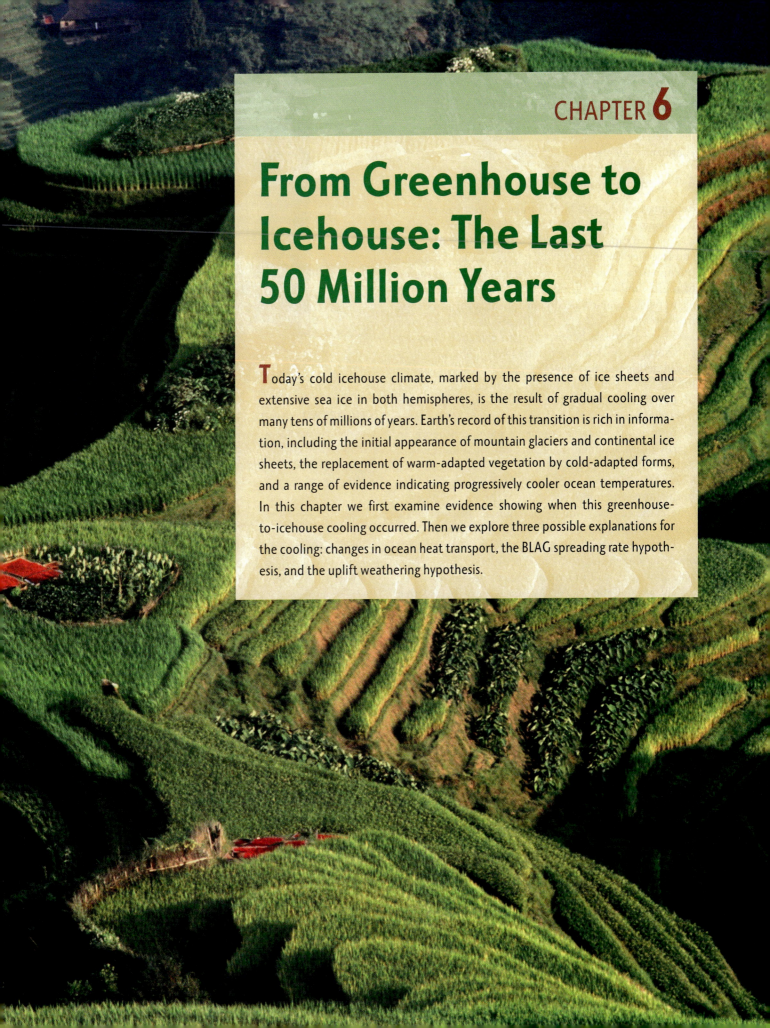

CHAPTER 6

From Greenhouse to Icehouse: The Last 50 Million Years

Today's cold icehouse climate, marked by the presence of ice sheets and extensive sea ice in both hemispheres, is the result of gradual cooling over many tens of millions of years. Earth's record of this transition is rich in information, including the initial appearance of mountain glaciers and continental ice sheets, the replacement of warm-adapted vegetation by cold-adapted forms, and a range of evidence indicating progressively cooler ocean temperatures. In this chapter we first examine evidence showing when this greenhouse-to-icehouse cooling occurred. Then we explore three possible explanations for the cooling: changes in ocean heat transport, the BLAG spreading rate hypothesis, and the uplift weathering hypothesis.

Global Climate Change since 50 Myr Ago

Earth has undergone a profound cooling at both poles and across the lower latitudes of both hemispheres during the last 50 Myr. Both ice and vegetation have left abundant evidence of this cooling (Figure 6-1).

6-1 Evidence from Ice and Vegetation

As climate cools, two kinds of glacial ice form on land (companion Web site, pp. 27–30). Small mountain glaciers and ice caps appear on the tops of high mountains, and large ice sheets cover much larger areas of the continents. Because average temperatures vary from region to region and with altitude, the conditions that permit ice to persist year-round do not appear at the same time in all areas.

In the southern hemisphere, no evidence exists for persistent ice on Antarctica until 35 Myr ago, when ice-rafted debris was first deposited in ocean sediments on the nearby continental margin. Since then, the size of the Antarctic ice sheet has increased irregularly toward the present, with a major growth phase near 13 Myr ago. Greater amounts of ice-rafted debris in nearby ocean sediments suggest additional increases in Antarctic ice during the last 10 Myr. Today more than 97% of Antarctica is buried under ice (Figure 6-2 left). In the lower and middle latitudes of the southern hemisphere, the earliest evidence of mountain glaciers in the high Andes is dated to between 7 and 4 Myr ago.

In the northern hemisphere, glacial ice first developed on Greenland sometime between 7 and 3 Myr ago, although small mountain glaciers may have existed locally around the North Atlantic Ocean before that

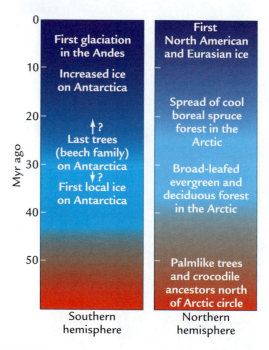

FIGURE 6-1 **Global cooling for 50 Myr** Gradual cooling during the last 50 Myr is demonstrated by the first appearance of mountain glaciers and continental-scale ice sheets and by a progressive trend toward cold-adapted vegetation in both hemispheres.

time (see Figure 6-1). The first evidence of glaciers in the high coastal mountains of southern Alaska dates to about 5 Myr ago. The first continental ice sheets of significant size appeared 2.75 Myr ago in North America and

FIGURE 6-2 **Cooling in Antarctica** (Left) Today an ice sheet up to 4 km thick covers most of Antarctica, although mountains locally protrude through the thinner cover around the margins. (Right) Until 30 Myr ago, *Nothofagus* trees, members of the beech family like those living today at the southern tip of South America, still existed in parts of Antarctica. (Left: Ward's Natural Science Establishment. Right: courtesy of Calvin Heuser, Tuxedo, NY.)

FIGURE 6-3 Cooling in the Arctic (Left) Warm-adapted breadfruit trees lived above the Arctic Circle in Canada until 60 Myr ago, but (right) land around the Arctic Ocean is now covered by scrubby tundra vegetation grazed by caribou. (Left: Swedish Museum of Natural History, photo by Yvonne Arremo, Stockholm. Right: Corbis.)

Eurasia. These ice sheets grew and melted in repeated cycles, and their maximum size increased after 0.9 Myr ago. Although these northern ice sheets developed more than 30 Myr later than the ones in Antarctica, they are a response to the same overall global cooling trend.

Fossil remains of vegetation also indicate a progressive cooling over the last 50 Myr. A form of beech tree called *Nothofagus* (Figure 6-2 right) lived on Antarctica before 40 Myr ago, along with several types of ferns. This vegetation disappeared as climate became more frigid and ice spread across Antarctica. Today the only vegetation on Antarctica is lichen and algae found in summer melt water ponds in ice-free regions of a few coastal valleys.

The same kind of long-term cooling trend is evident in north polar regions (Figure 6-3). Palmlike and other broad-leafed evergreen vegetation existed in the Canadian Arctic at 80° N from 60 to 50 Myr ago (Figure 6-3 left), as did the ancestors of modern alligators that would presumably have been ill-adapted to extreme cold. Sea ice was apparently absent, even along the coastal Arctic margins.

Gradually the warm conditions in the Arctic gave way to today's cold. The development of conifer forests of spruce and larch by 20 Myr ago indicates cooling, and the ring of tundra that has encircled the Arctic Ocean in the last few million years indicates deepening cold. Tundra is scrubby grasslike or shrublike vegetation that lives on thawed layers lying above **permafrost,** ground frozen each winter by intense cold (Figure 6-3 right). The appearance of tundra and permafrost is probably linked to frigid winters brought on by expanding sea ice.

The shapes of tree leaves can be used to reconstruct past climate (Figure 6-4). Leaves of trees living today in the warm tropics tend to have smoothly rounded margins, while leaves of trees in cooler climates generally

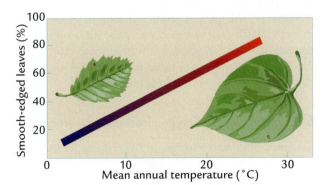

FIGURE 6-4 Leaf outlines indicate temperature Trees with smooth-edged leaves flourish today in the tropics, while trees with more jagged-edged leaves grow in colder climates. (Adapted from S. Stanley, *Earth System History,* © 1999 by W. H. Freeman and Company, after J. A. Wolfe, "A Paleobotanical Interpretation of Tertiary Climates in the Northern Hemisphere," *American Scientist 66* [1978]: 994–1003.)

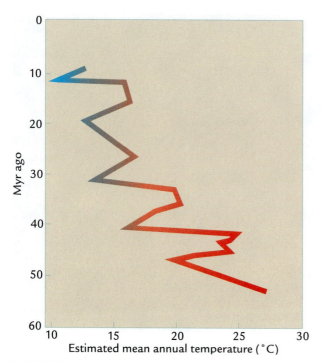

FIGURE 6-5 Cooling in western North America
Temperature trends estimated from the outline shapes of fossil leaves indicate an erratic but progressive cooling of northern middle latitudes during the last 55 Myr. (Adapted from J. A. Wolfe, "Tertiary Climatic Changes in Western North America," *Palaeogeography, Palaeoclimatology, Palaeoecology 108* [1994]: 195–205.)

have irregular edges, jagged or serrated in outline. The reason for this relationship is not known, but the correlation with temperature in the modern vegetation is strong enough that climate scientists have used this relationship to estimate past temperatures from assemblages of fossil leaves preserved in sedimentary rocks.

One record derived from leaf-margin evidence in western North America shows an ongoing cooling over the last 50 Myr (Figure 6-5). Although interrupted by small warm intervals, the trend toward ever-lower temperatures persists over the long term.

6-2 Evidence from Oxygen Isotope Measurements

Evidence of climate change on the continents over the last 50 Myr is incomplete. The first occurrences of ice sheets on land and their subsequent fluctuations in size are difficult to define. In addition, lakes that accumulate remains of past continental vegetation in their muddy sediments rarely persist for millions of years. In contrast, parts of the deep ocean have accumulated a continuous climatic record with quantitative information about climate change across the entire 50-Myr interval.

The most important climatic record in the ocean is the oxygen isotope signal (see Appendix 1 for a full summary). Most of the oxygen in nature occurs as the very abundant ^{16}O isotope or as the much less abundant ^{18}O isotope. Scientists refer to changes in the relative amounts of these two isotopes as variations in $\delta^{18}O$, measured as changes in parts per thousand (‰).

Typical modern $\delta^{18}O$ values are 0 to –2‰ for the surface ocean, +3 to +4‰ for the deep ocean, and –30 to –55‰ for ice sheets (Figure 6-6). Changes occur through time in the $\delta^{18}O$ values of water in the ocean and of ice in the glaciers. In this chapter the main focus is on changes in $\delta^{18}O$ values in the ocean. Foraminifera living both in surface waters and on the seafloor use HCO_3^- ions dissolved in seawater as the source of carbon and oxygen for their $CaCO_3$ shells. Because the oxygen in HCO_3^- comes directly from seawater, it gives climate scientists information on past variations in the two isotopes of oxygen in the ocean, as quantified by the $\delta^{18}O$ signal.

In the past, $\delta^{18}O$ variations have been produced by two climatic factors: (1) changes in the temperature of ocean water and (2) changes in the size of ice sheets on the continents. Changes in $\delta^{18}O$ values measured in the foraminiferal shells decrease by 1‰ for each 4.2°C increase in the temperature of ocean water at the location where the foraminifera lived. The same relationship holds if seawater cools but in the reverse sense ($\delta^{18}O$ values become heavier).

Changes in size of the ice sheets also alter $\delta^{18}O$ values in foraminiferal shells. Ice sheets are formed from water vapor evaporated from the oceans and later precipitated as snow (Appendix 1). Because the snow and ice are enriched in the lighter ^{16}O isotope, more of the heavier ^{18}O isotope is left behind in the oceans. This enrichment process is called **fractionation.** As a result, the $\delta^{18}O$ value of ocean water becomes more positive as ice sheets grow.

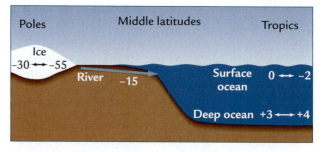

FIGURE 6-6 Typical $\delta^{18}O$ values in the modern world
$\delta^{18}O$ values in the ocean vary from 0 to –2‰ in warm tropical surface waters today to as much as +3 to +4‰ in cold deep-ocean waters. In today's ice sheets, typical $\delta^{18}O$ values reach –30‰ in Greenland and –55‰ in Antarctica.

The opposite is true if ice melts. If all the ice present on Antarctica and Greenland today melted and flowed back to the ocean, it would deliver a large volume of ^{16}O-rich meltwater that would shift the ocean's average $\delta^{18}O$ value ($\delta^{18}O_w$) from its present value of 0‰ to a value of –1‰.

A simple equation summarizes these processes:

$$\Delta\delta^{18}O_c = \Delta\delta^{18}O_w \times 0.23\Delta T$$

where Δ means "change in." This equation indicates that measured changes in $\delta^{18}O$ values in the shells of foraminifera ($\Delta\delta^{18}O_c$) result from changes in the mean $\delta^{18}O$ value of the oceans ($\Delta\delta^{18}O_w$) and from variations in the temperature of the water in which the shell formed (ΔT). The value 0.23 results from inverting the 4.2 value noted earlier (1/4.2 = 0.23).

A record of $\delta^{18}O_c$ over the last 70 Myr has been compiled from benthic foraminifera living on the ocean floor (Figure 6-7). Although the trend is shown as a single line, it is actually derived from hundreds of individual analyses scattered around the line because of local temperature conditions specific to each site and because of short-term changes. This signal begins to trend erratically toward more positive values near 50 Myr ago, and intervals of fastest change occur near 35 Myr ago, 13 Myr ago, and within the last 3 Myr.

These changes toward more positive $\delta^{18}O_c$ values are caused by some combination of (1) cooling of the deep ocean and (2) growth of ice sheets on land. Both factors are critical aspects of the transition from a greenhouse to an icehouse climate.

We can disentangle the effects of temperature and ice volume on this signal to some extent. No evidence exists of significant amounts of ice on Antarctica or anywhere else on Earth prior to 40 Myr ago. During the interval between 50 and 40 Myr ago, the $\delta^{18}O_c$ values increased from –0.75‰ to +0.75‰, a net change of +1.5‰. The temperature/$\delta^{18}O_c$ relationship tells us that deep waters must have cooled by more than 6°C (1.5‰ × 4.2°C/‰) during this interval before major ice sheets appeared.

Between 40 Myr ago and today, the deep-ocean $\delta^{18}O_c$ values increased from about +0.75‰ to +3.5‰, a further increase of 2.75‰ (see Figure 6-7). Both long-term cooling and growth of ice sheets contributed to this trend. The first substantial amounts of ice appeared near 35 Myr ago on Antarctica but not until just after 3 Myr ago on Greenland. Together, the modern Antarctic and Greenland ice sheets make the $\delta^{18}O$ value of ocean water about 1‰ heavier than it would otherwise be. Consequently, ~1‰ of the 2.75‰ $\delta^{18}O_c$ increase between 40 Myr ago and today can be explained by the growth of ice sheets within that interval. The remainder of the $\delta^{18}O_c$ increase—1.75‰—must have been caused by an

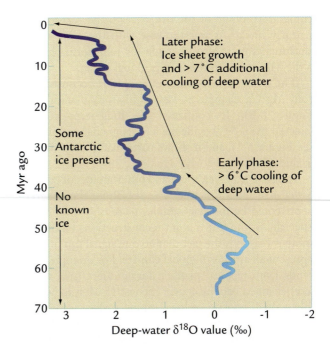

FIGURE 6-7 Long-term $\delta^{18}O$ trend Measurements of $\delta^{18}O$ in benthic foraminifera show an erratic long-term trend toward more positive values. From 50 to 40 Myr ago, the increase in $\delta^{18}O$ was caused by cooling of the deep ocean. After 40 to 35 Myr ago, it reflects further ocean cooling and the formation of ice sheets. (Adapted from K. G. Miller et al., "Tertiary Oxygen Isotope Synthesis: Sea Level History and Continental Margin Erosion," *Paleoceanography* 2 [1987]: 1–19.)

additional cooling of the deep ocean by more than 7°C between 40 Myr ago and today (1.75‰ × 4.2°C/‰). This additional cooling of 7°C in the last 40 Myr brings the total deep-ocean cooling since 50 Myr ago to more than 14°C.

Because the temperature of today's deep ocean averages about 2°C and has cooled by at least 14°C over the last 50 Myr, the deep-ocean temperature must have been near 16°C before 50 Myr ago. If deep water formed mainly in high latitudes as it does today, the polar climates that sent such warm water into the deep ocean must have been much warmer than they are today (Chapter 5).

6-3 Evidence from Mg/Ca Measurements

Another valuable index of the climatic response of the ocean comes from analyzing the ratio of the elements magnesium (Mg) and calcium (Ca) in the shells of foraminifera. The process by which Mg substitutes for Ca in the foraminiferal shells depends on the temperature of the waters in which the shells form. Across the temperature range of the deep ocean, the relationship is nearly linear. As was the case with the δ¹⁸O signal, adjustments must also be made for long-term changes in the Mg concentration of the global ocean.

The trend in deep-water temperature reconstructed from Mg/Ca changes is very similar to that of δ¹⁸O (Figure 6-8). It confirms the long-term cooling detected from δ¹⁸O evidence and from changes on land (see Figure 6-1), and it shows particularly large steps at or near 35, 13, and 3 Myr ago.

In detail, however, these and other indices of past climate can at times disagree. One reason for the disagreements could be too few samples to resolve finer detail, especially in the leaf-margin reconstructions of climate. Another reason is the fact that no one signal can possibly represent all aspects of global climate. For example, δ¹⁸O trends from tropical planktic foraminifera (not shown here) changed very little over the last 15 Myr, indicating only a small cooling of low-latitude surface waters. Yet ice sheets were growing and both high-latitude and deep-ocean temperatures were cooling markedly during the last 15 Myr.

A wide and convincing array of evidence documents the progressive cooling of both poles and of mid-latitude areas during the last 50 Myr. Several hypotheses have been put forward to explain this gradual greenhouse-to-icehouse transition.

Do Changes in Geography Explain the Cooling?

As noted in Chapter 5, the polar position hypothesis cannot explain this slow cooling over the last 50 Myr. Antarctica was located at the South Pole during the greenhouse climate of 100 Myr ago and it is still at the pole during the icehouse climate of today. Subsequently, the largest latitudinal shift of the continents during the last 50 Myr has been the northward movement of India and Australia into tropical latitudes (Figure 6-9). These movements away from the poles are unlikely to have produced the onset or intensification of glaciation.

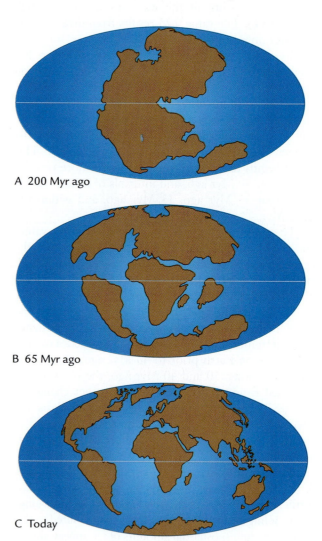

A 200 Myr ago

B 65 Myr ago

C Today

FIGURE 6-9 Continental movements since 200 Myr ago
Since (A) the time of Pangaea, 200 Myr ago, (B, C) the Atlantic Ocean has widened, the Pacific Ocean has narrowed, and India and Australia have separated from Antarctica and moved northward to lower latitudes. (Modified from F. Press and R. Siever, *Understanding Earth*, 2nd ed., © 1998 by W. H. Freeman and Company.)

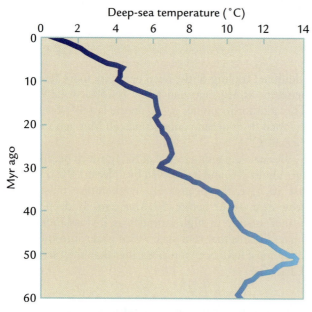

FIGURE 6-8 Long-term Mg/Ca trend Measurements of Mg/Ca ratios in benthic foraminifera indicate a progressive cooling of deep water over the last 50 Myr. (Adapted from C. H. Lear et al., "Cenozoic Deep-Sea Temperatures and Global Ice Volumes from Mg/Ca in Benthic Foraminifera," *Science* 287 [2000]: 269–72.)

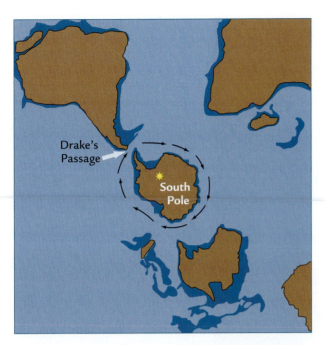

FIGURE 6-10 Opening of Drake's Passage Opening of an ocean gap between South America and Antarctica near 25 to 20 Myr ago allowed a strong Antarctic circumpolar current (arrows) to flow uninterrupted around the Antarctic continent. The passageway between Australia and Antarctica had opened 10 Myr earlier. (Adapted from E. J. Barron et al., "Paleogeography: 180 Million Years Ago to the Present," *Ecologae Geologicae Helvatiae* 74 [1981]: 443–70.)

6-4 Gateway Hypothesis

Some climate scientists have called on the opening or closing of **ocean gateways** to explain the onset of both southern and later northern glaciation during the last 50 Myr. These hypotheses focus on narrow passages that allow or impede exchanges of ocean water between ocean basins. They propose that changes in key gateways caused glaciations by altering the poleward transport of heat or salt.

Case Study 1: Antarctica During the last 50 Myr, the last of the Gondwana continents connected to Antarctica split off and moved north, leaving Antarctica isolated and surrounded by a circumpolar ocean (Figure 6-10). In the late 1970s, marine geologist James Kennett proposed that this breakup caused the onset of glaciation on Antarctica. Before the continents separated, oceanic flow around Antarctica had been impeded by the land connections with South America and Australia. Kennett hypothesized that these barriers had diverted warm ocean currents poleward from lower latitudes and had delivered enough heat to Antarctica to prevent glaciation.

After the continents separated, a strong, unimpeded eastward flow developed around Antarctica. Kennett proposed that the loss of the warm poleward flow of heat caused the continent to cool and glaciation to begin. Recent investigations indicate that Australia separated from Antarctica around 37 to 33 Myr ago, the same interval as the first glaciation on Antarctica. The opening of Drake's Passage between South America and Antarctica occurred near 25 to 20 Myr ago (see Figure 6-10). This final establishment of unimpeded circulation around the Antarctic continent falls roughly midway between the initial onset of glaciation and the large increase in Antarctic ice near 13 Myr ago.

Climate scientists have used sensitivity tests with O-GCMs to evaluate this hypothesis. Drake's Passage was closed in one experiment and left open in another while all other features of Earth's geography were kept the same. The model results suggested that opening Drake's Passage did not significantly alter ocean (or atmospheric) temperatures near Antarctica. Instead, the model simulated a frigid climate over Antarctica regardless of the kind of ocean flow. The combined heat transport by the ocean and the atmosphere remained about the same in both experiments. Apparently, the opening of the circum-Antarctic gateway was not a critical factor in the onset and development of Antarctic glaciation.

Case Study 2: Central American Seaway During the last 10 Myr, uplift in Central America gradually closed a deep ocean passage that had previously separated North and South America in the region of Panama. The last stages of uplift created the Central American part of the Cordilleran mountain chain. Final closure of this open passage occurred just before 4 Myr ago with the emergence of the Isthmus of Panama, and the first large-scale glaciation of North America followed at 2.75 Myr ago.

Several climate scientists have speculated that these two episodes are linked. They hypothesized that construction of the Isthmus of Panama blocked the strong westward flow of warm, salty tropical water that had previously been driven westward out of the tropical Atlantic Ocean and into the eastern Pacific by trade winds. The newly formed isthmus should have redirected this flow into the Gulf Stream and toward the high latitudes of the Atlantic. They further hypothesized that this strengthened northward flow of warm, salty water would have suppressed the formation of sea ice in north polar regions because saltier waters resist freezing better than fresher water (companion Web site, pp. 41–42). According to this hypothesis, the reduced cover of sea ice would have made more moisture from the ocean available to nearby landmasses and triggered the growth of ice sheets.

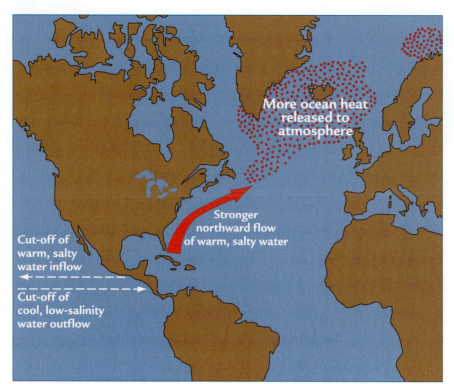

FIGURE 6-11 Closing of the Isthmus of Panama Simulations with ocean models indicate that gradual closing of the Central American isthmus between 10 and 4 Myr ago redirected warm, salty water northward into the Atlantic Ocean, reduced the extent of high-latitude sea ice, and handed off additional heat to the atmosphere. (Adapted from E. Maier-Reimer et al., "Ocean General Circulation Model Sensitivity Experiment with an Open Central American Isthmus," *Paleoceanography* 5 [1990]: 349–66.)

Climate modelers have used ocean general circulation models (O-GCMs) to test this hypothesis by running pairs of experiments with the Panama region configured both as an open gateway passage and as a closed-off isthmus. These two configurations roughly correspond to the end points of the tectonic changes that occurred between about 10 and 4 Myr ago. These simulations confirmed the prediction that warm, salty water would have been retained within the Atlantic Ocean and redirected toward northern latitudes (Figure 6-11). Closing of the Panama Isthmus also cut off a return flow of low-salinity Pacific water into the Atlantic. Blockage of the low-salinity return flow by the isthmus further increased the salinity of the northward-flowing Atlantic water.

In a critical respect, however, the model simulations contradicted the gateway hypothesis. The stronger northward flow of salty water in the Atlantic and the resulting reduction of sea ice caused by closing the Panama Isthmus did not greatly alter precipitation patterns around the high-latitude North Atlantic. As a result, the hypothesized increase in moisture needed to grow ice sheets did not occur. Instead, the stronger northward flow of water from the tropics and subtropics transferred a large amount of heat to the atmosphere and warmed the regions where ice sheets were eventually to form. This warming increased summer melting of snow and opposed the conditions needed for the inception of glaciation, contrary to the gateway hypothesis.

6-5 Assessment of Gateway Changes

These two case examples invoke very different roles for the ocean in glacial inception. For Antarctica, a *reduced* poleward flow of warm ocean water was proposed to have caused a cooling and subsequent glaciation. For the Isthmus of Panama, an *increased* poleward flow of warm ocean water was invoked as the cause of an increase in moisture flux that promoted glaciation.

These differing assumptions reflect disagreement among climate scientists about how the ocean affects ice sheets. Some climate scientists emphasize that a warmer ocean will release more *latent* heat (water vapor) to the atmosphere and thereby supply more moisture (snow) to aid ice growth. Most climate scientists, however, emphasize the fact that a warmer ocean will release more *sensible* heat to the overlying atmosphere and thereby potentially melt more ice (companion Web site, pp. 14–15).

In any case, neither set of modeling experiments supports the hypothesis that changes in poleward flow of warm ocean water tied to gateway changes would have had a large enough effect on climate to initiate the growth of ice sheets. This criticism needs to be tempered by the realization that O-GCMs are still at a relatively early stage of development and that precipitation patterns are often rather poorly simulated in GCMs. A second criticism of the gateway hypothesis is that each opening or closing of a gateway occurs over perhaps 10 Myr of time, while the cooling has lasted far longer.

IN SUMMARY, it seems unlikely that such discontinuous gateway episodes could have driven a progressive climatic cooling for 50 Myr.

Whether or not gateway changes affect climate on a global scale, they certainly have the potential to alter the flow of deep and bottom water. Major gateway changes redistribute heat and salt at the high-latitude sites where deep waters form, and changes in these surface-water properties may affect formation of deep water. For example, the model experiments indicated that closing the Isthmus of Panama increased the formation of deep water in the high latitudes of the North Atlantic. Formation of deep water increased in this simulation because of the increased salinity of northward-flowing Gulf Stream waters (see Figure 6-11). Higher-salinity surface waters promote stronger deep-water formation because the water is already dense when it encounters cold winter air masses at high latitudes. The results from this experiment agree with independent evidence that formation of deep water in the North Atlantic increased between 10 and 4 Myr ago, the interval over which the Central American seaway was gradually reduced and finally closed off by the emergence of the Central American isthmus.

Hypotheses Linked to Changes in CO_2

The evidence presented here requires a mechanism that has cooled climate at both poles over 50 Myr. Many climate scientists regard falling CO_2 levels as the most obvious way to do this, whether by slower CO_2 input (BLAG) or by faster removal (uplift weathering). A wide range of proxy indicators have been used to try to estimate changes in atmospheric CO_2 concentrations during the last 50 Myr. All the methods show a general trend of decreasing CO_2 levels, but they show large disagreements about the size and timing of the major CO_2 drops. The operating assumption used in this section is simply that CO_2 concentrations fell during the last 50 Myr.

6-6 Evaluation of the BLAG Spreading Rate Hypothesis

To explain the global cooling of the last 50 Myr, the spreading rate hypothesis must pass one critical test. A slowing of global mean spreading and subduction rates must have occurred through the interval, leading to slower rates of CO_2 input to the atmosphere.

It might seem initially that the spreading rate hypothesis has already passed this test. We saw in Chapter 5 that spreading rates 100 Myr ago were probably

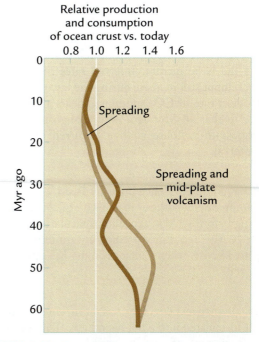

FIGURE 6-12 Changes in spreading rates The average rate of seafloor spreading slowed until 15 Myr ago, but it has since increased slightly. Adding the effects of generation of new crust by volcanism at hot spots away from plate margins does not change this basic trend. (Adapted from L. R. Kump and M. A. Arthur, "Global Chemical Erosion During the Cenozoic," in *Tectonic Uplift and Climate Change,* ed. W. F. Ruddiman [New York: Plenum Press, 1997].)

faster than they are today, although some scientists question whether or not this is so. Slower spreading today is consistent with the colder modern climate. But when we look more closely at the last 50 Myr, this explanation is not completely satisfactory. Before 15 Myr ago, global mean spreading (and subduction) rates show a decrease consistent with the spreading rate hypothesis (Figure 6-12). But since 15 Myr ago, the mean rates of spreading and subduction have slightly increased to a present value that is equal to the one that existed almost 30 Myr ago.

The trend since 15 Myr ago does not support the spreading rate hypothesis. An increase in rates of spreading during that time should have put more CO_2 into the atmosphere and warmed the global climate. Instead, the climate continued to cool, with a substantial increase in size of the Antarctic ice sheet and the first appearance of northern hemisphere ice (mountain glaciers and ice sheets).

Could additional volcanic input of CO_2 at sites away from ocean ridges and subduction zones explain this discrepancy? During some intervals in the past, extra CO_2 was released at oceanic and continental hot spots located well away from plate margins. Some of these episodes of volcanism have been radiometrically dated, and the estimated volume of volcanic rock produced can be added to the amount produced by spreading and subduction (see Figure 6-12). This adjustment does not change the basic picture. Inferred rates of volcanism and CO_2 input still increase during the last 15 Myr, and the modern rate of CO_2 addition appears almost comparable to that 40 Myr ago, even though most of the greenhouse-to-icehouse cooling occurred during this time interval.

IN SUMMARY, the evidence indicates that the spreading rate hypothesis may have explained global cooling before 15 Myr ago, and particularly before 30 or 40 Myr ago. But it predicts a warming during the last 15 Myr, when a major cooling has actually occurred.

An alternative possibility has been considered but not yet formulated into a full hypothesis. The concept is that the amount of carbon carried down into ocean trenches may have varied because of changes in the type of sediments being subducted, even in the absence of changes in spreading rates. Today, most subduction occurs on the margins of the Pacific Ocean, where sediments are $CaCO_3$-poor because of strong dissolution on the seafloor by corrosive deep waters (Chapter 2). Much of the global carbonate total is now being deposited on the Atlantic seafloor where dissolution is less intense. If a future change in the plate tectonic regime were to initiate subduction in the Atlantic Ocean, an enormous amount of carbonate would be available to be carried down into trenches for later melting and eventual release to the atmospheres through volcanoes. As a result, atmospheric CO_2 values would increase even in the absence of changes in spreading rates. At this point, this idea has not yet been tested.

6-7 Evaluation of the Uplift Weathering Hypothesis

To demonstrate that the uplift weathering hypothesis explains global cooling during the last 50 Myr, three main requirements must be met: (1) the amount of high-elevation terrain in existence today must be unusually large in comparison with earlier intervals; (2) this high terrain must be causing unusual amounts of rock fragmentation; and (3) the exposure of fresh debris must be causing unusually high rates of chemical weathering.

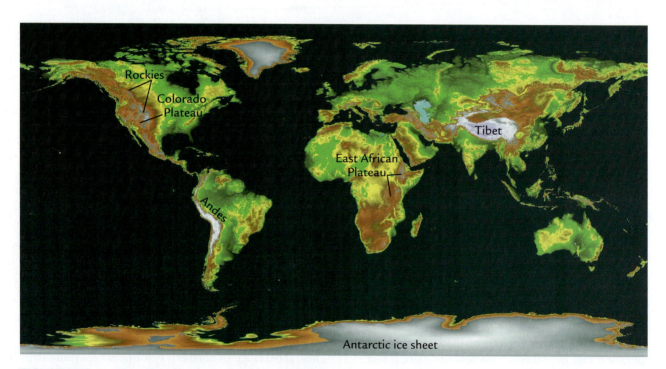

FIGURE 6-13 Earth's high topography Earth today has only a few regions where broad areas of land stand more than 1 km high (shown in brown, blue, and white). Except for the high ice domes on Antarctica and Greenland, the highest bedrock surfaces are the Tibetan Plateau and other high terrain in southern Asia, the Andes of South America, the Rocky Mountains and Colorado Plateau of North America, and the volcanic plateaus of eastern and southern Africa. (Courtesy of Peter Schloss, National Geophysical Data Center, Boulder, CO.)

To determine whether or not these requirements are met, we have to compare the present time with some interval in the past. The last half of the Cretaceous interval, from 100 to 65 Myr ago, is a useful basis for comparison for two reasons: (1) abundant evidence is still left in the geologic record and (2) it is an interval of full greenhouse climate (Chapter 5).

Prediction 1: Extensive High Terrain At first glance it might seem obvious that uplift has been unusually active in most mountain ranges during the last few tens of millions of years. Marine sediments deposited at or below sea level 100 to 65 Myr ago are now found at high elevations in Tibet and the Himalaya of Asia, the South American Andes, the North American Rocky Mountains, and the European Alps (Figure 6-13). These sediments have been uplifted from sea level to their present heights in the last 70 Myr or less.

Although much of today's high topography is geologically youthful, this evidence alone does not prove that the modern elevations are uniquely high or extensive. Plate tectonic processes continually cause uplift in many regions throughout geologic time, while erosion continually attacks the highest topography and wears it down. As a result, the highest topography during any interval of geologic time is always recent in origin, just as it is today.

The strongest evidence that the amount of high terrain is indeed more massive today than it was in earlier geologic eras is the existence of the Tibetan Plateau, some 2.5 million km² (1 million mi²) in area at an average elevation above 5 km. This plateau has slowly risen since the initial collision of India and Asia 55 Myr ago (Figure 6-14A and B). Uplift occurred earliest in the south-central parts of the plateau and later in the northeastern and southeastern sectors.

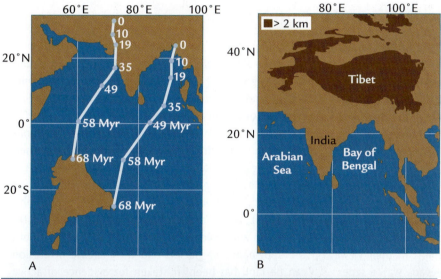

FIGURE 6-14 India-Asia collision and Tibet (A) Collision of India and Asia produced (B) the Tibetan Plateau, the largest high-elevation rock feature on Earth today. (C) The Himalaya Mountains tower over the Indian subcontinent to the south (in the foreground). Behind the Himalayas to the north lies the vast Tibetan Plateau at an average elevation above 5000 m. (A and B: Adapted from P. Molnar et al., "Mantle Dynamics, the Uplift of the Tibetan Plateau, and the Indian Monsoon," *Review of Geophysics* 31 [1993]: 357–96. C: Emil Muench/Photo Researchers.)

In contrast, no major continental collisions occurred from 100 to 65 Myr ago, and no massive plateaus existed then or for the preceding 150 Myr. The presence of the Tibetan Plateau and Himalayan complex (Figure 6-14B and C) is a strong argument for an unusually massive amount of high topography at the present time.

Most other high-elevation regions on Earth (see Figure 6-13) have been formed by subduction of ocean crust beneath continental margins. Because subduction is an ongoing process, mountain terrain has existed continuously through time, in contrast to plateaus produced by sporadic continental collisions. The modern Andes and narrow central plateau called the Altiplano are the result of subduction along the west coast of South America. Because subduction has been under way

there for more than 100 Myr, a mountain range has long existed in the western Andes, but much of the high topography of the central Altiplano and the eastern Andes was created within the last 15 Myr.

Subduction has also occurred along western North America for some 200 Myr. Scientists are sharply divided about the history of uplift in this region over the last 50 Myr, but it seems unlikely that the high topography in this region today is unique (Box 6-1).

Another kind of high terrain is the extensive low plateau in eastern and southern Africa at an elevation of 1 km (see Figure 6-13). This plateau results from deep-seated heating that causes a broad upward doming and outpouring of volcanic lava. Much of the East African plateau was built in the last 30 Myr, but similar

BOX 6-1   CLIMATE DEBATE

Timing of Uplift in Western North America

Scientists disagree about the age of uplift of high terrain across western North America. All agree that mountains of some size have existed in Nevada and eastern California for 200 Myr or more because ongoing subduction occurred along the west coast until about 30 Myr ago. They also agree that the Rocky Mountain West (the modern High Plains, Colorado Plateau, and both the U.S. and Canadian Rocky Mountains) was flooded by an inland sea from 100 to 70 Myr ago and has since been uplifted to its present height. But just about everything else about the timing and amount of uplift in this region is in dispute.

One group emphasizes recent uplift. In their view, the earlier mountain terrain near Nevada was a series of discontinuous low-elevation peaks, not a major topographic feature. This group infers that broad, large-scale upwarping of the entire West from the Sierra Nevada of California to the High Plains of Colorado and Wyoming began about 20 Myr ago because of some kind of deep-seated heating process. Visitors to parks in the American West will notice that this view is widely promoted by the U.S. Park Service.

The other group interprets major uplift as occurring at an earlier time, followed by a more recent loss of elevation in many regions. In their view, the mountain belt that existed in the Far West before 100 Myr ago was a continuous feature at 3–4 km elevation like the modern Andes, and it has subsequently dropped to its modern height of 1–2 km. This group acknowledges large-scale uplift of the

Rocky Mountain West since 70 Myr ago, but they infer that most of the uplift actually occurred between 70 and 45 Myr ago during an interval of heightened tectonic activity and that most of the Rocky Mountain area lost elevation once that activity ceased.

Geophysicist Peter Molnar has offered an explanation for this surprising difference in opinion. He proposed that the global cooling trend has increased the extent of glaciation on high-elevation mountains and plateaus by lowering the freezing line onto the high topography. As a result, glacial erosion has caused increases in rates of erosion and major incision of mountain valleys. Both the appearance of glaciers and the increase of erosion can be (and has been) mistaken for evidence of active uplift in mountain belts that are actually tectonically "dead."

Temperature reconstructions based on leaf-margin types have been used to test these opposing views. Temperatures are estimated from deposits of the same age both in higher-elevation mountain areas and in nearby coastal regions. Because temperature decreases with elevation in a known way, the estimated temperature differences between coastal and high-elevation regions can be converted to estimates of past elevation. This method suggests a small decrease in elevation of most parts of the Rocky Mountain West during the last 40 or 50 Myr. One exception during the last 10 or 20 Myr is the Yellowstone hot spot area, which has been domed upward by shallow subsurface heating.

amounts of plateau construction may have occurred on Africa near 100 Myr ago as the giant continent of Pangaea broke up. As a result, it is hard to argue that the present plateau exceeds features that may have existed during earlier intervals.

> **IN SUMMARY,** the existence of the massive Tibetan Plateau makes modern topography unusual, consistent with the uplift weathering hypothesis. Regions of high youthful terrain also exist along subducting plate margins and elsewhere, but they may be similar to features that existed in the past.

Prediction 2: Unusual Physical Weathering The second test of the uplift weathering hypothesis is whether or not today's high topography is causing higher rates of physical weathering and rock fragmentation than occurred in the past. The concentrations of suspended particles carried southward and eastward from the Himalaya and Tibet by Asian rivers are larger than any on Earth (Figure 6-15). The youthful topography of the eastern Andes drained by the Amazon River is another region of high particle concentrations. These measurements clearly indicate intense physical weathering at the present time in the two regions with highest terrain.

BOX 6-1 CLIMATE DEBATE

CONTINUED

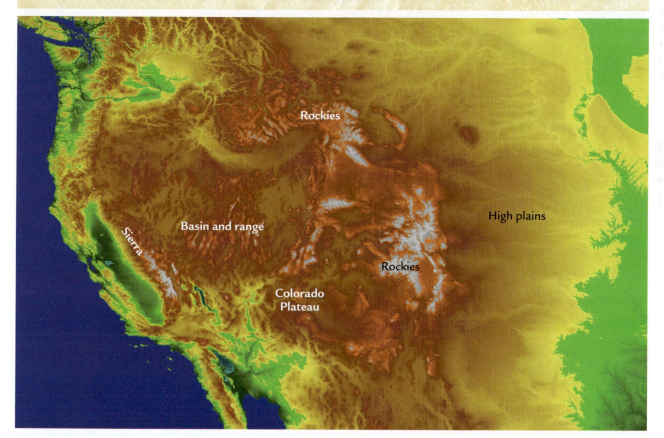

Topography in the American West A broad bulge of high topography reaches from the Sierra in the far western United States to the Rocky Mountains and High Plains farther east, with the Colorado Plateau and Basin and Range lying in between. Low-elevation regions a few hundred meters above sea level are shown in green, with progressively higher elevations in yellow, brown, pale orange, and white. (Courtesy of Peter Schloss, National Geophysical Data Center, Boulder, CO.)

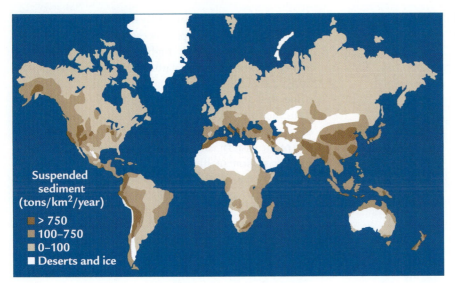

FIGURE 6-15 Sediments suspended in rivers The annual yield of suspended sediments is highest in two regions: the Himalayas of southeast Asia and the Andes of South America. (Adapted from D. E. Walling and B. W. Webb, "Patterns of Sediment Yield," in *Background to Paleohydrology*, ed. K. J. Gregory [New York: Wiley, 1983].)

The best record of rates of erosion lies in sediments deposited in ocean basins by rivers. By far the largest mass of young sediment in the ocean today is found on the Indian Ocean seafloor south of the Himalaya. Deposition of this pile of sediment began near 40 Myr ago, increased near 25 Myr ago, and accelerated rapidly near 10 Myr ago (Figure 6-16).

Climate model experiments indicate that this influx of sediment to the Indian Ocean is a result of two factors: (1) creation of steep terrain along the southern Himalayan margin of the Tibetan Plateau and (2) the fact that a plateau the size of Tibet in effect creates its own weather, including the powerful South Asian monsoon (Figure 6-17).

Monsoons result from different rates of heating of continents and oceans due to the different heat capacities of land and water (companion Web site, pp. 9–11, 17–18). In summer, the continents (including the surface of Tibet) warm up more rapidly than the oceans and heat the overlying air. The warm air rises and pulls in moist ocean air. The high Himalayas on the southern margins of the plateau form an obstacle to the incoming ocean air, forcing it to rise and its water vapor to condense in cooler temperatures at high altitudes. As a result, these steep slopes become a natural focal point for strong summer monsoon rains, which release latent heat and fuel even more powerful monsoons. The powerful south Asian monsoon came into existence in part as a direct result of the rise of the Tibetan Plateau.

Unfortunately, ocean sediments cannot give us a definitive estimate of *global* rates of physical weathering in the past. Much of the sediment eroded from coastal mountain ranges and deposited in the nearby ocean is soon subducted into nearby trenches. The large amount of sediment lost in this way cannot be quantified. In addition, some sediments deposited on the seafloor are eroded and redeposited, and this reworking skews compilations of sediment deposition rates through time toward younger ages. Rapid deposition of huge amounts of Himalayan sediment thus supports the

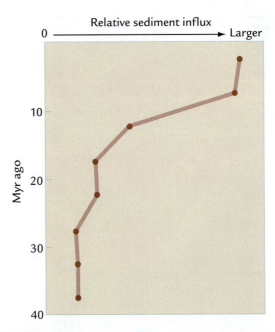

FIGURE 6-16 Himalayan sediments in the Indian Ocean The rate of influx of sediments from the Himalayas and Tibet to the deep Indian Ocean has increased almost tenfold since 40 Myr ago. (Adapted from D. K. Rea, "Delivery of Himalayan Sediment to the Northern Indian Ocean and Its Relation to Global Climate, Sea Level, Uplift, and Seawater Strontium," in *Synthesis of Results from Scientific Drilling of the Indian Ocean*, ed. R. A. Duncan et al. [Washington, DC: American Geophysical Union, 1992].)

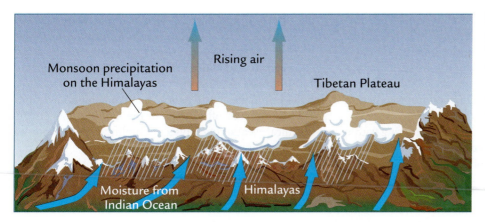

FIGURE 6-17 **Tibet and the monsoon** Heating of the Tibetan Plateau draws in moisture from the Indian Ocean and enhances the intensity of the warm, moist summer monsoon on its southern (Himalayan) margin

hypothesis that physical weathering is stronger today on a global basis than in earlier times, but this conclusion remains tentative because of the sediment lost to subduction.

Prediction 3: Unusual Chemical Weathering The final test of the uplift weathering hypothesis is whether or not the global average rate of chemical weathering is higher today than in the past. Unfortunately, chemical weathering rates are difficult to determine even on regional scales, much less for the entire Earth.

Climate scientists quantify modern rates of chemical weathering on a regional basis by measuring the total amount of ions dissolved and transported in rivers. This measure reflects the amount of chemical weathering within the watershed drained by each river, but modern disturbances of natural weathering processes by humans complicate such studies. In addition, it is difficult to distinguish between the ions provided by slow weathering of silicate rocks (hydrolysis) and those resulting from rapid dissolution of carbonate rocks. Only hydrolysis affects the CO_2 balance in the atmosphere (Chapter 3). Another limitation is strategic: it is impossible to study enough rivers to reach an accurate estimate of the global weathering rate because too many rivers contribute significantly to the global total.

It is even more difficult to reconstruct rates of chemical weathering during earlier intervals (Box 6–2). As a result, the case for unusual chemical weathering today rests only on a plausibility argument based on several observations: the unusual height and extent of the Tibetan-Himalayan complex, the unusual strength of the monsoon rains, and the unusual volume of sediments deposited in the nearby ocean. By inference, this combination of favorable factors should promote unusually rapid chemical weathering and cause CO_2 removal from the atmosphere. But inference is not proof.

> **IN SUMMARY**, the cause of global cooling during the last 50 Myr remains uncertain. The most likely culprit is a decrease in atmospheric CO_2, but it is unclear whether this change was driven by decreased input tied to seafloor spreading or increased removal tied to chemical weathering.

Future Climate Change at Tectonic Time Scales

Despite the wealth of evidence that climate has cooled over the last 50 Myr, it is impossible to predict the changes in climate that will be caused by tectonic processes. One reason is that scientists disagree about whether past cooling was driven by uplift, slower seafloor spreading, or other factors. In any case, future changes in plate tectonic processes are inherently unpredictable because the driving forces are not fully understood. If we cannot accurately predict the operation of tectonic processes in the future, we cannot predict their climatic effects. The safest prediction—that long-term (tectonic-scale) climate will continue to cool—is really not a prediction at all, just a forward projection of past trends.

The tectonic-scale climatic trends of the distant future will also be influenced by positive and negative feedback processes. We have already seen that negative feedback from chemical weathering is an integral part of the BLAG hypothesis (Chapter 4), but a rough calculation shows that it could also have acted to offset much of the increase in chemical weathering driven by uplift (Figure 6-18). In this calculation, we assume that uplift affected an area equivalent to 1% of the total area of continental crust, approximately the size of the Himalayas and the southern parts of the Tibetan Plateau. We also assume that uplift increased the rate of exposure of fresh bedrock in this small region by a factor of 50, broadly consistent with the relative differences in dissolved fluxes between the Andes Mountains and the lowlands in the Amazon Basin (Chapter 4). A 50-fold increase in weathering over 1% of Earth's land surface would increase global chemical weathering by 50%.

This localized increase in chemical weathering within the uplifted region would tend to be offset by a decrease in weathering across the rest of Earth's land surface. As climate cools, the negative feedback role of weathering should begin to moderate climate across the globe because chemical weathering is slower in cooler, drier, less vegetated conditions (Chapter 3). A global temperature decrease of 3° to 4°C would be enough to drop weathering rates by 50% over the remaining 99% of the land surface and offset most of the localized increase in chemical weathering caused by uplift. This amount of global cooling is well within the range estimated for the last 50 Myr. This calculation suggests that the moderat-

BOX 6-2   LOOKING DEEPER INTO CLIMATE SCIENCE

Organic Carbon: Monterey Hypothesis

The long-term cooling that produced the present icehouse climate was somewhat erratic, with interruptions by shorter intervals of warming and cooling lasting a few million years. One possible source of shorter-term climate changes at tectonic time scales is variations in the rate of burial and exposure of organic carbon. Organic carbon is a plausible driver of these climate changes because it accounts for 20% of the carbon cycling into and out of Earth's sediments and rocks (Chapter 3). Organic carbon also has the potential to affect climate relatively rapidly because large amounts can be quickly buried in the sedimentary record, causing rapid reductions of atmospheric CO_2 levels.

Several kinds of climatic and tectonic changes could favor rapid increases in burial of organic carbon (companion Web site, pp. 32–33): changes in wind direction along a coastal region that cause increased upwelling and carbon production, an increase in the total amount of organic carbon and nutrients delivered to the ocean, or a change toward a wetter climate on continental margins, where flat topography naturally favors development of swamps and deposition of organic matter.

An increase in the rate of burial of organic carbon has been proposed as the cause of the cooling trend near 13 Myr ago. The large increase in deep-ocean $\delta^{18}O$ values at this time indicates some combination of deep-water cooling and increase in size of the ice sheet on Antarctica. These changes followed an interval when carbon-rich sediments were deposited in shallow waters around the margins of the Pacific Ocean, including the Monterey coast of California. The marine geologists Edith Vincent and Wolfgang Berger suggested that a major increase in coastal upwelling, perhaps driven by stronger winds caused by long-term climate cooling, buried enough organic carbon along the margins of the Pacific to reduce atmospheric CO_2, cool the global climate, and allow ice to build up on Antarctica. They called this the **Monterey hypothesis.**

The Monterey hypothesis has been criticized because of a lag of 2 to 3 Myr between the onset of increased carbon burial and the time of fastest cooling shown by the $\delta^{18}O$ record, although the fastest rates of carbon burial appear to have occurred closer to the cooling. Other scientists have suggested that the increased carbon burial in the Pacific could be linked to the supply of carbon eroded from older sedimentary rocks on land (in the Himalayas).

Burial of organic carbon on shallow continental margins also tends to produce its own negative feedback. Carbon-rich sediments deposited in shallow areas are later re-exposed to the atmosphere if sea level falls because ice sheets grow. Exposure of this buried organic carbon allows it to be oxidized back to CO_2 and returned to the atmosphere. The return of CO_2 then causes the climate to warm.

Some climate scientists speculate that changes in rates of weathering on land and burial of organic carbon in the ocean could also be an important cause of longer-term cooling over tens of millions of years. If imbalances between these rates persist for many millions of years, the result could be increases or decreases in the total amount of carbon in the ocean-atmosphere system and of the level of CO_2 in the atmosphere. Scientists are investigating whether the organic carbon subcycle has been adding or removing carbon (and CO_2) from the ocean-atmosphere system over long intervals.

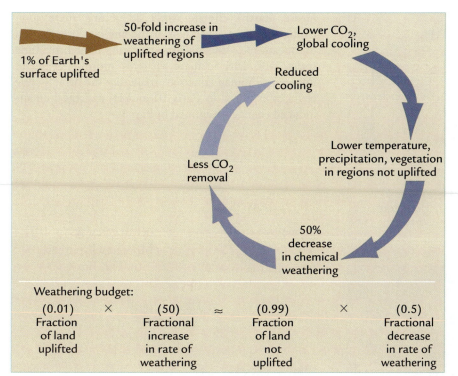

FIGURE 6-18 Negative feedback from chemical weathering? If increased chemical weathering in localized regions of uplift causes global climate cooling, the resulting reduction of chemical weathering in other regions may act as a negative feedback that moderates much of the cooling driven by uplift.

Labels within the figure:

1% of Earth's surface uplifted

50-fold increase in weathering of uplifted regions

Lower CO_2, global cooling

Reduced cooling

Less CO_2 removal

Lower temperature, precipitation, vegetation in regions not uplifted

50% decrease in chemical weathering

Weathering budget:

(0.01)	×	(50)	≈	(0.99)	×	(0.5)
Fraction of land uplifted		Fractional increase in rate of weathering		Fraction of land not uplifted		Fractional decrease in rate of weathering

ing effects of the chemical weathering thermostat may have balanced (or nearly balanced) the effects of uplift-driven weathering during the last 50 Myr.

But the possibility also exists that the process of global cooling could produce positive feedbacks that keep driving climate toward even colder conditions. These feedbacks result from the increased amounts of freshly fragmented rock generated by ice (Figure 6-19A). If the mountain glaciers grind large volumes of bedrock as climate cools, the rate of exposure of freshly fragmented rock should increase, thereby speeding up rates of chemical weathering, pulling more CO_2 out of the atmosphere, and causing additional cooling (Figure 6-19B). This kind of feedback can happen both in regions of active uplift and in high terrain that is no longer being uplifted (see Box 6-1).

An increase in weathering is particularly likely if the lower limits of the mountain glaciers move up and down the sides of mountains, alternately grinding fresh rock and then exposing it to the atmosphere for chemical weathering (Figure 6-19B). These kinds of fluctuations result from the orbital-scale cycles of warming and cooling we will examine in Part III.

A similar positive feedback may occur in connection with large continental-scale ice sheets (Figure 6-19C). As global climate cools and vast ice sheets appear, they erode the preexisting cover of already weathered soils, expose fresh underlying bedrock, and grind the rock down to finer particles. This debris is then deposited in extensive moraine ridges at the ice margins.

If the ice sheets simply remain at or near their maximum extents for millions of years (as the Antarctic ice sheet seems to have done), they would probably slow the overall rate of chemical weathering. Weathering may increase along the ice margins where piles of fragmented debris are exposed, but these localized increases would probably be overwhelmed by the larger decrease in weathering caused by ice cover across much of the continent.

If, however, the ice sheets fluctuate in size (as they have done in the northern hemisphere), these variations should cause a net increase in chemical weathering. When shorter-term climatic variations cause the ice sheets to melt back from their maximum extents or even disappear (as has happened in North America and Europe since 20,000 years ago), vast areas of finely ground debris are exposed (Figure 6-19C). This fresh debris can then be rapidly weathered during the warmer climates during intervals between glaciations. The net result is an increase in chemical weathering.

Further research is obviously needed to assess the role of tectonic forcing and internal feedbacks, both

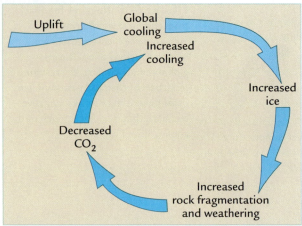

A Positive weathering feedback

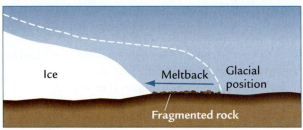

B Mountain glaciers

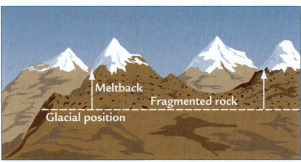

C Continental ice sheets

FIGURE 6-19 Positive feedback from ice? (A) Global cooling produces more ice on Earth, and the ice increases rock fragmentation (B) in high mountain terrain and (C) near ice sheets. Chemical weathering of this fragmented debris may cause further cooling by positive feedback.

positive and negative. In any case, Earth's long-term "forecast" over tectonic time scales calls for colder temperatures and more ice, assuming that the current plate tectonic regime does not change. This cooling is hardly imminent. All tectonic-scale processes and feedbacks operate at extremely slow rates, and the changes they produce become evident only over millions of years. Even though we seem to be headed toward a colder

future, Earth's climate won't be getting there soon enough for it to cause you or me any concern. And in the meantime, other factors operating on shorter time scales will drive climate changes more relevant to immediate human concerns, including the orbital-scale changes explored in Parts III and IV and the human-induced changes examined in Part V.

Key Terms

permafrost (p. 99)

$\delta^{18}O$ (p. 100)

fractionation (p. 100)

ocean gateways (p. 103)

Monterey hypothesis
(p. 112, Box 6–2)

Review Questions

1. What kinds of changes in vegetation and ice show that Earth has cooled in the last 50 Myr?

2. What two things do changes in $\delta^{18}O$ values in the last 50 Myr tell us about climate changes?

3. How well does the spreading rate (BLAG) hypothesis explain the last 50 Myr of cooling?

4. How well does the uplift weathering hypothesis account for the last 50 Myr of cooling?

5. Explain how chemical weathering could either moderate or deepen long-term cooling.

6. The volume of water in the world ocean is 48.5 times larger than the amount stored in the two largest ice sheets. The average $\delta^{18}O$ value of the ocean is near zero and the mean $\delta^{18}O$ value of ice on Antarctica and Greenland is –50‰. Show a calculation indicating how much the mean $\delta^{18}O$ value of ocean water would decrease if the two ice sheets melted.

7. Over the last 15 Myr, ice volume has increased, $\delta^{18}O_c$ values measured in deep-ocean foraminifera have increased, and long-term $\delta^{18}O_c$ values in planktonic foraminifera from the tropical Pacific Ocean have remained almost constant. What explains the difference between the two $\delta^{18}O_c$ trends?

Additional Resources

Basic Reading/Viewing

Companion Web site at www.whfreeman.com/ruddiman2e, pp. 9–11, 14–15, 17–18, 27–30, 32–33.

Cracking the Ice Ages. 1996. NOVA Video. Boston: WGBH.

Ruddiman, W. F., and J. E. Kutzbach. 1991. "Plateau Uplift and Climatic Change." *Scientific American* (March), 66–75.

Advanced Reading

Berner, R. A. 1999. "A New Look at the Long-Term Carbon Cycle." *GSA Today* 9: 1–6.

Kennett, J. P. 1977. "Cenozoic Evolution of Antarctic Glaciation, the Circum-Antarctic Ocean, and Their Impact on Global Paleoceanography." *Journal of Geophysical Research* 82: 3843–60.

Mikolajewicz, U., T. Maier-Reimer, T. J. Crowley, and K.-Y. Kim. 1993. "Effect of Drake and Panamanian Gateways on the Circulation of an Ocean Model." *Paleoceanography* 8:409–26.

Miller, K. G., R. A. Fairbanks, and G. S. Mountain. 1987. "Tertiary Oxygen Isotope Synthesis, Sea-Level History, and Continental Margin Erosion." *Paleoceanography* 2: 1–19.

Molnar, P., and P. England. 1990. "Late Cenozoic Uplift of Mountain Ranges and Global Climate Change: Chicken or Egg?" *Nature* 346: 29–34.

Raymo, M. E., and W. F. Ruddiman. 1992. "Tectonic Forcing of Late Cenozoic Climate." *Nature* 359: 117–22.

Ruddiman, W. F. (1997). *Tectonic Uplift and Climate Change*. New York: Plenum Press.

Wolfe, J. A. 1994. "Tectonic Climate Change at Middle Latitudes of Western North America." *Palaeogeography, Palaeoclimate, Palaeoecology* 108: 195–205.

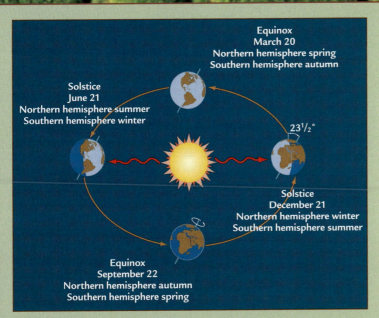

Orbital changes All aspects of Earth's present orbit have changed with time: the tilt of its axis, the shape of its path around the Sun, and the positions of the seasons on this path. These changes in orbit have driven climatic changes on Earth. (Adapted from F. K. Lutgens and E. J. Tarbuck, *The Atmosphere* [Englewood Cliffs, NJ: Prentice-Hall, 1992].)

In Part III we move from tectonic-scale climate changes to *orbital-scale* changes during the last several million years, a time when the continents and oceans were reaching their present positions. During this interval, changes in Earth's orbit have been the major driver of climate by altering the amount of solar radiation received by season and by latitude (Chapter 7). Three aspects of Earth's orbit have varied over cycles ranging in length from roughly 20,000 to 400,000 years: the tilt of its axis, the shape of its yearly path of revolution around the Sun, and the changing positions of the seasons along that path.

Orbital-scale changes have occurred throughout Earth's history, but our focus here is on the last 3 Myr because well-dated climate records are available from far more sites than in earlier times. The resulting increase in regional coverage provides greater insight into the operation of the climate system. Most climate

Orbital-Scale Climate Change

records containing orbital changes over this time span come from ocean sediments, which are dated by radiometric methods and by orbital "tuning." A wide array of records covering the last 650,000 years also comes from ice cores, which are dated by counting annual layers and by orbital tuning. These techniques make it possible to resolve time to within a few thousand years in both kinds of records. Because ocean sediments and ice cores are multichannel recorders that carry several kinds of climate signals side by side, scientists can also determine the relative timing of climatic responses in the oceans and on land, including the ice sheets.

The major orbital cycles have been detected in records of several important climatic responses on Earth: the strength of the low-latitude African and Asian monsoons (Chapter 8), the size of north polar ice sheets (Chapter 9), and the concentrations of important greenhouse gases (CO_2 and CH_4) through time

(Chapter 10). Because climate scientists now have in hand accurate records of both the orbital forcing and the internal climatic responses, the mechanisms of orbital-scale changes in ice sheets are gradually becoming clearer.

In Part III we address the following major questions:

- **How do orbital variations drive the strength of tropical monsoons?**

- **How do changes in Earth's orbit affect the size of northern hemisphere ice sheets?**

- **What controls orbital-scale fluctuations of atmospheric greenhouse gases?**

- **What is the origin of the 41,000-year ice-age cycle between 2.75 and 0.9 Myr and the variations at ~100,000 years during the last 0.9 Myr?**

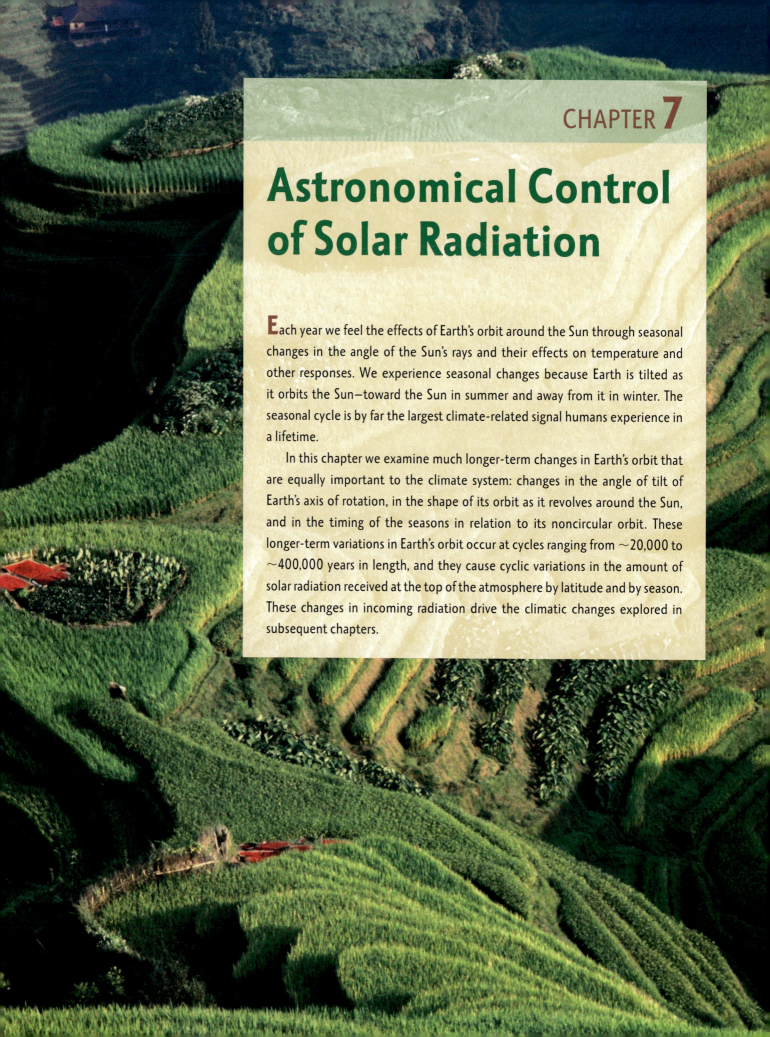

Astronomical Control of Solar Radiation

Each year we feel the effects of Earth's orbit around the Sun through seasonal changes in the angle of the Sun's rays and their effects on temperature and other responses. We experience seasonal changes because Earth is tilted as it orbits the Sun—toward the Sun in summer and away from it in winter. The seasonal cycle is by far the largest climate-related signal humans experience in a lifetime.

In this chapter we examine much longer-term changes in Earth's orbit that are equally important to the climate system: changes in the angle of tilt of Earth's axis of rotation, in the shape of its orbit as it revolves around the Sun, and in the timing of the seasons in relation to its noncircular orbit. These longer-term variations in Earth's orbit occur at cycles ranging from ~20,000 to ~400,000 years in length, and they cause cyclic variations in the amount of solar radiation received at the top of the atmosphere by latitude and by season. These changes in incoming radiation drive the climatic changes explored in subsequent chapters.

Earth's Orbit Today

The geometry of Earth's present solar orbit is the starting point for understanding past changes in Earth-Sun geometry. Much of our knowledge of Earth's orbit dates back to investigations in the seventeenth century by the astronomer Johannes Kepler. The larger frame of reference for understanding Earth's present orbit is the plane in which it moves around the Sun, the **plane of the ecliptic** (Figure 7-1).

7-1 Earth's Tilted Axis of Rotation and the Seasons

Two fundamental motions describe today's orbit. First, Earth spins on its axis once every day. One result is the daily "rising and setting" of the Sun, but of course that description is inaccurate. Days and nights are caused by Earth's rotational spin, which carries different regions of Earth's surface into and out of the Sun's direct radiation every 24 hours.

Earth rotates around an axis (or line) that passes through its poles (see Figure 7-1). This axis is tilted at an angle of 23.5°, called Earth's "obliquity," or **tilt.** This tilt angle can be visualized in either of two ways: (1) as the angle Earth's axis of rotation makes with a line perpendicular to the plane of the ecliptic or (2) as the angle that a plane passing through Earth's equator makes with the plane of the ecliptic.

The second basic motion in Earth's present orbit is its once-a-year revolution around the Sun. This motion results in seasonal shifts between long summer days, when the Sun rises high in the sky and delivers stronger radiation, and short winter days, when the Sun stays low in the sky and delivers weaker radiation. These seasonal differences culminate at the summer and winter **solstices,** which mark the longest and shortest days of the year (June 21 and December 21 in the northern hemisphere, the reverse in the southern hemisphere).

If we move outside our Earthbound perspective, we find that the cause of the seasons, the solstices, and the changes in length of day and angle of incoming solar radiation actually lies in the changing *position* of the tilted Earth with respect to the Sun. During each yearly revo-

lution around the Sun, Earth maintains a constant *angle* of tilt (23.5°) and a constant *direction* of this tilt in space. When the northern or southern hemisphere is tilted directly toward the Sun, it receives the more direct radiation of summer. When it tilts directly away from the Sun, it receives the less direct radiation of winter. But at both times and at all times of year it keeps the same 23.5° tilt.

If we switch back to our Earthbound perspective, we see the overhead Sun appearing to move back and forth through the year between the north tropic (Cancer) at 23.5°N and the south tropic (Capricorn) at 23.5°S. But again, this apparent movement is actually the result of Earth's revolution around the Sun with a constant 23.5° tilt. Earth's 23.5° tilt also defines the 66.5° latitude of the Arctic and Antarctic circles: 90° − 23.5° = 66.5°. Because of the 23.5° tilt away from the Sun in winter, no sunlight reaches latitudes poleward of 66.5° on the shortest winter day (winter solstice).

Midway between the extremes of the winter and summer solstices, during intermediate positions in Earth's revolution around the Sun, the lengths of night and day become equal in each hemisphere at the **equinoxes** (which means "equal nights"—that is, nights equal in length to days). Again, Earth's tilt angle remains at 23.5° during the equinoxes, and its direction of tilt in space stays the same. The only factor that changes is Earth's position in respect to the Sun. The two equinoxes and two solstices are handy reference points for describing distinctive features of its orbit.

7-2 Earth's Eccentric Orbit: Distance between Earth and Sun

Up to this point, everything that has been described would be true whether Earth's orbit was perfectly circular or not. But Earth's actual orbit (Figure 7-2) is not a perfect circle: it has a slightly eccentric or elliptical shape. The noncircular shape of Earth's orbit is the result of the gravitational pull of other planets on Earth as it moves through space.

Basic geometry shows that ellipses have two focal points rather than the single focus (center) of a circle. In Earth's case, the Sun lies at one of the two focal points in its elliptical orbit, as required by the physical laws of gravitation. The other focus is empty (see Figure 7-2).

Earth's distance from the Sun changes according to its position in this elliptical orbit. Not surprisingly, these changes in Earth-Sun distance affect the amount of solar radiation Earth receives, especially at two extreme positions in the orbit. The position in which Earth is closest to the Sun is called **perihelion** (the "close pass" position, from the Greek meaning "near the Sun"), while the position farthest from the Sun is called **aphelion** (the "distant pass" position, from the Greek meaning "away from the Sun"). On average, Earth lies 149.6 million

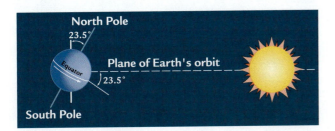

FIGURE 7-1 Earth's tilt Earth's rotational (spin) axis is currently tilted at an angle of 23.5° away from a line perpendicular to the plane of its orbit around the Sun.

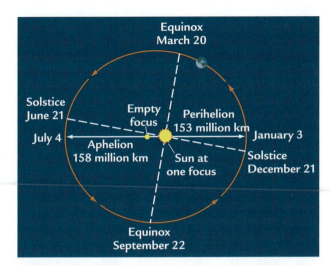

FIGURE 7-2 Earth's eccentric orbit Earth's orbit around the Sun is slightly elliptical. Earth is most distant from the sun at aphelion, on July 4, just after the June 21 solstice, and closest to the Sun at perihelion, on January 3, just after the December 21 solstice. (Modified from J. Imbrie and K. P. Imbrie, *Ice Ages: Solving the Mystery* [Short Hills, NJ: Enslow, 1979].)

kilometers from the Sun, but the distance ranges between 146 million kilometers at perihelion and 152 million at aphelion. This difference is equivalent to a total range of variation of slightly more than 3% around the mean value.

Earth is now in the perihelion position (closest to the Sun) on January 3, near the time of the December 21 winter solstice in the northern hemisphere and summer solstice in the southern hemisphere (see Figure 7-2). The fact that the close-pass position occurs in January causes winter radiation in the northern hemisphere and summer radiation in the southern hemisphere to be slightly stronger than they would be in a perfectly circular orbit.

Conversely, Earth lies farthest from the Sun on July 4, near the time of the June 21 summer solstice in the northern hemisphere and winter solstice in the southern hemisphere. The occurrence of this distant-pass position in July makes summer radiation in the northern hemisphere and winter radiation in the southern hemisphere slightly weaker than they would be in a circular orbit.

The effect of Earth's elliptical orbit on its seasons is small, enhancing or reducing the intensity of radiation received by just a few percent. Remember that the main cause of the seasons is the direction of tilt of Earth's axis in its orbit around the Sun (see Figure 7-1).

Another consequence of Earth's eccentric orbit is that the time intervals between the two equinoxes are not exactly equal: there are seven more days in the long part of the orbit, between the March 20 equinox and the

September 22 equinox, than in the short part of the orbit, between September 22 and March 20. The greater length of the interval from March 20 to September 22 tends to compensate for the fact that Earth is farther from the Sun on this part of the orbit and thus is receiving less solar radiation.

Long-Term Changes in Earth's Orbit

Astronomers have known for centuries that Earth's orbit around the Sun is not fixed over long intervals of time. Instead, it varies in a regular (cyclic) way because of the mass gravitational attractions among Earth, its moon, the Sun, and the other planets and their moons. These changing gravitational attractions cause cyclic variations in Earth's angle of tilt, its eccentricity of orbit, and the relative position of the solstices and equinoxes around its elliptical orbit (Box 7–1).

7-3 Changes in Earth's Axial Tilt through Time

If we assume for simplicity that Earth has a perfectly circular orbit around the Sun, we can examine two hypothetical cases that show the most extreme differences in tilt. For both cases, we look at the summer and winter solstices, the two seasonal extremes in Earth's orbit.

For the first case, Earth's axis is not tilted at all (Figure 7-3A). Incoming solar radiation is directed straight at the equator throughout the year, and it always passes by the poles at a 90° angle. Without any tilt, no seasonal changes occur in the amount of solar

A No tilt

B 90° tilt

FIGURE 7-3 Extremes of tilt (A) If Earth's orbit were circular and its axis had no tilt, solar radiation would not change through the year and there would be no seasons. (B) For a 90° tilt, the poles would alternate seasonally between conditions of day-long darkness and day-long direct overhead Sun. (Adapted from J. Imbrie and K. P. Imbrie, *Ice Ages: Solving the Mystery* [Short Hills, NJ: Enslow, 1979].)

BOX 7-1   TOOLS OF CLIMATE SCIENCE

Cycles and Modulation

Slow changes in Earth's orbit around the Sun occur in a cyclic or rhythmic way, as do the changes in amount of incoming solar radiation they produce. The science of wave physics provides the terminology needed to describe these changes. The length of a cycle is referred to as its **wavelength.** Expressed in units of time, the wavelength of a cycle is called the **period,** the time span between successive pairs of peaks or valleys.

The opposite (or inverse) of the period of a cycle is its **frequency,** the number of cycles (or in this case fractions of one cycle) that occur in one year. If a cycle has a period of 10,000 years, its frequency is 0.0001 cycle per year (one cycle every 10,000 years). In this book, we will refer to cycles in terms of their periods.

Another important aspect of cycles is their **amplitude,** a measure of the amount by which they vary around their long-term average. Low-amplitude cycles barely depart from the long-term mean trend; high-amplitude cycles fluctuate more widely.

Not all cycles are perfectly regular. Commonly the sizes of peaks and valleys oscillate irregularly around the long-term mean value through time. Behavior in which the amplitude of peaks and valleys changes in a repetitive or cyclic way is called **modulation,** a concept that lies behind the principle of AM (amplitude modulation) radio. Modulation creates an envelope that encompasses the changing amplitudes that occur at a specific cycle. Note that *modulation of a cycle is not in itself a cycle;* it simply adds amplitude variations to an actual cycle.

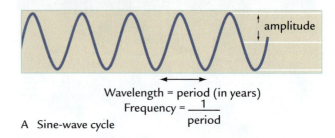

$$\text{Wavelength} = \text{period (in years)}$$
$$\text{Frequency} = \frac{1}{\text{period}}$$

A   Sine-wave cycle

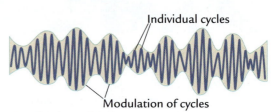

B   Amplitude modulation

Description of wave behavior (A) Perfectly cyclic behavior can be represented by a sine wave with a particular period and amplitude. (B) Cycles may show regular variations in amplitude, or modulation.

If variations in a particular signal are regular in *both* period and amplitude, it is appropriate to use the term "cycle." For the case of perfect cyclicity, this behavior is described as "sinusoidal" or **sine waves.** If the variations are irregular in period, the term "cyclical" is technically incorrect; "quasi-cyclical" or "quasi-periodic" is preferable. In the case of orbital-scale changes, we will informally use the term "cyclic" or "periodic" for climatic signals that are nearly regular but vary slightly in wavelength or amplitude.

---

radiation received at any latitude. As a result, solstices and equinoxes do not even exist because every day has the same length. A tilted axis is necessary for Earth to have seasons.

Next consider the opposite extreme with a maximum tilt of 90° (Figure 7-3B). Solar radiation is directed straight at the summer-season pole, while the winter-season pole lies in complete darkness. Six months later, the two poles have completely reversed position. The difference between these two extreme configurations shows that tilt is an important control on solar radiation at polar latitudes.

The angle of Earth's tilt has varied through time within a narrow range, between values as small as 22.2° and as large as almost 24.5° (Figure 7-4). The French astronomer Urbain Leverrier discovered these variations in the 1840s. Today Earth's tilt (23.5°) is near the middle

of this range, and the angle is currently decreasing. Cyclic changes in tilt angle occur mainly at a period of 41,000 years, the time interval that separates successive peaks or successive valleys (see Box 7–1). The cycles are fairly regular, both in period (wavelength) and in amplitude.

Changes in tilt amplify or suppress the strength of the seasons, especially at high latitudes (Figure 7-5). Larger tilt angles turn the summer hemisphere poles more directly toward the Sun and increase the amount of solar radiation received. The increase in tilt that turns the North Pole more directly toward the Sun at its summer solstice on June 21 also turns the South Pole more directly toward the Sun at its summer solstice six months later (December 21). On the other hand, the increased angle of tilt that turns each polar region more directly toward the Sun in summer also turns each winter season pole away from the Sun.

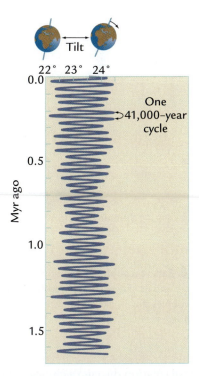

FIGURE 7-4 Long-term changes in tilt Changes in the tilt of Earth's axis have occurred at a regular 41,000-year cycle.

Decreases in tilt have the opposite effect: they diminish the amplitude of seasonal differences. Smaller tilt angles put the Earth slightly closer to the configuration shown in Figure 7-3A, which has no seasonal differences at all.

IN SUMMARY, changes in tilt mainly amplify or suppress the seasons, particularly at the poles.

7-4 Changes in Earth's Eccentric Orbit through Time

The shape of Earth's orbit around the Sun has also varied in the past, becoming at times more circular and at other times more elliptical (or "eccentric") than it is

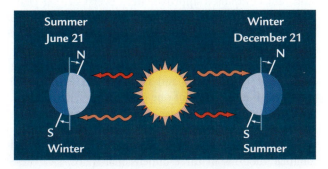

FIGURE 7-5 Effects of increased tilt on polar regions
Increased tilt brings more solar radiation to the two summer season poles and less radiation to the two winter season poles.

today. Leverrier discovered these variations in the 1840s. The shape of an ellipse can be described by reference to its two main axes: the "major" (or longer) axis and the "minor" (or shorter) axis (Figure 7-6). The degree of departure from a perfectly circular orbit can be described by

$$\epsilon = \frac{\sqrt{a^2 - b^2}}{a}$$

where ϵ is the **eccentricity** of the ellipse and a and b are half of the lengths of the major and minor axes (called the "semimajor" and "semiminor" axes).

The eccentricity of the elliptical orbit increases as these two axes become more unequal in length. At the extreme where the two axes become exactly equal ($a = b$), the eccentricity drops to zero because the orbit is circular ($a^2 - b^2 = 0$). Eccentricity (ϵ) has varied over time between values of 0.005 and 0.0607 (Figure 7-7). The present value (0.0167) lies toward the lower (more circular) end of the range.

Changes in orbital eccentricity are concentrated mainly at two periods. One eccentricity cycle shows up as variations at intervals near 100,000 years (see Figure 7-7). This cycle actually consists of four cycles of nearly equal strength and periods ranging between 95,000 and 131,000 years, but these cycles blend into a cycle near 100,000 years.

The second eccentricity cycle has a wavelength of 413,000 years. This longer cycle is not as obvious, but it shows up as alternations of the 100,000-year cycles between larger and smaller peak values. Larger amplitudes can be seen near 200,000, 600,000, 1,000,000, and 1,400,000 years ago (see Figure 7-7). A third eccentricity cycle also exists at a period of 2.1 Myr, but this cycle is much weaker in amplitude.

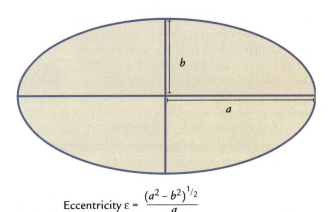

Eccentricity $\varepsilon = \dfrac{(a^2 - b^2)^{1/2}}{a}$

FIGURE 7-6 Eccentricity of an ellipse The eccentricity of an ellipse is related to half of the lengths of its longer (major) and shorter (minor) axes.

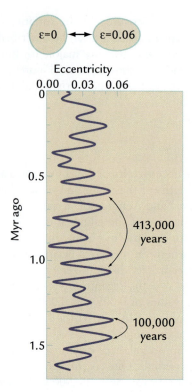

FIGURE 7-7 Long-term changes in eccentricity The eccentricity (ϵ) of Earth's orbit varies at periods of 100,000 and 413,000 years.

7-5 Precession of the Solstices and Equinoxes around Earth's Orbit

The positions of the solstices and equinoxes in relation to the eccentric orbit have not always been fixed at their present locations (see Figure 7-2). Instead, they have slowly shifted through time with respect to the eccentric orbit and the perihelion (close-pass) and aphelion (distant-pass) positions. Although Hipparchus in ancient Greece first noticed these changes, the French mathematician, scientist, and philosopher Jean Le Rond d'Alembert was the first to understand them in the eighteenth century.

The cause of these changes lies in a long-term wobbling similar to that of a top. Tops typically move with three superimposed motions (Figure 7-8). They spin very rapidly (rotate) around a tilted axis. They also revolve with a slower near-circular motion across the surface on which they spin, with many spins (rotations) for each complete revolution. Finally, tops also wobble, gradually leaning in different directions through time. This wobbling motion in not caused by changes in the *amount* by which the top leans (its angle of tilt), but rather by changes in the *direction* in which it leans.

Earth's wobbling motion, called **axial precession**, is caused by the gravitational pull of the Sun and Moon on the slight bulge in Earth's diameter at the equator. Axial precession can also be visualized as a slow turning of

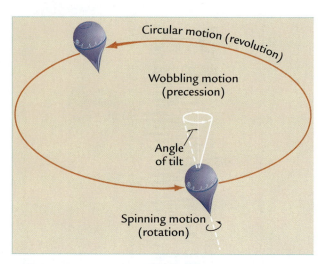

FIGURE 7-8 Earth's wobble In addition to its rapid (daily) rotational spin and its slower (yearly) revolution around the Sun, Earth wobbles slowly, like a top, with one full wobble every 25,700 years.

Earth's axis of rotation through a circular path, with one full turn every 25,700 years. Today Earth rotates around an axis that points to the North Star (Polaris), but over time the wobbling motion causes the axis of rotation to point to other celestial reference points (Figure 7-9). Earth wobbles very slowly; it revolves 25,700 times around the Sun and rotates almost 10 million times on its axis during the time it takes to complete just a single wobble.

A second kind of precessional motion is known as **precession of the ellipse.** In this case, the entire elliptically shaped orbit of the Earth rotates, with the long and short axes of the ellipse turning slowly in space (Figure 7-10). This motion is even slower than the wobbling motion of axial precession.

The combined effects of these two precessional motions (wobbling of the axis and turning of the ellipse) cause the solstices and equinoxes to move around Earth's orbit, with one full 360° orbit around the Sun completed approximately every 22,000 years (Figure 7-11). This combined movement, called the **precession of the equinoxes,** describes the absolute motion of the equinoxes and solstices in the larger reference frame of the universe. It consists of a strong cycle near 23,000 years and a weaker one near 19,000 years, with an average of one cycle every 21,700 years. For the rest of this book, we will concentrate mainly on the strong precession cycle near 23,000 years.

The precession of the equinoxes involves complicated angular motions in three-dimensional space, and these motions need to be reduced to a simple, easy-to-use mathematical form that can be plotted against time like the changes in tilt shown in Figure 7-4. To accomplish

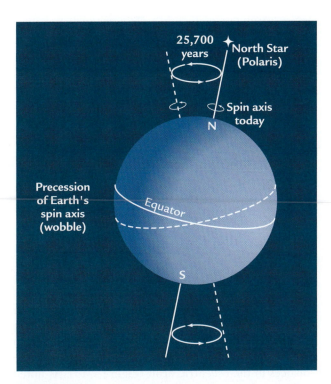

FIGURE 7-9 Precession of Earth's axis Earth's slow wobbling motion causes its rotational axis to point in different directions through time, sometimes (as today) toward the North Star, Polaris, but at other times toward other stars. (Adapted from J. Imbrie and K. P. Imbrie, *Ice Ages: Solving the Mystery* [Short Hills, NJ: Enslow, 1979].).

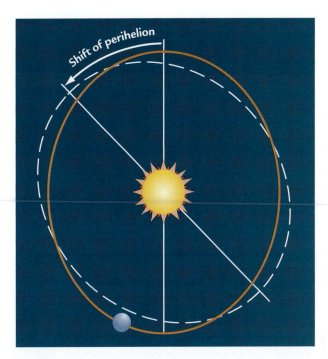

FIGURE 7-10 Precession of the ellipse The elliptical shape of Earth's orbit slowly precesses in space so that the major and minor axes of the ellipse slowly shift through time. (Adapted from N. Pisias and J. Imbrie, "Orbital Geometry, CO_2, and Pleistocene Climate," *Oceanus* 29 [1986–87]: 43–49.)

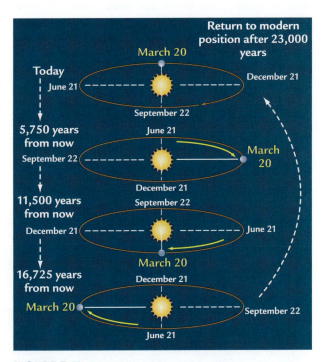

FIGURE 7-11 Precession of the equinoxes Earth's wobble and the slow turning of its elliptical orbit combine to produce the precession of the equinoxes. Both the solstices and equinoxes move slowly around the eccentric orbit in cycles of 23,000 years. (Adapted from J. Imbrie and K. P. Imbrie, *Ice Ages: Solving the Mystery* [Short Hills, NJ: Enslow, 1979].)

this goal, we make use of two basic geometric characteristics of precessional motion.

The first characteristic has to do with the angular form of Earth's motion with respect to the Sun. We define ω (omega) as the angle between two imaginary lines (Figure 7-12A): (1) a line connecting the Sun to Earth's position at perihelion (its closest pass to the Sun) and (2) a line connecting the Sun to Earth's position at the March 20 equinox. The first line is tied to the elliptical shape of Earth's orbit and the second to the varying positions of the seasons within the orbit. As a result, the slow change in the angle ω is a measure of Earth's wobbling motion—the very slow changes in the positions of the seasons with respect to the elliptical orbit.

The changing angle ω slowly sweeps out a 360° arc, starting at 0° (where the March 20 equinox coincides with the perihelion position), increasing to 90°, then to 180° (where the March 20 equinox occurs on the other side of the orbit, coincident with the aphelion position), later to 270°, and finally to 360°, at which point the cycle is complete and the angle returns to 0° (Figure 7-12B).

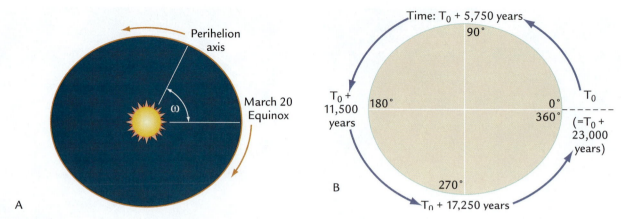

A

B

FIGURE 7-12 Precession and the angle ω (A) The angle between lines marking Earth's perihelion axis and the vernal equinox (March 20) is called ω. (B) The angle ω increases from 0° to 360° with each full 23,000-year cycle of precession.

This complicated angular motion can be represented in a simplified mathematical form by using basic geometry and trigonometry to convert the *angular* motions in Figure 7-12 to a *rectangular* coordinate system. Box 7–2 shows how the mathematical sine wave function projects the motion of a radius vector sweeping around a circle onto a vertical coordinate. This conversion allows the circular motion to be represented as an oscillating sine wave on a simple *x-y* plot. The amplitude of sinω moves from a value of +1 to –1 and back again over each 23,000-year precession cycle.

The second aspect of Earth's orbital motion that needs to be considered is its eccentricity. If Earth's orbit were perfectly circular, the slow movements of the solstices and equinoxes caused by precession would not alter the amount of sunlight received on Earth because the distance to the Sun would remain constant through time. Because the orbit is not circular, however, movements of the solstices and equinoxes (see Figure 7-11) cause long-term changes in the amount of solar radiation received on Earth.

These gradual movements of precession bring the solstices and equinoxes (and all other times of the year) into orbital positions that vary in distance from the Sun. Consider the two extreme positions of the solstices in the eccentric orbit (Figure 7-13). As noted earlier, in the present orbit, the position of the June 21 solstice (northern hemisphere summer and southern hemisphere winter) occurs very near aphelion, the most distant pass from the Sun (Figure 7-13 top). This greater Earth-Sun distance on June 21 slightly reduces the amount of solar radiation received during those seasons. Conversely, with the December 21 solstice (northern hemisphere winter and southern hemisphere summer) currently occurring near perihelion, the closest pass to the Sun, solar radiation is higher at those seasons than it would be in a perfectly circular orbit. Approximately 11,000 years ago, half of a precession cycle before now, this configuration was reversed (Figure 7-13 bottom). The June 21 solstice occurred at perihelion, and the December 21 solstice occurred at aphelion.

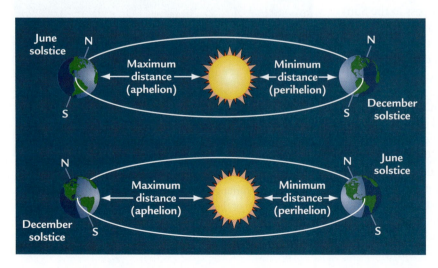

FIGURE 7-13 Extreme solstice positions Slow precessional changes in the attitude (direction) of Earth's spin axis produce changes in the distance between Earth and Sun as the summer and winter solstices move into the extreme (perihelion and aphelion) positions in Earth's eccentric orbit. (Modified from W. F. Ruddiman and A. McIntyre, "Oceanic Mechanisms for Amplification of the 23,000-Year Ice-Volume Cycle," *Science* 212 [1981]: 617–27.)

BOX 7-2   LOOKING DEEPER INTO CLIMATE SCIENCE

Earth's Precession as a Sine Wave

For a right-angle triangle (A), the sine of the angle ω is defined as the length of the opposite side over the length of the hypotenuse (the longest side). Consider a circle whose radius is a vector r that sweeps around in a 360° arc in an angular motion measured by the changing angle ω (B). Note that the circular motion described by the angle ω is analogous to actual changes in Earth-Sun geometry.

The angular motion of the radius vector r around the circle can be converted into changes in the dimensions of a triangle lying within the circle such that the sides of this triangle can be measured in a rectangular (horizontal and vertical) coordinate system (C). In this conversion, the hypotenuse of the triangle is also the radius vector r of the circle.

The sweeping motion of the radius vector r around the circle causes the shape of the internal triangle to change. The radius vector r always has a value of +1 because its length stays the same and its sign is defined within the angular coordinate system as a positive value.

But the length of the opposite side of the triangle (y) is defined within the rectangular coordinate system, and it can change both in amplitude and in sign (positive or negative). As the radius vector r sweeps around the circle, y increases and decreases along the vertical scale, cycling back and forth between values of +1 and –1. When r lies in the top half of the circle, y has values greater than 0. When it lies in the lower half, y is negative.

The angular motion of r can be converted to a linear mathematical form by plotting changes in sinω as the radius vector r sweeps out a full 360° circle, with the angle ω increasing from 0° to 90°, 180°, 270°, and back to 360° (= 0°). As before, sinω is defined as the ratio of the length of the opposite side y over the hypotenuse r.

The mathematical function sinω cycles smoothly from +1 to –1 and then back to +1 for each complete revolution of the radius vector r. At the starting point (ω = 0°), the length of the opposite side y is 0 and the radius is +1, so the value of sinω is 0/1, or 0. As the angle ω increases, the length of the opposite side of the triangle (y) increases in relation to the (constant) radius of the circle. When ω reaches 90°, sinω = +1 because the lengths of the opposite side and the radius (hypotenuse) are identical (1/1). At 180°, sinω has returned to 0 because the length of the opposite side (y) is again 0. For angles greater than 180°, the sinω values become negative because the opposite side of the triangle y now falls in negative rectangular coordinates (values below 0 on the vertical axis). Sinω values reach a minimum value of –1 at ω = 270° (–1/1). After that, sinω again begins to increase, returning to a value of 0 at ω = 360° (= 0°).

Converting angular motion to a sine wave   (A) The sine of an angle is the length of the opposite side of a triangle over its hypotenuse. (B) This concept can be applied to a circle where the hypotenuse is the radius (amplitude = 1) and the length of the opposite side of the triangle varies from +1 to –1 along a vertical coordinate axis. (C) As the radius vector sweeps out a full circle and ω increases from 0° to 360°, the sine of ω changes from +1 to –1 and back to +1, producing a sine wave representation of circular motion and of Earth's precessional motion.

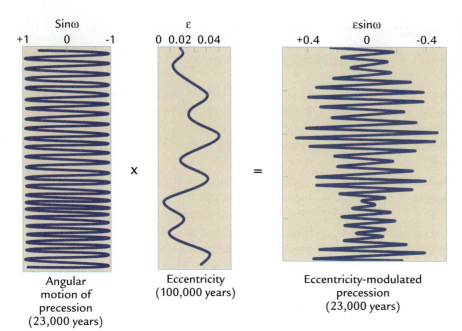

FIGURE 7-14 The precessional index The precessional index, $\varepsilon\sin\omega$, is the product of the sine wave function ($\sin\omega$) caused by precessional motion and the eccentricity (ϵ) of Earth's orbit.

The two solstice positions shown in Figure 7-13 are extreme points in a continuously changing orbit. Precession also moves the solstices through orbital positions with intermediate Earth-Sun distances like those shown in Figure 7-11. In the next 11,000 years, the solstices will move from their present positions back to those shown at the bottom of Figure 7-13.

Eccentricity plays an important role in the effect of precession on the amount of solar radiation received on Earth. The full expression for this impact is $\varepsilon\sin\omega$, the **precessional index** (Figure 7-14). The $\sin\omega$ part of this term is the sine wave representation of the movement of the equinoxes and solstices around the orbit (see Box 7-2). The eccentricity (ε) acts as a multiplier of the $\sin\omega$ term.

As noted earlier, the present value of ϵ is 0.0167. If this value remained constant through time, the $\varepsilon\sin\omega$ index would cycle smoothly between values of +0.0167 and −0.0167 over each precession cycle of ~23,000 years. As shown in Section 7-4, however, the eccentricity of Earth's orbit varies through time, ranging between 0.005 and 0.06 (see Figure 7-7). These changes in ϵ cause the $\varepsilon\sin\omega$ term to vary in amplitude (see Figure 7-14).

Long-term variations in the precessional index have two major characteristics (Figure 7-15). First, they occur at a cycle with a period near 23,000 years because of the regular angular motion of precession at that cycle (see Figure 7-14). Second, the individual cycles vary widely in amplitude because changes in eccentricity modulate the 23,000-year signal (see Box 7-1). At times the 23,000-year cycle swings back and forth between extreme maxima and minima; at other times the amplitude of the changes is small.

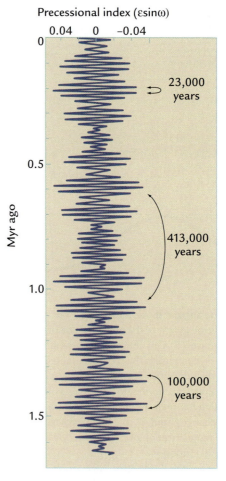

FIGURE 7-15 Long-term changes in precession The precessional index ($\varepsilon\sin\omega$) changes mainly at a cycle of 23,000 years. The amplitude of this cycle is modulated at the eccentricity periods of 100,000 and 413,000 years.

The changing values of $\epsilon\sin\omega$ affect the extreme perihelion and aphelion positions shown in Figure 7-13 by altering the distance between Earth and the Sun. With greater eccentricity, the differences in distance between a close pass and a distant pass are magnified. With a nearly circular orbit, differences in distance nearly vanish.

> **IN SUMMARY,** changes in eccentricity magnify or suppress contrasts in Earth-Sun distance around the orbit at the 23,000-year precession cycle. These changes in distance to the Sun in turn alter the amount of solar radiation received on Earth (more radiation at the perihelion close-pass position, less at the distant-pass aphelion position).

The modulation of the $\epsilon\sin\omega$ signal by eccentricity is not a real cycle (see Box 7–1), even though this statement probably goes against your intuition. You have learned that eccentricity varies at cycles of 100,000 and 413,000 years (see Figures 7–7 and 7–14), and you can see that the upper and lower envelopes of the $\epsilon\sin\omega$ signal vary at these periods (see Figure 7-15). But the offsetting effects of the upper and lower envelopes cancel each other out.

For example, when the 23,000-year cycle is varying between large minima and large maxima, these adjacent minima and maxima are approximately equal in size. Over the longer (100,000-year) wavelengths of the eccentricity variations, the amplitudes of the shorter-term (23,000-year) oscillations cancel each other out, leaving a negligible amount of net variation. Similarly, short-term variations between small-amplitude maxima and minima at other times also offset each other. The importance of this point will become obvious in Chapters 9 and 11.

> **IN SUMMARY,** the combined effects of eccentricity and precession cause the distance from the Earth to the Sun to vary by season, primarily at a cycle of 23,000 years. Times of high eccentricity produce the largest contrasts in Earth-Sun distance within the orbit, and conversely. As Earth precesses in its orbit, the changes in Earth-Sun distance are registered as seasonal changes in arriving radiation.

Changes in Insolation Received on Earth

Changes in Earth's orbit alter the amount of solar radiation received by latitude and by season. Climate scientists refer to the radiation arriving at the top of Earth's atmosphere as **insolation**. Some of this incoming insolation does not arrive at Earth's surface because clouds and other features in the climate system alter the amount that actually penetrates the atmosphere (companion Web site, pp. 2-4). Still, these calculations of insolation are the best guide to the effects of orbital changes on Earth's climate.

7-6 Insolation Changes by Month and Season

The long-term trends of tilt (see Figure 7-4) and $\epsilon\sin\omega$ (see Figure 7-15) contain all the information needed to calculate the amount of insolation arriving at any latitude and season. By convention, climate scientists usually show the amount of insolation (or the departures of insolation from a long-term average) during the solstice months of June and December in watts per square meter (W/m²). Some studies use an alternate form, calories per square centimeter per day.

June and December insolation values over the last 300,000 years show a strong dominance of the 23,000-year precession cycle at lower and middle latitudes and also at higher latitudes during the summer season (Figure 7-16). Just like the $\epsilon\sin\omega$ precessional index, individual insolation cycles at lower latitudes occur at wavelengths near 23,000 years, but their amplitudes are modulated at periods of 100,000 and 413,000 years. The June and December monthly insolation curves at each latitude in Figure 7-16 are also opposite in sign. Both can vary by as much as 12% (40 W/m²) around the long-term mean value for each latitude.

The 41,000-year cycle of tilt (obliquity) is not evident at lower latitudes but is visible in the low-amplitude variations of winter-season insolation at higher mid-latitudes (northern hemisphere January and southern hemisphere June at 60°). Summer season insolation changes at the tilt cycle are actually larger than those in winter, although this excess is not evident in these precession-dominated plots. One example is two precession cycles that are evident near 50,000 years ago in the June insolation signal for latitude 20°N but gradually blend and merge into a single tilt cycle at latitude 80°N (see Figure 7-16).

Changes in annual mean insolation at the 41,000-year tilt signal at high latitudes have the same sign as the summer insolation anomalies, but they are lower in amplitude. The lesser significance of winter season changes in tilt at full-polar latitudes results from the fact that no insolation at all arrives during long stretches of polar winter.

> **IN SUMMARY,** monthly seasonal insolation changes are dominated by precession at low and middle latitudes, with the effects of tilt evident only at higher latitudes.

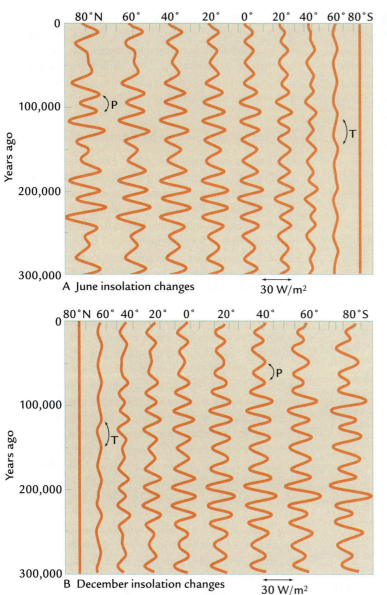

FIGURE 7-16 **June and December insolation variations** June and December monthly insolation values show the prevalence of precessional changes at low and middle latitudes and the presence of tilt changes at higher latitudes. Cycles of tilt and precession are indicated by T and P. The double arrows indicate variations of 30 W/m² for these signals.

A June insolation changes

B December insolation changes

As noted earlier, cycles of insolation change at 100,000 or 413,000 years are *not* evident in these signals because eccentricity is not a source of seasonal insolation changes. Actually, very small variations in received insolation do occur in connection with Earth's eccentric orbit around the Sun, but these appear only as changes in the total energy received by the entire Earth, not as seasonal variations. These changes are governed by the term $(1 - \epsilon^2)^{1/2}$. We have already seen that ϵ varies through time between 0.005 and 0.0607. Substituting these values for ϵ in the term above reveals that changes in total insolation received because of changes in eccentricity have varied by at most 0.002 (0.2%) around the long-term mean. Compared to changes in seasonal insolation of 10% or more at the tilt and precession

cycles, these annual eccentricity changes are negligible (smaller by a factor of about 50).

The pattern of insolation changes for tilt and precession can be compared by season and by hemisphere (northern versus southern). Insolation variations at high latitudes caused by changes in tilt are *in phase* between the hemispheres from a seasonal perspective: tilt maxima in the northern winter solstice of December match tilt maxima in the southern winter solstice of June. With increased tilt (Figure 7-17A), summer (June) insolation maxima in the northern hemisphere occur at the same time in the 41,000-year cycle as summer (December) insolation maxima in the southern hemisphere on the opposite side of the orbit. Higher tilt produces more insolation at both poles in their respective summers

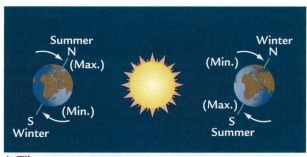

A Tilt

B Precession

FIGURE 7-17 Phasing of insolation maxima and minima
(A) Tilt causes in-phase changes for polar regions of both hemispheres in their respective summer and winter seasons. (B) Precession causes out-of-phase changes between hemispheres for their summer and winter seasons.

because both poles are turned more directly toward the Sun. For the same reason, more pronounced insolation minima also occur at both winter poles for a higher tilt: the two winter poles are tilted away from the Sun during the same orbit.

If we compare the North Pole with the South Pole *at a particular month* in the orbit, however, the two hemispheres are exactly out of phase (see Figure 7-17A). The increased tilt angle that turns north polar regions more directly toward the Sun in northern hemisphere summer also tilts the southern polar regions farther away from the Sun at that same place in the orbit (southern hemisphere winter). As a result, tilt causes opposite insolation effects at the North and South poles for a given point in the orbit.

For precession, the relative sense of phasing between seasons and hemispheres is exactly reversed from that of tilt (Figure 7-17B). Because Earth-Sun distance is the major control on these changes in insolation, a position close to the Sun (at perihelion) produces higher insolation than normal over Earth's *entire* surface. A precessional-cycle insolation maximum occurring at June 21 (or December 21) will be simultaneous everywhere on Earth. Distant-pass positions (at aphelion) will simultaneously diminish insolation everywhere on Earth.

An important fact to remember about precession is that the seasons are reversed across the equator. As a

result, an insolation maximum at June 21 is a *summer* insolation maximum in the northern hemisphere, but it is a *winter* insolation maximum in the southern hemisphere, where June 21 is the winter solstice. As a result of the seasonal reversal at the equator, insolation signals considered in terms of the season of the year are *out of phase* between the hemispheres for precession. This pattern is exactly opposite in sense to the in-phase pattern for tilt at high latitudes of both hemispheres.

Another way of looking at the relative phasing of precessional insolation is to track changes between seasons within a single hemisphere. The orbital position on the left in Figure 7-17B, which produces minimum summer (June 21) insolation in the northern hemisphere because it occurs at a distant position from the Sun (aphelion), must six months later cause maximum winter (December 21) insolation in the same hemisphere when Earth revolves around to the perihelion position (see Figure 7-17B right). As a result, precessional variations in insolation at any one location always move in opposite directions for the summer versus winter seasons.

Precessional changes in insolation have an additional characteristic not found in changes caused by tilt: an entire family of insolation curves exists for each season and month (and even day) of the year. As a matter of convention, insolation changes are typically shown only for the extreme solstice months of June and December, but in fact every season and month precesses into parts of the eccentric orbit that are alternately farther from the Sun and closer to the Sun at the same 23,000-year cycle.

As a result, each season and month experiences the same 23,000-year cycle of increasing and decreasing insolation values relative to the long-term mean, but the anomalies (departures from the mean) are offset in time from the preceding month or season. These offsets produce an entire family of monthly (and seasonal) insolation curves (Figure 7-18). Each successive month passes through perihelion (or aphelion) roughly 1916 years later than the previous month did ($1/12 \times 23{,}000 = 1916$).

7-7 Insolation Changes by Caloric Seasons

Calculations of monthly insolation are complicated by an additional factor related to the eccentricity of Earth's orbit. Earth gradually moves through a 360° arc in its orbit around the Sun, but this angular motion does not result in a constant rate of motion in space. Instead, Earth speeds up as it nears the extreme perihelion position and slows down near aphelion. As a result, as the solstices move slowly around the eccentric orbit, they gradually pass through regions of faster or slower movement in space.

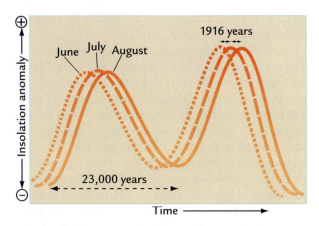

FIGURE 7-18 Family of monthly precession curves Because all seasons change position (precess) around Earth's orbit, each season (and month) has its own insolation trend through time. Monthly insolation curves are offset by slightly less than 2000 years (23,000 years divided by 12 months).

These changes in speed cause changes in the lengths of the months and seasons in relation to a year determined by "calendar time" (day of the year). The net effect is that changes in the amplitude of insolation variations in the monthly signals tend to be canceled by opposing changes in the lengths of the seasons. For example, times of unusually high summer insolation values at a perihelion position are also times of shorter summers. It is not obvious to scientists how to balance these two offsetting factors.

One way of minimizing these complications is to calculate the changes in insolation received on Earth within the framework of **caloric insolation seasons**. The summer caloric half-year is defined as the 182 days of the year when the incoming insolation exceeds the amount received during the other 182 days. Caloric seasons are not fixed in relation to the calendar because the insolation variations caused by orbital changes are added to or subtracted from different parts of the calendar year (see Figure 7-18). As a result, the caloric summer half-year falls during the part of the year we think of as summer, but it is not precisely centered on the June 21 summer solstice.

Changes in insolation viewed in reference to the half-year caloric seasons put a somewhat different emphasis on the relative importance of tilt and precession. Although low-latitude insolation anomalies are still dominated in both seasons by the 23,000-year precession signal, the 41,000-year tilt rhythm is much more obvious in high-latitude anomalies during the summer caloric half-year (Figure 7-19) than it is in the monthly insolation curves (see Figure 7-16). Another aspect of caloric season calculations is that the insolation values

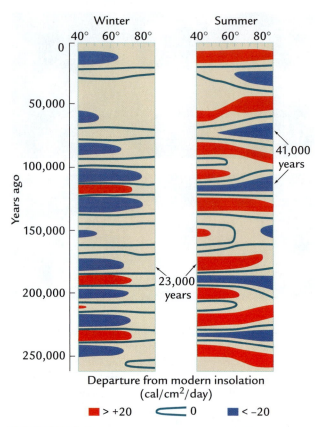

FIGURE 7-19 Caloric season insolation anomalies Plots of insolation anomalies for the summer and winter caloric half-year show a larger influence of tilt in relation to precession at higher latitudes than do the monthly anomalies. (Adapted from W. F. Ruddiman and A. McIntyre, "Oceanic Mechanisms for Amplification of the 23,000-Year Ice-Volume Cycle," *Science* 212 [1981]: 617–27.)

vary by a maximum of only ~5% around the mean, compared to variations as large as 12% for the monthly insolation changes.

Searching for Orbital-Scale Changes in Climatic Records

In the next four chapters we will explore abundant evidence that orbital-scale cycles are recorded in Earth's climate records. Many records contain two or even three superimposed orbital-scale cycles, and it can often be difficult to disentangle them visually.

For example, consider the three cycles shown in Figure 7-20A, with periods of 100,000 years, 41,000 years, and 23,000 years. These three cycles are equivalent to the three most prominent cycles of orbital change, but for simplicity they are shown as perfect sine waves rather than the more complex forms of the actual variations.

We can combine these three cycles by adding them together in various ways (Figure 7-20B). When the 23,000-year and 100,000-year cycles are combined, the resulting signal is obviously a simple addition of the two separate cycles. The two cycles are easy to distinguish because they differ in period by a factor of more than 4 (100,000 divided by 23,000 = 4.3).

It becomes more difficult to detect the two original signals when only the 23,000-year and 41,000-year cycles are combined. Because the periods of these two cycles are more similar, they reinforce and cancel each other in somewhat complicated ways. The task becomes even more difficult when all three cycles are combined, as in the bottom curve of Figure 7-20B. It is not at all obvious to the eye that this signal is a simple addition of three perfect sine waves.

In the case of Earth's actual climate records, the situation is even more complex because the three cycles are not only superimposed on each other but also

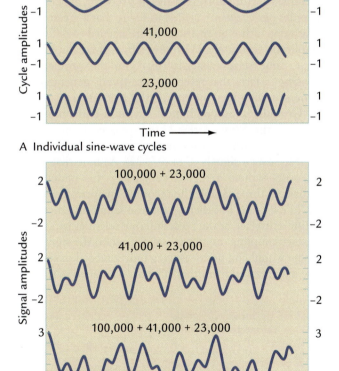

A Individual sine-wave cycles

B Combination of cycles

FIGURE 7-20 Complications from overlapping cycles If perfect sine wave cycles with periods of 100,000, 41,000, and 23,000 years are added together so that they are superimposed on top of one another, the original cycles are almost impossible to detect by eye in the combined signal.

change in amplitude through time (see Figures 7-4 and 7-15). Obviously, it will be impossible to disentangle all this information simply by eye.

7-8 Time Series Analysis

To simplify analyses of cyclic variations in climate changes, scientists use **time series analysis.** The term "time series" refers to records plotted against age (time). These techniques extract rhythmic cycles embedded within records of climate.

The first step in time series analysis is to convert climatic records to a time framework. After individual measurements of a climatic indicator have been made (for example, across an interval in a sediment core), all available sources of dating are used to define the ages of particular levels within the sequence. A complete time scale for the sequence is then created by interpolating the ages of all sediment depths between the dated levels. This time scale can then be used to plot the climatic record against time for further analysis.

One technique is **spectral analysis.** Modern techniques of spectral analysis are beyond the scope of this book, but we need at least a basic sense of how this technique detects cycles in records of past climate change. One way to visualize what happens in spectral analysis is to imagine taking a climate record plotted on a time axis and gradually sliding a series of sine waves of different periods across it. As this is done, the correlation between each sine wave and the full climatic signal is measured for each point in the sliding process. If the climate record that is being examined contains a strong cycle at one of the sine wave periods, the climate record will show a strong correlation with that sine wave at some point in the sliding process. The strong correlation indicates that the climatic signal contains a strong cycle at that period. As this process is repeated for different sine waves with different periods, other cycles may emerge.

Now we return to the example of the three superimposed cycles in the bottom part of Figure 7-20B. A spectral analysis run on this signal will extract the three component (orbital) cycles, which can be displayed on a plot called a **power spectrum** (Figure 7-21). The horizontal axis shows a range of periods plotted on a log scale, with the shorter periods to the right. The vertical axis represents the amplitude of the cycles (see Box 7-1), also known as their "power." The height of the lines plotted on the power spectrum is related to the square of the amplitude of the cycle at that period.

For the example shown in Figure 7-21, all three cycles detected by spectral analysis plot as narrow "line spectra," with their power concentrated entirely at the periods shown by the solid vertical lines. In this idealized example, no power occurs anywhere else in the spectrum than at these three cycles.

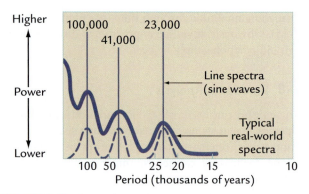

FIGURE 7-21 Spectral analysis Spectral analysis reveals the presence of cycles within complex climate signals. In this example, the original sine wave cycles from Figure 7–20A form line spectra (vertical bars) whose heights indicate their amplitudes. Actual climate records have peaks that are spread over a broader range of periods (dashed line and curving solid line).

In actual studies of climate, however, power spectra are never this simple. One reason is that even the most regular-looking orbital cycles such as the tilt changes in Figure 7-4 are not perfect sine waves but instead vary over a small range of periods. In addition, errors in dating records of climate change or in measuring their amplitude also have the effect of spreading power over a broader range of periods than would be the case for perfectly measured and dated signals. As a result of these complications, the total amount of power associated with each cycle looks like the area under the dashed curves in Figure 7-21.

Still another reason that real-world spectra are more complicated is that random noise exists in the climate system, consisting of irregular climatic responses not concentrated at orbital or other cycles. In most records, the effect of noise is spread out over a range of periods in the spectrum. In general, the amount of power tends to be larger at longer periods. As a result, spectra from real-world climatic signals tend to look like the thick curved line in Figure 7-21. The spectral peaks that rise farthest above the baseline of the trend are the most significant (believable) ones in a statistical sense.

A second useful time series analysis technique is called **filtering**. This technique extracts individual cycles at a specific period (or narrow range of periods) from the complexity of the total signal. This process is often referred to as "band-pass filtering" (filtering of a narrow band or range of the many periods present in a given signal). Filtering is analogous to using glasses with colored lenses to filter out all colors of the light spectrum except the one color (wavelength) we wish to see.

Filters are constructed directly from well-defined peaks in power spectra like those in Figure 7-21. The highest point on the spectral peak defines the central period of the filter, and the sloping sides of the spectral peak define the shape of the rest of the filter.

To understand the importance of filtering, consider again the three hypothetical sine waves in Figure 7-20. We can create filters for these three cycles based on the peaks in the power spectrum shown in Figure 7-21. If we pass these filters across the combined signal at the bottom of Figure 7-20B, the filters will extract the original form of all three individual cycles (at 23,000, 41,000, and 100,000 years). In effect, the filtering operation extracts the time-varying shapes of individual cycles embedded in the complexities of actual climate records.

7-9 Effects of Undersampling Climate Records

The technique of spectral analysis can be used only for a specific range of cycles within any climate record. Confident identification of a cycle by time series analysis requires that the cycle be repeated at least four times in the original record (the record must be at least four times longer than the cycle analyzed). At the other extreme (for the shortest cycles in a record), at least two samples per cycle are needed to verify that a given cycle is present, although many more are needed to define its amplitude accurately. With fewer than two samples per cycle, time series analysis runs into the problem of **aliasing**, a term that refers to false trends generated by undersampling the true complexity in a signal.

Consider the hypothetical case of a climatic signal that has the form of the 23,000-year cycle of orbital precession, with the wide range of amplitude variation typical of such signals (Figure 7-22). Assume that three scientists sample a record containing this underlying signal. All three sample the record at an average spacing of 23,000 years, but each begins the sampling process at a different place in the record. If one scientist happened to start sampling exactly at a maximum in the signal, he or she would end up measuring only a record of successive maxima, but if another scientist happened to start at

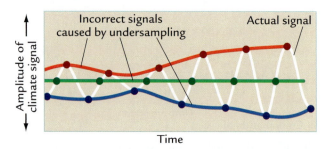

FIGURE 7-22 Aliasing (undersampling) of climate signals Undersampling of a climate signal (in this case one that is a direct response to changes in orbital precession) can produce aliased climate signals completely unlike the actual one.

a minimum, the record would show only successive minima. These sampling attempts give completely different results because they are persistently biased toward different sides of a highly modulated cycle. If the third scientist happened to start sampling exactly at a crossover point between minima and maxima, the scientist might extract a record suggesting that no signal exists at all.

These differences show the danger of aliasing. Although this example is obviously chosen to show the worst possible effect of aliasing, undersampling is a problem in most climate records.

7-10 Tectonic-Scale Changes in Earth's Orbit

Over time scales of hundreds of millions of years, some of Earth's orbital characteristics slowly evolved, as shown by evidence in ancient corals. Corals are made of banded $CaCO_3$ layers caused by changes in environmental conditions. The primary annual banding reflects seasonal changes in sunlight and water temperature (Chapter 2). A secondary banding follows the tidal cycles created by the Moon and Sun. The tidal cycles also affect water depth and other factors in the reef environment that influence coral growth.

Corals from 440 Myr ago show 11% more tidal cycles per year than modern corals do, implying that Earth spun on its rotational axis 11% more times per year than at present. As a result, each year had 11% more days. Gradually over the last 440 Myr, the spin rate and number of days decreased to their current levels. This gradual slowing in Earth's rate of rotation was caused by the frictional effect of the tides.

Other changes in Earth's orbit that can be inferred from this kind of information, such as changes in Earth-Moon distance, are thought to have affected the wavelengths of tilt and precession over tectonic-scale intervals. One estimate of the slow, long-term increases in the periods of tilt and precession toward their present values is shown in Figure 7-23.

Key Terms

plane of the ecliptic (p. 120)	precession of the ellipse (p. 124)
tilt (p. 120)	precession of the equinoxes (p. 124)
solstices (p. 120)	precessional index (p. 128)
equinoxes (p. 120)	
perihelion (p. 120)	insolation (p. 129)
aphelion (p. 120)	caloric insolation seasons (p. 132)
wavelength (p. 122)	
period (p. 122)	time series analysis (p. 133)
frequency (p. 122)	
amplitude (p. 122)	spectral analysis (p. 133)
modulation (p. 122)	power spectrum (p. 133)
sine waves (p. 122)	filtering (p. 134)
eccentricity (p. 123)	aliasing (p. 134)
axial precession (p. 124)	

Review Questions

1. Why does Earth have seasons?

2. When is Earth closest to the Sun in its present orbit? How does this "near-pass" position affect the amount of radiation received on Earth?

3. Describe in your own words the concept of modulation of a cycle.

4. Earth's tilt is slowly decreasing today. As it does so, are the polar regions receiving more or less solar radiation in summer? In winter?

5. How is axial precession different from precession of the ellipse?

6. How does eccentricity combine with precession to control a key aspect of the amount of insolation Earth receives?

7. Do insolation changes during summer and winter have the same or opposite timing (sign) at any single location on Earth? Why or why not?

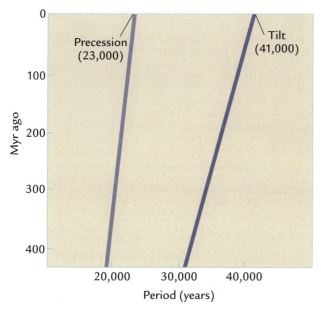

FIGURE 7-23 Tectonic-scale orbital changes Gradual changes in Earth's orbit over long tectonic time scales have caused a slow increase in the periods of the tilt and precession cycles. (Adapted from A. Berger et al., "Pre-Quaternary Milankovitch Frequencies," *Nature* 342 [1989]: 133–34.)

8. Do the following changes occur at the same time (same year) in Earth's orbital cycles?

   a. summer insolation maxima at both poles caused by changes in tilt

   b. summer insolation maxima in the tropics of both hemispheres caused by precession

Additional Resources

Basic Reading

Companion Web site at www.whfreeman.com /ruddiman2e, pp. 2–4.

Imbrie, J., and K. P. Imbrie, 1979. *Ice Ages: Solving the Mystery*. Short Hills, NJ: Enslow.

Ruddiman, W. F. 2005. *Plows, Plagues and Petroleum*, Chapter 3. Princeton, NJ: Princeton University Press.

www.classzone.com/book/earth_science/terc /content/visualization, Chapter 15: Earth's Orbital Motions.

Advanced Reading

Berger, A. L. 1978. "Long-Term Variations of Caloric Insolation Resulting from the Earth's Orbital Elements." *Quaternary Research* 9: 139–167.

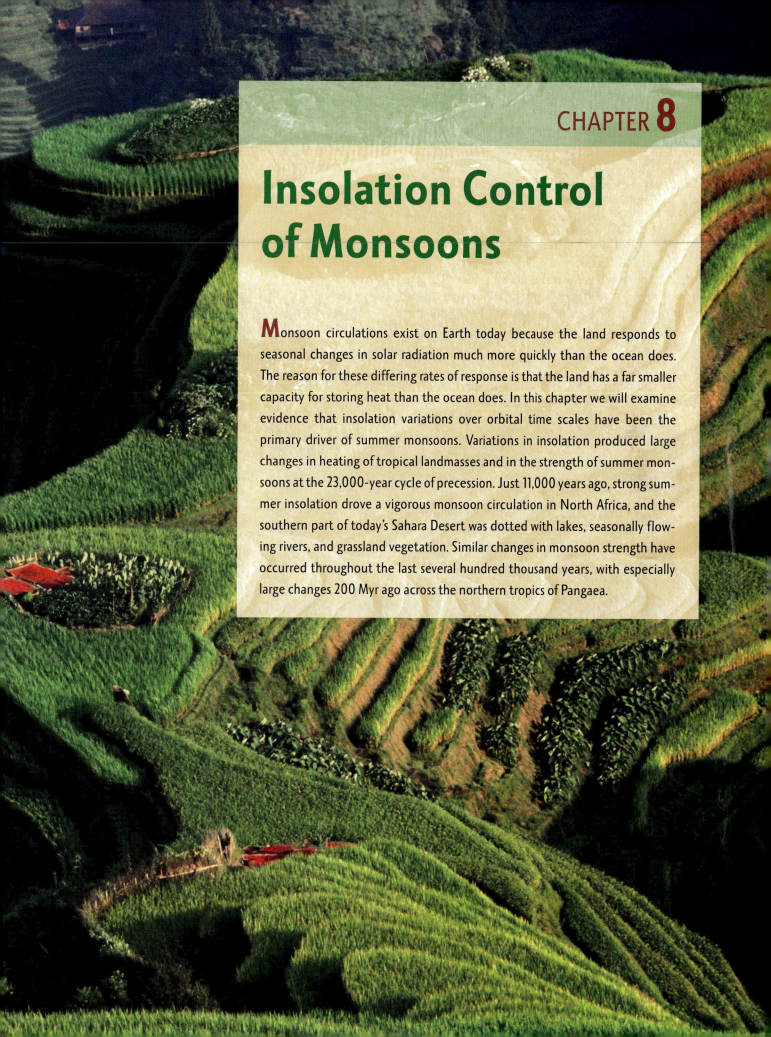

CHAPTER 8

Insolation Control of Monsoons

Monsoon circulations exist on Earth today because the land responds to seasonal changes in solar radiation much more quickly than the ocean does. The reason for these differing rates of response is that the land has a far smaller capacity for storing heat than the ocean does. In this chapter we will examine evidence that insolation variations over orbital time scales have been the primary driver of summer monsoons. Variations in insolation produced large changes in heating of tropical landmasses and in the strength of summer monsoons at the 23,000-year cycle of precession. Just 11,000 years ago, strong summer insolation drove a vigorous monsoon circulation in North Africa, and the southern part of today's Sahara Desert was dotted with lakes, seasonally flowing rivers, and grassland vegetation. Similar changes in monsoon strength have occurred throughout the last several hundred thousand years, with especially large changes 200 Myr ago across the northern tropics of Pangaea.

Monsoon Circulations

In summer, strong solar radiation causes rapid warming of the land but slower and much less intense warming of the ocean. Rapid heating over the continents causes air to warm, expand, and rise, and the upward movement of air creates an area of low pressure at the surface (companion Web site, pp. 9, 11, 15–18, 21). The flow of air toward this low-pressure region brings in air from the ocean, which also warms and rises (Figure 8-1A). This inflow carries water vapor evaporated from nearby oceans and contributes to monsoonal rainfall. Regions of the continents far from the ocean or protected by intervening mountain ranges may lie beyond the reach of the oceanic moisture and bake under the strong summer radiation.

During winter, when solar radiation is weaker, air over the land cools off rapidly, becomes denser than the air over the still-warm ocean, and sinks from higher levels in the atmosphere. This downward movement creates a region of high pressure over the land in contrast to the lower pressure over the still-warm oceans. The overall atmospheric flow in winter is a down-and-out movement of cold, dry air from the land to the sea (Figure 8-1B).

Most of the strong summer monsoons occur in the northern hemisphere, where landmasses are large (Asia and North Africa) and elevations are high (especially the Tibetan-Himalayan region of Southeast Asia). Monsoons are weaker in the southern hemisphere, where landmasses at tropical and subtropical latitudes are smaller and high topography is more limited in extent.

Here we focus initially on past variations of the North African monsoon for two reasons. First, North Africa lies far from the high-latitude ice sheets that might complicate the direct response of land surfaces to solar heating. In addition, nearby oceans yield a rich variety of climate records showing monsoon-related signals on North Africa.

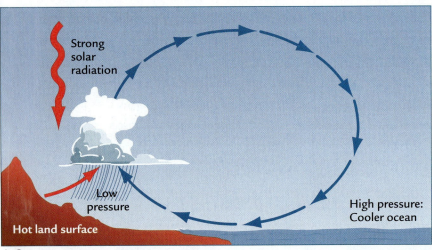

A Summer monsoon

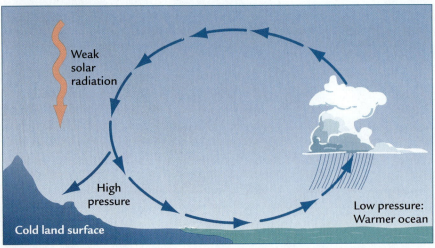

B Winter monsoon

FIGURE 8-1 Seasonal monsoon circulations Seasonal changes in the strength of solar radiation affect the surface of the land more than the ocean. (A) In summer, intense solar heating of the land causes an in-and-up circulation of moist air from the ocean. (B) In winter, weak solar radiation allows the land to cool off and creates a down-and-out circulation of cold dry air.

Africa is a deceptively large landmass compared to its appearance on Mercator maps. It stretches from 37°N to 35°S, with far more of its land area north of the equator—in fact almost twice the area of the U.S. mainland. Because the huge North African land surface is situated at tropical and lower subtropical latitudes, it is strongly influenced by the overhead Sun.

As a result of strong solar heating during northern hemisphere summer, a low-pressure region that develops over west-central North Africa draws moisture-bearing winds in from the tropical Atlantic (Figure 8-2A). During typical summers, this monsoonal rainfall penetrates northward to about 17°N latitude (the southern edge of the Sahara Desert) before retreating southward later in the year.

During northern hemisphere winter the overhead Sun moves to the southern hemisphere, and solar radiation over North Africa is weaker. Cooling of the North African land surface by back radiation causes sinking of air from above, and a high-pressure cell develops at the surface over the northwestern Sahara Desert (Figure 8-2B). Strong and persistent trade winds associated with this high-pressure cell and with similar circulation over the adjacent North Atlantic blow southwestward from North Africa across the tropical Atlantic.

Because the trade winds of the winter monsoon carry little moisture, winter precipitation is rare in North Africa. Only two areas receive much rain in this season: the northernmost Mediterranean margin, where storms occasionally form over the nearby ocean, and the tropical southwest coast (the Ivory Coast), where the moist intertropical convergence zone (ITCZ) remains over the land.

Because most of the rainfall in North Africa occurs in association with the summer monsoon, the distribution of major vegetation types reflects the monsoonal delivery of precipitation from the south (Figure 8-3). Rain forest in the year-round wet climate near the equator gives way northward to a sequence of progressively drier vegetation, first the tree-and-grass savannas of the Sahel region and then the desert scrub of the arid Sahara.

8-1 Orbital-Scale Control of Summer Monsoons

The idea that changing insolation could control the strength of monsoons over orbital time scales was proposed by the meteorologist John Kutzbach in the early 1980s, although it had been anticipated to some extent by Rudolf Spitaler late in the nineteenth century. This concept is called the **orbital monsoon hypothesis.**

The orbital monsoon hypothesis is a direct logical extension of factors at work in the present monsoon circulations (Figure 8-4). Because seasonal monsoon circulations are driven by changes in the strength of solar radiation, orbital-scale changes in summer and winter insolation (Chapter 7) should have produced a similar response. If summer insolation was higher in the past than it is today, the summer monsoon circulation should have been stronger, with greater heating of the land, stronger rising motion, more inflow of moist ocean air, and more rainfall (Figure 8-4B). Conversely, summer insolation levels lower than those today should have driven a weaker summer monsoon in the past.

The same kind of reasoning applies to the winter monsoon. Winter insolation minima weaker than the one today should have enhanced the cooling of the land surface, which should have driven a stronger down-and-out flow of dry air from land to sea (Figure 8-4C).

Recall from Chapter 7 that more intense summer insolation maxima and deeper winter insolation minima *always* occur together at any one location. As a result, stronger in-and-up monsoon flows in summer should

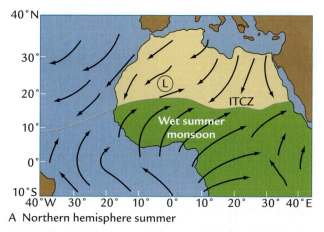

A Northern hemisphere summer

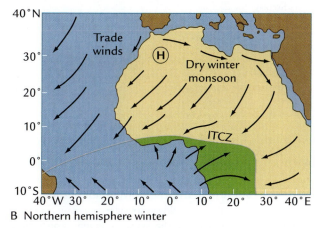

B Northern hemisphere winter

FIGURE 8-2 Monsoon circulations over North Africa Seasonal changes cause (A) a moist inflow of monsoonal air toward a low-pressure center over North Africa in summer and (B) a dry monsoonal outflow from a high-pressure center over the land in winter. (Adapted from J. F. Griffiths, *Climates of Africa* [Amsterdam: Elsevier, 1972].)

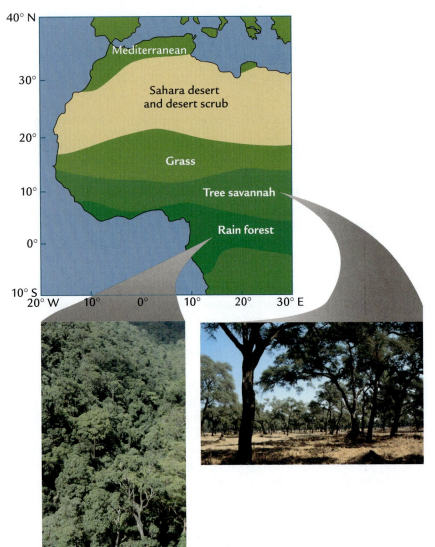

FIGURE 8-3 Vegetation in North Africa Vegetation across the northern part of Africa ranges from rain forest near the equator to savanna and grassland in the Sahel to desert scrub vegetation in the Sahara. This pattern reflects the diminishing northward reach of summer monsoon moisture from the tropical Atlantic. (Adapted from J. F. Griffiths, *Climates of Africa* [Amsterdam: Elsevier, 1972]. Inset photos courtesy of Tom Smith, University of Virginia.)

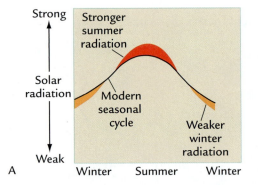

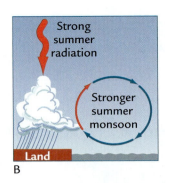

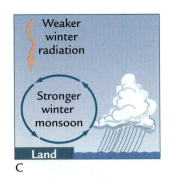

FIGURE 8-4 The orbital monsoon hypothesis (A) Departures from the modern seasonal cycle of solar radiation have driven stronger monsoon circulations in the past. (B) Greater summer radiation intensified the wet summer monsoon, while (C) decreased winter insolation intensified the dry winter monsoon. (Adapted from J. E. Kutzbach and T. Webb III, "Late Quaternary Climatic and Vegetational Change in Eastern North America: Concepts, Models, and Data," in *Quaternary Landscapes,* ed. L. C. K. Shane and E. J. Cushing [Minneapolis: University of Minnesota Press, 1991].)

occur at the same times in the past as stronger down-and-out monsoon flows in winter.

At first it might seem that the climatic effects of these opposed insolation trends in the two seasons might cancel each other, but this is not the case for the annual precipitation produced by the monsoons. Monsoonal winters are always dry, regardless of the amount of insolation, because the air descending from higher in the atmosphere holds very little moisture (see Figure 8-4C). As a result, orbital-scale changes in winter insolation have essentially no effect on annual rainfall. In contrast, because summer monsoon winds coming in from the ocean carry abundant moisture, orbital-scale changes in summer insolation have a large affect and in fact the dominant impact on annual rainfall.

This imbalance is an example of a **nonlinear response** of the climate system to insolation: the amount of rainfall is highly sensitive to insolation change in one season (summer) but not sensitive to changes in the other (winter). As a result, the system has a strong annual response even though the equal-and-opposite insolation trends in the two seasons might have been expected to cancel each other.

Orbital-Scale Changes in North African Summer Monsoons

The orbital monsoon hypothesis can be tested against a wide range of evidence, including changes in lake levels across arid regions like North Africa. At low and middle latitudes, changes in the amount of incoming solar insolation follow the 23,000-year rhythm of orbital precession (Chapter 7). A June insolation curve from latitude 30°N covering the last 140,000 years clearly shows this 23,000-year tempo (Figure 8-5). Note that today's June insolation level is well below the longer-term average. The orbital monsoon hypothesis predicts that stronger summer monsoons should have occurred at those times in the past when summer insolation values were significantly larger than the modern value.

The most recent instance when summer insolation values were substantially higher than today's occurred near 11,000 years ago. Evidence we will examine in Part IV of this book shows that lakes across tropical and subtropical North Africa were at much higher levels during this insolation maximum and for some time afterward. In fact, many lakes that were filled to high levels at that time are completely dry today. This evidence indicates that a strengthened summer monsoon circulation reached much farther northward into North Africa 11,000 years ago than it does today, in agreement with the orbital monsoon hypothesis.

But this interval was only one brief period in a long history of rainfall changes in North Africa. What about the much longer-term behavior of the summer monsoon?

We can use the June 30°N insolation curve in Figure 8-5 to construct a simple conceptual model that predicts how the summer monsoon should have varied with time if the orbital monsoon hypothesis is valid. The predicted monsoon response is based on three assumptions.

First, we assume that a **threshold level** of insolation exists, below which the monsoon response will be so weak that it will leave little or no evidence in the geologic record (see Figure 8-5). A good example is the level of lakes in arid regions. The orbital monsoon hypothesis predicts that the higher the level of summer insolation, the stronger the summer monsoon rainfall and the higher the lake levels. But if the insolation value falls below a certain level, the weakened summer monsoon may produce so little rain that the lakes dry up completely. A dried-out lake can no longer register changes toward even drier climates because it has reached the threshold limit of its ability to record climate.

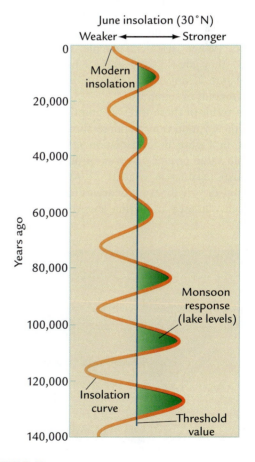

FIGURE 8-5 Conceptual model of monsoon response to summer insolation Increases in summer insolation heating above a critical threshold value drive a strong monsoon response at the 23,000-year tempo of orbital precession. The amplitude of this strong monsoon response is related to the size of the increase in summer insolation forcing.

This assumption has a good basis in fact. Many lakes in North Africa that existed 11,000 years ago during the strong monsoon interval are dry today, even though a weak summer monsoon still occurs at present in response to today's low levels of summer radiation. Apparently it takes a threshold insolation value well above the present level to bring most North African lakes into existence.

Second, we assume that the strength of the monsoon response (such as the water level of the North African lakes) is directly proportional to the amount by which summer insolation exceeds the threshold value. This assumption has a reasonable physical basis: stronger insolation should drive stronger monsoons and fill lakes to higher levels.

Third, we assume that the strength of the monsoon in the past as recorded in lake level records is a composite of the average monsoon strength over many individual summers. Actually, this is more fact than assumption: the wet monsoon circulations that develop every summer inevitably cease during the following winter. The lakes fill because of the integrated effect of many wet summers. When scientists sample records of these changes, they are looking at year-by-year responses blended into a longer-term average over hundreds or even thousands of summers.

As a result of these three assumptions, we arrive at the predicted monsoon response shown by the green shading in Figure 8-5. Insolation maxima above the threshold value produce a series of pulselike monsoon (and lake) maxima at regular 23,000-year intervals. These pulses vary in strength according to the amount by which summer insolation exceeds the threshold

value. Insolation levels below the threshold value leave no monsoon evidence in the geologic record.

The strongest predicted monsoon peaks in Figure 8-5 occur between 85,000 and 130,000 years ago, when the summer insolation curve reached its largest maxima because of modulation of the precession signal by orbital eccentricity (Chapter 7). In contrast, the weaker insolation maxima near 35,000 and 60,000 years ago should have produced less powerful monsoons. All long-term summer insolation minima (such as the one we are in today) fall below the critical threshold.

We can examine actual climate records for evidence of this predicted monsoon response. Because most of North Africa is arid and because erosion of sediment is much more prevalent than deposition, its climate history is sparse and difficult to date. Fortunately, the nearby seas and oceans contain continuous and well-dated records.

8-2 "Stinky Muds" in the Mediterranean

The water that fills the Mediterranean Sea today has a relatively high oxygen content. Near-surface waters are well oxygenated because they exchange oxygen-rich air with the atmosphere and because photosynthesis by marine organisms produces O_2 (companion Web site, p. 32). The high oxygen content of deep waters results from sinking of oxygen-rich surface water during winter (Figure 8-6A).

This sinking motion results from two factors (companion Web site, pp. 24–25): (1) the high salt content of the Mediterranean Sea, caused by the excess of summer evaporation over precipitation, and (2) winter chilling

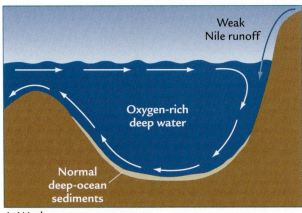

A Weak summer monsoon

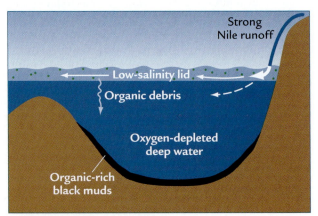

B Strong summer monsoon

FIGURE 8-6 Mediterranean circulation and monsoons (A) In today's Mediterranean circulation, salty surface water chilled by cold air in winter sinks and carries dissolved oxygen to deeper layers. (B) At intervals in the past, strong summer monsoons in tropical Africa caused an increased discharge of Nile River freshwater into the eastern Mediterranean, creating a low-density surface-water lid that inhibited sinking of surface water and caused the deep ocean to lose its oxygen and deposit organic-rich black muds.

of salty water along the northern margins of the Mediterranean during incursions of cold air from the north. These two factors make surface waters dense enough to sink to great depths. The dense waters that sink deep into the Mediterranean Sea eventually exit westward into the Atlantic Ocean. As a result of this flow, the floor of today's Mediterranean Sea is covered by sediments typical of well-oxygenated ocean basins: light tan silty mud containing shells of plankton that once lived at the sea surface and benthic foraminifera that once lived on the seafloor.

Mediterranean sediments also contain occasional distinct layers of black organic-rich muds, called **sapropels.** Their high organic carbon content indicates that they formed at times when the waters at the seafloor were **anoxic:** they lacked the oxygen needed to convert (oxidize) organic carbon to inorganic form. The lack of oxygen led to stagnation of the deep waters and deposition of iron sulfides, giving the sediments a "stinky" (rotten-egg) odor. The lack of oxygen also kept benthic foraminifera and other creatures from living on the seafloor.

Paleoecologist Martine Rossignol-Strick proposed that the deep Mediterranean basin was deprived of oxygen during intervals when sinking of oxygen-rich surface waters was cut off by a cap of low-density freshwater brought in by rivers (Figure 8-6B). Even though the surface waters were still chilled by cold air masses at these times, the low-salinity lid kept them from becoming dense enough to sink deep into the basin. As a result, the deep Mediterranean basin lost its supply of oxygen.

At the same time, production of planktic organisms continued at the surface and probably even increased as the stronger river inflow delivered extra nutrients (food) to the Mediterranean. The high productivity at the surface continually sent organic-rich remains of dead plankton toward the seafloor. Sinking and oxidation of this organic carbon continually depleted the oxygen levels in the deep Mediterranean and produced the stinky muds on the seafloor.

The most recent sapropel in the eastern Mediterranean dates to 10,000 to 8000 years ago, an interval when summer insolation levels were higher than today, the African summer monsoon was stronger, and African lakes were at higher levels. Earlier layers of organic-rich mud deeper in Mediterranean sediment cores occur at regular 23,000-year intervals during times when summer insolation was higher than it is today. The sapropels were best developed (thickest and most carbon-rich) near the time of the strongest summer insolation maxima, but they were poorly developed during weaker insolation maxima and were absent the rest of the time. This history of sapropel deposition matches very well the pattern predicted by the orbital monsoon

hypothesis (see Figure 8-5). The close match indicates some kind of connection to the low-latitude monsoon over North Africa.

Initially some climate scientists questioned this explanation. The Mediterranean Sea lies at high subtropical latitudes (30°–40°N), beyond even the greatest northward expansions of past summer monsoons indicated by lake-level evidence across North Africa. If climate within the confines of the Mediterranean region never became truly monsoonal, how could the stinky muds deposited in that basin be a response to the North African monsoon?

The critical link turned out to be the Nile River (Figure 8-7), which gathers most of its water from the highlands of eastern North Africa at tropical latitudes. Even today these highlands receive summer rains during the relatively weak tropical monsoon, and the Nile delivers the water to the Mediterranean Sea far to the

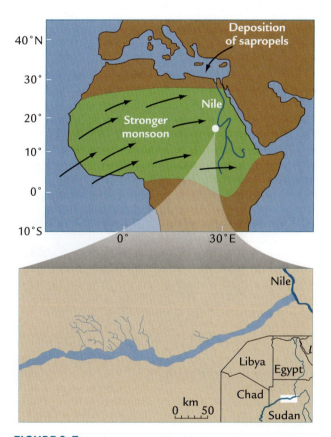

FIGURE 8-7 Monsoons and the Nile River Strong summer monsoons in tropical North Africa periodically produced large discharges of Nile freshwater into the Mediterranean Sea. Satellite sensors have detected riverbed sediments deposited during strong monsoons but now buried beneath sheets of sand in the hyperarid eastern Sahara Desert (inset). (Inset adapted from H.-J. Pachur and S. Kroplein, "Wadi Howar: Paleoclimatic Evidence from an Extinct River System in the Southeastern Sahara," *Science* 237 [1987]: 298–300.

north. At times when summer insolation was much stronger than it is today, the strengthened summer monsoon expanded northward and eastward, bringing much heavier rainfall to these high-elevation regions. In effect, rainfall in the North African tropics exerts a remote control on the salinity of the subtropical Mediterranean Sea via the Nile River.

Satellite sensors have detected the buried remnants of streams and rivers that once flowed across Sudan (see Figure 8-7) but are now covered by sheets of sand blowing across the hyperarid southeastern Sahara Desert. The fact that these streams once flowed eastward and joined the Nile River indicates that lower-elevation regions also contributed to the Nile's stronger flow during major monsoons.

8-3 Freshwater Diatoms in the Tropical Atlantic

Evidence that North African lakes fluctuate at the 23,000-year tempo of orbital precession can also be found in sediment cores from the north tropical Atlantic Ocean (Figure 8-8). These sediments contain layers with high concentrations of the opaline ($SiO_2 \cdot H_2O$) shells of the freshwater diatom *Aulacoseira granulata*. Because these diatoms could not have lived in

the ocean and because these ocean cores are hundreds and even thousands of kilometers away from the land, the only explanation for the presence of the diatoms is wind transport. In arid and semiarid regions, winds scoop out ("deflate") sediment from the beds of dried-out lakes and blow the fine debris far away, some of it to the nearby oceans.

The lake diatoms in these tropical Atlantic cores must have come from North Africa, which lies directly upwind in the prevailing northeasterly flow of winter trade winds. The intervals in the Atlantic cores containing freshwater diatoms mark times in the past when North African lakes were drying out and their muddy lakebeds were becoming exposed to strong winter trade winds.

The records in the Atlantic cores show that lake diatoms were delivered in distinct pulses separated by 23,000 years. As was the case for the Mediterranean sapropels, this 23,000-year tempo in diatom influxes is a direct indication of a connection to the tropical monsoon fluctuations in North Africa. In this case, however, each diatom pulse occurs later than the summer insolation maxima by 5000 to 6000 years (Figure 8-9).

This delay makes physical sense as part of the sequence of events during a typical monsoon cycle.

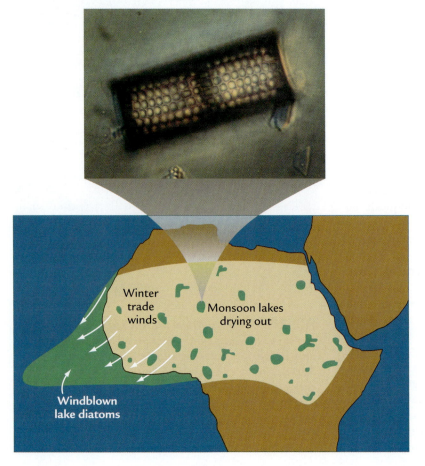

FIGURE 8-8 Drying of monsoonal lakes
North African lakes filled by strong monsoon rains later dried out and were exposed to erosion by winds. Lake muds containing the freshwater diatom *Aulacoseira granulata* (inset) were carried by winds to the tropical Atlantic. (Inset courtesy of Bjørg Stabell, University of Oslo.)

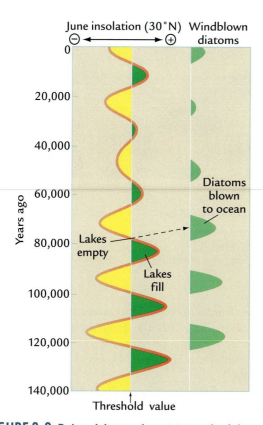

FIGURE 8-9 Delayed diatom deposition in the Atlantic
Diatoms from North African lakes were deposited in the
tropical Atlantic Ocean some 5000 years after the intervals of
strongest monsoons, as the lakes dried out. (Adapted from W.
F. Ruddiman, "Tropical Atlantic Terrigenous Fluxes Since 25,000
Years B.P.," *Marine Geology* 136 [1997]: 189–207; based on E. M.
Pokras and A. C. Mix, "Earth's Precession Cycle and Quaternary
Climatic Changes in Tropical Africa," *Nature* 326 [1987]: 486–7.)

Lakes in North Africa filled to maximum size during
the summer insolation maxima that drove the strong
monsoons. These high lake levels deposited lakebeds
rich in diatoms, but the high water levels in the lakes for
a time kept these sediments from being exposed to
winds, eroded, and blown to the oceans.

Then, as summer insolation began to decrease toward
the next insolation minimum, the monsoon weakened,
summers became drier, and the lake levels began to drop.
The fall in lake levels exposed the diatom-bearing silts
and clays to winter winds, which blew them out to the
ocean. Once the lakes had dried out completely and most
of the diatom-bearing sediments had been blown away,
transport of diatoms to the ocean slowed or stopped, even
though the monsoon continued to weaken as summer
insolation continued to fall toward the next minimum.
As a result, the diatom pulses sent to the ocean should lag
well behind the summer monsoon maxima because of the
time needed to dry the lakes, but they should precede the
subsequent summer insolation minima because many of

the lakebeds were already fully exposed partway into the
drying trend.

Another indication of a link to the North African
summer monsoon comes from the amplitude of the
diatom peaks. Each 23,000-year diatom pulse has the
same relative strength as the immediately preceding summer insolation maximum (see Figure 8-9). This pattern
is consistent with a scenario in which stronger insolation
maxima drove stronger summer monsoon maxima, which
created bigger lakes, which provided larger sources of
diatom-bearing sediments for subsequent transport to
the ocean.

8-4 Upwelling in the Equatorial Atlantic

Atlantic sediments contain additional evidence consistent
with the hypothesis that the North African monsoon
fluctuates at the 23,000-year tempo of orbital precession.
Cores in the eastern Atlantic just south of the equator
show that the structure of the upper water layers has
varied with a prominent 23,000-year rhythm. Part of
the reason for this response is that the North African
summer monsoon imposes an atmospheric circulation
pattern that overrides the local circulation.

When the North African summer monsoon is relatively weak (as it is today), trade winds along the equator have a strong east-to-west flow (Figure 8-10A). The
strongest trade winds occur in southern hemisphere
winter (July and August) and blow from the South
Atlantic toward the equator. Part of this flow crosses the
equator, turns to the northwest, and enters North
Africa in the summer monsoon flow, but this part of the
flow is not strong when the monsoon is weak, as it is
today. Instead, strong trade winds blow mainly toward
the west and drive warm surface waters away from
the equator (companion Web site, pp. 22–24). This
upper-ocean flow causes a shallowing of the seasonal
thermocline, a subsurface region of steep temperature
gradients between the warm surface waters and much
cooler temperatures below. As the thermocline shal-
lows, cooler waters rich in nutrients rise toward the sea
surface just south of the equator.

In contrast, at times when summer insolation was
higher than it is today, the stronger summer monsoon
flow altered this circulation pattern (Figure 8-10B). A
much larger portion of the southeast trade-wind flow
crossed the equator, turned to the northeast, and was
drawn into North Africa in the monsoon circulation.
This strengthening of the monsoon flow into Africa
weakened the westward trade-wind flow along the
equator, and the weaker trade winds reduced the
upwelling of cold waters, leaving the surface waters
poorer in nutrients from below.

Changes between these two circulation patterns
over time can be measured by examining variations in

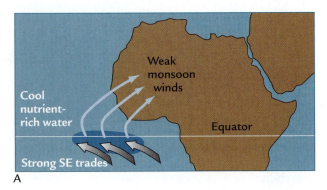
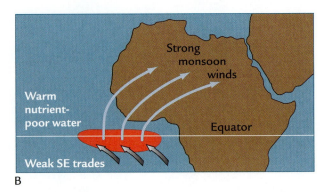

FIGURE 8-10 Effect of monsoons on southeast trade winds (A) When monsoonal circulation over North Africa is weak, strong southeasterly trade winds in the eastern tropical Atlantic cause cool, nutrient-rich waters to rise close to the surface. (B) When a strong monsoon circulation over North Africa weakens the trade winds, tropical waters are warm and depleted in nutrients.

the relative amounts of planktic organisms that inhabit near-surface waters and leave shells in the sediments below. In equatorial Atlantic sediments, planktic foraminifera and coccoliths are the most common shelled organisms (Chapter 2). Different species of these two kinds of plankton prefer different environmental conditions near the sea surface—either warmer waters with fewer nutrients or cooler waters rich in nutrients. Sediment cores from the Atlantic Ocean just south of the equator show 23,000-year cycles of alternating abundances in these two types of plankton, still another indication of the effects of the North African summer monsoon.

Orbital Monsoon Hypothesis: Regional Assessment

The evidence from North Africa and surrounding seas supports the orbital monsoon hypothesis, but monsoon circulations also exist on other continents. The south Asian monsoon is the most powerful monsoon of all because of the size of the land mass and the extent of the high topography of Tibet (Chapter 6). As a result, a large area of low pressure and heavy monsoonal precipitation covers southern Asia in summer (Figure 8-11).

8-5 Cave Speleothems in China and Brazil

Stalactites and stalagmites are constructed of calcite ($CaCO_3$) deposited by groundwater dripping in caves. These deposits build up layer by layer over thousands to tens of thousands of years, and they can be very accurately dated by radiometric analysis of small amounts of thorium and uranium (Chapter 2). In addition, the relative amount of the isotopes of oxygen in the calcite layers varies though time because of several factors, one of

which is changes in the surface climate that provides the precipitation that feeds the groundwater. As a result, changes in overlying air masses are an important control on variations in the $\delta^{18}O$ signal of cave calcite (Appendix 1).

Records from caves in southern China show $\delta^{18}O$ variations so large that they can be explained only by

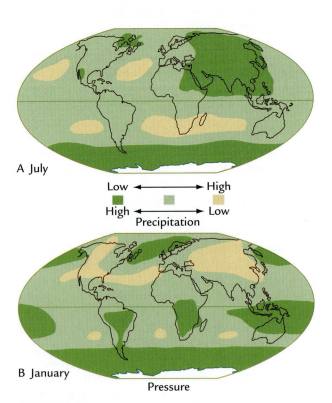

FIGURE 8-11 Seasonal patterns of pressure and precipitation In the summer season of each hemisphere, heating of the continents produces low pressures and heavy monsoonal rainfall.

changes in air masses linked to the Asian monsoon (Figure 8-11A). Highly negative $\delta^{18}O$ values indicate a stronger monsoon flow from the ocean along with greater fractionation of the oxygen isotopes (Figure 8-12). These $\delta^{18}O$ changes correlate closely with changes in midsummer (July) insolation at 25°N, especially during the interval prior to 100,000 years ago where both the insolation and $\delta^{18}O$ changes were largest. This record provides unambiguous evidence in support of the orbital monsoon hypothesis: the Asian monsoon has varied at 23,000-year intervals and the changes had a midsummer (July) phase. During times of smaller variations in insolation and $\delta^{18}O$, other changes are evident in the $\delta^{18}O$

signal, but they occur over shorter intervals than the orbital-scale oscillations.

Records of $\delta^{18}O$ in cave calcite have also been recovered from southeastern Brazil, a subtropical location that is under the influence of a summer monsoon circulation (see Figure 8-11). Again, the 23,000-year cycle is obvious in the $\delta^{18}O$ signal, but in this case it has the phase of February insolation rather than July (Figure 8-12). This result is a particularly elegant confirmation of the orbital-monsoon hypothesis. Because February is midsummer in the southern hemisphere, monsoon variations on southern continents should have this February phase at the precession cycle, and they do.

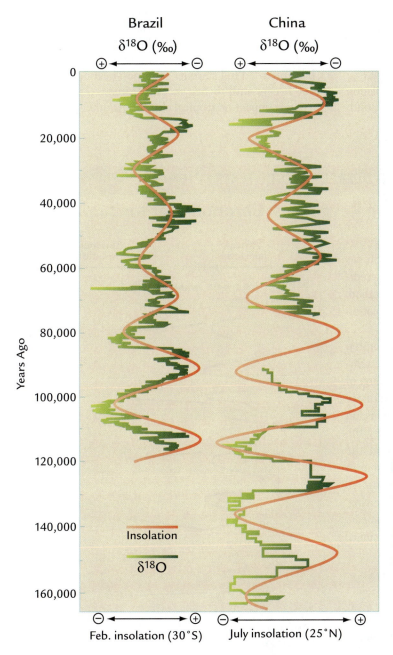

FIGURE 8-12 Monsoon $\delta^{18}O$ signals from caves
Calcite from cave deposits in China and Brazil shows $\delta^{18}O$ changes produced by variations in the strength of monsoonal air masses at the 23,000-year precession cycle. The $\delta^{18}O$ variations have the phase of midsummer insolation in each hemisphere. (Adapted from D. Yuan et al., "Timing, Duration and Transitions of the Last Interglacial Asian Monsoon," *Science* 304 [2004]: 575–579, and from F. W. Cruz et al., "Insolation-Driven Changes in Atmospheric Circulation over the Past 116,000 Years in Subtropical Brazil," *Nature* 434 [2005]: 63–66.

IN SUMMARY, evidence from several continents fully supports John Kutzbach's orbital monsoon hypothesis. At this point, the hypothesis has passed so many tests that it merits the higher status of a theory (Chapter 1).

8-6 The Phasing of Summer Monsoons

The evidence examined in this chapter has shown that peak development of past summer monsoons at the 23,000-year cycle occurred in midsummer (July north of the equator, February to the south). This evidence has been interpreted in two ways. To understand the difference in these interpretations, we need to return to the effects of precession on insolation. Recall that Earth's precessional motion produces a family of monthly insolation curves, each offset from the preceding month by one-twelfth of a 23,000-year cycle, or slightly less than 2000 years. For example, the insolation signal for the month of July lags that for the month of June by this amount (see Figure 7–18). Because an entire family of insolation curves is available as possible monsoon drivers, any month could be the source of the "true" forcing of summer monsoons at the 23,000-year cycle. The problem is to provide a specific physical justification for the one chosen.

One view is that peak monsoon responses should have the same phase as the peak in annual insolation forcing at the June 21 summer solstice. Additions to this already strong solstice forcing at the 23,000-year cycle could then drive the strongest monsoon response. In this view, the later (July) phase indicated by a range of evidence noted earlier results from lags in the climate system that retard the peak monsoon development. The size of this lag can be calculated as slightly less than 2000 years, the difference between a phase of June 21 and a phase of middle or late July at the 23,000-year cycle.

Ice sheets have been proposed as one possible source of such a retarding effect. Strong monsoons developed during intervals when northern hemisphere ice sheets were present but melting rapidly. In this view, the large-scale cooling caused by the lingering ice sheets could have retarded full summer monsoon heating for 1000 to 2000 years until the ice sheets became too small to have a major impact on tropical climate. Another possibility is

---

BOX 8-1   LOOKING DEEPER INTO CLIMATE SCIENCE

Insolation-Driven Monsoon Responses: Chronometer for Tuning

The clearly demonstrated link between summer insolation forcing and monsoon responses at low latitudes has become part of the basis for a new way of dating sedimentary records on land and in the oceans. This method, **orbital tuning,** can provide even better time resolution than radiometric methods.

The tuning method is based on the relationship between the insolation signal (the forcing) and the summer monsoon changes (the response). The timing of orbital insolation changes is known with great accuracy from astronomical calculations, and a range of monsoon responses can be measured in sediments. By making the simple assumption that the insolation driver and the monsoon responses have kept the same relationship in the past, the monsoon responses in the sediments can be dated with nearly the same accuracy as the orbital forcing.

This method is most easily applied in ocean sediments because deposition in the ocean tends to be continuous.

Tuning sediment sequences to orbital variations If lavas or magnetic reversal boundaries provide radiometrically dated levels in terrestrial or marine sediments, the age of intervening sediment intervals can be determined by tuning monsoon-driven sediment responses to insolation changes at the 23,000-year precession cycle.

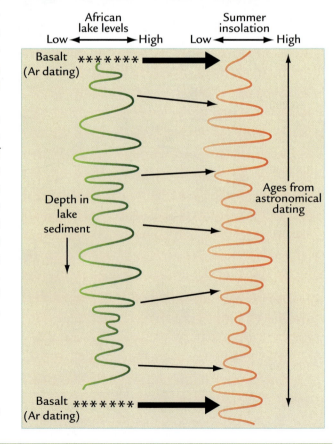

that the tropical ocean may have delayed full development of the summer monsoon by controlling the supply of latent heat that fuels the monsoons.

An alternative view is that the peak monsoon response should have the phase of insolation forcing during the midsummer month of July (or February). This view has a reasonable physical basis because modern summer monsoons reach their peak development during these hot midsummer months. With the continents heated to their maximum summer temperatures, the temperature contrast with the cool oceans also reaches a maximum, and this temperature difference drives the strongest monsoon circulations of the year. In this view, precession-cycle insolation changes aligned with this intense midsummer heating should have the greatest impact in boosting continental temperatures and in driving the strongest monsoons. From this perspective, the midsummer monsoon responses are forced by midsummer insolation with no significant lags in the climate system.

At this point, climate scientists cannot prove which of these views is correct. In addition, the phasing of peak summer monsoons at orbital time scales is not yet fully known for all regions. In any case, the evidence supports Kutzbach's original hypothesis that summer monsoon variations every 23,000 years should have a July phase in the north (and February in the south).

Monsoon Forcing Earlier in Earth's History

The concept of insolation forcing of summer monsoons can be used to investigate environments that existed in the distant geologic past. Orbital-scale changes in summer insolation at the precession cycle drive changes in monsoonal precipitation, and precipitation is a key control of processes that leave evidence in ancient climate records, such as sediment-laden runoff and lake depth. As a result, many ancient sedimentary rocks contain valuable information about varying monsoonal precipitation.

If high-quality time control is available for ancient deposits, we can look for evidence of the kind of monsoon signature shown in Figure 8-5. Because many of these deposits extend over millions of years, we can expect to see records that look like those in Figure 8-13. A wide range of sediment indicators linked to precipitation, erosion, runoff, transport, and deposition may have this appearance. In some cases, this relationship can even be used to refine ("tune") time scales initially determined by radiometric dating (Box 8–1).

As in the case of North African lakes, we expect the monsoon signature to show clusters of two or three strong maxima separated by clusters of two or three weak maxima, with these clusters repeating in the record at intervals of about 100,000 years because of control of

BOX 8-1   LOOKING DEEPER INTO CLIMATE SCIENCE

CONTINUED

Assume that a sediment core contains two magnetic reversal boundaries that have been dated by correlation to the global magnetic stratigraphy established by dating basalt layers on land (Chapter 4). The ages of these reversals constrain the intervening sequence of sediments to a particular interval of time. Also assume that this sediment sequence contains a record that is directly tied to the strength of the tropical monsoons, such as a sequence of sapropel layers or of changes in the composition of marine plankton.

In many cases, the monsoon-related response measured in the sediments will show an obvious correlation to the summer insolation forcing. Both the insolation changes and the monsoon responses recorded will show cycles near 23,000 years and obvious modulation of these cycles at eccentricity periods near 100,000 and 400,000 years. The tuning process (matching maxima and minima in the monsoon response to correlative features in the insolation signal) allows the ages within the sedimentary sequence to be assigned to specific precession cycles in the past at a finer resolution than the 23,000-year length of each cycle.

This method has also been applied to sediments deposited in long-lived lakes on continents. In regions like

East Africa, volcanic eruptions deposit beds of basalt (lava) or volcanic ash that can be radiometrically dated by K/Ar methods (see Chapter 2). The volcanic deposits provide the initial time framework for the tuning process, analogous to the use of magnetic reversal boundaries in marine sediments. The monsoon-driven variations in lake size during the sequence lying between the basalt layers can then be tuned to the astronomically dated record of summer insolation. Because records on land are generally much more difficult to date than those in ocean sediments, tuning of these lake sequences provides an enormous improvement over other dating methods.

For sedimentary records that contain climatic responses at the cycles of both precession and obliquity, the tuning method can be tested even more rigorously. In this case, the "tuned" time scale must match not just the amplitude-modulated precession cycle at 23,000 years but also the tilt cycle at 41,000 years. This added requirement makes the tuning process an even more demanding exercise but one that is even more likely to yield a unique time scale.

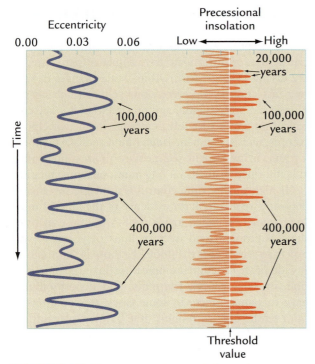

FIGURE 8-13 Monsoon signals recorded in sediments
Monsoonal influences can be detected in older sediment
sequences. High orbital eccentricity values (left) should
amplify individual 23,000-year precession cycles approximately
every 100,000 and 400,000 years (right). The monsoon signal
in the sediments could resemble the red-shaded area to the
right of the threshold insolation value.

the amplitude of precession by orbital eccentricity
(Chapter 7). In this case, because we are looking at much
longer records, we should also see clusters of monsoon-
driven maxima at the longer eccentricity period of about
400,000 years (see Figure 8-13). The truncation of the
summer monsoon response pattern at a critical threshold value is called **clipping.** As a result of this truncation,
many monsoon responses register only one side of each
23,000-year precession cycle, with modulation of this
one-sided response at 100,000 and 400,000 years.

8-7 Monsoons on Pangaea 200 Myr ago

Just before 200 Myr ago, a chain of basins formed in a
region that is now the eastern United States but at that
time lay deep in the interior of the giant supercontinent
Pangaea (Figure 8-14). These deep depressions in a
region of generally high terrain were formed by pre-
cursors of the forces that would eventually pull the
Pangaean continent apart and create the Atlantic
Ocean. But this part of Pangaea would not break up
until tens of millions of years later (Chapter 5).

Sediments deposited in one of these depressions, the
Newark Basin in modern New Jersey, have been extensively investigated. The fossil compasses provided by

magnetic evidence from volcanic rocks indicate that
the Newark Basin was located in the tropics 200 Myr
ago, about 10° of latitude north of the equator (see
Figure 8-14). Because of its tropical location, the Newark
Basin was dominated by precessional insolation changes,
similar to those in modern North Africa and southern
Asia. Because the basin was far from the ocean, its climate
was relatively arid, but enough moisture arrived to create
a lake that varied greatly in size over time.

Evidence preserved in a thick (> 7000 m) sequence of
lake sediments shows that the size of this lake fluctuated
at a tempo near 20,000 years. Several layers of molten
magma that intruded into the lakebed sequence and
quickly cooled have been dated by radiometric methods.
These dates show that the lakebed sequence was
deposited over an interval of at least 20 Myr centered
near 200 Myr ago.

This estimate is confirmed by the presence of
fine laminations (varves) in parts of the sequence. The
varves are tiny (0.2–0.3 mm) couplets of alternating light

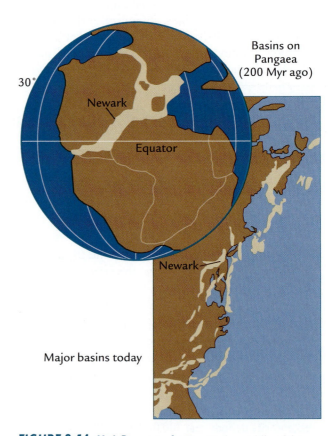

FIGURE 8-14 Mid-Pangaean basins In the middle of the
Pangaean supercontinent 200 Myr ago (top left), the Newark
Basin developed in what is now New Jersey as one of a chain of
basins of equivalent age (bottom right). (Adapted from P. E.
Olsen and D. V. Kent, "Milankovitch Climate Forcing in the
Tropics of Pangaea During the Late Triassic," *Palaeogeography,
Palaeoclimatology, Palaeoecology* 122 [1996]: 1–26.)

drying up of the lakes. Individual layers in these sequences are continuous over large areas, indicating that the wet-dry variations in climate affected the entire basin.

Extensive investigations show that these fluctuations in lake depth over millions of years were cyclic (Figure 8-16). The shortest cycles occur over rock thicknesses averaging 4–5 m, equivalent to about 20,000 years in time based on the average thickness of each annual varve (0.2–0.3 mm). These cycles were driven by precession. Monsoons filled and emptied these Pangaean lakes (Chapter 5) in response to orbital precession in the same way that North African lakes have filled and emptied during much more recent times. Because we are looking much farther back in time, the periods of the orbital cycles were slightly shorter than they are today (Chapter 7).

Two larger-scale groupings of cycle peaks are also evident. The amplitude of individual 20,000-year peaks

FIGURE 8-15 Evidence of changing lake levels Dinosaur footprints in lake muds that have since hardened into rock show that the Pangaean lakes occasionally dried out completely. These footprints are from a basin in Connecticut formed at the same time as the Newark Basin in New Jersey. (Dinosaur State Park, Rocky Hill, CT.)

and dark layers, with one light/dark pair deposited each year. Darker organic-rich layers were deposited in summer, lighter mineral-rich layers in winter. Dissolved oxygen concentrations must have been low or zero in the deeper levels of the lake when the organic-rich layers accumulated to prevent destruction of the delicate varves by animals moving across and within the sediments. Use of these varves as an internal chronometer to count elapsed time (Chapter 2) confirms that the total time of lake-sediment deposition was about 20 Myr.

The types of sediment deposited in the Newark Basin varied widely in response to changes in lake depth. When the lake was deep (100 m or more), the sediments tended to be gray or black muds with large amounts of organic carbon. These sediments contain finely laminated varves and are rich in well-preserved remains of fish. Sediments deposited when the lake was shallower or entirely dried out tend to be red or purple because they were oxidized (rusted) by contact with air, and they often contain mud cracks due to exposure to the dry air. Dinosaur footprints and the remains of plant roots are also common in sediments from the dried-out, vegetated parts of the lakebeds (Figure 8-15).

The thick sequences preserved in the Newark Basin repeatedly fluctuate between sediments typical of deep lakes and those that indicate a shallowing or complete

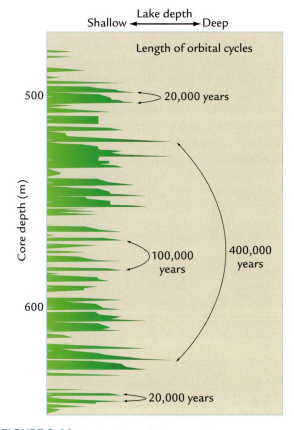

FIGURE 8-16 Fluctuations of Pangaean lakes Newark Basin lake sediments varied in depth from very shallow to over 100 m deep at three tempos. Individual cycles in lake depth every 4–5 m occur at a period of 20,000 years, clusters of larger deep-lake maxima every 20–25 m occur at intervals near 100,000 years, and unusually large deep-lake clusters at intervals of 90–100 m occur every 400,000 years. (Adapted from P. E. Olsen and D. V. Kent, "Milankovitch Climate Forcing in the Tropics of Pangaea During the Late Triassic," *Palaeogeography, Palaeoclimatology, Palaeoecology* 122 [1996]: 1–26.)

in lake depth rises and falls roughly every five or six cycles separated by 20–25 m of sediment, or a little less than 100,000 years. An even larger-scale change in amplitude of the monsoon-cycle peaks occurs between approximately 530 and 620 meters depth in the core (see Figure 8-16), or over a time interval of about 400,000 years.

These two longer-term patterns match the expected monsoon signature shown in Figure 8-13 remarkably well. They reflect a modulation of the strength of the 20,000-year precession cycles by eccentricity changes at intervals of about 100,000 and 400,000 years. The full imprint of ancient monsoons is amazingly clear in the sediments of this basin despite the passage of 200 Myr.

8-8 Joint Tectonic and Orbital Control of Monsoons

We saw in Chapter 6 that tectonic changes affect the intensity of monsoon circulations. Large landmasses such as Pangaea intensify monsoons by offering a larger area for the Sun to heat. Positioning of landmasses at lower latitudes is important because solar radiation is more direct and albedos are much lower than at higher, snow-covered latitudes. Topography is a key control over monsoon strength at tectonic time scales because high-elevation regions focus strong monsoonal rains on their margins.

The processes that control monsoon intensity over tectonic time scales interact with those at orbital scales. Tectonic-scale processes alter the average strength of the monsoon over millions of years, while the orbital-scale insolation changes drive shorter-term monsoon strength at a cycle of about 20,000 years. One way the tectonic and orbital factors might interact is suggested in Figure 8-17. On the left is a schematic version of a low-latitude summer insolation curve, with individual maxima and minima at the 20,000-year precessional cycle and modulation of this cycle every 100,000 and 400,000 years. The smooth curve in the center represents gradually changing tectonic-scale processes, such as the slow uplift that gradually intensifies the average strength of the monsoon over millions of years. This slow tectonic-scale increase in monsoon strength combines with the orbital-scale monsoon cycles to produce the response shown on the right—a *slow increase in the amplitude of the orbital-scale cycles* caused by tectonic amplification.

We can hypothesize the existence of a threshold value above which key climatic indices record monsoon responses but below which no response is registered (as in Figure 8-13). In the changes shown on the right in Figure 8-17, the tectonic influence may have been weak enough during the earlier intervals that the orbital-scale

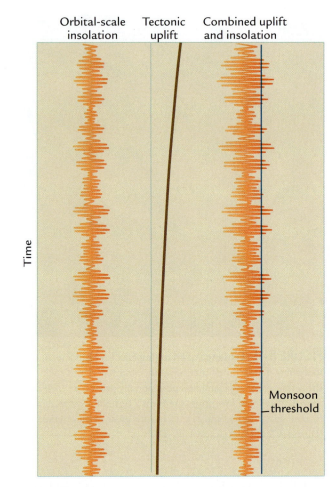

FIGURE 8-17 Combined tectonic and orbital forcing of monsoons Monsoons are driven by orbital-scale variations in insolation (left) and by slower-acting tectonic factors such as plateau uplift (center). The combined tectonic and orbital forcing causes the amplitude of orbital-scale monsoon responses to increase and gradually exceed critical thresholds (right). (Adapted from W. F. Ruddiman et al., "Late Miocene to Pleistocene Evolution of Climate in Africa and the Low-Latitude Atlantic: Overview of Leg 108 Results," Ocean Drilling Program Initial Reports 108B [1989]: 463–84.)

monsoon cycles never exceeded this threshold. Later, as tectonic processes created conditions more favorable to monsoons, peaks in summer insolation would have driven monsoons that began to exceed the threshold by small amounts and then later by steadily increasing amounts.

Something like this kind of evolving climatic response is thought to have occurred in Southeast Asia over the last 30 or 40 Myr. A long-term tectonic increase in monsoon intensity due to uplift progressively intensified the amplitude of orbitally driven monsoon cycles in this region. Simulations run on general circulation models indicate that the combined effects of orbital-scale insolation and uplift are not additive in a simple linear

way. Instead, it appears that plateau uplift sensitizes the monsoon system to insolation forcing in such a way that the combined monsoon response to uplift and insolation is stronger than a simple linear combination of the two effects.

Key Terms

orbital monsoon
  hypothesis (p. 139)
nonlinear response
  (p. 140)
threshold level (p. 141)
sapropels (p. 143)

anoxic (p. 143)
thermocline (p. 145)
clipping (p. 148)
orbital tuning (p. 149)

Review Questions

1. In what way is the orbital monsoon hypothesis an extension of processes driving modern monsoons?

2. Why does the intensity of 23,000-year monsoon peaks vary at intervals of 100,000 and 413,000 years?

3. How did the Mediterranean Sea acquire a freshwater lid during times when very little precipitation was falling in that region?

4. Explain how the opposed July/February timing of past monsoon changes in China and Brazil lends strong support to the orbital monsoon hypothesis.

5. Does peak monsoon strength lag behind summer insolation forcing?

6. What similarities exist between monsoon changes in Pangaea 200 Myr ago and those in North Africa during the last several hundred thousand years?

7. How do tectonic uplift and orbital variations combine to affect the long-term intensity of monsoons?

Additional Resources

Basic Reading

Companion Web site at www.whfreeman.com /ruddiman2e, pp. 9–11, 15–18, 21–24, 32.

Kutzbach, J. E. 1981. "Monsoon Climate of the Early Holocene: Climate Experiment with Earth's Orbital Parameters for 9000 Years Ago." *Science* 214: 59–61.

Ruddiman, W. F. 2005. *Plows, Plagues and Petroleum*, Chapter 5. Princeton, NJ: Princeton University Press.

www.classzone.com/book/earth_science/terc/content /visualization, Chapter 24: Monsoon Circulations.

Advanced Reading

Olsen, P. E. 1986. "A 40-Million-Year Lake Record of Early Mesozoic Orbital Climatic Forcing." *Science* 234: 842–48.

Pokras, E. M., and A. C. Mix. 1987. "Earth's Precession Cycle and Quaternary Climatic Changes in Tropical Africa." *Nature* 326: 486–87.

Rossignol-Strick, M., W. Nesteroff, P. Olive, and C. Vergnaud-Grazzini. 1982. "After the Deluge: Mediterranean Stagnation and Sapropel Formation." *Nature* 295: 105–10.

Ruddiman, W. F., 2006. "Viewpoint: What Is the Timing of the Orbital Monsoon?" *Quaternary Science Reviews* 25: 657–58.

Yuan, D., et al. 2004. "Timing, Duration, and Transitions of the Last Interglacial Monsoon." *Science* 304: 575–78.

CHAPTER 9

Insolation Control of Ice Sheets

Ice sheets covered northern North America and Europe 20,000 years ago. The present locations of Toronto, New York, Chicago, Seattle, and London were buried under hundreds of meters of ice. Later, the ice melted, and the last remnants disappeared by 6000 years ago, near the time human civilizations came into existence. The fact that ice sheets first appeared in the northern hemisphere within the last 3 Myr can be explained by very slow tectonic-scale cooling (Part II), but the evidence that ice sheets grew and melted over much shorter intervals of time requires a different explanation.

The driver of these shorter-term variations in the amount of ice is orbitally driven insolation changes. In this chapter, we investigate how changes in summer insolation control the size of ice sheets by determining the rate of ice melting or accumulation. We explore two lags that are important to understanding the ice response: the lag of slow-responding ice sheets behind the insolation changes and the delayed depression of bedrock beneath the weight of the overlying ice. Both these lags are thousands of years in length. Then we examine past changes in ice volume based on evidence from oxygen isotopes and coral reefs. Finally, we analyze how the ice sheet history over the last 3 Myr compares with predictions from the theory of orbital control of ice volume.

Milankovitch Theory: Orbital Control of Ice Sheets

Continental ice sheets exist in regions where the overall rate of snow and ice accumulation across the entire mass of ice equals or exceeds the overall rate of ice loss or **ablation** (companion Web site, pp. 27–30). Snow accumulates as ice at high latitudes and altitudes where temperatures are cold enough to permit frozen pre-cipitation and to prevent melting in summer. For continent-sized ice sheets that reach sea level, tempera-tures sufficiently cold to sustain ice occur today only at high latitudes.

Rates of ice accumulation and ablation vary with tem-perature, but the two relationships differ in a critical way. Ice (snow) accumulates at mean annual temperatures below 10°C, but rates of accumulation remain below 0.5 meter per year regardless of temperature (Figure 9–1A). At higher temperatures, ice accumulation is limited by the fact that more of the precipitation falls as rain. At extremely low temperatures, all the precipitation is snow, but frigid air carries so little water vapor that rates of ice accumulation are low.

In contrast, ablation of ice accelerates rapidly when temperatures warm. Melting begins at mean annual temperatures above –10°C, equivalent to summer tem-peratures above 0°C, and can reach rates equivalent to several meters of ice per year, much larger than maxi-mum rates of accumulation.

Ice ablation can occur as a result of incoming solar radiation, by uptake of sensible or latent heat delivered by warm air masses (and rain), and by shedding of ice-bergs to the ocean or to lakes (called **calving**). Calving differs from other ablation processes because icebergs leave the main ice mass and move elsewhere to melt, often in a warmer environment than the one in or near the ice sheet.

The net balance between accumulation and ablation over an entire ice sheet is called the **ice mass balance** (Figure 9–1B). The mass balance at very cold tempera-tures (below –20°C) is positive but small because so little snow falls. The mass balance at mean annual tem-peratures near –15° to –10°C is more positive because snow accumulation rates are more rapid but ablation is not strong. The mass balance turns sharply negative at temperatures above –10°C because ablation accelerates and overwhelms accumulation. The boundary between positive and negative mass balance is called the **equilib-rium line**.

If net accumulation and ablation are in balance over an entire ice sheet, the ice sheet is said to be in a condi-tion of stable equilibrium. Net accumulation high on the ice sheet is exactly balanced by ablation at lower eleva-tions, and no net change in total ice volume occurs. Ice flows within the ice sheet from areas of accumulation to

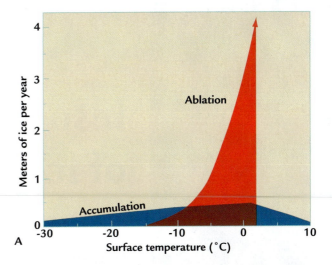

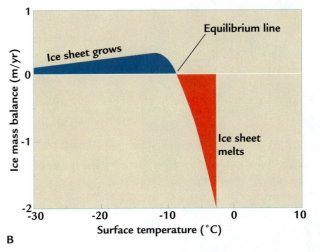

FIGURE 9-1 Temperature and ice mass balance
Temperature is the main factor that determines whether ice sheets are in a regime of net ablation (negative mass balance) or accumulation (positive mass balance). Ablation increases sharply at higher temperatures. (Modified from J. Oerlemans, "The Role of Ice Sheets in the Pleistocene Climate," *Norsk Geologisk Tidsskrift* 71 [1991]: 155–61.)

areas of ablation, but the total mass of the ice sheet remains unchanged. Here, our focus is not on ice sheets that remain in equilibrium. We seek to understand what makes ice sheets grow and shrink.

Beginning with the Belgian mathematician Joseph Adhemar in the 1840s, scientists suspected that orbitally driven changes in solar insolation might be linked in some way to the growth and melting of continent-sized ice sheets. Because orbital changes alter the amount of insolation received on Earth in all seasons (Chapter 7), scientists faced an important question: Which season is critical in controlling the size of ice sheets?

Winter would seem to be the obvious choice, because snow falls mainly in winter. Colder winters

caused by lower solar radiation at that time of year should help to accumulate larger amounts of snow and promote glaciation. But this seemingly reasonable idea turned out to be wrong. One problem is that ice sheets grow at high latitudes where temperatures are *always* cold in winter, even during intervals of relatively warm climate like the one we live in today. In addition, the Sun at these latitudes always lies low in the winter sky, regardless of ongoing orbital changes, and incoming solar radiation in winter is never strong. Winter is not the critical season.

The opposite idea—*summer insolation control of ice sheets*—was proposed by several scientists working in the late nineteenth and early twentieth centuries, including Rudolf Spitaler (who was also the first to realize that summer insolation changes might drive monsoons), Wladimir Köppen, and Alfred Wegener (who also proposed the theory that continents drift). Their reasoning was simple: no matter how much snow falls during winter, it can all be easily melted if the following summer is warm and ablation is rapid (see Figure 9–1).

As a result, these scientists reasoned that low summer insolation is critical in producing summers cool enough for snow and ice to persist from one winter to the next. This idea gained popularity during the early and mid-twentieth century from work by the Serbian astronomer Milutin Milankovitch, who first calculated in a systematic way the impact of astronomical changes on insolation received on Earth at different latitudes and in different seasons. This idea is now known as the **Milankovitch theory.**

Milankovitch proposed that ice growth in the northern hemisphere occurs during times when summer insolation is reduced. Low summer insolation occurs when Earth's orbital tilt is small and its poles are pointed less directly at the sun (Figure 9–2A). Low insolation also occurs when the northern summer solstice occurs with Earth farthest from the Sun (in the aphelion or distant-pass position) and when the orbit is highly eccentric (further increasing the Earth-Sun distance). Milankovitch reasoned that the most sensitive latitude for low insolation values is 65°N, the latitude at which ice sheets first accumulate and last melt. He also proposed that ice melts during the stronger summer insolation resulting from the opposite orbital configuration (Figure 9–2B).

The amount of summer insolation arriving at the top of Earth's atmosphere at 65°N can vary by as much as ±12% around the long-term mean value (Chapter 7). We have no way of knowing how much of this incoming solar radiation actually makes it through the atmosphere to Earth's high-latitude ice sheets because of the complicating effects of regional changes in atmospheric circulation, clouds, and water vapor. Milankovitch noted these complications but assumed that the amount of

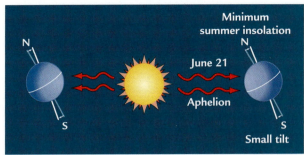

A Northern hemisphere ice growth

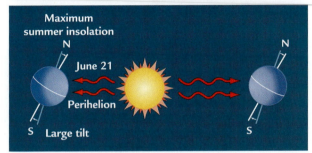

B Northern hemisphere ice decay

FIGURE 9-2 Orbital changes and ice sheets (A) According to the Milankovitch theory, ice sheets grow in the northern hemisphere at times when summer insolation is low, because tilt is low and Earth lies in the aphelion position farthest from the Sun. (B) Ice melts when summer insolation is high because tilt is high and Earth lies in the perihelion position closest to the Sun. (Adapted from W. F. Ruddiman and A. McIntyre, "Oceanic Mechanisms for Amplification of the 23,000-Year Ice-Volume Cycle," *Science* 212 [1981]: 617–27.)

radiation penetrating to Earth's surface is closely related to the amount arriving at the top of the atmosphere.

IN SUMMARY, the Milankovitch theory proposes that when summer insolation is strong, more radiation is absorbed at Earth's surface at high latitudes, making the climate in those regions warmer. Warming accelerates ablation, melts more snow and ice, and either prevents glaciation or shrinks existing ice sheets (Figure 9–3 top). Conversely, when summer insolation is weak, less radiation is delivered to high latitudes, and the reduction in radiation cools the regional climate. This cooling reduces the rate of summer ablation and allows snow to accumulate and ice sheets to grow (Figure 9–3 bottom).

Modeling the Behavior of Ice Sheets

To gain more insight into summer insolation control of ice sheets in the northern hemisphere, climate scientists have developed numerical models based on an idealized

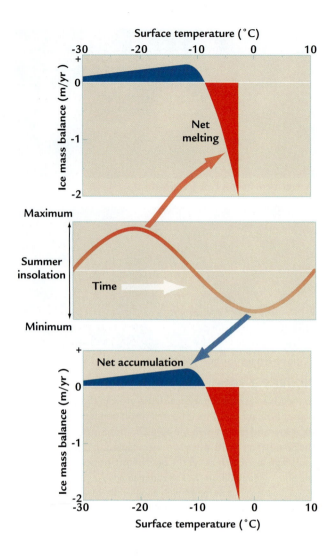

FIGURE 9-3 Milankovitch theory According to the Milankovitch theory, (top) high summer insolation heats the land and results in greater ice ablation, while (bottom) low summer insolation allows the land to cool and ice sheets to form. (Modified from J. Oerlemans, "The Role of Ice Sheets in the Pleistocene Climate," *Norsk Geologisk Tidsskrift* 71 [1991]: 155–61.)

(simplified) representation of the ice sheets surrounding the Arctic Ocean. These models portray the changes in ice sheets across a single (idealized) high-latitude continent near the Arctic, but they ignore possible differences between various sectors of the Arctic (Europe versus North America). This simplification is reasonably well justified because ice sheets seem to have grown on all the continents around the Arctic, at least during the last glacial maximum 20,000 years ago (Figure 9–4).

9-1 Insolation Control of Ice Sheet Size

The mechanism by which changes in summer insolation control the size of ice sheets on northern landmasses in these models follows directly from Milankovitch's theory. Changes in summer insolation drive regional temperature responses that alter both melting rates and ice mass balance.

One kind of model represents this relationship as changes in mass balance along a north-south line (Figure 9–5). These transects have just two dimensions, one in a vertical direction (altitude) and the other in a north-south direction (latitude). Changes in the other horizontal dimension (longitude) are ignored to allow the models to simulate changes over longer intervals of time than would be possible with full three-dimensional models, which are computationally more demanding.

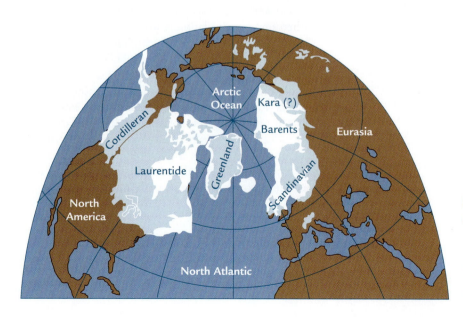

FIGURE 9-4 Ice sheets around the Arctic Ocean At the last glacial maximum, 20,000 years ago, ice sheets surrounded much of the Arctic Ocean. (Modified from G. Denton and T. Hughes, *The Last Great Ice Sheets* [New York: Wiley, 1981].)

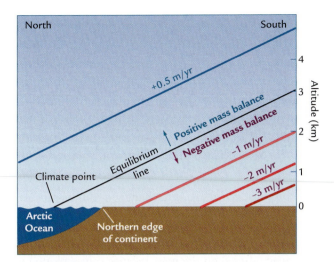

FIGURE 9-5 Ice sheet models Two-dimensional models represent northern hemisphere ice sheets along a north-south line. The equilibrium line separates northern (and higher) regions of net accumulation from southern (and lower) regions of net ablation, and it intersects Earth's surface at the climate point. (Adapted from J. Oerlemans, "Model Experiments of the 100,000-Year Glacial Cycle," *Nature* 287 [1987]: 430–32.)

The equilibrium line in these models, the boundary between areas of net ice ablation and accumulation, slopes upward into the atmosphere toward the south at a low angle. This slope is consistent with conditions today: temperatures are colder toward higher latitudes and altitudes and warmer toward lower latitudes and altitudes. As a result, subfreezing temperatures occur today only at high latitudes and altitudes. Long distance air travel commonly occurs at these subfreezing altitudes.

Parallel to this moving equilibrium line are lines of ice mass balance. These lines show the thickness in meters of ice that accumulates or melts each year (as before, snowfall is converted to an equivalent thickness of ice). Ice accumulates above the equilibrium line and in the north because of the colder temperatures, and it melts in the warmer temperatures below the equilibrium line and toward the south. The rates of ice melting are more closely spaced in the warmer areas because of the rapid melting rates shown in Figure 9–1.

The equilibrium line intercepts Earth's surface in the higher latitudes at the **climate point** (Figure 9–5). Ice sheet models use orbital-scale changes in summer insolation to move this climate point (and the equilibrium line attached to it) north and south across the landmasses (Figure 9–6). The amount of north-south shift of the equilibrium line is set proportional to the amount of change in summer insolation. These shifts can cover 10° to 15° of latitude.

Gradual changes in the amount of summer radiation received at high latitudes cause the areas of net snow

accumulation and net ice ablation to shift back and forth across the land. Strong summer insolation warms the high-latitude landmasses in summer, moves the climate point northward over the Arctic Ocean, and puts northern landmasses in an ablation regime which melts all winter snow each summer and does not allow ice to accumulate (see Figure 9–6 top). Weak summer insolation allows the landmasses to cool, shifts the climate point southward over the land, and sets up a positive mass balance over the northern edge of the continents so that permanent ice can accumulate (see Figure 9–6 bottom).

Once ice sheets begin to form, their vertical dimension (altitude) comes into play in a powerful way (Figure 9–7). As the ice sheets thicken, their upper surfaces

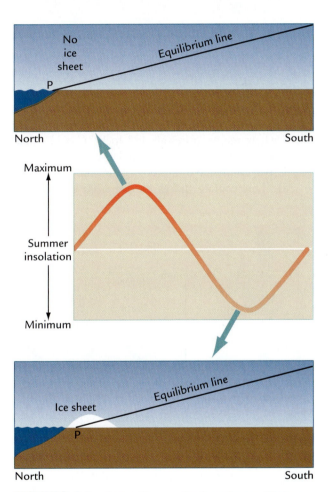

FIGURE 9-6 Insolation changes displace the equilibrium line (Top) When the equilibrium line is driven north by high values of summer insolation, the continents lie in a regime of net ablation and no ice can accumulate. (Bottom) When it is driven south by summer insolation minima, the northern landmasses lie in a regime of net accumulation and ice sheets can grow. (*P* = climate point.) (Modified from J. Oerlemans, "The Role of Ice Sheets in the Pleistocene Climate," *Norsk Geologisk Tidsskrift* 71 [1991]: 155–61.)

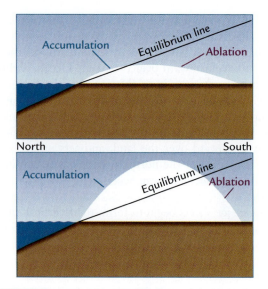

FIGURE 9-7 Ice elevation feedback As ice sheets grow higher, more of their surface lies above the equilibrium line in a regime of net accumulation, even if the equilibrium line does not move.

reach altitudes where temperatures are much colder. The tops of the ice sheets can reach elevations of 2 to 3 km, where temperatures are 12° to 19°C cooler than those at sea level (using an average lapse-rate cooling of 6.5°C per kilometer of altitude (companion Web site, p. 15). This cooling makes the ice mass balance still

more positive and increases the accumulation of snow and ice. In effect, once ice sheets start to grow, they contribute to their own positive mass balance by growing upward into a net accumulation regime. Eventually the tops of the ice sheets reach elevations at which snowfall is lower because the frigid air contains very little water vapor and this positive feedback effect weakens.

9-2 Ice Sheets Lag Behind Summer Insolation Forcing

The response of ice sheets to changes in summer insolation is far from immediate. Imagine what would happen if climate suddenly cooled enough to permit snow to fall throughout the year and accumulate rapidly over all of Canada. Based on the mass balance values plotted in Figure 9–1, about 0.3 m of ice might accumulate each year. But even under this unrealistically favorable condition, a full-sized ice sheet 3000 m thick would take 10,000 years to form. In reality, the ice would take much longer to accumulate because the initial cooling would not be instantaneous.

The geologist John Imbrie and his colleagues have led the exploration of the link between insolation and ice volume. They use as an analogy a conceptual model based on variations through time in the intensity of the flame in a Bunsen burner shown in Figure 9–8 (top); also see Chapter 1. Because water has a high heat capacity

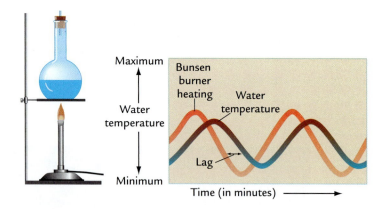

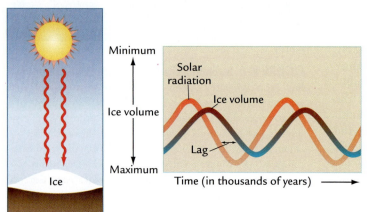

Figure 9-8 Ice volume lags insolation (Top) As the flame of a Bunsen burner is alternately turned higher and lower, the water heats and cools but with a short time lag behind the changes in heating. (Bottom) Similarly, long-term increases and decreases in summer insolation heating cause ice sheets to melt and grow but with lags of thousands of years. (Adapted from J. Imbrie, "A Theoretical Framework for the Pleistocene Ice Ages," *Journal of the Geological Society* (London) 142 [1985]: 417–32.)

and reacts slowly to changes in the heating applied, changes in water temperature lag behind changes in intensity of the heat source.

Ice sheets have the same lagging response to summer insolation (Figure 9–8 bottom) but on a much longer time scale. As summer insolation declines from a maximum value, ice begins to accumulate. The rate at which ice volume grows reaches a maximum when summer insolation has fallen to its lowest value because summer ablation is at a minimum at that time. Later, as summer insolation begins to increase, the ice sheet continues to grow for a while longer because insolation values are still relatively low. The ice sheet does not reach its maximum size until insolation crosses the value that initiates net ice ablation. As a result, the maximum in ice volume lags thousands of years behind the minimum in summer insolation.

As summer insolation continues to increase, the rate of ablation increases (see Figure 9–8 bottom). When insolation reaches its maximum value, the *rate* of ablation also reaches its maximum. Again, however, the minimum in ice volume does not occur at the insolation maximum because falling insolation values still remain high enough to melt even more ice for thousands of years. Later, about halfway through the drop in insolation, ice volume reaches its maximum value and then starts to decrease in a regime of increasing ablation. This relationship can be described by the equation shown in Box 9–1.

The persistent delay in ice volume relative to summer insolation shown in Figure 9–8 (bottom) is the **phase lag** between the two cycles. Remember that the period of a cycle is the interval of time separating successive peaks or successive valleys. In this example, the phase lag of ice volume behind summer insolation represents one-quarter of the cycle length.

The same relationship can be applied to the separate ice volume responses driven by the insolation cycles of orbital tilt and precession. At the orbital tilt cycle, the ice volume response would have the same regular sine wave shape as the summer insolation signal, but it would lag behind it by one-quarter of the 41,000-year wavelength or about 10,000 years (Figure 9–9A). At the precession cycle, the ice volume response would again lag insolation by one-quarter of its 23,000-year length, or just under 6000 years. The ice volume response at this cycle also shows the same amplitude modulation as the precession insolation signal (Figure 9–9B). As we will see later, the actual lags of the ice sheets behind the solar forcing are large but not quite a full quarter wavelength.

9-3 Delayed Bedrock Response beneath Ice Sheets

As ice sheets grow, so does the pressure of their weight on the underlying bedrock. Although the density of solid ice (just under 1g/cm^3) is much lower than that of the underlying rock (about 3.3 g/cm^3), ice sheets can reach thicknesses in excess of 3000 m, equivalent to the weight of 1000 m of rock. This load is enough to depress the underlying bedrock far beneath its level when no ice sheets existed.

One way to visualize this process is a thought experiment in which an ice sheet 3.3 km thick is instantaneously

---

BOX 9-1   LOOKING DEEPER INTO CLIMATE SCIENCE

Ice Volume Response to Insolation

The dependence of ice volume on summer insolation can be expressed by this equation:

$$\frac{d(I)}{d(t)} = \frac{1}{T}(S-I)$$

where I is ice volume, $\frac{d(I)}{dt}$ is the rate of change of ice volume per unit of time (t), T is the response time of the ice sheet (in years), and S is the summer insolation signal.

This equation specifies that the rate of change of ice volume with time is a function of two factors. One factor is T, the response time of the ice sheets, measured in thousands of years. The equation specifies that the larger the time constant T of ice response, the slower will be the resulting rate of ice volume change, which depends on the inverse of the value of T, or $1/T$.

The second factor controlling the rate of ice volume change is the term $S - I$, which is a measure of the degree of disequilibrium (offset) between the summer insolation forcing signal (S) and the ice volume response (I). Conceptually, the disequilibrium can be thought of in this way: the ice volume response (I) is constantly chasing after the insolation forcing signal (S), but it never catches up to it. For example, when insolation is at a minimum, ice volume is growing at its fastest rate toward its maximum value, but has not yet gotten to that value. When ice volume finally does reach its maximum size, the insolation curve has already turned and risen halfway toward its next maximum, causing the ice volume curve to reverse direction and shrink.

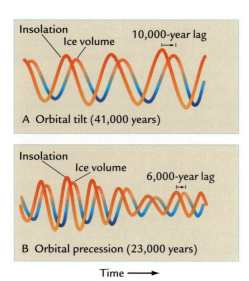

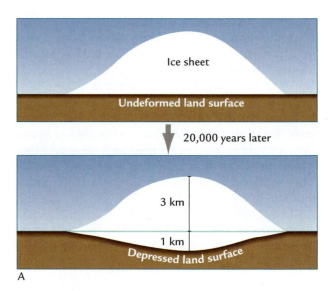

FIGURE 9-9 Ice volume lags tilt and precession
(A) At the 41,000-year cycle of orbital tilt, the lag of ice sheet size behind changes in summer insolation approaches one-quarter wavelength, or 10,000 years. (B) At the 23,000-year cycle of orbital precession, ice sheets lag roughly one-quarter wavelength (6,000 years) behind changes in summer insolation and show the same modulation of amplitude.

(18,000 years), only a tiny fraction of the eventual bedrock depression remains unrealized.

Bedrock behavior would work in the same sense but in the opposite direction if the ice load were abruptly removed. The rock surface would rebound toward the level that is in equilibrium with the absence of an ice load. The initial rapid elastic rebound would be followed by a slow viscous rebound lasting thousands of years. Today parts of Canada (in the Hudson Bay region) and Scandinavia (around the Baltic Sea) are still undergoing a delayed slow viscous rebound in response to ice melting that occurred many thousands of years ago.

Actual ice sheets in nature grow and melt much more slowly than these idealized (instantaneous) examples.

loaded onto bedrock (Figure 9–10A). In time, the 3.3-km ice sheet would eventually depress the underlying bedrock by 1 km. To put this bedrock change in a climatic context, 1 km of elevation change is equivalent to a 6.5 °C change in temperature at Earth's prevailing lapse rate. For this reason, these large changes in bedrock elevation can translate into significant effects on temperature and mass balance at the surface of the overlying ice sheet.

Bedrock responds to the ice load in two phases (Figure 9–10B). The initial reaction is a quick sagging beneath the weight of the ice. This **elastic response** represents about 30% of the total vertical change in the bedrock. Over the next several thousand years, the bedrock continues to sink in a much slower (and larger) **viscous response** caused by the extremely slow flow of rock in a relatively "soft" layer of the upper mantle between 100 and 350 km depth (see Chapter 4).

This viscous response slows progressively as the bedrock adjustment moves toward a final state of equilibrium. Viscous behavior has a response time (see Chapter 1) of about 3000 years: that is, about half of the remaining response needed to reach final equilibrium is achieved every 3000 years. The rate of change of the curve gradually slows through time because each successive 3000-year response time eliminates half of the remaining (unrealized) response (1 > 1/2 > 1/4 > 1/8, and so on). After six response times of 3000 years each

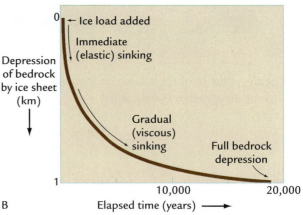

FIGURE 9-10 Bedrock sinking (A) If an ice sheet 3.3 km thick were suddenly placed on the land, the bedrock would sink almost 1 km under the load. (B) The initial sinking would be elastic and immediate, but the later response would be viscous and slower, with about half of the remaining sinking occurring every 3000 years.

When ice begins to accumulate on the land, the immediate (elastic) sagging of the bedrock depresses the land and promotes ice sheet melting. But the slower (and larger) viscous bedrock response keeps the growing ice sheet at higher elevations, where temperatures are colder, ablation is slower, and the ice mass balance is more positive. Overall, the delay in bedrock sinking provides a positive feedback to the growing ice sheet (Figure 9–11 top).

Bedrock plays the same overall positive feedback role during times when the ice is melting (Figure 9–11 bottom). The weight of a large ice sheet that exists for thousands of years creates a deep depression in the underlying bedrock. As the ice begins to melt, the (smaller) elastic part of the rebound quickly lifts the bedrock and eliminates part of the depression. But the (larger) viscous part of the rebound leaves the ice sheet at lower elevations in the depression it created and in warmer air that causes further ice melting.

9-4 Full Cycle of Ice Growth and Decay

In this section, we explore the interactions of insolation, ice volume, and bedrock responses during a typical cycle of ice sheet growth and decay (Figure 9–12). Because of the long lags inherent in the responses, the factors interact in an intricate way to create and destroy ice sheets.

We start with an interglacial maximum with the climate point *P* located in the Arctic Ocean and with no ice sheet present on the northern continent (Figure 9–12A). As summer insolation begins to decrease from a previous maximum, the equilibrium line shifts to the south and the climate point gradually moves onto the land. Some snow survives summer ablation on the far northern part of the continent, and a small ice sheet begins to form (Figure 9–12B).

As the ice sheet slowly grows, it reaches higher, colder elevations where accumulation dominates over ablation (Figure 9–12C). The ice also advances southward, partly because the equilibrium line is moving south and partly because internal flow from the area of ice accumulation in the north carries ice to the south. The thickening ice sheet slowly begins to weigh down the bedrock, but most of the bedrock depression lags several thousand years behind ice accumulation. This delay in bedrock sagging helps to keep the surface of the ice sheet at higher and colder elevations where accumulation exceeds ablation.

The highest rate of ice accumulation occurs when summer insolation reaches a minimum value and the equilibrium line is displaced farthest south (Figure 9–12C). At this point the ice sheet has not yet reached maximum size because of the lag of ice volume behind the insolation driver. The rapid growth of new ice continues to weigh down the bedrock even more, with a lag of thousands of years for each new increment of ice.

Summer insolation then begins to increase and shift the equilibrium line slowly back to the north, but the ice sheet continues to grow to its maximum size for several thousand years (Figure 9–12D). Ice growth continues because insolation levels are still relatively low and because most of the surface of the ice sheet lies above the equilibrium line, protected from the slowly increasing levels of ablation.

At some point, the combined effects of the ongoing northward shift of the equilibrium line along with the

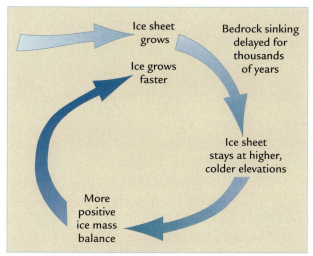

A Low summer insolation

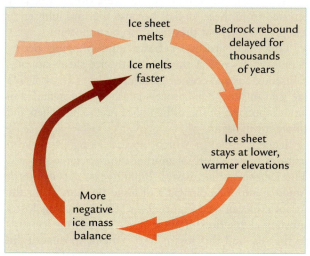

B High summer insolation

FIGURE 9-11 Bedrock feedback to ice growth and melting
(A) Delayed bedrock sinking during ice accumulation and (B) delayed rebound during ice melting provide positive feedback to the growth and decay of ice sheets.

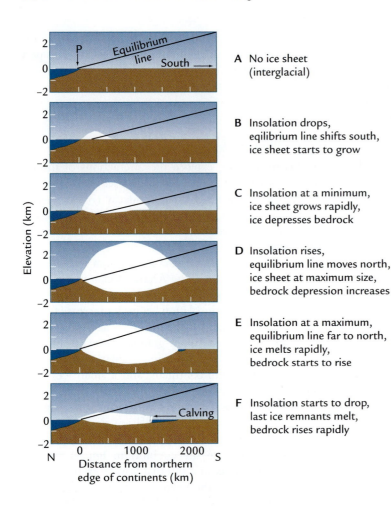

FIGURE 9-12 Cycle of ice sheet growth and decay Typical cycles of ice sheet growth and decay incorporate both the delayed response of the ice sheets to summer insolation forcing and the delayed response of the bedrock to the ice sheet loading and unloading. (Adapted from J. Oerlemans, "The Role of Ice Sheets in the Pleistocene Climate," *Norsk Geologisk Tidsskrift* 71 [1991]: 155-61.)

A No ice sheet (interglacial)

B Insolation drops, eqilibrium line shifts south, ice sheet starts to grow

C Insolation at a minimum, ice sheet grows rapidly, ice depresses bedrock

D Insolation rises, equilibrium line moves north, ice sheet at maximum size, bedrock depression increases

E Insolation at a maximum, equilibrium line far to north, ice melts rapidly, bedrock starts to rise

F Insolation starts to drop, last ice remnants melt, bedrock rises rapidly

increasing amount of bedrock depression bring a large area of the southern end of the ice sheet below the equilibrium line (Figure 9–12E). With a larger surface area of the ice sheet now undergoing ablation, the overall mass balance turns negative and ice volume begins to decrease. Melting is aided by the delayed rebound of bedrock from the weight of the earlier (larger) ice sheet. In effect, the southern edge of the ice sheet is now sagging into the bedrock hole it had previously created.

Eventually, rising summer insolation drives the equilibrium line far enough north to move the climate point back over the Arctic Ocean (Figure 9–12F). Most of the remaining ice now lies in an ablation regime, and the last remnants may disappear several thousand years later. But if a small amount of ice survives intact through a summer insolation maximum, it can serve as a nucleus from which the next large ice sheet can grow.

9-5 Ice Slipping and Calving

Ice is transferred within the body of ice sheets by slow flow from colder, higher regions of net accumulation to lower, warmer regions of net ablation (companion Web site, pp. 27–30). In two-dimensional ice models, this flow is usually represented in a simplified way as a slow diffusion (spreading) of ice from higher to lower elevations. Several other types of ice behavior are usually omitted from these models because they are inherently less predictable, but they may also be important.

One such process is **basal slip.** Slipping occurs because meltwater at the base of the ice sheet saturates soft sediments and creates a lubricated layer across which the ice can slide. This process is usually not included in models because of the difficulty of predicting when and where it will occur.

Iceberg calving, which occurs along the ocean margins of ice sheets, is another unpredictable process. The ice sheets that existed 20,000 years ago in North America and Scandinavia had large borders along the Atlantic Ocean (see Figure 9–4). These margins lost a substantial fraction of their mass by calving icebergs to the sea, and this loss is also difficult to quantify in models because it is irregular and unpredictable.

One method for modeling the long-term evolution of ice sheets is to couple a two-dimensional (altitude/latitude) ice sheet model like the one shown in Figure 9–5 to a simplified two-dimensional physical model of the atmosphere and ocean. Similar to the ice sheet models, the 2-D atmosphere-ocean models have one vertical dimension (altitude in the atmosphere, depth in the ocean) and one horizontal dimension (latitude); changes in the other horizontal dimension (longitude) are omitted. The goal of these coupled models is to simulate the linked changes in ice sheets and the atmosphere-ocean system.

Northern Hemisphere Ice Sheet History

The history of glaciation in the northern hemisphere has been reconstructed during the last four decades. The two most definitive kinds of evidence of this history have come from the ocean.

9-6 Ice Sheet History: $\delta^{18}O$ Evidence

At first thought, it might seem that the best records of past glaciations would be found on continents where the ice sheets actually existed. Ice erodes underlying sediments and bedrock and deposits long moraine ridges containing unsorted sediment called *till* (Chapter 2). Unfortunately, these deposits are of little use in reconstructing long-term glacial history because each successive glaciation erodes and destroys most of the sediment left by the previous ones. The few undisturbed deposits that remain are isolated fragments beyond the reach of radiocarbon dating.

Continuous records of glacial history come from ocean basins where sediment deposition is uninterrupted. Ocean sediments contain two key indicators of past glaciations: (1) ice-rafted debris, a mixture of coarse and fine sediments delivered to the ocean by melting icebergs that calve from ice sheet margins; and (2) $\delta^{18}O$ records from the shells of foraminifera, which provide a quantitative measure of the combined effects of changes in ice volume and in the temperature of ocean water (Appendix 1). These signals accumulate layer by layer in sediments on the ocean floor.

Decades ago, the marine scientists Cesare Emiliani and Nick Shackleton pioneered the use of oxygen-isotope ratios recorded in the shells of marine foraminifera to study past climates. In the 1950s and 1960s Emiliani analyzed $\delta^{18}O$ records extending back a few hundred thousand years and interpreted the $\delta^{18}O$ variations primarily as a record of past temperature changes. In the late 1960s Shackleton proposed instead that the $\delta^{18}O$ signals were mostly a record of changing global ice volume, with only a small overprint from temperature changes. (Current thinking is that the effect of ice volume on $\delta^{18}O$ signals lies somewhere between these two views and is slightly larger than the temperature effect in most regions). Using new mass spectrometers capable of analyzing very small samples, Shackleton published detailed $\delta^{18}O$ signals based on bottom-dwelling (benthic) foraminifera and extended our knowledge of glacial history much further into the past.

In 1976 James Hays and John Imbrie joined with Shackleton to write a landmark paper that conclusively linked changes in $\delta^{18}O$ to changes in orbital insolation. They found that orbital periods were clearly present in $\delta^{18}O$ changes over the last 300,000 years and that the $\delta^{18}O$ changes lagged behind changes in summer insolation forcing by several thousand years, equivalent to the lag of ice volume behind summer insolation predicted by Milankovitch.

The first continuous and detailed $\delta^{18}O$ record of the entire 2.75 Myr of northern hemisphere glacial history was compiled in the late 1980s by isotopic analysis of benthic foraminifera from the North Atlantic Ocean (Figure 9–13). The core from which this $\delta^{18}O$ record was taken also contains ice-rafted debris from ice sheets on the adjacent continents.

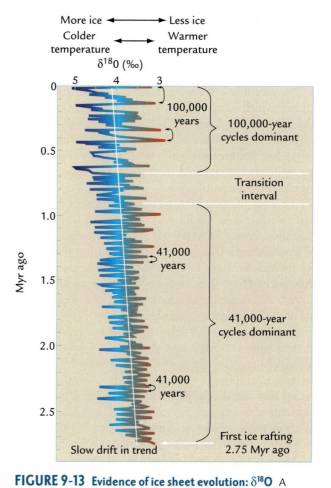

FIGURE 9-13 Evidence of ice sheet evolution: $\delta^{18}O$ A sediment core from the North Atlantic Ocean reveals a long $\delta^{18}O$ record of ice volume and deep-water temperature change. No major ice sheets existed before 2.75 Myr ago, after which small ice sheets grew and melted mainly at a cycle of 41,000 years until 0.9 Myr ago. Since that time, large ice sheets grew and melted at intervals near 100,000 years. The diagonal white line shows a gradual long-term $\delta^{18}O$ trend toward more ice and colder temperature. (Adapted from M. E. Raymo, "The Initiation of Northern Hemisphere Glaciation," *Annual Reviews of Earth and Planetary Sciences* 22 [1994]: 353-83.)

This long $\delta^{18}O$ record shows two trends: (1) a gradual drift toward more positive values and (2) numerous cyclic-looking oscillations between positive and negative values. Both features reflect some combination of changes in temperature and fluctuations in ice volume (Appendix 1). Changes toward more positive $\delta^{18}O$ values indicate more ice on the land and/or a cooling of deep-ocean temperatures. More negative $\delta^{18}O$ values indicate smaller ice sheets and/or warmer deep-ocean temperatures.

Before 2.75 Myr ago, the $\delta^{18}O$ values were relatively negative (+3.5‰ or larger) and no ice-rafted debris was present. During this interval, northern hemisphere ice sheets either did not exist or never reached the size needed to send large numbers of icebergs to the North Atlantic south of Iceland. The smaller variations in $\delta^{18}O$ during this interval probably reflect temperature changes in the deep waters.

Beginning 2.75 Myr ago, significant amounts of ice-rafted debris appeared in the record, an indication that ice sheets were now present at least sporadically. This debris accumulated during intervals of positive $\delta^{18}O$ values, which occurred mainly at a regular cycle of 41,000 years (see Figure 9–13). This part of the record suggests that ice sheets were now forming during intervals of low summer insolation but that all or most of the ice probably disappeared during the subsequent summer insolation maxima.

This regime of 41,000-year cycles persisted for the first two-thirds of the interval of northern hemisphere glaciation from 2.75 to 0.9 Myr ago. The forty or more $\delta^{18}O$ oscillations that can be detected during this interval indicate at least forty episodes of glaciation. The slow background shift of the $\delta^{18}O$ signal toward more positive values during this interval also indicates a gradual underlying drift into a colder world.

Beginning near 0.9 Myr ago and becoming more obvious after 0.6 Myr ago, the character of the $\delta^{18}O$ record changes (see Figure 9–13). Maximum $\delta^{18}O$ values increase in amplitude but are spaced farther apart, indicating that ice sheets persisted for longer intervals of time and grew larger in a colder world. These glacial intervals come to an end during abrupt $\delta^{18}O$ decreases that indicate rapid ice melting and ocean warming. Over the last 0.6 Myr, there have been six of these large $\delta^{18}O$ maxima, each followed by an abrupt deglaciation (called a **termination**) at an average spacing near 100,000 years. Almost hidden in the highly compressed record shown in Figure 9–13 are smaller 41,000-year and 23,000-year $\delta^{18}O$ oscillations that persist during the last 0.9 Myr as secondary cycles superimposed on the larger oscillations near 100,000 years.

To get a clearer sense of the character of these later cycles, we zoom in on the most recent part of a record

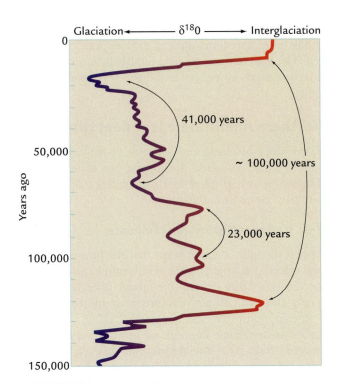

FIGURE 9-14 **Ice sheet $\delta^{18}O$ changes over the last 150,000 years** A multicore combined $\delta^{18}O$ record covering the last 150,000 years shows 23,000-year and 41,000-year oscillations in addition to the larger oscillation near 100,000 years. (Adapted from D. Martinson et al., "Age Dating and the Orbital Theory of the Ice Ages: Development of a High-Resolution 0 to 300,000-Year Chronostratigraphy," *Quaternary Research* 27 [1987]: 1–29.)

that begins during the major glaciation near 150,000 years ago (Figure 9–14). Near 130,000 years ago, an abrupt shift occurred into an interglacial interval that lasted until 120,000 years ago. Like the modern interglaciation, this interval had no ice-rafted debris in North Atlantic sediments, because northern ice sheets were not present except on Greenland.

Between 125,000 and 80,000 years ago, the $\delta^{18}O$ signal oscillated several times between values that indicate more or less ice and colder or warmer temperatures. The spacing of these oscillations at approximately 23,000 years confirms the presence of the orbital precession signal in this record. The two later glacial maxima near 63,000 and 21,000 years ago are separated by about 42,000 years, an indication that the 41,000-year orbital tilt signal is also present in this record.

The rapid transition between 17,000 and 10,000 years ago marks a second abrupt deglaciation, the first since 130,000 years ago. These terminations are the most prominent marker of the longer-period oscillations near a period of 100,000 years.

IN SUMMARY, evidence from $\delta^{18}O$ signals indicates that ice sheets have fluctuated at orbital cycles of approximately 23,000, 41,000, and 100,000 years during the long history of northern hemisphere glaciation. Unfortunately, these $\delta^{18}O$ signals also contain a large temperature overprint that makes it difficult to constrain the actual size of the ice sheets. For this reason, independent confirmation is needed to confirm that $\delta^{18}O$ signals provide a reasonably accurate history of ice volume.

9-7 Confirming Ice Volume Changes: Coral Reefs and Sea Level

The oceans provide an independent measure of ice volume—the fossil remains of coral reefs. Today most coral reefs grow in warm tropical ocean water and prefer clear water near small islands (Figure 9–15) rather than water muddied by river runoff from large continents. Coral reefs grow near sea level, and the species most useful to climate scientists (such as *Acropora palmata*) grow at or just below sea level. Reefs have strong structural frameworks that remain intact long after individual coral organisms have died and that preserve records of past sea level positions.

As sea level rises and falls, coral reefs migrate upslope and downslope. In effect, ancient reefs function as dipsticks that measure the past level of water in the world ocean. Over orbital cycles of tens to hundreds of thousands of years, fluctuations in sea level result mainly from changes in the amount of water extracted from the ocean and stored in ice sheets on land. As a result, the sea level history recorded by the coral reef dipsticks is a record of ice volume on land.

Old coral reefs can be dated by radiometric decay methods. Their skeletons contain small amounts of ^{234}U, which slowly decays to ^{230}Th (Chapter 2). This dating technique is well suited for use during the last several hundred thousand years. The sea-level record from dated coral reefs can be compared with changes in the $\delta^{18}O$ signal covering the last 150,000 years (see Figure 9–14).

Ocean islands in tectonically stable regions like Bermuda have a prominent fossil coral reef that dates to 125,000 years ago and lies about 6 m above modern sea level. Reefs of this age are unique during the last 150,000 years; they are the only indication of a sea level higher than today. This evidence agrees with the marine $\delta^{18}O$ record in Figure 9–14: the only $\delta^{18}O$ minimum comparable to the modern value within the last 150,000 years dates to between 130,000 and 120,000 years ago. Both types of evidence agree that this was the only interval in the last 150,000 years when the amount

FIGURE 9-15 Coral reefs Coral reefs form in clear, shallow waters in warm tropical seas. (Ian Cartwright/PhotoDisc.)

of ice on Earth was as small as it is now. In fact, the amount was slightly smaller, because the "extra" 6 m of sea level requires that some of today's ice on Greenland or Antarctica or both had melted at that time.

Unfortunately, all other coral reefs that grew during the last 150,000 years formed when sea level was below the modern position because of the greater amount of seawater tied up in ice sheets. These other reefs must today lie submerged deep on the underwater slopes of these islands and beyond easy reach. To circumvent this problem, marine scientists have turned to ocean islands in a different tectonic setting: areas where tectonic uplift has raised older reefs that formed below today's sea level and has exposed them above sea level (Figure 9–16).

The two most intensively studied of these islands are Barbados in the eastern Caribbean Sea and New Guinea in the western Pacific Ocean. Both islands are rimmed by prominent coral reef terraces that have been

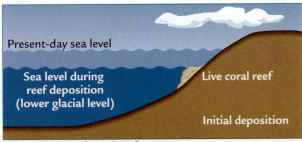

A Deposition of coral reef

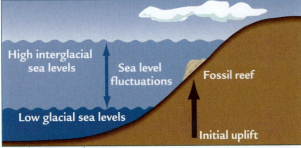

B Subsequent changes

C Present

FIGURE 9-16 Gradual uplift of coral reefs (A) Coral reefs may initially form on the edge of an uplifting island at times when sea level lies below its modern position. (B) As time passes, uplift steadily raises the island and the fossil reef toward higher elevations, while sea level moves up and down against the island in response to changes in ice volume. (C) Today old fossil reefs can lie well above sea level as a result of uplift.

lifted out of the ocean since they formed (Figure 9–17). One of the major reefs exposed on both Barbados and New Guinea dates from 125,000 years ago, the same age as the last interglacial reefs on tectonically stable islands. Two other reefs date from about 104,000 and 82,000 years ago. The ages of the two younger reefs on Barbados and New Guinea match the ages of two prominent minima in the $\delta^{18}O$ signal in Figure 9–14. These reefs record oscillations toward higher sea levels because of ice melting, but the melting was not as complete as today and sea levels were not as high.

Recent technological advances that permit drilling of deeply submerged reefs have shown that the lowest global sea level reached in the last 125,000 years was somewhere between –110 m and –125 m below the modern level near 20,000 years ago. The timing of this sea level minimum correlates with the largest $\delta^{18}O$ maximum in Figure 9–14, again confirming that $\delta^{18}O$ is a good index of ice volume.

The sea-level minimum from 20,000 years ago and the +6-m maximum level from 125,000 years ago can serve as anchor points to remove the effect of uplift on these tectonically active islands and to calculate changes in sea level during the intervening interval (Box 9–2). This method reveals that the reefs at 82,000 and 104,000 years ago were formed when sea level was lower than today by an estimated 17 m. This estimate falls about 15% of the way from full interglacial to full glacial sea levels, about the same as the relative (proportional) change in $\delta^{18}O$ between the minimum value 125,000 years ago and the maximum value 20,000 years ago (see Figure 9–14). This agreement provides even more confirmation that $\delta^{18}O$ is a good index of ice volume, despite the temperature overprint known to be present. Each 10-m change in global sea level results in an isotopic ($\delta^{18}O$) change of 0.8–1.1‰.

IN SUMMARY, coral reefs that formed during the last 150,000 years confirm that the $\delta^{18}O$ signal is a reasonable proxy for ice sheet size. The ages of the most prominent $\delta^{18}O$ minima correspond to the ages of coral reefs formed during high stands of sea level caused by reduced ice volume, and the amplitudes of the sea level changes estimated from the reefs correspond to the relative changes in ice volume inferred from the $\delta^{18}O$ signal.

Is Milankovitch's Theory the Full Answer?

As noted earlier, the 2.75-Myr history of northern hemisphere glaciation shown in Figure 9–13 has two defining characteristics. One feature is the long sequence of $\delta^{18}O$ oscillations, each of which lasted for tens of thousands of years. As noted previously, this aspect of the signal is linked to orbital forcing of glacial cycles. Milutin Milankovitch, who died in 1958, did not live to see these remarkable records of glacial history. During his lifetime, the only records available were those from glacial moraines on land, and most of those were deposited during or after the most recent glaciation 20,000 years ago. Until well into the 1960s, most glacial geologists thought that only four or five glaciations had occurred, but the marine $\delta^{18}O$ record in Figure 9–13 revealed some fifty glacial maxima. The other obvious feature in this $\delta^{18}O$ record is the slow, persistent increase in $\delta^{18}O$ values. This trend is part of the gradual cooling that has been underway for 55 Myr and that led to northern hemisphere glaciation (Chapter 6).

FIGURE 9-17 Uplifted coral reef terraces Terraces formed by erosion-resistant coral reefs lie well above sea level on the island of New Guinea in the western Pacific. (Courtesy of Arthur Bloom, Cornell University.)

These longer-term tectonic and shorter-term orbital controls on ice sheets are combined here in an initial interpretation of the history of northern hemisphere glaciation (Figure 9–18). The first component of this conceptual model is a signal representative of changes in summer insolation at high northern latitudes. This curve incorporates the combined influence of the cycles at 23,000 years (precession) and at 41,000 years (tilt). Because insolation changes at these cycles have varied

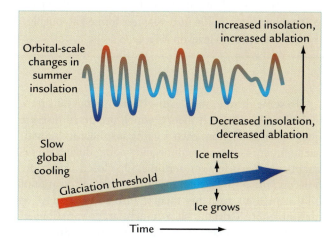

FIGURE 9-18 Factors in long-term evolution of ice sheets The long-term evolution of ice sheets reflects the interaction of two factors: (top) cyclic changes in summer insolation that drive shorter-term changes in ice sheet mass balance and (bottom) a much more gradual global cooling represented by a slowly changing glaciation threshold. Ice sheets accumulate when summer insolation falls below the critical glaciation threshold and melt when it rises above it.

around a constant long-term mean value for millions of years, their basic character has stayed nearly the same, and the same curve is repeated throughout this analysis.

The second component of the model is the assumption of a threshold temperature below which ice sheets can form. This threshold is represented as an equilibrium line that separates temperatures cold enough to permit net ice accumulation (glaciation) from those warm enough to cause net melting of snow and ice. The position of this threshold relative to the insolation curve changes very slowly through time as the gradual cooling of the last several million years proceeds.

The ice sheet response in this model results from the interaction between the equilibrium line threshold and the summer insolation signal. When insolation values fall below the equilibrium line threshold, ice sheets grow. When they rise above it, ice sheets melt. Both the growth and melting of ice sheets lag several thousand years behind the insolation forcing.

This conceptual model is used to examine stages in the development of northern hemisphere glaciation (see Figure 9–19). During each stage, the threshold line is shown as flat, although it is actually changing very slowly with respect to the insolation signal. Changes in the position of the equilibrium line during intervals of maximum and minimum summer insolation are also shown on a north-south profile.

Preglaciation Phase For millions of years prior to 2.75 Myr ago, ice sheets of substantial size did not exist on North America or Eurasia. Small amounts of ice formed on Greenland (but are ignored here). Even the deepest summer insolation minima failed to reach the critical threshold necessary for glaciation to develop over the major continents (Figure 9–19A), and the climate

BOX 9-2 LOOKING DEEPER INTO CLIMATE SCIENCE

Sea Level on Uplifting Islands

Coral reefs on islands that are slowly being lifted out of the ocean by tectonic processes can be used as sea level dipsticks if the overprint caused by uplift can be removed. Ancient coral reefs now sit on the emerged flanks of islands at elevations defined relative to modern sea level. The present elevation is the result of two factors: (1) the position of sea level at the time the reef formed (the difference from the modern position), and (2) the amount of uplift of the island since the reef formed.

The amount of subsequent uplift of each reef depends on its age (older reefs have undergone greater uplift) and on the rate of uplift of the island (reefs of a given age have undergone more uplift on fast-rising islands than on slow-rising ones). Although the present elevations of these older reefs are measured from present sea level, modern sea level is just one position in a pattern of continual rises and falls through time.

First the average rate of uplift (U) is calculated for the island under study by using the last interglacial reef on the island as a reference point. Both the age of this reef (125,000 years) and the sea level at the time it formed (6 m higher than today) are known. The calculation is

$$U = \frac{(Ht - 6)}{125}$$

where U is the mean uplift rate in meters per 1000 years and Ht is the present elevation (in meters) of the 125,000-year reef on the island under study.

This calculation removes the 6-m difference in sea level between today and 125,000 years ago to isolate the average effect of the gradual uplift between the two times. With the mean uplift rate determined (and assumed to have remained constant over the last 125,000 years), the amount of uplift that has occurred since a reef of any intermediate age formed can be corrected in order to derive an estimate of sea level at the time that it formed:

$$S = h - (U)(t)$$

where S is the relative sea level at the time the older reef formed (in meters), h is the present elevation of the older reef (in meters), U is the mean uplift rate in meters per 1000 years, and t is the time elapsed since the reef formed (in thousands of years).

With these equations, the differing effects of uplift that occurred on any island can be removed. For example, New Guinea is being uplifted at a rate close to 2 m per 1000 years, while Barbados is rising at a rate of 0.3 m per 1000 years. A coral reef dated to 82,000 years ago will have been uplifted by 150 m on New Guinea since it formed, but only by 25 m on Barbados. With the correction for local uplift rates, the 82,000-year reef formed on Barbados must have formed when global sea level was about 17 m lower than it is today. The same kind of calculation shows that the 104,000-year reef formed at roughly the same relative sea level on New Guinea.

---

point remained over the Arctic Ocean at all times (Figure 9–19B). North America and Eurasia remained too warm for ice sheets to form.

Small Glaciation Phase (2.75–0.9 Myr ago) By 2.75 Myr ago, global cooling had altered the position of the equilibrium-line threshold which now began to interact with the summer insolation curve. At intervals of 41,000 or 23,000 years, summer insolation minima crossed the equilibrium line threshold (Figure 9–19C), the climate point moved southward over the continents (Figure 9–19D), and ice sheets began to grow. But before these ice sheets could grow very large, summer insolation had already begun to increase toward the next maximum, and the equilibrium line had moved northward off the continents. As a result, the ice sheets melted away.

To some degree, this small-glaciation interval agrees with the Milankovitch theory, which predicted discrete intervals of glaciation, each lagging just behind individual summer insolation minima and ending during subsequent summer insolation maxima. On closer inspection, however, the $\delta^{18}O$ oscillations during this period are not fully consistent with his theory. Insolation changes at high northern latitudes show much larger variations at the 23,000-year precession cycle than at the 41,000-year tilt cycle, both in monthly changes and in the caloric summer index used by Milankovitch (Chapter 7). Yet the $\delta^{18}O$ signal during the small glaciation phase is dominated by very strong 41,000-year variations, with much weaker changes at 23,000 years (see Figure 9–13). A plot of the relative amplitude of these orbital cycle variations using the power spectrum method (see Chapter 7)

BOX 9-2 LOOKING DEEPER INTO CLIMATE SCIENCE

CONTINUED

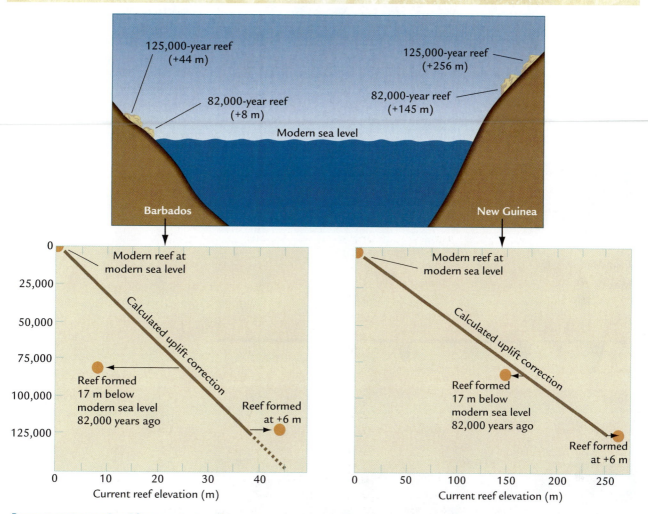

Reconstructing sea level from ancient reefs By subtracting the effects of slow tectonic uplift of the islands of Barbados and New Guinea, scientists can reconstruct sea level at earlier times when coral reefs formed.

highlights the mismatch between the rhythms present in the insolation signal and those found in the $\delta^{18}O$ (ice volume) response (Figure 9–20 top, center).

Large Glaciation Phase (0.9 Myr ago to the Present) As global cooling continued, the glaciation threshold line again shifted relative to the insolation curve. In this new regime, conditions favorable for ice accumulation rather than ablation prevailed most of the time (Figure 9–19E). Through much of this interval, the climate point remained over the land and allowed ice growth. It retreated to the Arctic Ocean only during unusually strong insolation maxima (Figure 9–19F). Because the land remained in a regime of ice accumula-

tion even during many of the smaller insolation maxima, ice sheets did not disappear as easily but persisted until a sufficiently strong maximum occurred to cause complete melting. With a longer time span in which to grow, these ice sheets became larger than the ones during the small glaciation phase.

This large glaciation phase corresponds in a general way with the interval of $\delta^{18}O$ changes after 0.9 Myr ago (see Figure 9–13). During the last 900,000 years, larger ice sheets have grown over longer intervals of time. Milankovitch would probably have been surprised by this part of the $\delta^{18}O$ record. Although the continued presence of 23,000-year and 41,000-year glacial cycles during this

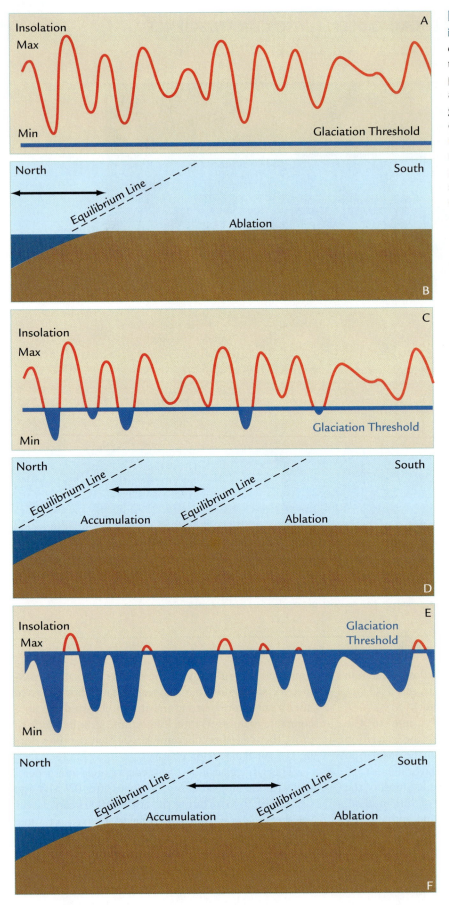

FIGURE 9-19 Conceptual phases of ice sheet evolution The long-term evolution of ice sheets should pass through three phases. (A, B) In the preglaciation phase, no ice accumulates. (C, D) In the small glaciation phase, ice accumulates during summer insolation minima but melts entirely during the next insolation maximum. (E, F) In the large glaciation phase, ice persists through weak summer insolation maxima and melts only during larger maxima.

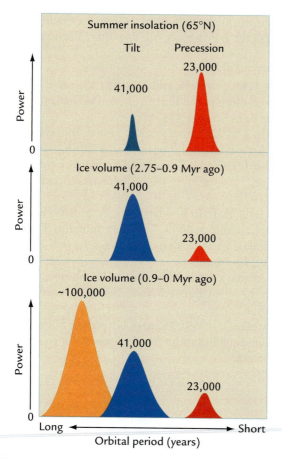

FIGURE 9-20 Spectral analysis: insolation and ice volume
(Top) Summer insolation changes at 65 °N are strong at
23,000 years, weaker at 41,000 years, and negligible at 100,000
years. (Center) In contrast, the δ¹⁸O signal of northern
hemisphere ice volume (and deep-ocean temperature) between
2.75 and 0.9 Myr ago was very strong at 41,000 years but much
weaker at 23,000 years. (Bottom) Since 0.9 Myr ago, the δ¹⁸O
signal has been strongest at or near 100,000 years.

If the global cooling trend of the last 55 Myr were to
persist for tens of millions of years more, it is possible
that the northern hemisphere would enter a different
phase in which the equilibrium line remains permanently
on the continents and ice sheets never melt. During the
last 500,000 years, we have avoided this state only during
relatively brief 10,000-year interglaciations that represent 10% of the entire time span. We have moved very
close to a state in which ice sheets are present permanently on the northern continents. Antarctica has been in
a state of permanent glaciation for many millions of
years, Greenland for at least the last million years.

IN SUMMARY, the Milankovitch theory is a useful
starting point for understanding the history of
northern hemisphere glaciation. It explains the
presence of ice volume responses at 41,000 and 23,000
years and the reason they lag several thousand years
behind the summer insolation forcing. But it fails to
explain the dominance of the 41,000-year cycle for
almost 2 Myr and the emergence of large oscillations
near 100,000 years within the last million years.
Something more complicated must be happening to
make ice sheets grow and melt at these rhythms. We
will return to this problem in Chapter 11.

Key Terms

ablation (p. 156)

calving (p. 156)

ice mass balance (p. 156)

equilibrium line (p. 156)

Milankovitch theory
  (p. 157)

climate point (p. 159)

phase lag (p. 161)

elastic response (p. 162)

viscous response (p. 162)

basal slip (p. 164)

termination (p. 166)

Review Questions

1. What is the equilibrium line and why is it
   important?

2. Why are northern ice sheets likely to be more
   responsive to insolation changes compared to ice
   in Antarctica?

3. Why does the size of a growing or melting ice
   sheet lag well behind changes in insolation?

4. How does the delay in bedrock response to ice
   loading act as a positive feedback on ice volume?

5. If the average amplitude of a 41,000-year δ¹⁸O
   cycle in the deep North Atlantic prior to 0.9 Myr
   ago was 1.0‰ and if changes in ice volume
   account for half of that total, how large was the
   average change in deep-water temperature?

interval is in part consistent with his theory, he did not
anticipate the much larger oscillations at a period near
100,000 years (see Figure 9–20 bottom). He would also
have been surprised by the saw-toothed character of
these oscillations, with relatively gradual build-up of ice
volume followed by rapid melting (see Figure 9–14).

The fact that these large oscillations in ice volume
occurred at a cycle near 100,000 years might at first
seem to be explained by changes in orbital eccentricity
at that period. But remember that changes in orbital
eccentricity act only indirectly as a multiplier effect on
the size of the precession cycle, not as a direct driver of
ice volume cycles. Incoming insolation is the mechanism that drives ice sheets, and the direct effect of the
100,000-year eccentricity variations on the insolation
signals is trivial. The large oscillations in ice volume at
or near a period of 100,000 years are not explained by
the Milankovitch theory.

6. How do corals provide an indication of the volume of water tied up in ice on land?

7. What has been the average rate of uplift of a coral reef formed during the last interglaciation 125,000 years ago and now lying 131 m above sea level?

8. Which part of the Milankovitch theory has proven accurate? Which parts are insufficient?

9. If a reef on New Guinea that has been dated to 104,000 years formed at 17 m below modern sea level, and if the average uplift rate on that island has been 2 m per thousand years, what is the present elevation of that reef?

Additional Resources

Basic Reading

Companion Web site at www.whfreeman.com /ruddiman2e, pp. 15, 27–30.

Imbrie, J., and K. P. Imbrie. 1979. *Ice Ages: Solving the Mystery*. Short Hills, NJ: Enslow.

Ruddiman, W. F. 2005. *Plows, Plagues and Petroleum*, Chapter 4. Princeton, NJ: Princeton University Press.

Advanced Reading

Hays, J. D., J. Imbrie, and N. J. Shackleton. 1976. "Variations in the Earth's Orbit: Pacemaker of the Ice Ages." *Science* 194: 1121–32.

Imbrie, J. 1982. "Astronomical Theory of the Ice Ages: A Brief Historical Review." *Icarus* 50: 408–22.

Imbrie, J. 1985. "A Theoretical Framework for the Ice Ages." *Journal of the Geological Society* (London) 142: 417–32.

Mesolella, K. J., R. K. Matthews, W. S. Broecker, and D. L. Thurber. 1969. "The Astronomical Theory of Climate Change: Barbados Data." *Journal of Geology* 77: 250–74.

Oerlemans, J. 1991. "The Role of Ice Sheets in the Pleistocene Climate." *Norsk Geologisk Tidsskrift* 71: 155–61.

Weertman, J. (1964). "Rate of Growth or Shrinkage of Non-equilibrium Ice Sheets." *Journal of Glaciology* 5: 145–58.

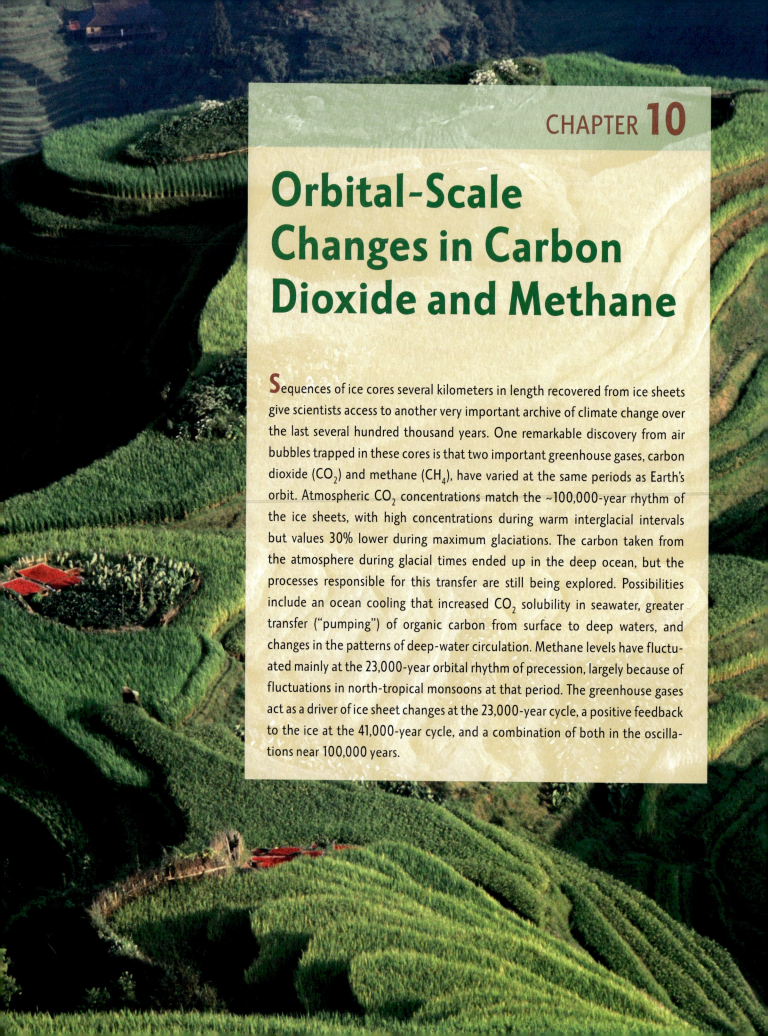

Orbital-Scale Changes in Carbon Dioxide and Methane

Sequences of ice cores several kilometers in length recovered from ice sheets give scientists access to another very important archive of climate change over the last several hundred thousand years. One remarkable discovery from air bubbles trapped in these cores is that two important greenhouse gases, carbon dioxide (CO_2) and methane (CH_4), have varied at the same periods as Earth's orbit. Atmospheric CO_2 concentrations match the ~100,000-year rhythm of the ice sheets, with high concentrations during warm interglacial intervals but values 30% lower during maximum glaciations. The carbon taken from the atmosphere during glacial times ended up in the deep ocean, but the processes responsible for this transfer are still being explored. Possibilities include an ocean cooling that increased CO_2 solubility in seawater, greater transfer ("pumping") of organic carbon from surface to deep waters, and changes in the patterns of deep-water circulation. Methane levels have fluctuated mainly at the 23,000-year orbital rhythm of precession, largely because of fluctuations in north-tropical monsoons at that period. The greenhouse gases act as a driver of ice sheet changes at the 23,000-year cycle, a positive feedback to the ice at the 41,000-year cycle, and a combination of both in the oscillations near 100,000 years.

Ice Cores

Cores from today's ice sheets span the last several hundred thousand years and contain invaluable archives of climatic signals that are not available from other sources. Two of the most important records are those of the greenhouse gases carbon dioxide (CO_2) and methane (CH_4).

10-1 Drilling and Dating Ice Cores

Scientists searching for the oldest ice in an ice sheet drill down from the top of the highest ice domes (Figure 10–1). They avoid the steeper edges because the ice there flows relatively quickly toward the ice margins and melts. In contrast, the ice that accumulates on the highest domes flows slowly down into the interior of the ice sheet, where it is stretched and thinned but not melted. As a result, the oldest ice sits under the middle of an ice sheet.

Because winter weather on ice sheets is inhospitable, drilling is done in the "summer" season (Figure 10–2). On Greenland a warm summer day may reach temperatures a few degrees below freezing (0°C), but on the Antarctic ice sheet temperatures rarely warm to –20°C in summer. Drilling takes place in structures that provide protection from the elements but keep the ice cores frozen. Hundreds of ice cores, each a few meters in length, are retrieved as drilling proceeds through several

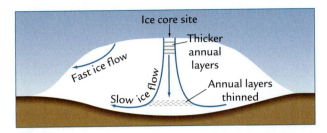

FIGURE 10-1 Ice coring The best place on an ice sheet to take ice cores is at the top of the ice dome because ice flows slowly down into the ice sheet and old ice is preserved at the bottom.

kilometers of ice. Drilling all the way through an ice sheet takes more than a single summer.

Some ice cores can be dated by counting annually deposited layers (Chapter 2). Annual layering is recorded in several properties within the ice, including layers of dust that may be visible to the eye. The dust is deposited at the end of cold, dry, windy winters. The count starts with the top layer, the year the coring operation began, and it proceeds downward as far as annual layers remain detectable. Eventually the natural stretching of ice layers by flow deeper in the ice sheet blurs the layers beyond recognition.

The layer-count method works best for ice sheets where snow is deposited rapidly. For the Greenland ice sheet, where ice accumulates at 0.5 m or more per year,

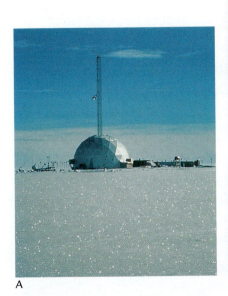

A   B   C

FIGURE 10-2 Ice coring operations (A, B) Ice drilling during cold "summer" conditions retrieves sequences of ice cores thousands of meters thick. (C) Scientists may also examine upper ice layers in pits dug into the ice. (Courtesy of Paul Mayewski, University of Maine.)

annual layers can be detected thousands of years into the past. In contrast, on the moisture-starved central domes of the Antarctic ice sheet, where ice accumulates at less than 5 cm per year (the length of a finger), annual layering is barely detectable even in the surface snow and not apparent in the stretched and thinned ice below.

For cores without annual layering, one common technique for dating the ice is to construct an **ice flow model** based on the physical properties of the ice sheet and the assumption of a smooth steady flow of ice below the surface, like that shown in Figure 10–1. These models produce fairly good estimates of the age of the ice, but they are not as accurate as annual layer counts.

Snow that falls on the high central domes of ice sheets is light and fluffy, and air circulates easily among its upper layers (Figure 10–3). As additional snow accumulates, the underlying layers are gradually buried and compressed by the weight of the overlying snow and ice, and the pressure eventually transforms the snow into crystals of ice. Little by little, the flow of air in the subsurface is reduced, and the air diffuses only slowly through the ice crystals. At a depth of about 50 m below the surface, air can no longer circulate at all and is trapped in place as small bubbles, a process called **sintering.** These air bubbles form a permanent record of the past atmosphere.

At the time the air bubbles are trapped, their slow diffusion beneath the surface makes their average age greater than that of the overlying atmosphere and the snow falling on the ice sheet. But the age of the bubbles is younger than that of the ice in which they are trapped because the surrounding ice was deposited many years earlier. The difference in age between the air bubbles and the surrounding ice depends on the rate at which the ice accumulates. If deposition of ice is fast (0.5 m/yr), the age difference between the bubbles and the ice enclosing them will only be a few hundred years. If deposition is slow (0.05 m/yr), the age offset can be as large as 2000 years or more.

10-2 Verifying Ice Core Measurements of Ancient Air

Before interpreting records of greenhouse gases trapped in ice cores, scientists first needed to verify that the techniques they use to extract and measure the gas concentrations were reliable. To do this, they measured air bubbles deposited in the upper layers of ice in cores taken from sheltered places where snow accumulates relatively rapidly. Ice cores from these sites provide measurements of CO_2 and methane values from recent centuries, with each analysis representing an average of 5 to 10 years.

Records obtained from such ice cores indicate that CO_2 concentrations in the atmosphere were about 280 ppm (parts per million by volume) near the middle of the nineteenth century. By late in the century, the concentrations began to rise toward (and past) 300 ppm (Figure 10–4A). The rising CO_2 trend in the ice core measurements then merges smoothly with a record based on instrumental analyses of air samples taken at the Mauna Loa Observatory in Hawaii beginning in 1958 by the atmospheric chemist David Keeling. The instrumental measurements show CO_2 levels accelerating from 315 ppm in 1958 to today's value of more than 385 ppm. The smooth merging of the ice core and instrumental trends shows that the CO_2 measurements in the ice cores are reliable.

Ice core measurements of methane over the last few centuries blend in a similar way with instrumental measurements made later in the twentieth century (Figure 10–4B). Ice core CH_4 concentrations were between 750 and 800 ppb (parts per billion by volume) before the 1800s but then began to accelerate, eventually merging with the rapid increase measured by instruments during and since the 1980s. The modern

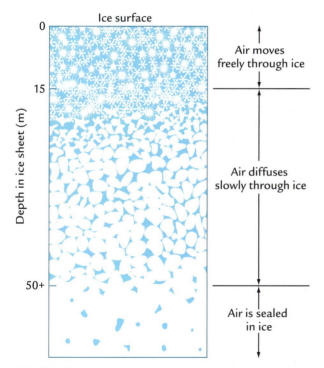

FIGURE 10-3 Sintering: Sealing air bubbles in ice Air moves freely through snow and ice in the upper 15 m of an ice sheet, but flow is increasingly restricted below this level. Bubbles of old air are eventually sealed off completely in ice 50 m or more below the surface. (Adapted from D. Raynaud, "The Ice Core Record of the Atmospheric Composition: A Summary, Chiefly of CO_2, CH_4, and O_2," in *Trace Gases in the Biosphere,* ed. B. Moore and D. Schimel [Boulder, CO: UCAR Office for Interdisciplinary Studies, 1992].)

Figure labels: Ice surface; Depth in ice sheet (m); 0; 15; 50+; Air moves freely through ice; Air diffuses slowly through ice; Air is sealed in ice

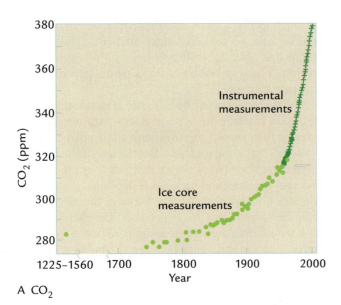

A CO_2

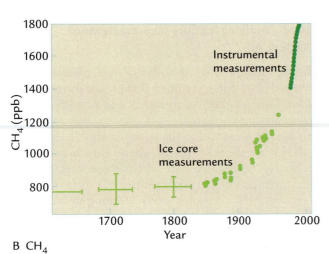

B CH_4

FIGURE 10–4 Ice core and instrumental CO_2 and CH_4 measurements Measurements of (A) carbon dioxide and (B) methane from bubbles in ice cores merge perfectly with measurements of the atmosphere in recent decades. (A: Adapted from H. H. Friedli et al., "Ice Core Record of the $^{13}C/^{12}C$ Ratio of Atmospheric CO_2 in the Past Two Centuries," *Nature* 324 [1986]: 237–38. B: Adapted from M. A. K. Khalil and R. A. Rasmussen, "Atmospheric Methane: Trends over the Last 10,000 Years," *Atmospheric Environment* 21 [1987]: 2445–52.)

methane level is near 1800 ppb. With the validity of both the CO_2 and CH_4 records in ice cores proven by their excellent match with instrumental measurements, the longer-term records of these gases extracted from ice cores can be accepted as reliable.

10-3 Orbital-Scale Carbon Transfers: Carbon Isotopes

Before we explore the long-term CO_2 and CH_4 records in ice cores, we need to revisit the distribution of carbon

in Earth's natural reservoirs and the transfers among them (Figure 10–5). In Part II, we focused on the very slow exchanges between the carbon buried in Earth's sediments and rocks and the carbon stored in Earth's surface reservoirs (the atmosphere, vegetation and ocean). Over many millions of years, the cumulative effects of these slow exchanges caused large changes in CO_2.

In this chapter, we are interested in orbital-scale changes in CO_2 and CH_4 that occur over thousands to tens of thousands of years. These faster changes can be explained only by rapid exchanges of carbon among the surface reservoirs (see Figure 3–3). Large amounts of carbon must have moved among these reservoirs during the length of an orbital cycle.

To explore how carbon has moved among these reservoirs, we need a quantitative way to track its movement. Fortunately, two **carbon isotopes** exist in nature, and different types of carbon in the climate system have distinctive carbon isotope ($\delta^{13}C$) ratios that give scientists a way of tracking how carbon has moved among these reservoirs (Appendix 2).

Most carbon occurs in oxygen-rich environments in the atmosphere, oceans, and vegetation. Carbon moves among these reservoirs in one of two forms: organic carbon, which includes both living and dead organic matter, and inorganic carbon, which consists mainly of ions dissolved in water (HCO_3^{-1} and CO_3^{-2}) but also includes CO_2 in the atmosphere (companion Web site, pp. 30–33). Abundances and typical $\delta^{13}C$ values of organic and inorganic carbon in the major reservoirs are shown in Figure 10–5.

The $\delta^{13}C$ values of inorganic and organic carbon differ mainly because of changes that occur during photosynthesis, a process by which plants create organic carbon from inorganic sources (companion Web site, pp. 30–31). During photosynthesis, plants preferentially incorporate the ^{12}C isotope rather than the ^{13}C isotope into their living tissue. This discrimination (fractionation) in favor of the ^{12}C isotope shifts the $\delta^{13}C$ composition of the organic matter produced toward more negative (^{12}C-enriched) values compared with the inorganic carbon source (Appendix 2).

Ocean **phytoplankton** (plant plankton) take inorganic carbon with $\delta^{13}C$ values near 0‰ from seawater and convert it to organic carbon with $\delta^{13}C$ values near –22‰, a net fractionation of –22‰. Overall, organic carbon forms a small fraction of the ocean reservoir, and inorganic carbon is the predominant form. Terrestrial plants use atmospheric inorganic carbon (CO_2) with a $\delta^{13}C$ value near –7‰ and convert it to organic carbon with values that range between –11 and –28‰ for different kinds of plants. This fractionation of carbon by living systems is different from the physical fractionation of oxygen isotopes by evaporation and condensation (Chapter 6, Appendix 1).

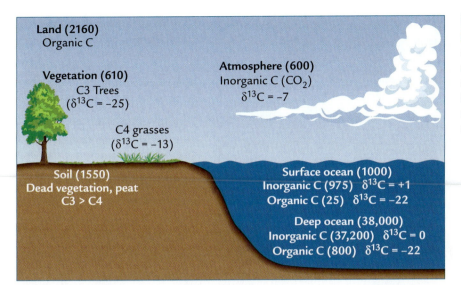

FIGURE 10-5 Carbon reservoir $\delta^{13}C$ **values** The major reservoirs of carbon on Earth have varying amounts of organic and inorganic carbon (shown in parentheses as billions of tons of carbon), and each type of carbon has characteristic carbon isotope ($\delta^{13}C$) values.

Both the organic and inorganic forms of carbon are preserved in the geologic record. $CaCO_3$ shells of marine foraminifera are formed from inorganic carbon dissolved in seawater, typically with values near 0‰. Many kinds of geologic deposits on land and in the ocean also contain small amounts of organic carbon with more negative $\delta^{13}C$ values, typically averaging −25‰. These deposits are useful for tracking past transfers of carbon among the surface reservoirs over orbital (and shorter) time scales.

Orbital-Scale Changes in CO_2

The longest records of past CO_2 changes come from sequences of cores extracted from the Antarctic ice sheet. The CO_2 record from one of these sites (Vostok Station) shows oscillations between values as high as 280–300 ppm and as low as 180–190 ppm over the last 400,000 years (Figure 10–6). These CO_2 oscillations line up well with variations in marine $\delta^{18}O$, which indicate changes in ice volume. When the ice sheets were large, CO_2 concentrations were low (190 ppm or less). When ice sheets were small or absent in the northern hemisphere, CO_2 concentrations were high (280 to 300 ppm). Ice cores extracted in recent years have extended this close link of CO_2 and ice volume back to more than 650,000 years ago.

10-4 Where Did the Missing Carbon Go?

The observation that a general match exists between CO_2 and ice volume tells us that some kind of cause-and-effect relationship exists between these two components of the climate system. One way to begin exploring this relationship is to find out the fate of the carbon that was removed from the atmosphere. The 90-ppm reductions in CO_2 values mean that almost one-third of the carbon in the atmosphere during interglacial times (~180 billion tons) moved to some other reservoir during glaciations (Figure 10–7).

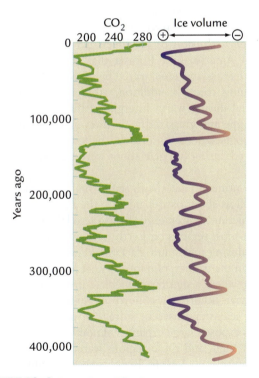

FIGURE 10-6 Long-term CO_2 changes A 400,000-year record of CO_2 from Vostok ice in Antarctica shows four large-scale cycles near a period of 100,000 years that are similar to those in the marine $\delta^{18}O$ (ice volume) record. (Adapted from J. R. Petit et al., "Climate and Atmospheric History of the Past 420,000 Years from the Vostok Ice Core, Antarctica," *Nature* 399 [1999]: 429–36.)

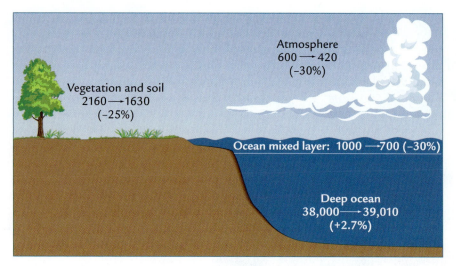

FIGURE 10-7 Interglacial-to-glacial changes in carbon reservoirs During the glacial maximum 20,000 years ago, large reductions in carbon biomass occurred in the atmosphere, in vegetation and soils on land, and in the surface ocean. The total amount of carbon removed from these reservoirs (more than 1000 billion tons) was added to the much larger reservoir in the deep ocean.

One place to look for this "missing" CO_2 carbon is the vegetation-soil reservoir. Scientists have abundant information on the amount of carbon stored on land in vegetation and soils during the last glacial maximum, 20,000 years ago. The evidence shows that the continents had less net vegetation cover and held less carbon during glaciations than they did during warm interglacial intervals like today.

The main reason for the decrease in glacial carbon was the expansion of ice sheets across large areas of North America and Eurasia that during interglacial times like today were covered by forests of conifers and deciduous trees. In addition, other regions that were forested during warm interglacial times were covered by steppe and grassland with lower amounts of carbon during glaciations. As a result, the forested regions that survived during full glaciations were smaller than those today.

Continental lakes with sediments containing pollen can tell us about past changes of the nearby vegetation. The picture that emerges from these lake core records is that most regions on Earth were drier and less vegetated during maximum glaciations than they are today. Even in the tropics, rain forests were less extensive. One region where a significant increase in glacial vegetation may have occurred was north of Australia. There, the fall of sea level caused by storage of water in glacial ice sheets exposed large expanses of now-submerged continental shelf, and these regions were probably covered by tropical rain forests. But this regional increase was not enough to offset losses elsewhere.

From this evidence, climate scientists estimate that the total amount of vegetation on land was reduced by roughly 25% (from 2160 to 1630 billion tons) during the last glacial maximum 20,000 years ago (see Figure 10–7). These estimated reductions are uncertain, and the actual reduction could be anywhere between ~15%

and 30%, but in any case the decrease was substantial. Clearly, the carbon removed from the atmosphere did not go into land vegetation, and now we face the added problem of explaining the carbon missing from not just the atmosphere but also the land.

The only remaining place where the missing carbon could have been stored is in the ocean. But which ocean reservoir took up the carbon, the small surface reservoir or the much larger deep reservoir? The surface ocean is not the answer. Ocean surface waters exchange all their carbon with the atmosphere within just a few years. Because of this rapid exchange of CO_2 gas, most areas of the surface ocean today have CO_2 values within 30 ppm of the value in the overlying atmosphere. Such rapid exchanges mean that if CO_2 values were 30% lower in the glacial atmosphere, they must have been lower by nearly the same average amount in the glacial surface ocean. This estimate adds another 300 billion tons to the growing list of carbon that was "missing" during glacial times (Figure 10–7).

The only available carbon reservoir left is the deep ocean. The missing glacial carbon removed from the atmosphere, vegetation, and surface waters must have been stored or sequestered there. The total amount of carbon missing from the other reservoirs adds up to about 1000 billion tons (see Figure 10–7). This amount must have ended up in the deep ocean during the last glaciation. But the deep ocean is such an enormous carbon reservoir (~38,000 billion tons) that this additional carbon would have increased the amount present during interglacial times by only about 2.7%.

10-5 $\delta^{13}C$ Evidence of Carbon Transfer

The amount of carbon transferred from the land to the deep ocean can be estimated independently from $\delta^{13}C$ measurements of foraminifera. Organic carbon in terrestrial vegetation is tagged with a negative $\delta^{13}C$ value averaging –25‰, whereas the large amount of inorganic carbon in the ocean has an average value near 0‰. Although a small fraction of the terrestrial organic carbon delivered to the ocean remains as organic matter, most of it is converted to inorganic carbon that retains the very negative –25‰ $\delta^{13}C$ composition.

During glacial times (Figure 10–8A), ^{12}C-rich organic carbon transferred from the land to the deep ocean and converted to inorganic form should have made the $\delta^{13}C$ value of the inorganic carbon in the ocean slightly more negative. These lower $\delta^{13}C$ values should correlate with the more positive $\delta^{18}O$ values resulting from more glacial ice and colder deep-ocean temperatures. The opposite pattern should have occurred during interglacial times (Figure 10–8B), with more positive $\delta^{13}C$ values because of the return of the negative carbon to the land and more negative $\delta^{18}O$ values because of the melting of ^{16}O-rich ice.

We can use a mass balance calculation to estimate the effect of adding very negative (^{12}C-enriched) carbon to the deep sea during maximum glaciations:

$$\underset{\substack{\text{Inorganic}\\\text{C in ocean}}}{(38,000)}\ \underset{\substack{\text{Mean}\\\delta^{13}C}}{(0‰)}\ +\ \underset{\substack{\text{C added}\\\text{from land}}}{(530)}\ \underset{\substack{\text{Mean}\\\delta^{13}C}}{(-25‰)}\ =\ \underset{\substack{\text{Glacial inorganic}\\\text{carbon total}}}{(38,530)}\ \underset{\substack{\text{Mean}\\\delta^{13}C}}{(x)}$$

where x is the $\delta^{13}C$ value of glacial inorganic carbon in the ocean and the sizes of the carbon reservoirs shown are in billions of tons of carbon. Solving for x, we find that the mean $\delta^{13}C$ value of inorganic carbon in the glacial ocean should have shifted from ~0‰ to –0.34‰ because of the addition of ^{12}C-rich carbon transferred from the land.

This estimated –0.34‰ shift has been tested by comparing it to $\delta^{13}C$ values in the $CaCO_3$ shells of glacial-age benthic foraminifera in cores distributed across the world ocean over a range of water depths. The measurements show the $\delta^{13}C$ composition of the deep waters region by region. With all these analyses combined into a global average value, the estimated $\delta^{13}C$ change for the entire ocean is –0.35‰ to –0.4‰. In view of the uncertainties involved, the good agreement between the two methods indicates that carbon isotopes are useful tracers of past carbon shifts among Earth's reservoirs.

This evidence suggests that we should also be able to trace carbon transfers during earlier glacial cycles using $\delta^{13}C$ changes measured in the shells of benthic foraminifera. Because the Pacific contains by far the largest volume of water of Earth's oceans, scientists have used changes in its deep-water $\delta^{13}C$ values through time as the best single "quick" index of average $\delta^{13}C$ changes in the global ocean. The longer-term $\delta^{13}C$ record in Figure 10–9 shows the basic relationship expected from the transfers shown in Figure 10–8: large amounts of terrestrial carbon with negative (–25‰) $\delta^{13}C$ values were transferred into the ocean every time ice sheets grew, and then were returned to the land when the ice melted.

The most negative $\delta^{13}C$ values are spaced at intervals close to 100,000 years during the last 0.9 Myr and at cycles of 41,000 years prior to 0.9 Myr ago. Although the correlation between the two signals is far from perfect, time series analysis confirms that the $\delta^{13}C$ (carbon transfer) and $\delta^{18}O$ (ice volume) signals have varied at the same periods and with the same approximate timing over millions of years.

On closer inspection, many of the $\delta^{13}C$ variations shown in Figure 10–9 exceed the –0.35 to –0.4‰ shift that can easily be attributed to transfers of carbon from land to sea. This unexpectedly large $\delta^{13}C$ response tells us that additional factors must have been affecting oceanic $\delta^{13}C$ values through time. The source of these variations is not known.

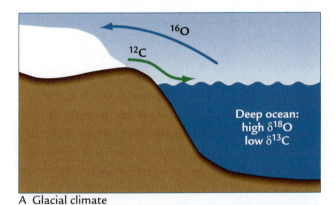

A Glacial climate

B Interglacial climate

FIGURE 10-8 Glacial transfers of ^{12}C and ^{16}O (A) During glaciations, ^{12}C-enriched organic matter is transferred from the land to the ocean at the same time that ^{16}O-enriched water vapor is extracted from the ocean and stored in ice sheets. (B) During interglaciations, ^{12}C-rich carbon returns to the land as ^{16}O-rich water flows back into the ocean.

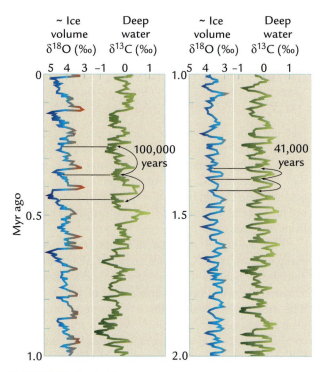

FIGURE 10-9 Carbon transfer during glaciations A sediment core from the deep Pacific Ocean shows more negative $\delta^{13}C$ values during glaciations ($\delta^{18}O$ maxima), mainly because ^{12}C-enriched carbon is transferred from the land into the ocean. (Adapted from D. W. Oppo et al., "A $\delta^{13}C$ Record of Upper North Atlantic Deep Water During the Past 2.6 Million Years," *Paleoceanography* 10 [1995]: 373–94.)

How Did the Carbon Get into the Deep Ocean?

The answer to this question is complicated and not yet resolved. Scientists are actively investigating several factors that seem likely to have played a role in this transfer.

10-6 Increased CO_2 Solubility in Seawater

Changes in the average temperature of the ocean during glaciations would have altered the chemical solubility of CO_2 in seawater and thereby affected the amount of CO_2 left in the atmosphere. Because CO_2 dissolves more readily in colder seawater, atmospheric CO_2 levels will drop by ~10 ppm for each 1 °C of ocean cooling. As we will see in Chapter 12, surface-ocean temperatures in the tropics cooled by an estimated average of 2 °–4 °C during the last glacial maximum 20,000 years ago and by larger amounts in many high-latitude regions. The much larger volume of water in the deep ocean cooled by an average of 2 °–3 °C. As a result of

this cooling, the atmospheric CO_2 concentration should have fallen by ~20–30 ppm because of increased CO_2 solubility.

The altered salinity of the glacial ocean would also have affected atmospheric CO_2, but in the opposite direction. CO_2 dissolves more easily in seawater with a lower salinity, but the average glacial ocean was saltier than it is today because of the amount of freshwater taken from the ocean and stored in ice sheets. Although some high-latitude ocean surfaces (such as the North Atlantic) became less salty during glaciations, the average salinity of the entire ocean increased by about 1.1‰, enough to cause an estimated glacial CO_2 increase of 11 ppm.

The 11-ppm CO_2 increase caused by higher salinity would have offset just under half of the 20–30 ppm decrease caused by ocean cooling for a net CO_2 drop of ~14 ppm. Most of the observed CO_2 decrease of 90 ppm must have occurred by means of other mechanisms.

10-7 Biological Transfer from Surface Waters

One way to move carbon from surface to deep waters is through biological activity. Photosynthesis occurs in sunlit surface waters where sufficient nutrients (phosphorus and nitrogen) are available to stimulate growth of marine phytoplankton (Figure 10–10). CO_2 is extracted

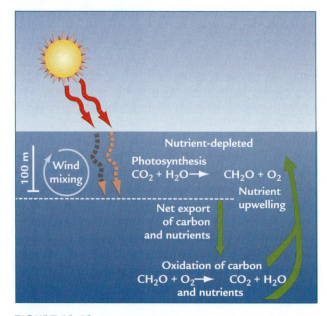

FIGURE 10-10 Photosynthesis in the ocean Sunlight that penetrates the surface layers of the ocean causes photosynthesis in microscopic marine plankton. After the plankton die, their organic tissue sinks to the seafloor and is oxidized to nutrient form. Upwelling returns the nutrients (including organic carbon) to the ocean surface.

BOX 10-1   A CLOSER LOOK AT CLIMATE SCIENCE
Using $\delta^{13}C$ to Measure Carbon Pumping

As photosynthesis sends carbon to the deep ocean, it also preferentially removes the ^{12}C isotope from surface waters and sends it to the seafloor in dead organic matter. Fractionation and removal of ^{12}C-rich carbon leaves the inorganic carbon in surface waters enriched in ^{13}C, and inorganic $\delta^{13}C$ values become more positive, ranging as high as +1 to +2‰. In the deep ocean, where organic carbon is oxidized back to inorganic form, the deep waters become enriched in ^{12}C, and the $\delta^{13}C$ values become more negative.

These opposing $\delta^{13}C$ trends provide a way to measure the strength of the ocean carbon pump: by analyzing changes in carbon isotopic values in foraminifera. The shells of planktic foraminifera record changes in surface-water $\delta^{13}C$ values through time, while the shells of benthic foraminifera record ongoing changes in deep-water $\delta^{13}C$ values. The changing difference between these two trends through time is a monitor of the changing strength of the carbon pump. If nutrient delivery to surface waters increased during glacial times, productivity and carbon pumping should also have increased. The result should have been an increase in the difference between the $\delta^{13}C$ values of surface and deep waters: more positive values in surface waters and more negative values below.

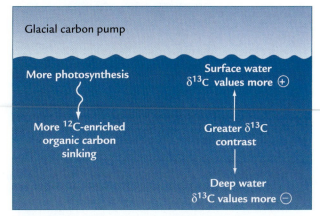

Measuring changes in the ocean carbon pump Greater photosynthesis in surface waters during glaciations would pump more organic carbon to the deep sea and reduce atmospheric CO_2 levels. Past changes in this process can be measured by the increased difference between ^{13}C-enriched carbon (higher $\delta^{13}C$ values) in the shells of planktic foraminifera living in surface waters and ^{12}C-enriched carbon (lower $\delta^{13}C$ values) in the shells of benthic foraminifera living on the deep ocean floor.

from surface waters and incorporated in organic tissue (CH_2O), some of which sinks to the deep ocean:

$$CO_2 + H_2O \rightarrow CH_2O + O_2$$

If the rate of photosynthesis increases, CO_2 concentrations in surface waters will decrease because of greater downward export of carbon in dead organic matter. To increase this transfer, sometimes referred to as the **carbon pump,** more nutrients must become available to the plant plankton in surface waters to stimulate greater productivity. If greater productivity occurred during glacial times, the increased downward transfer of organic carbon would have reduced CO_2 values in surface waters and in the overlying atmosphere. One way to track past changes in the carbon pump is to use paired $\delta^{13}C$ measurements of foraminifera that lived in surface and deep waters (Box 10–1).

Most ocean surface waters have very small amounts of nutrients, which are quickly consumed by planktic organisms. As a result, annual productivity in many mid-ocean regions is very low in the current interglacial climate (Figure 10–11). Because a range of evidence indicates that many low-productivity regions remained nutrient-depleted during glaciations, increased carbon pumping could not have occurred in these regions.

In contrast, the deep ocean is loaded with nutrients because the nitrogen and phosphorus and carbon contained in organic matter falling from surface waters are oxidized back to mineral form (see Figure 10–10). But these nutrients remain unused in the darkness of the deep sea unless they are delivered back to the surface waters by upwelling or other processes to stimulate surface productivity.

High productivity occurs today in several kinds of regions: coastal areas, where rivers and upwelling deliver nutrients; narrow regions of upwelling near the equator, especially in the Pacific Ocean; and high latitudes of the North Pacific Ocean and the Southern Ocean around Antarctica (see Figure 10–11). Large-scale changes in biological pumping of carbon to the deep sea can occur in these productive regions.

The high-latitude oceans are particularly promising locations for increased carbon pumping because they contain large amounts of unused nutrients in the

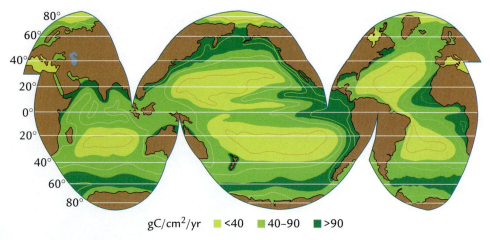

gC/cm²/yr ■ <40 ■ 40–90 ■ >90

FIGURE 10-11 Annual carbon production in the modern surface ocean Primary production of carbon (grams per square centimeter per year) is highest in shallow coastal regions, in high-latitude oceans (especially the Southern Ocean), and across equatorial upwelling belts, but it is lower in central ocean gyres. (Adapted from W. H. Berger et al., "Ocean Carbon Flux: Global Maps of Primary Production and Export Production," in *Biogeochemical Cycling and Fluxes between the Deep Euphotic Zone and Other Oceanic Realms,* National Undersea Research Program Report 88–1 [Asheville, NC: NOAA, 1987].)

current interglacial climate. Solar radiation in these areas is relatively weak, and active photosynthesis is limited to a relatively brief summer season. As a result, photosynthesis does not utilize most of the available nutrients, and high concentrations persist even during the productive season. This modern excess of nutrients means that increased productivity during glacial intervals could have transferred additional carbon to deep waters compared to today, leaving surface waters with reduced CO_2 levels.

Scientists are investigating several mechanisms that might have increased glacial productivity and downward transfer of carbon. One intriguing explanation is linked to an increase in delivery of nutrients from the continents by stronger glacial winds (Figure 10–12). The marine scientist John Martin proposed that iron, a trace element, is critical to marine life, as it is to humans. Because erosion of the land is the main source of iron for the oceans, he suggested that ocean regions should receive an iron "boost" if extra dust is blown in by winds. This concept is called the **iron fertilization hypothesis.**

Stronger glacial winds blowing from semiarid and arid continental areas should have carried greater amounts of iron to both coastal and mid-ocean areas than they do today, stimulating greater productivity over broad areas. Iron fertilization might have stimulated productivity in two regions with modern excesses of nutrients: the high latitudes of the North Pacific Ocean, which receive enormous dust influxes from central Asia; and the Atlantic sector of the Southern Ocean,

which receives smaller influxes from the Patagonian tip of South America. Extra iron arriving in these two areas during glaciations could have stimulated greater productivity and carbon pumping to the deep ocean.

Martin's hypothesis is still being debated. Oceanic field tests have shown that adding iron to surface waters can stimulate greater short-term productivity of ocean phytoplankton. Estimates of the effect of iron fertilization on atmospheric CO_2 concentrations range all the way from negligible (a few parts per million) to sizeable (several tens of parts per million). Whether or not iron fertilization stimulated a large enough increase in carbon

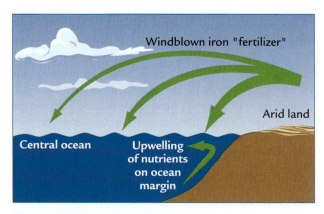

FIGURE 10-12 Iron fertilization of ocean surface waters
Dust rich in trace elements such as iron is blown from continental interiors to the ocean during glaciations. The addition of this and other nutritional supplements stimulates productivity across broad regions.

pumping and glacial productivity to alter atmospheric CO_2 concentrations remains unclear. Some scientists have suggested that the dust from the land may deliver other key elements that stimulate ocean productivity.

10-8 Changes in Deep-Water Circulation

Another mechanism for transferring more carbon to the deep ocean is to change the pattern of deep-ocean circulation. Today, most of the water that fills the deeper oceans forms in the subpolar North Atlantic Ocean or in the Southern Ocean (companion Web site, pp. 24–25). North Atlantic Deep Water flows southward through the Atlantic basin, while colder and denser Antarctic Bottom Water fills the much larger volume in the Pacific and Indian Oceans.

Evidence from $\delta^{13}C$ measurements in bottom-dwelling foraminifera indicates that the circulation pattern during glacial times was different. We saw earlier that the average $\delta^{13}C$ composition of the entire ocean became more negative during glacial times, but in this case, our focus is on *regional $\delta^{13}C$ variations in the ocean*. Varying degrees of photosynthesis (see Box 10–1) in different polar areas of the ocean give inorganic carbon distinctively different $\delta^{13}C$ values in north polar versus south polar areas. As a result, the deep waters that form from surface waters in these regions begin their downward trip with distinctively different $\delta^{13}C$ values (Figure 10–13).

At one extreme are the relatively positive $\delta^{13}C$ values (> 0.8‰) in the North Atlantic Ocean near 2000–4000 m depth. These values are positive because North Atlantic Deep Water is formed from surface waters that have been greatly enriched in ^{13}C by photosynthesis (see Box 10–1). A plume of positive $\delta^{13}C$ values at depths of 2000–4000 m defines the core of this southward flow (Figure 10–14A).

In contrast, the waters that form in the Antarctic region and flow to a wide range of depths have $\delta^{13}C$ values lower than 0.5‰. Because photosynthetic fractionation of carbon isotopes in the Southern Ocean is incomplete, extremely positive $\delta^{13}C$ values do not develop in the surface waters that feed the deeper flow. The contrast between the low-$\delta^{13}C$ water from the Antarctic and the high-$\delta^{13}C$ water from the North Atlantic makes them relatively easy to trace.

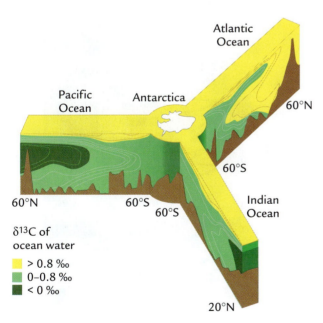

FIGURE 10-13 Modern deep-ocean $\delta^{13}C$ patterns In today's ocean, photosynthesis and carbon isotope fractionation drive $\delta^{13}C$ values higher in surface waters compared to deep waters. (Adapted from C. D. Charles and R. G. Fairbanks, "Evidence from Southern Ocean Sediments for the Effect of North Atlantic Deepwater Flux on Climate," *Nature* 355 [1992]: 416-19.)

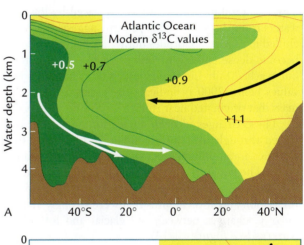

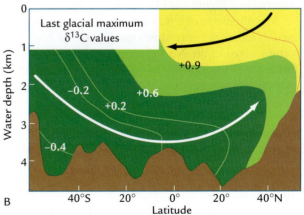

FIGURE 10-14 Change in deep Atlantic circulation during glaciation In contrast to (A) the modern distribution of $\delta^{13}C$ in the Atlantic Ocean, (B) the axis of high-$\delta^{13}C$ water formed in the north flowed south at shallower levels during the last glacial maximum. (Modified from J.-C. Duplessy and E. Maier-Reimer, "Global Ocean Circulation Changes," in *Global Changes in the Perspective of the Past*, ed. J. A. Eddy and H. Oeschger [New York: Wiley, 1993].)

A second process that affects the regional $\delta^{13}C$ pattern is called **$\delta^{13}C$ aging.** As deep waters flow from their source regions to other parts of the world ocean, their $\delta^{13}C$ values gradually become more negative. The gradual shift results from the continual downward rain of ^{12}C-rich carbon ($\delta^{13}C = -22$‰) from surface waters along the path of deep-water flow (see Box 10–1).

This aging effect is evident mainly in the Pacific Ocean, where the flow of deep water is slow enough to give the water time to age significantly (see Figure 10–13). In contrast, deep water circulates so quickly through much of the Atlantic Ocean that it has little time to age. The variations in $\delta^{13}C$ values within the deep Atlantic are determined mainly by physical mixing of water from the North Atlantic and Antarctic sources.

During the maximum glaciation 20,000 years ago, the $\delta^{13}C$ pattern in the deep Atlantic Ocean changed considerably (Figure 10–14B). The tongue of high-$\delta^{13}C$ water that today lies at 2000–4000 m shifted upward to a depth of 1500 m or less and was replaced by low-$\delta^{13}C$ waters from the south. This pattern suggests that water chilled in northern latitudes during glacial times did not become dense enough to sink as deep as it does today but reached only intermediate depths. The lower $\delta^{13}C$ values below 2000 m indicate the presence of a water mass that originated near Antarctica and flowed north.

These changes in deep circulation patterns can be traced through time by examining $\delta^{13}C$ records from benthic foraminifera in tropical Atlantic sediment cores. The equatorial Atlantic seafloor at 3000–4000 m is a good place for this kind of study because it lies in the region and at the depths that were strongly affected by $\delta^{13}C$ changes between the glacial and interglacial patterns (see Figure 10–14).

Measurements of $\delta^{13}C$ values from this region can be used to calculate the relative contributions of high-$\delta^{13}C$ water formed in the North Atlantic and low-$\delta^{13}C$ water formed in the Southern Ocean. Changes in $\delta^{13}C$ values in these source areas through time can be determined by analyzing benthic foraminifera in cores from those regions. The $\delta^{13}C$ values measured in the tropical foraminifera can then be expressed as percentage contributions from these two sources. This analysis removes the whole-ocean $\delta^{13}C$ change caused by carbon transfers from land to sea and isolates the relative contributions from the northern and southern sources through time.

A long record from the tropical Atlantic shows oscillations in the percentage of northern-source and southern-source water (Figure 10–15). Northern-source waters generally dominated during $\delta^{18}O$ minima (interglaciations), while Antarctic-source waters dominated during $\delta^{18}O$ maxima (glaciations). The measured fluctuations in deep-water sources occurred mainly at a period near 100,000 years after 0.9 Myr ago. This $\delta^{13}C$ trend indicates that a link exists between the size of

northern hemisphere ice sheets and the pattern of deep-water flow in the North Atlantic Ocean.

These changes in the pattern of deep-water circulation could have affected CO_2 concentrations in the atmosphere in several ways. One possibility is that stronger overturn of Antarctic-source waters at intermediate depths could have delivered more nutrients to surface waters and stimulated more biological productivity and downward pumping of carbon. The carbon sent to the deep ocean could then reside there "hidden" from the atmosphere.

A second potential link to atmospheric CO_2 is through the carbonate chemistry of the ocean. Concentrations of dissolved CO_2 in the surface waters of the world ocean are linked to the amount of the carbonate ion CO_3^{-2} present. These ions are produced when corrosive deep water dissolves $CaCO_3$ on the seafloor. When the CO_3^{-2} is returned to surface waters by ocean circulation, it combines chemically with dissolved CO_2 in the surface waters to produce the bicarbonate ion HCO_3^{-1}. This process removes CO_2 from surface waters. As a result, if a larger amount of $CaCO_3$ is dissolved on the seafloor, the CO_2 concentrations will be reduced in surface waters and in the overlying atmosphere.

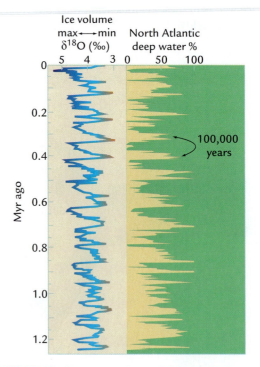

FIGURE 10-15   Changing sources of Atlantic deep water
The percentage of deep water originating in the North Atlantic and flowing to the equator during the last 1.25 Myr has been consistently lower during glaciations than during interglaciations. (Adapted from M. E. Raymo et al., "The Mid-Pleistocene Climate Transition: A Deep-Sea Carbon Isotopic Perspective," *Paleoceanography* 12 [1997]: 546–59.)

On a global average basis, this mechanism acts slowly (over thousands of years) because of the time required to dissolve deep-sea carbonates and because of the slow overturn of the deep ocean. Deep-ocean regions experienced different and often opposing changes in $CaCO_3$ dissolution during glaciations (increases in most of the Atlantic, decreases in most of the Pacific). Because these changes more or less canceled each other out at a global scale, little if any change in the average CO_3^{-2} content of deep or surface waters occurred.

Nevertheless, regional changes in CO_3^{-2} could also have altered atmospheric CO_2 concentrations. Today, deep water from North Atlantic sources comes to the surface in the Antarctic region (see Figure 10–14A). Because today's North Atlantic deep water is a relatively non-corrosive water mass, it dissolves relatively little $CaCO_3$ on the seafloor and delivers relatively small concentrations of CO_3^{-2} to the surface waters of the Southern Ocean. As a result, dissolved CO_2 concentrations can remain at relatively high levels in the Southern Ocean and the overlying atmosphere.

During glaciations, however, the water that formed in the North Atlantic sank to much shallower depths and did not intersect as much of the seafloor (see Figure 10–14B). Instead, an expanded area was bathed by southern-source water that was more corrosive, dissolved more $CaCO_3$, and eventually returned more CO_3^{-2} to Antarctic surface waters. The geochemists Wally Broecker and Tsung-Hung Peng proposed that this change in circulation would have reduced the concentration of dissolved CO_2 in the Southern Ocean along with the amount of CO_2 in the overlying atmosphere. They estimated that this mechanism, called the **polar alkalinity hypothesis,** might explain as much as 40 ppm of the observed 90-ppm decrease in atmospheric CO_2 during glacial times. Because the deep flow in the Atlantic Ocean is relatively rapid, this mechanism can influence atmospheric CO_2 concentrations within a few hundred years.

> **IN SUMMARY,** several factors probably contributed to the reduction in atmospheric CO_2 concentrations during glacial intervals: reduced CO_2 solubility in colder waters, greater biological pumping of carbon from surface to deep waters, and changes in deep-ocean circulation. The size of the latter two contributions is highly uncertain, and this problem is currently an area of intense investigation.

Orbital-Scale Changes in CH_4

In contrast to the oxidized carbon common in most of Earth's environments, the carbon in methane (CH_4) is in a reduced form that is produced when oxygen is absent. Most natural methane originates in wetlands, where decaying plant matter uses up the available oxygen and creates the necessary reducing conditions. Most natural wetlands are located in the tropics and in boreal (circum-Arctic) regions.

Methane concentrations in ice cores show a series of cyclic variations between maxima of 650–700 parts per billion (ppb) and minima of 350–450 ppb (Figure 10–16). Ice flow models show that the peaks in methane concentration fall very close to times of maxima in northern hemisphere summer insolation at the 23,000-year cycle of orbital precession. Small adjustments in estimated ages within the ice cores bring the CH_4 peaks into full alignment with July insolation maxima. This midsummer timing for methane peaks is supported by the age of the most recent CH_4 maximum in

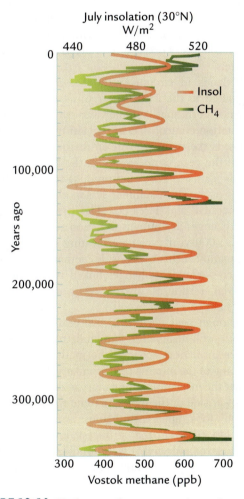

FIGURE 10-16 Methane and monsoons The methane record from Vostok ice in Antarctica shows regular cycles at intervals of 23,000 years and closely resembles the monsoon response to low-latitude insolation forcing. (Left: Adapted from W. F. Ruddiman and M. E. Raymo, "A Methane-Based Time Scale for Vostok Ice: Climatic Implications," *Quaternary Science Reviews* 21 [2003]: 141–55.)

Greenland ice cores of 11,000 to 10,500 years ago, coincident with the time of the most recent July insolation maximum.

The methane signal in Figure 10–16 closely resembles the monsoon response signal (see Chapter 8). Peak methane values match predicted peaks in monsoon intensity not just in timing but also in amplitude, with the largest methane peaks lining up with the strongest insolation maxima. This match suggests that a connection exists between past CH_4 concentrations and changes in monsoon strength.

Monsoon fluctuations determine the amount of precipitation that falls in southern Asia. Heavy rainfall saturates the ground, reduces its ability to absorb water, and increases the amount of standing water in swampy areas. Vegetation that grows and decays each summer uses up the oxygen dissolved in the water and creates the reducing conditions needed to generate methane. The extent of these boggy areas expanded during wet monsoon maxima and shrank during monsoon minima at the 23,000-year precession cycle, as did the methane emissions.

Wetlands in circum-Arctic regions probably also contributed to the 23,000-year variations in methane. In far-northern bogs, summer warmth, rather than the precipitation control that prevails in tropical monsoon regions, is the main control on methane releases. For most of the year, temperatures in the far north are too cold for plants to grow and produce methane, but the warmth of the short summer season allows methane to be generated and released. The primary tempo of orbital changes in summer insolation across Asia occurs at the 23,000-year cycle, and heating of the continent varies mainly at this cycle, as do methane releases from northern wetlands. (Changes linked to ice sheets and ice-driven responses also affect northern methane releases but to a lesser extent.)

Although peaks in the 23,000-year CH_4 signal are generally well matched in amplitude to the strength of summer insolation maxima, the methane minima are not. Many of the minima reach similar values despite the fact that the insolation minima vary widely in amplitude. The lower side of the methane trend has a truncated or "clipped" look, apparently because some tropical wetlands survive even the strongest insolation forcing toward extreme drying. Likely candidates are wetlands very near the equator, which remain in the tropical wet zone for any climate.

IN SUMMARY, prominent variations in methane at the 23,000-year cycle are linked primarily to changes in strength of the summer monsoon in tropical and subtropical regions and secondarily to variable heating of Asia and its northern wetlands.

Orbital-Scale Climatic Roles: CO_2 and CH_4

Because we know the basic timing of CO_2 and CH_4 variations during the last several hundred thousand years, we can attempt to assess the role these gases played in orbital-scale climatic variations. Milankovitch proposed that summer insolation drives ice sheets at the 41,000-year and 23,000-year cycles with a lag of approximately 5000 years (Figure 10–17A), but how do the greenhouse gases fit into this framework?

The answer to this question turns out to be different for the orbital cycles of precession and tilt. At the 23,000-year precession cycle (Figure 10–17B), both gases respond on or near the "early" tempo of the Sun (northern hemisphere July insolation). Because the processes that generate CO_2 and CH_4 have very short response times, the gases respond quickly to the forcing from the Sun. The 23,000-year methane response arises from year-by-year midsummer heating of northern continents, especially Asia. The 23,000-year CO_2 signal lags behind that of methane by ~1000 years, but it still has a phase much closer to that of the Sun than to that of the ice. The 23,000-year CO_2 response could result from a range of processes, some acting very quickly (yearly changes in carbon pumping) and others somewhat more slowly (the hundreds of years required for changes in deep-ocean circulation). Because the greenhouse gases have early phases like that of the Sun, they are a part of the forcing of the ice sheet response at 23,000 years, an addition to the initial forcing provided by summer insolation.

In contrast, the response of both gases at the 41,000-year tilt cycle falls on or close to that of the ice sheets (Figure 10–17C). This timing rules out any possibility that the gases are forcing the ice sheets at the tilt cycle. If they were forcing the ice, the slow-responding ice sheets would lag behind by thousands of years, but no such lag exists. The observed phasing at the 41,000-year cycle can be explained only if the ice sheets are controlling CO_2 and CH_4 with little or no lag in the gas response.

The ice-driven gas signals, particularly the signal of CO_2, provide immediate positive feedback to the ice sheets at the 41,000-year cycle. As ice sheets grow, they drive the CO_2 concentration in the atmosphere down, and the lower CO_2 concentration cools the climate and helps the ice grow even faster. When the ice melts, CO_2 concentrations rise, climate warms, and ice melting accelerates.

The greenhouse-gas role in the prominent oscillations at a period near 100,000 years is not as clear. The very small amount of insolation forcing at the 100,000-year eccentricity period has nearly the same phase as the changes in orbital eccentricity, but it has a negligible effect on the ice sheets (Chapter 9). In the absence of measurable insolation forcing, the only timing comparison we can

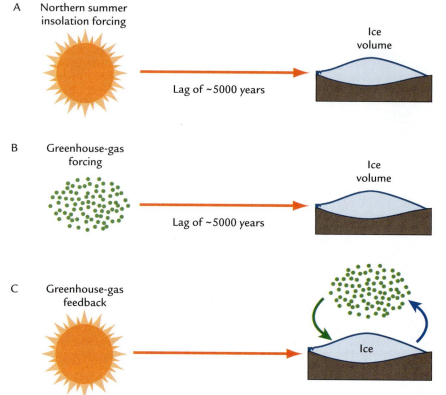

A   Northern summer insolation forcing

Ice volume

Lag of ~5000 years

B   Greenhouse-gas forcing

Ice volume

Lag of ~5000 years

C   Greenhouse-gas feedback

Ice

FIGURE 10-17 Greenhouse gas forcing or feedback? CO_2 and CH_4 join summer insolation as part of the ice sheet forcing at the 23,000-year cycle, but they act as a positive feedback at the 41,000-year cycle and play a mixed feedback/forcing role in the 100,000-year oscillations. (Adapted from W. F. Ruddiman, "Orbital Changes and Climate," *Quaternary Science Reviews* 24 [2006]: 3092–112.)

make is between the changes in the gas concentrations and the ice sheets (see Figure 10–17C).

Both the greenhouse gases and the ice sheets have phases that fall very close to that of eccentricity. CO_2 has at most a small lead relative to changes in eccentricity, while the ice sheets have a small lag. As a result, the gas responses lead that of the ice sheets by some 1000–3000 years.

This small difference in timing does not fit cleanly into the interpretation of greenhouse gases as either a "forcing" or a "feedback." The greenhouse gas lead of ~5000 years at the 23,000-year cycle is appropriate to a forcing-and-slow-response relationship, amounting to nearly one-quarter (25%) of the length of the cycle. In contrast, the greenhouse gas lead of ~2000 years at the 100,000-year period is only 2% of a much longer period. The only way that a 2000-year lead could be consistent with a forcing-and-response relationship is if the ice sheet response at the 100,000-year period is very fast, more than twice as fast as the response at the 23,000-year cycle. It is hard to imagine why (or how) this would be the case. On the other hand, the opposite possibility—that the ice sheets control the gas responses at ~100,000 years—also seems to be ruled out by the fact that the gases do lead the ice (by a small amount).

One possible resolution of this problem is that the small lead of the greenhouse gases ahead of the ice sheets at the ~100,000 year period reflects a combination of

both roles, one part gas forcing of the ice with a lead, one part ice control of the gases with no lead or lag. Because the gases and the ice sheets are so nearly in phase at the ~100,000-year period, the in-phase feedback role for the gases appears to be larger than the forcing role.

These various phase relationships between the greenhouse gases and the ice sheets at the orbital cycles are all apparent in changes that occurred during the intensively studied changes of the last 150,000 years (Figure 10–18). During isotopic stage 5, large insolation changes at the 23,000-year cycle produced large CH_4 variations that clearly led the $\delta^{18}O$ (~ice volume) signal, although the lead is not so readily apparent in the CO_2 signal. The lead of CH_4 ahead of $\delta^{18}O$ points to gas forcing of the ice sheets at the 23,000-year precession cycle.

In contrast, the strong glacial maxima near 21,000 and 63,000 years ago have the tempo of the 41,000-year cycle. Both maxima are accompanied by prominent and coincident CO_2 minima. This relationship suggests that the ice sheets controlled the in-phase CO_2 changes, and that the gases provided positive feedback to the ice sheets at the tilt cycle.

Both the CO_2 and $\delta^{18}O$ signals have saw-toothed shapes at a period near 100,000 years, with slow growth of ice and declines in CO_2, followed by rapid ice melting and CO_2 increases. The two signals have very similar overall timing, but a small CO_2 lead is evident at the deglacial terminations.

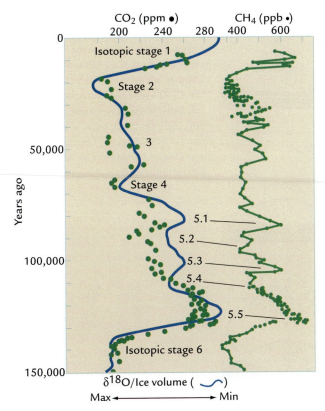

FIGURE 10-18 Ice/Gas Phasing The phasing of the greenhouse gases relative to ice volume at the three orbital periods is evident in the changes of the last 150,000 years. (Adapted from W. F. Ruddiman, "Ice-Driven CO_2 Feedback on Ice Volume," *Climate of the Past* 2 [2006]: 43–66.)

IN SUMMARY, the greenhouse gases act as a forcing of the ice sheets at the 23,000-year period, but they act as an ice-driven feedback at the 41,000-year period. The reason why the gases have these completely different roles at the two periods is not known. The role of the greenhouse gases in the major oscillations near ~100,000 years appears to be a combination of a large feedback role and a smaller forcing role.

Key Terms

ice flow model (p. 177)
sintering (p. 177)
carbon isotopes (p. 178)
phytoplankton (p. 178)
carbon pump (p. 183)

iron fertilization
   hypothesis (p. 184)
$\delta^{13}C$ aging (p. 186)
polar alkalinity
   hypotheseis (p. 187)

Review Questions

1. How are ice cores dated?

2. Why are air bubbles in ice cores younger than the ice in which they are sealed?

3. What features of the ice core CH_4 signal suggest a link to tropical monsoons?

4. To what extent does a cooler glacial ocean explain lower CO_2 levels in the glacial atmosphere?

5. Where did the carbon (CO_2) removed from the atmosphere go during glaciations?

6. What effect does fractionation have on carbon isotope values?

7. Explain how carbon isotopes ($\delta^{13}C$) trace shifts of carbon from the land to the ocean.

8. How does the oceanic carbon pump reduce CO_2 levels in the atmosphere?

9. What evidence indicates a different glacial deep-water flow pattern in the North Atlantic?

10. How could a change in deep circulation in the Atlantic Ocean alter atmospheric CO_2?

Additional Resources

Basic Reading

Ruddiman, W. F. 2005. *Plows, Plagues and Petroleum,* Chapter 5. Princeton, NJ: Princeton University Press.

Advanced Reading

Broecker, W. S., and T.-H. Peng. 1989. "The Cause of the Glacial to Interglacial Atmospheric CO_2 Change: A Polar Alkalinity Hypothesis." *Global Biogeochemical Cycles* 3: 215–39.

Kohfeld, K., C. Le Quere, S. P. Harrison, and R. F. Anderson. 2005. "Role of Marine Biology in Glacial-Interglacial Cycles." *Science* 308: 74–78.

Martin, J. H. 1990. "Glacial-Interglacial CO_2 Change: The Iron Hypothesis." *Paleoceanography* 5: 1–13.

Petit, J. R., et al. 1999. "Climate and Atmospheric History of the Past 420,000 Years from the Vostok Ice Core, Antarctica." *Nature* 399: 429–37.

Sigman, D. M., and E. A. Boyle. 2000. "Glacial/Interglacial Variations in Atmospheric Carbon Dioxide." *Nature* 407: 859–68.

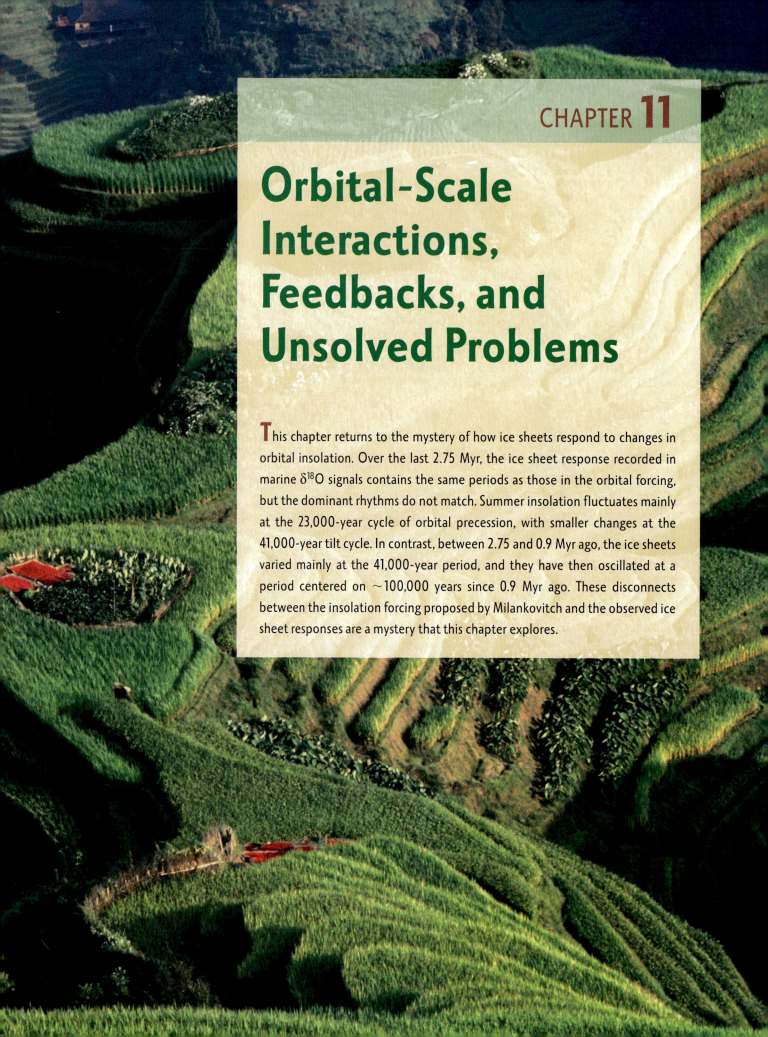

Orbital-Scale Interactions, Feedbacks, and Unsolved Problems

This chapter returns to the mystery of how ice sheets respond to changes in orbital insolation. Over the last 2.75 Myr, the ice sheet response recorded in marine $\delta^{18}O$ signals contains the same periods as those in the orbital forcing, but the dominant rhythms do not match. Summer insolation fluctuates mainly at the 23,000-year cycle of orbital precession, with smaller changes at the 41,000-year tilt cycle. In contrast, between 2.75 and 0.9 Myr ago, the ice sheets varied mainly at the 41,000-year period, and they have then oscillated at a period centered on ~100,000 years since 0.9 Myr ago. These disconnects between the insolation forcing proposed by Milankovitch and the observed ice sheet responses are a mystery that this chapter explores.

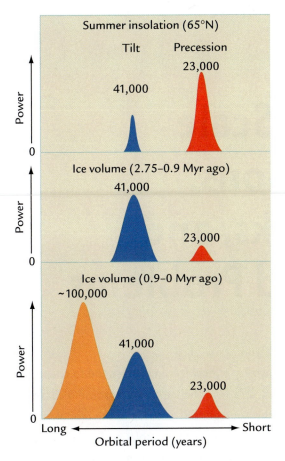

Summer insolation (65°N)
Tilt   Precession
23,000
41,000

Ice volume (2.75–0.9 Myr ago)
41,000
23,000

Ice volume (0.9–0 Myr ago)
~100,000
41,000
23,000

Long ←———→ Short
Orbital period (years)

FIGURE 11-1 Spectral analysis: Insolation and ice volume
(Top) Summer insolation changes at 65 °N are strongest at the 23,000-year period, weaker at 41,000 years, and negligible at 100,000 years. (Center) The δ¹⁸O signal of northern hemisphere ice volume (and deep-ocean temperature) between 2.75 and 0.9 Myr ago varied mainly at 41,000 years; little response occurred at 23,000 years. (Bottom) Since 0.9 Myr ago, a δ¹⁸O signal at or near 100,000 years has been dominant.

Climatic Responses Driven by the Ice Sheets

Chapter 9 ended with one of the major unsolved mysteries in all of climate science. The changes in summer insolation proposed by Milankovitch as the driver of ice sheets do not match the major rhythms of ice response between 2.75 and 0.9 Myr ago, nor those since 0.9 Myr ago (Figure 11–1). These unexpected ice sheet responses are widely thought to have emerged from responses internal to the climate system, and scientists have proposed many possible explanations, but with no consensus at this point. One approach to solving this problem is to look at the wide range of climatic behavior that occurs in different regions.

Ice sheets act as drivers of climate change from *inside* the climate system because of their physical

characteristics. One factor is their height: they protrude thousands of meters into the air and form massive obstacles to the free circulation of winds in the lower atmosphere, thereby rearranging the flow of air. In addition, their bright surfaces reflect much more incoming sunlight than do darker ice-free surfaces. This loss of solar heating in regions covered by ice cools the air above the ice sheets and in nearby areas.

Ice-driven responses are changes in climate that result directly from the presence or absence of ice sheets. Fast-responding parts of the climate system should register these changes with more or less the same timing as the ice sheet drivers. One such area is the high-latitude North Atlantic Ocean, which is rimmed by the great ice sheets of North America and Eurasia. Its temperature might be expected to track the changing size of the ice sheets with no perceptible lag (Figure 11–2).

Climate scientists have run sensitivity test experiments to isolate the effects of the massive domes of ice on atmospheric circulation by inserting ice sheets into general circulation models (GCMs) that otherwise have modern boundary conditions. These simulations show that the ice sheet over North America has a large influence on the nearby ocean temperatures. A clockwise flow of winds around the central dome of the North American ice sheet sends very cold winds blowing around the northern part of the ice sheet and out toward the southeast over the western North Atlantic Ocean.

As the ice sheets cool the North Atlantic Ocean, the climatic effects project farther east into Europe from the chilled sea surface. Other GCM sensitivity tests have been run to evaluate how changes in ocean temperature alter Europe's climate. In these model experiments, the only change in boundary condition input was a reduction of the temperature of the North

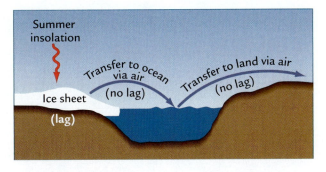
Summer insolation
Transfer to ocean via air (no lag)
Transfer to land via air (no lag)
Ice sheet
(lag)

FIGURE 11-2 Ice-driven responses Orbital-scale ice sheet rhythms may be quickly transferred to other parts of the climate system via the atmosphere and ocean. (Adapted from W. F. Ruddiman, "Northern Oceans," in *North America and Adjacent Oceans During the Last Deglaciation*, ed. W. F. Ruddiman and H. E. Wright, Geological Society of America DNAG vol. K–3 [Boulder, CO: Geological Society of America, 1987].)

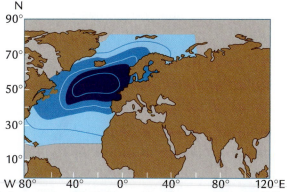

A Input to model: Colder sea surface temperatures (°C)

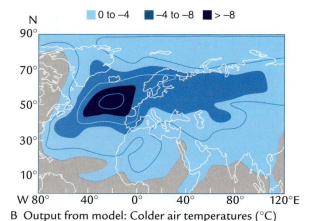

| 0 to –4 | –4 to –8 | > –8 |

B Output from model: Colder air temperatures (°C)

FIGURE 11-3 Surface-ocean sensitivity test Sensitivity tests with a GCM show that (A) inserting cold ocean temperatures into an interglacial world (B) produces colder air temperatures over Europe and Asia. (Adapted from D. Rind et al., "The Impact of Cold North Atlantic Sea-Surface Temperatures on Climate: Implications for the Younger Dryas," *Climate Dynamics* 1 [1986]: 3–33.)

Atlantic Ocean to near-glacial values north of 20°N (Figure 11–3). The rest of Earth's surface was left in its current state, with no ice sheets except the small one now on Greenland.

The model simulation shows that the cold North Atlantic sea surface projects very cold air temperatures downwind to the western maritime parts of Europe and

some cooling even as far east as the interior of Eurasia (Figure 11–3B). Precipitation in these chilled regions falls significantly because the colder ocean gives off less water vapor to the atmosphere. This experiment indicates that an ice sheet signal initially transferred to the North Atlantic would then be transferred to Europe and even Asia. Both signal transfers would have occurred with lags of no more than a few centuries.

How far east the ice sheet signal might penetrate into Asia is not entirely clear. The Siberian high-pressure center, which even now is the winter season center of cooling and powerful winds in northern Asia (companion Web site, p. 21), plays an important role. Model experiments indicate that the Siberian High would have been strengthened in winter and would have lasted for a longer part of the year in a world with ice sheets present.

These model experiments indicate that ice sheets have a large downstream climatic impact across a broad area of the northern hemisphere at high and middle latitudes (Figure 11–4). The effects reach farthest south during winter because of very strong wind flow during that season. In summer, wind strength drops and local heating of the land by the Sun becomes more important.

In the northern tropics and subtropics, the second fundamental tempo of global climate change exists—the orbitally driven monsoon circulations. These changes are dominated by the 23,000-year precession cycle, compared to the predominant ice volume responses at 41,000 and ~100,000 years. In summer, these local changes in solar radiation heating overwhelm the weaker message sent out by the ice sheets. Monsoon-related variations can also be found in the southern subtropics (Chapter 8).

The southern hemisphere lies entirely outside the region under the direct influence of northern ice sheets on atmospheric circulation. Tests with general circulation models show that the presence of large masses of ice in high northern latitudes has no significant effect on temperatures anywhere in the southern hemisphere. This conclusion pertains only to transfers that occur through the physical circulation of the atmosphere by winds and pressure fields. We will see later that other means of transferring northern signals southward exist.

FIGURE 11-4 Regions of ice-driven responses High and middle latitudes of the northern hemisphere show evidence of climate responses controlled by changes in the sizes of ice sheets.

Mystery of the 41,000-Year Glacial World

The evidence for 41,000-year variations in ice sheets between 2.75 and 0.9 Myr ago is based on variations in marine $\delta^{18}O$ values in benthic foraminifera (see Figure 9–13). Temperature overprints at the 41,000-year cycle are thought to account for somewhat less than half of the range of $\delta^{18}O$ variation found during this interval, but the remaining signal clearly shows that ice volume varied mainly at the 41,000-year tempo. Three explanations have been proposed for the mismatch between the strong insolation forcing at 23,000 years and the dominant ice volume response at 41,000 years.

11-1 Did Insolation Actually Vary Mainly at 41,000 Years?

The atmospheric scientist Peter Huybers proposed that climate scientists have been wrong in thinking that summer insolation changes in the northern hemisphere have been dominated by the 23,000-year period of precession. He acknowledged that large variations in the amplitude of insolation changes do occur at the 23,000-year period during summer, but he claimed that these changes are cancelled by reductions in the length of the summer season.

This explanation originates with work centuries ago by the astronomer Johannes Kepler. Kepler's second law states that planetary bodies moving in an elliptical orbit vary in angular speed with their distance from the Sun. When Earth is close to the Sun (at perihelion), it moves faster than it does at other times in the orbit. When the eccentricity of Earth's orbit is unusually high, Earth's precessional motion brings it even closer to the Sun at perihelion during the summer season, and it moves with even greater speed. At such times, the length of the summer season is reduced by Earth's greater speed.

Huybers proposed that the net effect is that higher-than-normal levels of insolation caused by Earth being unusually close to the Sun at perihelion are offset by the shorter-than-normal length of the summer season. The result is that no net insolation change occurs on Earth at the 23,000-year cycle (Figure 11–5). Insolation variations caused by changes in tilt are not affected by this factor. If this explanation is correct, no mismatch actually exists between the insolation forcing and the ice sheet responses, both of which vary mainly at 41,000 years.

One problem with this explanation is that the 23,000-year signal remains dominant even in the caloric summer half-year index used by Milankovitch (Chapter 7). This half-year index integrates the total insolation over those 182 days of the year for which insolation levels are higher than the other 182 days of the year. The specific 182 calendar days in each caloric season vary through time because of the insolation "boosts" introduced at different times of the year by Earth's precession. This shifting definition of "summer" avoids the problem of a fixed summer

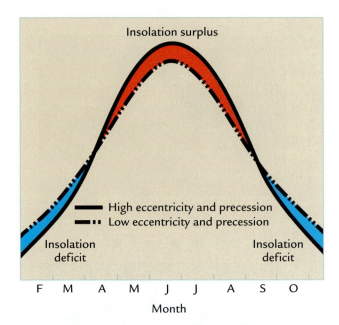

FIGURE 11-5 **Intensity versus length of season** Variations in the intensity of summer insolation at the 23,000-year precession cycle are balanced by changes in the duration of the summer season.

season, which would be subject to Kepler's second law. In addition, it seems likely that the 182 days of higher ("summer") insolation would span most of the part of the year when ice sheets lie in an ablation regime. If so, the fact that insolation variations at the 23,000-year precession signal are still larger than those at the 41,000-year tilt signal in the caloric half-year insolation index suggests that a mismatch still exists between the forcing and the response.

11-2 Interhemispheric Cancellation of 23,000-Year Ice Volume Responses?

The marine geologist Maureen Raymo proposed a different explanation for the mismatch. She assumed that northern hemisphere ice sheets have responded with a strong 23,000-year signal as expected from the insolation forcing, but she suggested that these changes were cancelled out in the global signal by a 23,000-year response of Antarctic ice with opposite timing (phasing).

Changes in precession occur at precisely opposite times in the two hemispheres from a seasonal point of view (Chapter 7). For example, the northern hemisphere summer currently has an insolation minimum because Earth is in its aphelion (distant-pass) position at that time of year. But six months later in early January, which is summer in the southern hemisphere, Earth has moved into its perihelion (close-pass) position. As a result, the southern hemisphere currently has a 23,000-year insolation maximum, exactly opposite the summer minimum in the northern hemisphere. Through time, the two hemispheres maintain these exactly opposite seasonal insolation trajectories at the 23,000-year cycle.

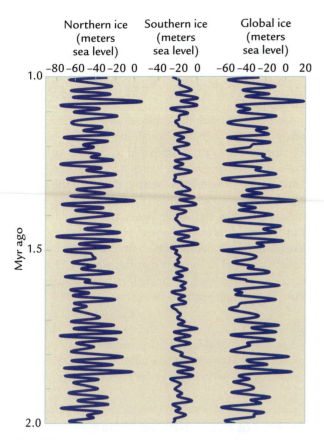

FIGURE 11-6 **Interhemispheric cancellation** The response of northern hemisphere ice sheets to summer insolation forcing at the 23,000-year cycle may have been offset by a response of Antarctic ice to summer insolation with opposite timing.

Raymo proposed that these oppositely phased insolation changes drove similarly opposed ice volume responses in the two hemispheres (Figure 11–6). When northern ice sheets were growing, the Antarctic ice sheet was shrinking; the converse is also true. As a result, the effects of the northern and southern ice sheets on the global $\delta^{18}O$ signal were also opposed. The $\delta^{18}O$ composition of the Antarctic ice sheet was also considerably more negative (–50 to –55‰) than that of the northern ice sheets (–30 to –35‰, the composition of Greenland ice). As a result, changes in the volume of Antarctic ice had to be only about 60% as large as those in the northern hemisphere to cancel them out in the global average value recorded in marine $\delta^{18}O$ records (32‰ divided by 52‰).

One problem with this explanation is a range of evidence that northern ice sheets did vary at the 41,000-year tempo, not at 23,000 years. Sea-surface temperature changes derived from assemblages of planktic foraminifera correlate peak for peak with the $\delta^{18}O$ signal in a core from the North Atlantic Ocean during the interval from 1.5 to 1.2 Myr ago (Figure 11–7). Because sea-surface temperatures in this region are considered to be an "ice-driven" response (see Figure 11–2), the

similarity of the two trends during this interval suggests that the northern ice sheets varied mainly at the 41,000-year tempo indicated by the $\delta^{18}O$ trend. Similar evidence comes from 41,000-year alternations between deposits of weathered soils (indicating warm wet conditions) and windblown loess (indicating cold dry conditions) in the loess plateau of southeast Asia (see Figure 2–3). This plateau is built of layered sequences of sediments hundreds of meters thick.

A second issue is whether the Antarctic ice sheet would have been able to respond to orbital forcing during the interval between 2.75 and 0.9 Myr. Even during the interglacial climates of the last several hundred thousand years (including today), Antarctica appears to have remained deeply refrigerated, with temperatures barely reaching the freezing point around the margins of the continent even in midsummer. General circulation modeling experiments suggest that a very large warming would be required to

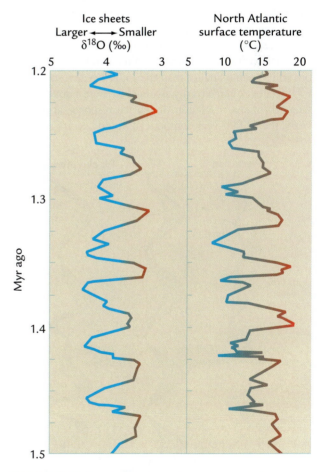

FIGURE 11-7 **North Atlantic surface response to ice**
Climate signals in a North Atlantic sediment core show that sea-surface temperatures between 1.5 and 1.2 Myr ago closely tracked $\delta^{18}O$ (~ice volume) fluctuations. (Adapted from W. F. Ruddiman et al., "Pleistocene Evolution: Northern Hemisphere Ice Sheets and North Atlantic Ocean Climate," *Paleoceanography* 4 [1989]: 353–412.)

produce melting conditions across the margins of the Antarctic ice sheet. Climate was presumably warmer than now during the interval between 2.75 and 0.9 Myr ago, but was it warm enough for major ice melting to have occurred during favorable orbital configurations?

11-3 CO₂ Feedback at 41,000 Years?

The marine geologist Bill Ruddiman has proposed that ice sheets varied mainly at a 41,000-year tempo because of positive feedback from CO_2. During the last 400,000 years, CO_2 variations at the 41,000-year cycles have been in phase with and therefore driven by the ice sheets (Chapter 10). Although no ice cores have yet reached back to the 41,000-year glacial regime, Ruddiman suggested that this same relationship would also have prevailed during the 2.75–0.9 Myr interval.

A conceptual model of this idea (Figure 11–8) starts with insolation variations at the 23,000-year cycle that are larger than those at the 41,000-year cycle. Because the ice sheets have almost twice as long to grow at the 41,000-year cycle as at the 23,000-year cycle (41,000/23,000 = 1.8), the 41,000-year component of the ice volume response to insolation forcing gains in relative strength but still remains smaller than the 23,000-year response. If CO_2 feedback is arbitrarily assumed to cause a doubling in strength of the insolation-driven ice response at 41,000 years, the 41,000-year component would become consid-

erably larger than the response at 23,000 years (which receives no positive feedback from CO_2). As in the "small glaciation" phase (Chapter 9), ice sheets would appear during summer insolation minima (mainly at the 41,000-year cycle) but melt away during the next insolation maximum.

One problem with this explanation is that climate scientists do not know how to weigh the effect on ice sheets of changes in insolation compared to changes in greenhouse gases. Expressed in units of Watts per square meter (W/m^2), the relative heating effects of the insolation changes are much larger than those of the greenhouse gases. On the other hand, the greenhouse-gas changes persist through the year, whereas the insolation changes trend in opposite directions during summer and winter. A second problem is that the reason CO_2 acts as an ice-driven feedback at the 41,000-year cycle, but not at the 23,000-year signal, is not known.

> **IN SUMMARY,** the reason for the dominance of the 41,000-year signal in δ¹⁸O (ice volume?) variations prior to 0.9 Myr ago is not well understood. Several possible explanations are under consideration.

Mystery of the ~100,000-Year Glacial World

After 0.9 Myr ago, the second great mystery of the ice ages emerged: oscillations centered near a period of 100,000 years. Both δ¹⁸O trends and coral reef positions confirm that the ice sheets fluctuated at this tempo (Chapter 9). Because more records from continental lakes reach back far enough in time to span several of these oscillations, more information on regional responses is available for this interval.

Climatic oscillations at a period near 100,000 years with the characteristic saw-toothed shape can be traced into nearby regions (Figure 11–9). Changes in sea-surface temperature in the North Atlantic reconstructed from assemblages of planktic foraminifera resemble the δ¹⁸O (ice volume) signal, just as they did in the interval of 41,000-year glacial changes (see Figure 11–7). Variations in pollen assemblages from European lakes also have the same general character. Pollen grains deposited during warm, moist interglacial climates come mainly from trees like those in today's European forests, while glacial-age pollen consists mainly of grasses and herbs, indicating much drier and colder (treeless) conditions. Both the SST and pollen records show the characteristic deglacial terminations that bring the glacial climates to an abrupt end. The basic correlation of these three records is a typical "ice-driven" response: a rapid transfer of the ice sheet signals to the ocean and to the land (see Figure 11–2).

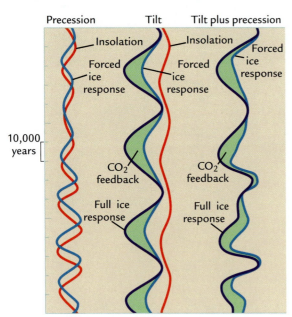

FIGURE 11-8 CO₂ feedback The ice volume response to insolation forcing at the 41,000-year cycle may have been amplified by CO₂ feedback. (Adapted from W. F. Ruddiman, "Ice-Driven CO₂ Feedback on Ice Volume," *Climate of the Past 2* [2006]: 43–78.)

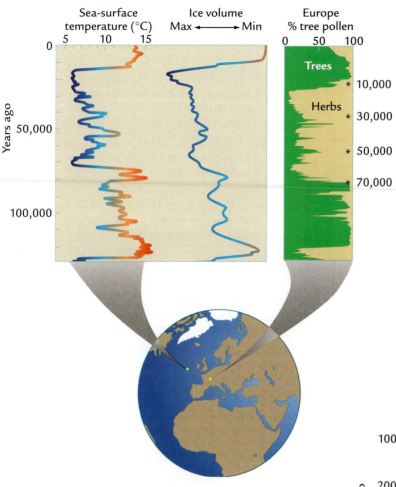

FIGURE 11-9 **North Atlantic sea-surface temperature and European vegetation** Variations in $\delta^{18}O$ (~ice volume) are similar to changes in North Atlantic sea-surface temperature (estimated from planktic foraminifera) and European vegetation (based on pollen assemblages). (Left: Adapted from C. Sancetta et al., "Climatic Record of the Past 130,000 Years in the North Atlantic Deep-Sea Core V23–82: Correlations with the Terrestrial Record," *Quaternary Research* 3 [1973]: 110–16. Center: Adapted from D. Martinson et al., "Age Dating and the Orbital Theory of the Ice Ages: Development of a High-Resolution 0 to 300,000-Year Chronostratigraphy," *Quaternary Research* 27 [1987]: 1–29. Right: Adapted from G. M. Woillard and W. G. Mook, "Carbon-14 Dates at Grande Pile: Correlation of Land and Sea Chronologies," *Science* 215 [1982]: 159–61.)

Similar-looking signals can be found much farther to the east, in the loess plateau of southern China. Over the last several hundred thousand years, interglacial soils and windblown glacial loess deposits have alternated at a period near 100,000 years (Figure 11–10). The correlation to the $\delta^{18}O$ record suggests that these loess/soil variations were ice-driven. The ice-driven imprint can even be tracked all the way around to the Greenland ice sheet. Dust particles in glacial-age layers of Greenland ice come mainly from Asia, where they were lifted by strong winds and blown eastward at jet stream elevations.

IN SUMMARY, large oscillations in ice volume at a period near 100,000 years drove climatic responses that can be traced across the high and middle latitudes of the northern hemisphere. These responses occurred in the region that model experiments indicate should be under the immediate influence of the northern ice sheets (see Figure 11–4). These ice-driven responses resulted from altered wind patterns that produced changes in surface-ocean temperatures and precipitation over land.

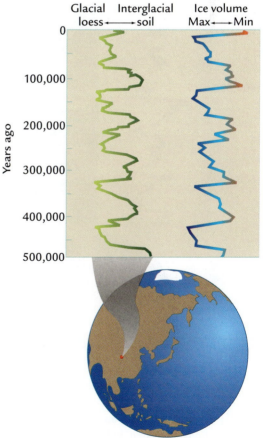

FIGURE 11-10 **Responses of windblown debris in East Asia to ice volume** Alternating layers of windblown loess and soils in Southeast Asia match variations in $\delta^{18}O$ (ice volume). (Adapted from G. Kukla et al., "Pleistocene Climates in China Dated by Magnetic Susceptibility," *Geology* 16 [1988]: 811–14.)

In contrast, most of the tropics and all of the southern hemisphere lie beyond the immediate reach of ice sheet rearrangements of atmospheric winds, according to general circulation model experiments. Yet many climatic records from these regions show climatic responses surprisingly similar to those near the ice sheets.

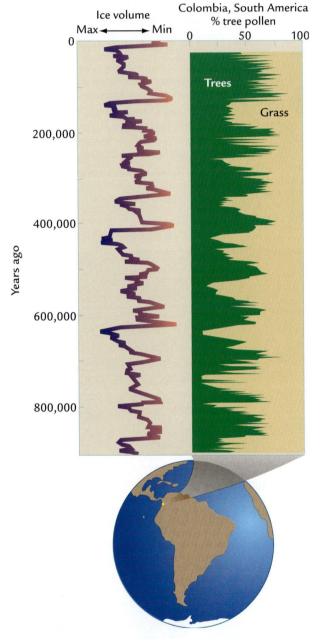

FIGURE 11-11 Vegetation response in South America A long lake core from the eastern Andes Mountains in Colombia shows major shifts between forest and grassland pollen that match 100,000-year glacial-interglacial ice volume changes in the northern hemisphere. (Adapted from H. Hooghiemstra et al., "Frequency Spectra and Paleoclimatic Variability of the High-Resolution 30–1450 Kyr Funza I Pollen Record," *Quaternary Science Reviews* 12 [1993]: 141–56.)

A long sediment core from a lake in the eastern Colombian Andes of South America shows cyclic alternations between pollen produced by trees and by high-mountain grasslands (Figure 11–11). Dating of this record by radiocarbon analysis, volcanic ash layers, and magnetic reversals indicates that the major pollen fluctuations occurred at a period near 100,000 years. Tree pollen increased during interglaciations, while grass pollen increased during glaciations, the same kind of pattern shown by pollen records in Europe. A marine sediment core from east of New Zealand contains a similar history of variations between tree pollen and mountain grassland types (Figure 11–12). In this case, because both the marine $\delta^{18}O$ signal and the terrestrial pollen signal are recorded in the same core, correlating them is not a problem. The southern pollen response resembles the northern ice sheet signal.

This persistence of a saw-toothed oscillation at ~100,000 years extending into the tropics and well into the southern hemisphere is a surprise. These responses cannot be "ice-driven," at least not in the sense of being created by changes in atmospheric winds. They also cannot be produced by fluctuations in the Antarctic ice sheet, which already covers 97% of the continent during the present warm interglacial interval and which cannot have covered a much larger area during full-glacial climates. Antarctic ice expanded to the continental margins uncovered by falling sea level (because of water trapped in northern ice sheets), but no major expansion in Antarctic ice volume occurred.

Another possibility that has been considered by several scientists is that these oscillations developed independently somewhere in the southern hemisphere or in the tropics and that they then played a role in forcing the northern ice sheet response. The most commonly cited mechanism is changes in carbon dioxide produced in some way by the ocean. The problem with this explanation is that the CO_2 lead relative to ice volume is only ~2000 years (Chapter 10), too small a lead to be interpreted in terms of a forcing-and-response relationship (Box 11–1).

11-4 How Is the Northern Ice Signal Transferred South?

The opposite possibility is that the saw-toothed oscillation at ~100,000 years originates in the northern ice sheets and is transferred south. But if this transfer cannot occur through changes in atmospheric circulation (see Figure 11–4), how does it happen?

One possible mechanism is changes in sea level tied to storage and release of ocean water in the northern ice sheets. Changes in sea level affect climate in coastal regions that are alternately flooded and exposed by vertical movements of shallow seas (Chapter 5). Climates in and near such areas may vary between

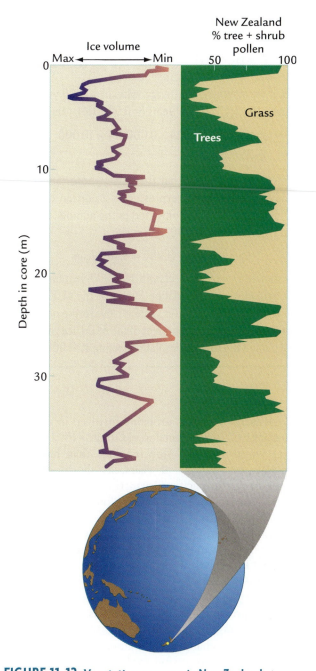

FIGURE 11-12 Vegetation response in New Zealand A marine sediment core from the east coast of New Zealand shows major 100,000-year shifts between forest and grassland pollen that match glacial-interglacial ice volume ($\delta^{18}O$) signals. (Adapted from L. E. Heusser and G. van der Geer, "Direct Correlation of Terrestrial and Marine Paleoclimatic Records from Four Glacial-Interglacial Cycles-DSDP Site 594, Southwest Pacific," *Quaternary Science Reviews* 13 [1994]: 275–82.)

relatively temperate maritime conditions and harsher, more continental environments as the ocean rises and falls. The problem with this explanation is that many of the regions that show the saw-toothed oscillations are located at high altitudes and well into continental interiors (see Figure 11–11). Climate in such regions would

not be highly sensitive to changes in distant coastal regions.

A second way of potentially linking the two hemispheres is by changes in deep-water circulation that transfer heat between their polar regions. The reductions in deep-water formation in the North Atlantic Ocean that occur during glaciations (see Figure 10–14) could reduce the amount of relatively warm and salty water sent to south polar latitudes. Again, however, while this mechanism might affect the temperature of the Southern Ocean, it would fail to explain changes at high elevations and in low-latitude or mid-latitude continental interiors of southern hemisphere continents.

A third way to transfer a northern signal to the southern hemisphere is through variations in atmospheric greenhouse gases. Sensitivity test experiments with general circulation models suggest that the 30% CO_2 reduction that occurred during the last glacial maximum (along with a 50% reduction in methane) would have cooled the southern hemisphere, especially in south polar regions where changes in sea ice amplify the response (Figure 11–13). Changes in greenhouse-gas concentrations would also affect the interior regions and mountainous areas of the continents.

The phasing between the greenhouse-gas changes and ice volume in the ~100,000-year oscillations is consistent with a link of this kind. The phasing of the gases and the northern ice sheets is so nearly the same that they argue for ice sheet forcing of the fast-responding gases rather than gas forcing of the slow-responding ice sheets. The small CO_2 lead indicates that the situation is not quite this simple and that some gas forcing of ice sheets is involved (probably tied to the 23,000-year cycle). But the overall similarity in timing of the gas and ice signals suggests that the main relationship is ice sheet control of the gases.

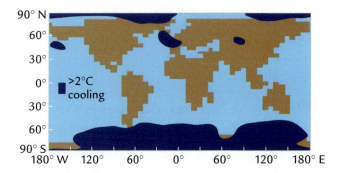

FIGURE 11-13 Southern Hemisphere response to CO₂ changes GCM experiments show that the southern hemisphere responds to lower CO_2 levels during glacial times, especially at higher latitudes. (Adapted from A. J. Broccoli and S. Manabe, "The Influence of Continental Ice, Atmospheric CO₂, and Land Albedo on the Climate of the Last Glacial Maximum," *Climate Dynamics* 1 [1987]: 87–100.)

The Link between Forcing and the Time Constants of Ice Response

The relationship between the forcing of ice sheets and their response can be quantified by the following equation, provided that the variations under investigation are sine waves:

$$\varphi = \arctan 2\pi f T$$

where *arctan* means "the angle whose tangent is . . .," φ is the lag of the ice response behind the forcing in degrees (out of a full 360° circle), f is the frequency under investigation in 1/year (f is the inverse of the period), and T is the response time (the time constant) of the ice sheets in years.

If φ is known, the equation can be used to solve for T and conversely. For example, consider the case of summer insolation forcing of ice sheets at the 41,000-year tilt cycle. Ice sheets are thought to have a time constant of response somewhere in the range of 5,000 to 15,000 years. If we assume a mid-range value of 10,000 years for T, and we use f = 1/41,000 years (= 0.000024), then

$$\varphi = \arctan 2 \times 3.14 \times 0.000024 \times 10,000$$

$$\varphi = \arctan 1.51$$

Geometry tables tell us that the angle whose tangent is 1.51 is 56.5°. This angle, expressed as a portion of a full 360° circle, needs to be converted to the number of years as a portion of a full 41,000-year cycle:

$$56.5°/360° \times 41,000 = 6,400 \text{ years}$$

The estimated lag of ice sheets with a 10,000-year time constant behind summer insolation forcing at the 41,000-year cycle is thus 6,400 years. This value is close to the observed lag of the 41,000-year component of the $\delta^{18}O$ ("ice volume") signal behind the summer insolation forcing at 41,000 years.

Ice sheets could have controlled CO_2 by any or all of several mechanisms (Chapter 10). If strong winds picked up dust formed along the ice margins or in regions of cold dry ice-driven climates and blew it out to the ocean, fertilization by iron or other trace elements could have strengthened the carbon pump mechanism and reduced atmospheric CO_2 concentrations. If ice sheets caused the decrease in depth of sinking of North Atlantic Deep Water, the resulting drop in the CO_3^{-2} ion concentration in south polar waters could have caused a CO_2 decrease. And if ice sheet growth helped to cool deep-water temperatures, the CO_2 solubility mechanism would have reduced atmospheric CO_2 concentrations. The evidence examined to this point suggests that ice sheets did drive dust concentrations in the northern hemisphere (see Figure 11–10) and also affected the flow of North Atlantic Deep Water (see Figure 10–15).

Why *Did* the Northern Ice Sheets Vary at ~100,000 Years?

This investigation has now led us full circle back to where we started—with the northern ice sheets. The ice sheets clearly fluctuated at a period near 100,000 years, and they probably sent this signal south by means of their control of the greenhouse-gas concentrations at ~100,000 years. But a major question remains unresolved: Why did the ice sheets fluctuate at a period near 100,000 years in the first place?

One factor that was probably involved in the shift from the 41,000-year glacial world to the ~100,000-year glacial world was the slow cooling that had been underway for millions of years, as shown by $\delta^{18}O$ records from the Pacific Ocean (Figure 11–14). This gradual polar cooling would have slowly reduced ablation in high northern latitudes to the point that some ice began to survive during relatively weak insolation maxima. This residual ice would then form a base for additional ice growth, as proposed for the "large glaciation" phase in Chapter 9.

A likely reason for the spacing of these larger glaciations at approximately 100,000 years is evident in the summer insolation trends in (Figure 11–15). Although these insolation changes are dominated by changes at the 23,000-year precession cycle, modulation of the precession signal by eccentricity at a period near 100,000 years is also obvious. This modulation produces clusters of unusually high insolation maxima at intervals of approximately 100,000 years. If ice sheets were able to grow to a relatively large size during intervals when insolation maxima were smaller, they could have been vulnerable to rapid melting when insolation maxima grew larger, at intervals of ~100,000 years.

This response is possible because ice melting is sensitive only to the "upper side" of the envelope of modulation by eccentricity. Insolation minima are equally prominent on the opposite side of this envelope, but they are irrelevant to ice melting, which occurs only during insolation maxima. In this way, major deglaciations could

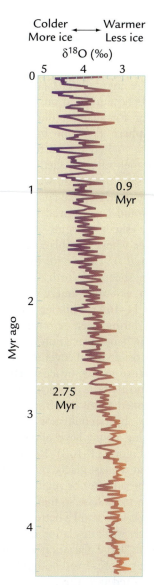

FIGURE 11-14 Changes in $\delta^{18}O$ in the last 4.5 Myr A $\delta^{18}O$ record from a sediment core in the eastern Pacific Ocean shows a slow increase in $\delta^{18}O$ values over the last 4.5 Myr. (Adapted from A. C. Mix et al., "Benthic Foraminifer Stable Isotope Record from Site 849 [0–5Ma]: Local and Global Climate Changes," *Ocean Drilling Program, Scientific Results* 138 [1995]: 371–412.)

be produced at intervals of ~100,000 years from a highly modulated 23,000-year cycle.

Although major deglaciations (terminations) occur at an average spacing of ~100,000 years, none of the gaps between terminations is actually that long. The spacing between the last five terminations was approximately 116,000, 117,000, 94,000, and 84,000 years. This somewhat irregular timing makes sense if the deglaciations were constrained to occur on or near one of the major insolation maxima at the precession cycle and only after

the ice sheets had grown to a large size. With these constraints, terminations would naturally tend to occur after either four or five 23,000-year insolation maxima, or at intervals of ~92,000 or ~115,000 years.

In any case, the origin of the ~100,000-year oscillations remains an area of intensive research. The many explanations that have been proposed fall into three major groups.

11-5 Ice Interactions with Bedrock

One suggestion is that Earth's climate system has a natural 100,000-year resonance, analogous to a bell that rings with a characteristic period or frequency, called a **resonant response.** In this view, any external force that disturbs Earth's climate (changes in tectonic configurations, orbital insolation, or something else) will produce a characteristic internal resonant response like a ringing bell.

The problem with this explanation is that the kinds of tectonic-scale changes that might create such a resonance have changed extremely slowly over the last few million years (Chapter 4). Plates move at rates of centimeters per year, which translate into a few tens of kilometers over a million years. Rates of net mountain uplift and erosion are even slower. No one has yet

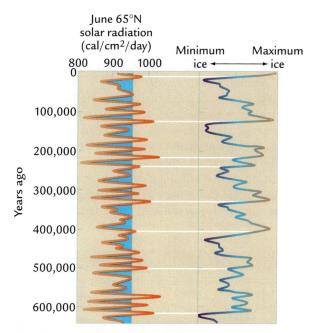

FIGURE 11-15 Strong summer insolation peaks pace rapid deglaciations (Left) Strong summer insolation peaks at the precession cycle resulting from eccentricity modulation match (right) rapid deglacial terminations indicated by $\delta^{18}O$ (ice volume) signals. Adapted from W. S. Broecker, "Terminations," in *Milankovitch and Climate*, ed. A. L. Berger et al. [Dordrecht: D. Reidel, 1984].)

proposed a specific mechanism by which such slow changes in bedrock configuration would have elicited a totally new resonant climatic response within the last million years.

A second, more credible explanation of the ~100,000-year oscillations focuses on delayed bedrock rebound during abrupt deglacial terminations caused by the weight of large ice sheets. When the south-central portions of large ice sheets began to melt under rising levels of summer insolation, the ice sheets would have retreated into the deep bedrock holes they had created (see Chapter 9). The delay in bedrock rebound would have kept the remaining ice in a warm environment and accelerated the rates of melting.

This explanation requires that larger ice sheets produce a slower rebound of the underlying bedrock than the smaller ice sheets that existed prior to 0.9 Myr ago. The rate of bedrock rebound depends on the viscosity (resistance to flow; see Chapter 4) of the material deep in the Earth that is "squeezed" outward from underneath the burden of the overlying ice sheets. Higher-viscosity rock would return slowly, while lower-viscosity material would flow back more rapidly. At this point, different models of the viscosity of Earth's mantle exist. Whether or not the larger and thicker ice sheets of the 100,000-year world would have tapped more of the high-viscosity (slow-flow) response is unclear.

Another explanation related to the underlying bedrock focuses on the character of the materials over which the ice sheets moved and the way their movement affected their thickness. Glacial geologists have found several ice-deposited moraines in Iowa and Nebraska that lie beyond the geographic limits of the large ice sheet that existed at the most recent glacial maximum (20,000 years ago). Layers of volcanic ash date these older moraines to about 2 Myr ago, early in the interval of the 41,000-year ice sheet cycles. These deposits prove that at least some of the smaller-volume 41,000-year ice sheets were already reaching maximum extents comparable to those of the later larger-volume

ice sheets. If these ice sheets were similar in extent but smaller in volume, they must have been thinner than the ones that varied at or near the 100,000-year cycle.

The glacial geologist Peter Clark proposed that the earlier ice sheets accumulated on top of soils that had been developing for many millions of years before northern hemisphere glaciation began. The weight of the overlying ice melted the bottom layers of the ice sheets, and the meltwater trickled down into the soils. Soils that are saturated with water are more easily deformed by the overlying ice and can cause the ice to slip. Slipping would have moved large amounts of ice toward the ice sheet margins and southward into warmer latitudes, where ice ablation rates were higher. Between 2.75 and 0.9 Myr ago, frequent sliding may have kept the ice sheet on North America low and thin, subject to high ablation, and consequently small in volume (Figure 11–16A). Thin ice sheets would also be easier to melt during even relatively weak insolation maxima (as in the "small glaciation" phase of Chapter 9).

None of the original soil cover is now left across central Canada because erosion by ice sheets has removed it. The surface is mostly bare bedrock, with scattered areas of coarse ice-eroded debris. With so little soft sediment, more recent ice sheets could not easily slide, and the absence of sliding could have allowed them to grow much thicker (Figure 11–16B). Reconstructions of ice sheet thickness at the last glacial maximum 20,000 years ago indicate a broad interior region where the ice was thick and frozen to its base so that sliding was unlikely. Thicker ice sheets also stand a better chance of surviving through relatively weak insolation maxima and growing to larger size (the "large glaciation" phase of Chapter 9).

Eroded material preserved both in old moraines and in sediments pushed into the ocean show that ancient soils were the main type of debris eroded prior to the last 1.5 to 1 Myr, while freshly pulverized debris has been predominant since that time. This evidence supports Clark's hypothesis that the ancient soils were gradually eroded by the early ice sheets. This

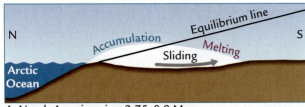

A North American ice 2.75–0.9 Myr

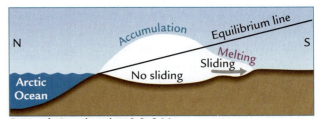

B North American ice 0.9–0 Myr

FIGURE 11-16 Ice slipping may control ice sheet volume (A) During earlier glaciations of North America, ice sheets may have been thin because they slid on water-saturated soils toward lower elevations and warmer temperatures. (B) Later, after ice sheets stripped off most of the underlying soil, their central regions could grow higher because they no longer slid.

mechanism does not depend on (but also does not contradict) the requirement of a long-term cooling to explain the transition from the 41,000-year to the ~100,000-year glacial world. It requires only a change in the nature of the material on which the ice sits.

11-6 Ice Interactions with the Local Environment

Another group of explanations of the oscillations at ~100,000 years looks to interactions between the ice sheets and the changes they impose on the nearby climate system. These ideas particularly focus on ways to explain the speed of deglacial terminations. Large insolation maxima are assumed to initiate and pace deglacial melting (see Figure 11–15), but processes within the climate system are invoked as the mechanism of accelerated melting.

One idea is that the cooling produced by northern ice sheets caused large amounts of sea ice to expand across nearby high-latitude oceans. This increased ice cover would reduce the extraction of moisture that could be delivered to the ice sheets, and the ice sheets would melt faster because of moisture starvation. Glacial geologists have criticized this idea because ice accumulation is a relatively weak factor in the mass balance of ice sheets compared with ablation (see Figure 9–1). As a result, moisture starvation should have had relatively little effect on the rapid reductions in ice volume during terminations.

Another proposal is that the windy, dusty glacial world produced a thin coating of dust on the lower southern margins of the ice sheets. Because dust has a lower albedo than ice, it would have absorbed more incoming solar radiation and warmed the surface of the ice. The absorbed heat would have promoted greater melting and more rapid deglaciation. One problem with this idea is that a slightly thicker coating of dust can have the opposite effect of insulating the ice surface from solar heating and actually reducing the rate of ablation.

Still another idea is that the large ~100,000-year ice sheets developed extensive margins that fronted on the ocean or rested on bedrock lying below sea level (similar to the modern West Antarctic Ice Sheet). These marine margins would have been vulnerable to rises in sea level that would have "lifted" them off the bedrock bases that otherwise stabilized their flow. As a result, they could have responded more quickly to climate changes (initiated by rising insolation levels) than the more sluggish land-based ice sheets deep in the continental interiors. Evidence from marine sediments suggests that ice sheets vulnerable to rising sea level first appeared along the Arctic margin north of Norway near 0.9 Myr ago. One of these ice sheets—the Barents Ice Sheet just north of Scandinavia (see Figure 9–4)—was also among the first northern ice to melt during the most recent deglaciation. The early disappearance of this ice sheet might have helped warm polar latitudes and hastened the melting of the slower-responding ice sheets within the northern continental interiors.

One problem with explanations that focus only on the rapid terminations is that they ignore the rest of the ~100,000-year cycles—the longer intervals of episodic ice growth that occurred during the other ~90,000 years. The accumulation of large ice sheets over these intervals cannot be explained as a linear one-for-one response to changes in summer insolation. Ice accumulation requires some kind of non-linear process just as much as rapid melting on deglaciations.

11-7 Ice Interactions with Greenhouse Gases

A final possibility is that greenhouse gases (particularly CO_2) play a key role in the ~100,000-year ice sheet oscillations. An early proposal was that a CO_2 response at the ~100,000-year period emerged independently from somewhere in the climate system and drove an ice sheet response at that period. As shown in Chapter 10, however, the CO_2 lead relative to ice volume is too small (~2,000 years) for CO_2 forcing to have been the primary relationship with the ice sheets (see Box 11–1).

The other possibility is that CO_2 is primarily a positive feedback on ice volume. In the 41,000-year glacial world, according to this hypothesis, ice sheets formed mainly at the 41,000-year period because of positive feedback from CO_2, but the ice melted during the next summer insolation maximum. In the new ~100,000-year glacial world that developed after 0.9 Myr ago, ice sheet growth again occurred mainly at the 41,000-year cycle, but complete ice melting did not.

The last interglacial-glacial oscillation (Figure 11–17) shows five intervals of ice growth marked by $\delta^{18}O$ increases. Neither of the two oscillations—those near 95,000 and 45,000 years ago—that were a response only to summer insolation changes at the 23,000-year cycle resulted in ice volumes larger than had been attained earlier in this climatic oscillation. All the net ice growth occurred during the three other oscillations—those near 115,000, 72,000, and 30,000 years ago. These ice growth episodes were separated by approximately 41,000 years, occurred during insolation minima at the tilt cycle, and were accompanied by large decreases in atmospheric CO_2 concentrations. Based on this evidence, net ice growth during this ~100,000-year climatic oscillation occurred during 41,000-year episodes and was aided by positive CO_2 feedback, much like the ice growth episodes in the earlier 41,000-year glacial world.

Unlike the earlier 41,000-year regime, however, only a fraction of the ice that grew during the intervals centered on 115,000 and 72,000 years ago melted during the following insolation maxima. Instead, most of the ice remained in place as a base for the next interval of growth to even larger volumes. Because climate had cooled by

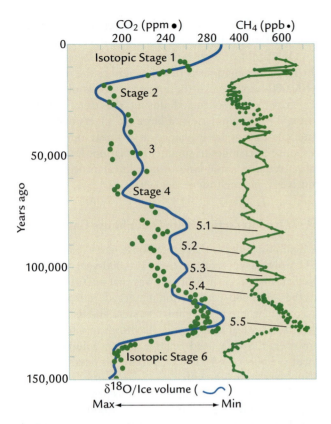

FIGURE 11-17 **Changes in ice volume, CO₂, and methane during the last 150,000 years** During the last full interglacial-glacial oscillation, sawtooth-shaped changes in CO_2 and ice volume ($\delta^{18}O$) occurred at similar times. (Adapted from W. F. Ruddiman, "Ice-Driven CO_2 Feedback on Ice Volume," *Climate of the Past* 2 [2006]: 43–78.)

150,000 years is again the summer half-year insolation signal of Milankovitch. The direct ice response to this forcing is the same as in the 41,000-year world, and the amplification of the 41,000-year ice volume response to Milankovitch forcing by CO_2 feedback is slightly (20%) larger than the doubling that was assumed for the 41,000-year glacial world (see Figure 11–8).

The major difference from the 41,000-year world is that only about one-third of the new ice that accumulated during ice growth episodes melted during the next summer insolation maximum. As a result, the three 41,000-year ice growth episodes combined to produce a longer-wavelength ice volume signal.

Ice growth came to an end when summer insolation rose toward a relatively strong maximum by 11,000 to 10,000 years ago (see Figure 11–15). Although rising insolation initiated the melting process, the reduction in ice volume allowed CO_2 concentrations to rise, and positive feedback from the increasing CO_2 levels kicked in as a powerful amplifier of ice melting. In effect, the same CO_2 feedback that built the large ice sheets was then readily available to help drive major deglaciations.

0.9 Myr ago, a reduction in ablation at high latitudes is the probable reason for the survival of larger amounts of ice (the "large glaciation" phase from Chapter 9).

In a conceptual model of these processes (Figure 11–18), the insolation forcing for the last

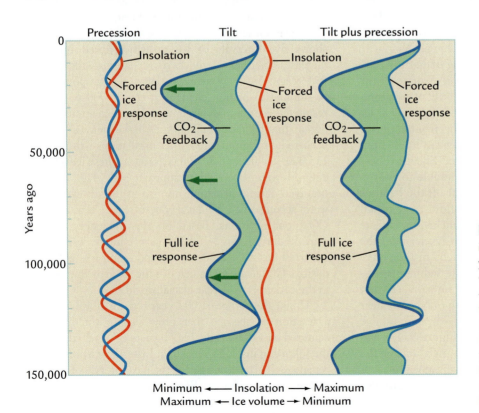

FIGURE 11-18 **CO₂ feedback** The ice volume response at ~100,000 years may have resulted from CO_2 feedback amplification of 41,000-year ice growth episodes, followed by CO_2 feedback amplification of rapid melting initiated by large peaks in 23,000-year insolation. (Adapted from W. F. Ruddiman, "Ice-Driven CO_2 Feedback on Ice Volume," *Climate of the Past* 2 [2006]: 43–78.)

Several factors contribute to the observed pacing of deglaciations at an average period of 100,000 years. One contribution noted earlier is the timing of clusters of high-insolation maxima at ~100,000 years see (Figure 11–15). These peaks are large mainly because of eccentricity modulation of the 23,000-year precession cycle and they tend to position major deglaciations at intervals of either ~92,000 years or ~115,000 years. Some insolation peaks are particularly large because they are closely aligned with high insolation caused by maxima at the tilt cycle.

Another factor that affects the timing of deglaciations is the net growth of ice sheets at 41,000-year intervals. A relatively large volume of ice obviously has to accumulate to create a major deglaciation. This constraint tends to position terminations at intervals of either 82,000 or 123,000 years, after either two or three intervals of ice growth. Together, these two constraints combine to position terminations within intervals of either 82,000–92,000 years or 115,000–123,000 years. The separations between the last five terminations all fall within one of these two time clusters. As a result, ice sheets that grew during 41,000-year episodes melted at intervals near 100,000 years.

> **IN SUMMARY,** several possible explanations for the ~100,000-year glacial world are being explored by climate scientists. In all the proposed explanations, the large ice sheets produce internal responses (either of bedrock or in the climate system) that hasten their own destruction during intervals whose timing is paced by changes in summer insolation. The internal processes that destroy the ice sheets act as positive feedbacks that accelerate ice melting initiated by rising insolation.

Key Terms

ice-driven responses (p. 192)

resonant response (p. 201)

Review Questions

1. In what sense are ice sheets both a climatic response and a source of climatic forcing?

2. Name an ice-driven response and explain its origin.

3. Summarize three possible explanations for the unexpected strength of the 41,000-year response of ice sheets between 2.75 and 0.9 Myr ago.

4. How could the northern hemisphere ice sheets affect climate in the southern hemisphere?

5. What evidence suggests that orbital-scale changes in northern hemisphere ice volume drive changes in atmospheric CO_2 rather than the opposite?

6. How could 100,000-year cycles in the size of northern hemisphere ice sheets be paced by summer insolation changes that occur only at cycles of 41,000 and 23,000 years?

7. If a sine wave CO_2 signal with a period of 100,000 years forces an ice sheet response at the same period and if the ice has a time constant of 10,000 years, what should be the size of the ice sheet lag behind the forcing?

Additional Resources

Basic Reading

Imbrie, J., and K. P. Imbrie. 1979. *Ice Ages: Solving the Mystery.* Short Hills, NJ: Enslow.

Ruddiman, W. F. 2005. *Plows, Plagues and Petroleum,* Chapters 3–5. Princeton, NJ: Princeton University Press.

Advanced Reading

Broecker, W. S. 1984. "Terminations." In A. L. Berger et al., eds., *Milankovitch and Climate,* pp. 687–98. Dordrecht: Reidel.

Broccoli, A. J., and Manabe, S. 1987. "The Influence of Continental Ice, Atmospheric CO_2, and Land Albedo on the Climate of the Last Glacial Maximum." *Climate Dynamics* 1: 87–100.

Huybers. P. 2006. "Early Pleistocene Glacial Cycles and the Integrated Summer Insolation Forcing." *Science* 313: 508–11.

Manabe, S., and A. J. Broccoli. 1985. "The Influence of Continental Ice Sheets on the Climate of an Ice Age." *Journal of Geophysical Research* 90: 2167–90.

Raymo, M. E., Lisiecki, L. E., and Nisancogliu, K. H. 2006. "Plio-Pleistocene Ice Volume, Antarctic Climate, and the Global $\delta^{18}O$ Signal." *Science* 313: 492–95.

Rind, D., D. Peteet, W. S. Broecker, A. McIntyre, and W. F. Ruddiman. 1986. "The Impact of Cold North Atlantic Sea-Surface Temperatures on Climate: Implications for the Younger Dryas Cooling (11–10K)." *Climate Dynamics* 1: 3–33.

Ruddiman, W. F., 2006. "Ice-Driven CO_2 Feedback on Ice Volume." *Climate of the Past* 2: 43–78.

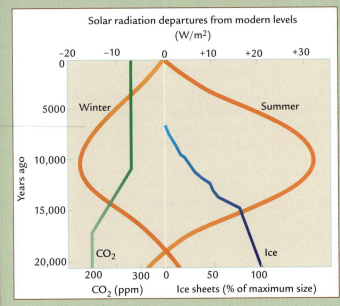

Solar radiation departures from modern levels
(W/m²)

Deglacial climatic forcing For the past 21,000 years, model experiments based on known climatic forcing (insolation, ice sheets, and CO_2) can be compared with observations from the geologic record. (Adapted from J. E. Kutzbach et al., "Climate and Biome Simulations for the Past 21,000 Years," *Quaternary Science Reviews* 17 [1998]: 473–506.)

The last several tens of thousands of years have seen enormous changes in climate, some of them within the time span of recorded human civilization. The largest change was the transition from the last glaciation near 20,000 years ago (Chapter 12) to the warmth of the present interglaciation.

Changes during this interval are recent enough to be dated by the radiocarbon method, and thousands of ^{14}C-dated records from the land can be added to hundreds of records from the ocean to provide coverage that is nearly global. Most terrestrial ^{14}C dates come from lakes or bogs, where watery environments preserve organic carbon. For radiocarbon dates to be used as true ages, they must first be converted to calendar years, which for most intervals are older than the radiocarbon dates by a few thousand years. Throughout these chapters, all ages are in calendar ("true") years unless ^{14}C ages are specified.

The Glacial World: More Ice, Less Gas

Even though the glacial world was icy, cold, dry, windy, and sparsely vegetated, from a tectonic perspective it was nearly identical to the world today. The continents had moved to essentially their modern positions, and plateaus and mountains were at very nearly the same elevations as they are today. In a sense, this glacial world represented an alternative version of today's Earth, the product of a giant experiment run by the climate system in response to changes in several forcing factors. As we saw in Parts II and III, the three factors with the greatest potential to account for differences from modern climate conditions were larger ice sheets, lower CO_2 levels, and changes in seasonal insolation.

Surprisingly, summer and winter insolation levels 21,000 years ago were close to those today. This seemingly counterintuitive fact results from the fact that intervals of lower summer insolation had helped to build the ice sheets thousands of years earlier, and the ice sheets had responded with their usual lags by slowly growing to maximum size by 21,000 years ago. By the time the ice sheets reached their maximum size, however, summer insolation had already risen close to today's level and was headed toward higher levels that would soon begin to melt the ice (Figure 12–1).

Because the seasonal insolation levels 21,000 years ago were close to those today, insolation cannot have been the major explanation of the differences in climate between these two intervals. Two factors are left as probable explanations of the colder and drier glacial maximum climates: the larger size of the ice sheets and the lower concentrations of greenhouse gases. This chapter focuses on the likelihood that these two factors account for the very different glacial-maximum world.

12-1 Project CLIMAP: Reconstructing the Last Glacial Maximum

Cooperative efforts to reconstruct past climates began in the 1970s, when a large interdisciplinary effort called the **CLIMAP (*Climate Mapping and Prediction*) Project** reconstructed the surface of Earth at the last glacial maximum. Led by the marine geologists John Imbrie and Jim Hays and the geochemist Nick Shackleton, CLIMAP drew on the expertise of scientists with specialized knowledge of ice sheets, windblown deposits, marine sediments, vegetation, and climate modeling.

In the years before CLIMAP, relatively few scientists worked on past climates, and the scientists who did usually worked on individual research projects using widely differing techniques. CLIMAP succeeded in bringing many of these scientists together and combining their skills into a single interdisciplinary effort. As a result, a generation of climate scientists in many disciplines learned to see Earth's climate as an integrated whole. Today interdisciplinary alliances are common in the study of climate science.

The CLIMAP group published a first map of the ice-age Earth in 1976 and then a revised version in 1981 (Figure 12–2). These maps show conditions at Earth's surface during typical seasons (in this case summer) at the last glacial maximum. Because individual years cannot be resolved in glacial-age records, the maps portray an average northern summer during the millennium or so centered on the glacial maximum.

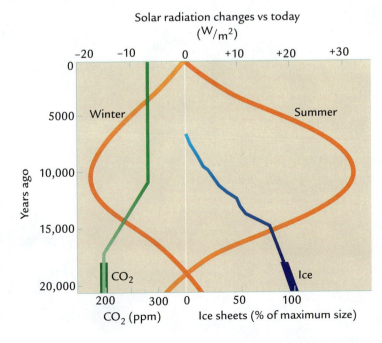

FIGURE 12-1 Boundary conditions for glacial maximum climate Climate models simulate glacial maximum climate by using larger ice sheets (thick blue line) and lower levels of greenhouse gas (thick green line) as boundary condition input. (Adapted From J. E. Kutzbach et al., "Climate and Biome Simulations for the Past 21,000 Years," *Quaternary Science Reviews* 17 [1998]: 473–506.)

CHAPTER 12

Last Glacial Maximum

When the most recent of the ~100,000-year glacial oscillations culminated 21,000 years ago, most of Earth's surface was very different from its current appearance. Ice sheets 2 or more km high covered Canada, the northern United States, northern Europe, and parts of Eurasia. Global sea level was 110–125 m lower, joining modern islands between Asia and Australia and connecting Britain to mainland Europe. South of the ice sheets, conditions were cold and windy, with dust blowing in many areas. The modern forests of North America, Europe, and Asia were regions of tundra or grasslands, and lower levels of atmospheric CO_2 and CH_4 caused cooling and drying across the tropics and the southern hemisphere.

In this chapter we focus on aspects of this glacial world such as the extent and thickness of the ice sheets and the debris they produced. Then we explore how changes in the distribution of life forms, especially land vegetation and ocean plankton, allow us to test climate simulations run on general circulation models. Finally, we examine a controversy that has implications for future changes in climate: How cold were the tropics at the last glacial maximum?

Deglacial Climate Change

The ability to reconstruct regional climate responses across Earth's entire surface also gives scientists the opportunity to look at climate change from a geographic perspective. Mapping climate changes at specific times ("time slices") gives researchers insights into specific regional processes that alter climate and allows comparisons of data with model simulations. Boundary conditions based on observations from the geologic record can be used to run model simulations of past climates, and the results (output) from these simulations can be compared with independent geologic data assembled into map form. The major boundary conditions that have driven climate changes during the last 21,000 years have been changes in the size of ice sheets, in seasonal insolation, and in the levels of greenhouse gases in the atmosphere (Chapter 13).

In climate archives with sufficiently high resolution, climatic oscillations lasting thousands of years are apparent, especially in and around the North Atlantic Ocean (Chapter 14). These fluctuations can rarely be dated accurately enough to permit firm correlations among different archives. These oscillations require an explanation other than orbital forcing, but their cause is as yet unknown.

In this part we address the following important questions:

- To what extent does the timing of ice melting in the last 21,000 years support the Milankovitch theory that ice sheets are forced by orbital insolation?

- What does the cooling of the tropics at the glacial maximum tell us about Earth's sensitivity to the concentration of atmospheric CO_2?

- To what extent do changes in tropical moisture during the last 21,000 years support the Kutzbach theory that orbital insolation controls the strength of summer monsoons?

- How has Earth's climate system responded to insolation changes during the several thousand years since the ice sheets melted?

- What is the origin of the brief oscillations of climate that occur at intervals much shorter than changes in Earth's orbital configurations?

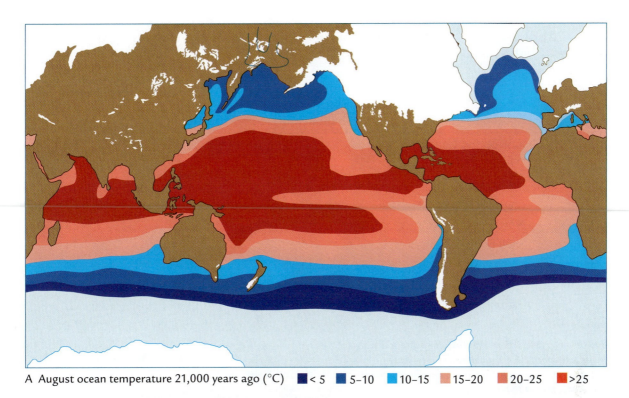

A August ocean temperature 21,000 years ago (°C)   ■ < 5   ■ 5–10   ■ 10–15   ■ 15–20   ■ 20–25   ■ >25

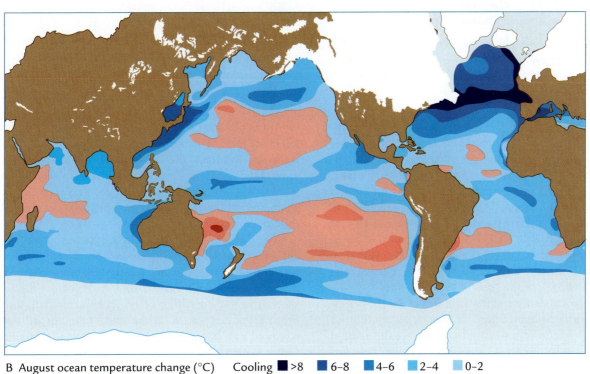

B August ocean temperature change (°C)   Cooling ■ >8   ■ 6–8   ■ 4–6   ■ 2–4   ■ 0–2
Warming ■ 0–2   ■ >2

FIGURE 12-2   CLIMAP reconstruction of glacial maximum ocean temperatures CLIMAP produced the first global-scale reconstruction of Earth's surface at the most recent glacial maximum, including ice sheet size and ocean surface temperature. These maps show (A) conditions during an average summer during the glacial maximum and (B) the changes in temperature between then and today. (Adapted from CLIMAP Project Members, *Seasonal Reconstruction of the Earth's Surface at the Last Glacial Maximum*, Map and Chart Series MC-36 [Boulder, CO: Geological Society of America, 1981].)

TABLE 12.1   Approximate Volumes of Ice and Amounts of Water Stored in Glacial Ice Sheets by Lowering Sea Level beneath Today's Position

Ice sheet	Location	Excess ice volume (million km³)	TSH:Sea level Amount (m)	Change (m)[a]
Laurentide	East-central Canada	25–34[b]	72–100	50–70
Cordilleran	Western North America	1.8	5	3.5
Greenland	Greenland	2.6[c]	7	5
Britain	England, Scotland, Ireland	0.8	2	1.5
Scandinavian	Northern Europe	7.3	21	15
Barents/Kara	Shelf north of Eurasia	6.9	20	14
East Antarctic	Eastern Antarctica	13.3[d]	9	6
West Antarctic	Western Antarctica	16.5[d]	18	13
Others	Various	1.2	3	2
All ice sheets		55–64	155–183	109–129

[a]Net sea level changes are 30% smaller than the volumes of seawater removed from the ocean because ocean bedrock rises when the weight of water is removed.

[b]The higher estimate shown is for a thick ice sheet like that in the CLIMAP maximum reconstruction; the lower estimate is for a thin ice sheet.

[c]Present volume of ice on Greenland is 3 million km³.

[d]Present volume of ice on Antarctica is 29 million km³.

Source: Adapted from G. H. Denton and T. J. Hughes, *The Last Great Ice Sheets* (New York: Wiley, 1981).

The most striking features of the CLIMAP reconstruction are the continent-sized ice sheets covering North America as far south as 37°N latitude, Scandinavia down to 48°N, and the Arctic margins of Eurasia. Today ice sheets on Antarctica and Greenland cover a combined area of about 14.2 million km², equivalent to just under 3% of Earth's surface area and about 10% of its land surface. The CLIMAP reconstruction of Earth at the last glacial maximum shows ice sheets covering an area of 35 million km², equal to 7% of Earth's total surface and 25% of the area of the continents.

The North American ice sheet was by far the largest of the northern hemisphere ice sheets, accounting for over 55% of the volume of ice in excess of the amount on Earth today (Table 12–1). This great ice sheet was roughly equivalent in volume to the ice sheet on modern Antarctica. Most of the ice in North America was in the **Laurentide ice sheet,** centered on east-central Canada, with the rest in the much smaller **Cordilleran ice sheet** over the Rockies in the American West. The **Scandinavian ice sheet** in northern Europe and the **Barents ice sheet** on the northern Eurasian continental shelf represented about 22% of the extra ice. The remainder was accounted for by expansion of ice sheets on Antarctica and Greenland across land exposed by the fall in sea level. Expansion of mountain glaciers dramatically transformed the appearance of high terrain around the world but contributed little to the change in ice volume. In some regions, such as Patagonia in southernmost South America, mountain glaciers merged into small ice caps.

The glacial world in the CLIMAP reconstruction was almost 4°C colder than Earth today. The North Atlantic cooled by 8°C or more, and sea ice was more extensive than it is today (see Figure 12–2B). Farther from the ice sheets, the North Pacific cooled by a lesser amount, with expanded winter sea ice in the northwest. Sea ice also advanced well beyond its present limits in the Southern Ocean, with a band of cooler ocean surface temperatures north of the expanded limit of sea ice around Antarctica. The CLIMAP reconstruction indicates low-latitude ocean temperatures only slightly cooler than today's and in some regions even a bit warmer.

In the decades since this reconstruction was published, several of its features have been challenged, and a few of them are almost certainly wrong. Yet the CLIMAP reconstruction of the ice-age Earth is still the standard against which challenges are directed.

12-2 How Large Were the Ice Sheets?

The glacial geologists George Denton and Mikhail Grosswald and the glaciologist Terry Hughes headed the CLIMAP ice sheet reconstructions. Of the two alternative reconstructions published, one broke with conventional thinking by portraying the ice sheets at their maximum plausible size (the "maximum reconstruction"). Three aspects of this reconstruction proved highly controversial.

One debate arose over the lateral extent of the ice sheets, which in most places were shown extending to the maximum limits that ice had reached at any time in the history of northern hemisphere glaciation (see Figure 12-2). The southern limits of most continental ice sheets at the last glacial maximum had not been in much doubt for many years prior to the CLIMAP reconstruction, but larger uncertainties remained about whether or not higher-latitude ice margins actually reached the ocean. Some glacial geologists argued that the ice sheets in the north were less extensive than they had been during earlier glaciations because they were starved for moisture. The main reason for this uncertainty was that the scarcity of organic carbon in the cold, dry Arctic made it difficult to find ^{14}C with which to date the deposits.

Estimates of the northern ice margins generally disagreed by a few hundred kilometers, amounts that were important to arguments about the physical state of the ice sheets in specific regions but actually rather small in comparison with the full lateral dimensions of ice sheets (thousands of kilometers) and with the size of grid boxes used in climate modeling (hundreds of kilometers on a side). As it turned out, careful ^{14}C dating along the northern margins of the ice sheets has confirmed that most of them were indeed at or near their maximum limits 21,000 years ago.

A second major disagreement about the extent of glacial maximum ice sheets centered on marine ice sheets, the ice sheets that formed on shallow continental shelves with their bases lying below sea level (companion Web site, p. 29). The CLIMAP reconstruction placed two marine ice sheets along the northern margin of Eurasia, a large one over the Barents Sea directly north of Scandinavia and a smaller one over the Kara Sea north of western Russia. This aspect of the reconstruction caused years of controversy. Subsequent surveys of these shallow seas collected evidence (glacial debris in sediment cores and depth soundings showing the extent of submerged moraine ridges) proving that at the glacial maximum, a large marine ice sheet did exist in the Barents Sea and perhaps in the westernmost Kara Sea but not in the eastern Kara Sea or east of that region.

Another controversial aspect of the CLIMAP reconstruction was the thickness (and height) of the ice sheets. Height and thickness are usually related in a simple way: as an ice sheet grows, in time it weighs down the underlying bedrock by an amount equal to about 30% of its thickness (Chapter 9). As a result, the height of the ice above the surrounding landscape usually represents about 70% of its total thickness. Unfortunately, no simple method exists to measure or reconstruct changes in either the thickness or height of ice sheets through time.

The CLIMAP maximum reconstruction showed relatively high (thick) ice sheets, especially over North America. This reconstruction was based on the assumption that the ice sheets had existed at or near their maximum extent long enough to build up slowly to full equilibrium thickness and that they consisted of stiff ice frozen to the underlying bedrock. These two assumptions produced ice sheets with profiles that rose relatively steeply from their outer margins to high elevations and great thicknesses in their central regions (Figure 12–3A, B).

This aspect of the CLIMAP reconstruction is still controversial. One line of evidence has come from measurements of the size of the fall in global sea level caused by water stored in the ice sheets. Coral reefs can be used as dipsticks to measure past changes in sea level (Chapter 9), and reefs in different areas indicate a sea level drop of somewhere between 110 and 125 m during the last glacial maximum. The higher estimate of sea level fall agrees with the large-volume ice sheets in the CLIMAP reconstruction, but the lower estimate falls close to the smaller estimate of ice volume (see Table 12–1).

If the ice sheets were as extensive as those CLIMAP proposed but held a smaller volume of water taken from the ocean, one or more of the ice sheets must have been thinner than the CLIMAP reconstruction indicated. Scientists naturally focused on the North American ice sheet because of its huge size: if this ice sheet were thinner by 30%, it would match the lower estimate of sea level change.

Independent evidence supports the possibility of thinner North American ice. Ice sheets slip and slide across soft water-laden sediments and unconsolidated sedimentary rock (Chapters 9 and 11). Only the central part of the North American ice sheet rested on hard bedrock, while the southern margins lay on easily deformed sediments at lower, warmer elevations. As a result, the southern margins of the ice sheet were probably thinned by frequent sliding (Figure 12–3C). This sliding could also have thinned the inner portion of the ice sheet by drawing ice out toward the margins.

A second line of evidence comes from the amount of bedrock rebound that has occurred since the ice sheets melted. Bedrock weighed down by ice sheets has a slow viscous response, and some of its rebound after the ice

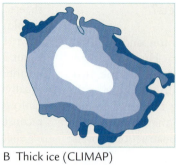

B Thick ice (CLIMAP)

Ice elevation (km)

■ <1 ■ 1–2 ■ 2–3 □ >3

C Thin ice

A Ice sheet extent

FIGURE 12-3 How thick were the ice sheets? (A) The limits of the North American ice sheet are well known, but (B) the CLIMAP reconstruction showed a thick high-elevation ice sheet (white) during the glacial maximum and (C) recent reconstructions favor a thinner ice sheet. (Adapted from P. Clark et al., "Numerical Reconstruction of a Soft-Bedded Laurentide Ice Sheet During the Last Glacial Maximum," *Geology* 24 [1996]: 679–82.)

melts is delayed for thousands of years later (Chapter 9). As a result, the bedrock retains a "memory" of how much ice was once on the land and when it melted. Scientists can exploit this behavior by examining regions in which the land is slowly rising out of the sea even now, leaving a trail of fossil beach shorelines exposed along the rising coasts. By radiocarbon dating the shells that formed when these beaches were at sea level and then measuring the present elevations of the old beaches above sea level, scientists can attempt to reconstruct the history of bedrock rebound and estimate how thick the ice sheets originally were. This technique has produced estimates of a thinner North American ice sheet than CLIMAP predicted, although doubts persist as to whether the bedrock "memory" of the ice sheets that existed 21,000 years ago has grown too dim to be useful.

12-3 Glacial Dirt and Winds

The ice sheets were prolific producers of debris in sizes ranging from large boulders to fine clay. Ice sheets grind across the landscape, scraping and dislodging soils and relatively unconsolidated sedimentary rocks. The weight of the ice sheets provides a pressure force that uses debris carried in the bottom layer of the ice to grind and gouge out small pieces of even the hardest bedrock. In areas where basal layers of ice alternately freeze and thaw, water trickles down into cracks in the rock when the ice melts and then expands when it freezes again,

breaking off large chunks. This freeze-thaw process quarries large slabs of bedrock and incorporates them in the ice for further grinding and fragmentation.

These and other processes erode large volumes of debris of all sizes. The ice sheets carry this material and deposit it along their margins when the ice melts. Much of the unsorted debris is piled into moraines (Chapter 2). Running water from melting ice or local precipitation reworks the debris, extracting finer sediments and producing **glacial outwash.** Ice margins usually have little vegetation because the constant supply of new debris buries new growth and because meltwater inundates the landscape in summer. The lack of vegetation exposes debris to further erosion.

Winds then rework these deposits, creating a gradation of grain sizes away from the ice margins. The coarsest debris (boulders, cobbles, and pebbles) remains in place, but strong winds can transport medium to fine sand over short distances. Winds also lift and carry finer silt-sized sediment farther from source regions, leaving loess deposits that become thinner and finer away from the glacial outwash (Figure 12–4). The loess patterns suggest that winds carried this debris mainly from the west-northwest to the east-southeast in both North America and Europe.

Winds can carry even finer (clay-sized) dust completely around the world. Glacial-age layers in the Greenland ice sheet contain ten times as much fine dust as interglacial layers. Chemical analysis of this dust indicates that the main source region was Asia rather

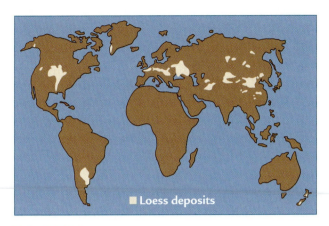

FIGURE 12-4 Glacial maximum loess Ice sheets and mountain glaciers eroded large amounts of debris of all sizes and carried it to their margins. Winds picked up silt-sized loess and deposited it downwind of these sources. (Adapted from K. Pye, "Loess," *Progress in Physical Geography* 8 [1984]: 176–217.)

than nearby North America. Glacial ice also contains more Na⁺ and Cl⁻ ions, an indication that far more salt was lifted from stormy glacial sea surfaces and deposited in the ice then than today.

Dust transport was also greater at lower latitudes during the last glacial maximum. Today the North African and Arabian deserts and their semiarid margins produce some of the largest dust storms on Earth. By comparison, sediment cores from the Indian Ocean east of the Arabian Peninsula show that dust accumulated five times faster during the last glacial maximum and during several previous glaciations than it does today. Cores from the equatorial Atlantic Ocean reveal that dust was also deposited in that region at higher rates during glaciations.

The extremely arid cores of the deserts were also affected (Figure 12–5). Regions of Arabia and North Africa identified as deserts on modern maps have actually varied in moisture level through time: moving sand dunes and loose soil are prevalent during extremely arid intervals, but the dunes are stabilized by sparse desert vegetation during monsoonal intervals that are moister (Chapter 8). In these areas and in Australia as well, moving sand dunes were much more extensive during the last glaciation than they are today because the winds were stronger and the climate was drier.

Even the South Pole was dustier. Glacial-age layers of ice cores from Antarctica contain more than ten times as much dust as modern interglacial layers. Geochemical fingerprinting of likely source areas suggests that this dust came from the southernmost tip of South America (Patagonia). The increased flux of dust probably resulted both from greater production of debris at the sources and from a more turbu-

lent atmosphere that carried dust farther south than today.

Almost everywhere we look, evidence shows that more debris was blowing across Earth's surface in the glacial world. One way of testing climate model simulations of the last glacial maximum is to examine the distribution of various kinds of debris carried by winds, ranging from desert sands to windblown silts (loess) and fine (clay-sized) dust. Because climate models can simulate the strength and direction of winds from the surface up to jet stream altitudes, the potential exists to compare model simulations with the observed patterns of windblown glacial debris.

Unfortunately, even though the current generation of climate models does a fairly good job of simulating the large-scale circulation of the atmosphere, the models do not do as good a job at the smaller scales needed to simulate dust transport. The models are less successful at simulating the processes that actually lift and transport silt and dust from Earth's surface, such as local wind gusts along frontal systems or small-scale eddies of wind.

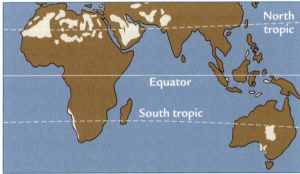

A Sand dunes active today

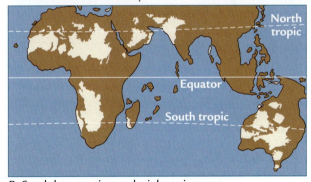

B Sand dunes active at glacial maximum

FIGURE 12-5 Glacial maximum sand dunes (A) Moving sand dunes occur today in Africa, Arabia, and Australia. (B) At the last glacial maximum, drier climates and stronger winds created more extensive sand dunes. (Adapted from M. Sarnthein, "Sand Deserts During Glacial Maximum and Climatic Optimum," *Nature* 272 [1978]: 43–46.)

Testing Model Simulations against Biotic Data

So far, we have examined only the physical aspects (ice and dirt) of the glacial maximum world. But living organisms also have a story to tell. They also allow us to test the performance of climate models on a world quite different from ours.

12-4 COHMAP: Data-Model Comparisons

During the 1980s, an interdisciplinary project called **COHMAP** (*Co*operative *H*olocene *Ma*pping *P*roject) used a combined data-model approach to examine the last glacial maximum and the subsequent changes to interglacial conditions. Led by the meteorologists John Kutzbach and Tom Webb, the paleoecologist Herb Wright, and the geographer Alayne Street-Perrott, COHMAP brought together scientists from countries around the world to pool information from hundreds of individual ^{14}C-dated records of lake levels and pollen in lake sediments for the purpose of examining regional-scale patterns.

The first step in the COHMAP approach was to assemble records of the changing boundary conditions that have driven climate over the last 21,000 years (Figure 12–6). As noted earlier, the largest differences in boundary conditions compared to conditions today were the larger ice sheets and the lower greenhouse-gas concentrations (see Figure 12–1).

The COHMAP researchers then ran model simulations of climate at intervals of several thousand years between the glacial maximum and the present to determine how changes in the major boundary conditions drove regional patterns of climate change. The COHMAP team focused on the role of orbital-scale changes in climate over intervals of thousands of years, rather than on shorter-term fluctuations superimposed on this gradual trend.

The climate data produced as output from these model simulations were then tested against climate reconstructions using ^{14}C-dated records of pollen from lake cores and plankton shells from ocean sediment cores. Modern relationships between the abundances of species and climatic variables can be measured, quantified, and used to reconstruct past climates from fossil organisms. By comparing these fossil-based estimates of climate with the changes simulated by the models, scientists can test the reliability of both approaches (see Figure 12–6).

12-5 Pollen: An Indicator of Climate on the Continents

Precipitation and temperature determine the larger-scale vegetation units such as forests, grasslands, and deserts and also the distribution of particular species within those units. Pollen is carried mainly by winds and to a lesser extent by water and insects. Some pollen comes to rest in lakes and settles into the mud, where its resistant outer layer aids preservation. The preserved pollen reflects the average composition of vegetation over a region extending tens of kilometers from the lake. The pollen percentages are generally similar to those of the actual vegetation, although "overproducers" such as pine trees leave disproportionately large amounts of pollen compared with "underproducers" such as maples. Climate scientists can adjust for this kind of disproportionate representation.

The northern midwestern states are a useful region for showing the climatic control on vegetation (Figure 12–7). The percentage of pollen from prairie grasses and herbs is higher in modern lake sediments west of the Mississippi River than in the wetter, tree-dominated area to the east. Within the eastern forest, cold-tolerant spruce pollen is more abundant in the north, while oak pollen is more abundant in warmer southern latitudes. These climatic controls can also be demonstrated by plotting pollen percentages against different combinations of seasonal and annual temperature and precipitation (Figure 12–8).

These modern relationships are a useful basis for understanding the past. The bottom layers of sediment in a ^{14}C-dated core from Minnesota are late glacial in age, dating from the time just after the North American

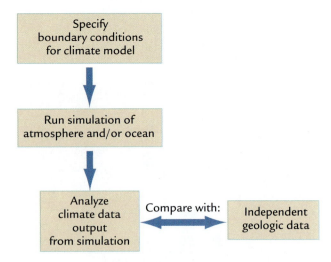

FIGURE 12-6 Data-model comparisons Past climates can be estimated by running climate model simulations with boundary conditions different from those of today and comparing the model output against estimates derived from pollen in lake sediments or other climatic data. (Adapted from J. Kutzbach et al., "Climate and Biome Simulations for the Past 21,000 Years," *Quaternary Science Reviews* 17 [1998]: 473–506.)

A Annual precipitation (cm)

FIGURE 12-7 Pollen distributions and climate Today the distribution of pollen in the northern midwestern United States reflects control by (A) precipitation and (B) temperature. (C) Prairie grasses and herbs are most abundant where rainfall is low, and tree pollen is more common in wetter eastern regions. (D) Spruce trees proliferate in the colder north, (E) oak in the warmer south. (Adapted from T. Webb III, "Holocene Palynology and Climate," in *Paleoclimate Analysis and Modeling*, ed. A. Hecht [New York: Wiley, 1985].)

B Annual temperature (°C)

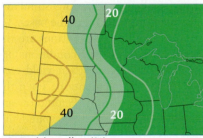

C Prairie pollen (%)

D Spruce pollen (%)

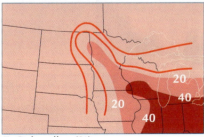

E Oak pollen (%)

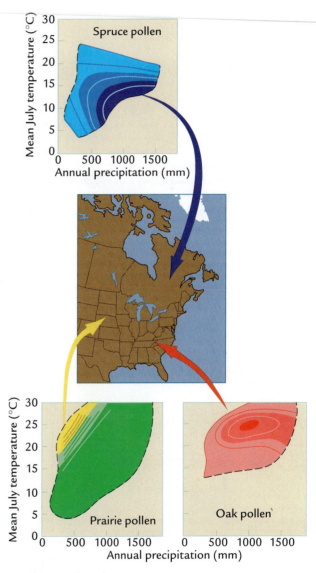

FIGURE 12-8 Pollen percentages and climate The abundances of spruce, oak, and prairie pollen follow distinct temperature and precipitation patterns. Colors indicate the same pollen abundances as in Figure 12-7. (Adapted from T. Webb III et al., "Climatic Change in Eastern North America During the Past 18,000 Years: Comparisons of Pollen Data with Model Results," in *North America and Adjacent Oceans During the Last Deglaciation,* ed. W. F. Ruddiman and H. E. Wright [Boulder, CO: Geological Society of America, 1987].)

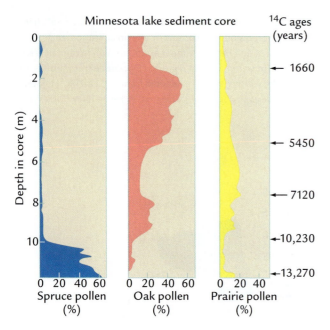

Minnesota lake sediment core

FIGURE 12-9 **Pollen in a lake core** A ^{14}C-dated sediment core from a Minnesota lake shows a transition in climate near 10,000 years ago from colder conditions (abundant spruce) to a warmer climate (abundant oak). High percentages of prairie grasses near 6000 years ago indicate a drier climate. (Adapted from H. E. Wright et al., "Two Pollen Diagrams from Southeastern Minnesota: Problems in the Late- and Postglacial Vegetation History," *Geological Society of America Bulletin* 74 [1963]: 1371–96.)

pollen to warm-tolerant oak pollen near 10,000 years ago indicates rapid warming in this region. Subsequently, changes to maximum values of dry-adapted herb and grass pollen indicate a climate drier than today's.

The Minnesota core is one of many hundreds examined in North America, along with additional hundreds in Europe and elsewhere in the world (see Chapter 2). Viewed together, these records provide a larger geographic perspective on the pattern of pollen (and vegetation) distribution at the last glacial maximum and during the deglaciation. This larger map perspective can be compared with map patterns produced by model simulations.

12-6 Using Pollen for Data-Model Comparisons

Data-model comparisons focus on the distribution of pollen at specific intervals in the past across geographic regions. Counts of pollen percentages in lake sediments within these regions produce mapped patterns of "observed" pollen abundance. These observed patterns are then compared with pollen distributions simulated by climate models for the same interval in the past.

These model-simulated pollen distributions are the result of several steps. First, boundary conditions are chosen and used in model simulations that yield estimates of past temperature and precipitation. Then the model estimates of temperature and precipitation are used to generate estimates of the percentage abundance of each type of pollen based on the modern relationship between climate and pollen (for example, see Figure 12–8). Each estimate of annual precipitation and mean July temperature simulated for a specific grid box in the model yields a specific estimate of the percentage of oak, spruce, and prairie pollen for that particular loca-

ice sheet melted back from this region, whereas the upper layers of mud record the postglacial climate of the present interglaciation (Figure 12–9). Most of the pollen in the older layers is from spruce trees, indicating conditions colder than today's. An abrupt switch from spruce

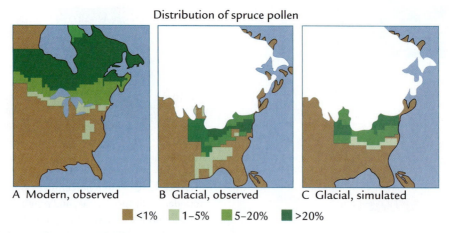

Distribution of spruce pollen

A Modern, observed   B Glacial, observed   C Glacial, simulated

■ <1%   ■ 1–5%   ■ 5–20%   ■ >20%

FIGURE 12-10 **Modern and glacial maximum spruce** (A) Today spruce pollen is most abundant in the cold climate of northeastern Canada. (B) At the glacial maximum, spruce pollen is found mainly in lake sediments from the northern United States. (C) Model simulations confirm that the large North American ice sheet produced temperatures cold enough for spruce to flourish in the northeastern United States. (Adapted from T. Webb III et al., "Late Quaternary Climate Change in Eastern North America: A Comparison of Pollen-Derived Estimates with Climate Model Results," *Quaternary Science Reviews* 17 [1998]: 587–606.)

tion. The map patterns of pollen abundance estimated in this way can then be compared directly with the map patterns derived from pollen counts from lake cores.

For example, observations today show maximum amounts of spruce pollen in northeastern Canada (Figure 12–10A). Counts of pollen in lake sediments during the last glaciation show spruce concentrated in the east-central United States just south of the ice sheet (Figure 12–10B). These analyses agree fairly well with climate model simulations of where spruce should have occurred at the glacial maximum (Figure 12–10C).

Comparisons with pollen data can also be made with **biome models.** To estimate the vegetation that would have been present in different regions, the biome method again makes use of climatic variables simulated by GCMs for times in the past when boundary conditions differed from those today. The first step in the method uses broad temperature and precipitation constraints to narrow the possible range of major vegetation types (for example, no trees can occur in model grid boxes for which hyperarid climates are simulated, but grass and desert scrub vegetation can).

In the second step, the surviving vegetation units within each grid box compete for the resources necessary for growth and reproduction, such as water, nutrients, and light. Both steps are based on today's relationships between vegetation and the environment. Because the first step in the biome method encompasses all the major vegetation groupings on Earth, this approach can simulate changing patterns of vegetation on any continent.

Data-Model Comparisons of Glacial Maximum Climates

In this section we examine the matches and mismatches between model simulations of the climate of the last glacial maximum and observations from the climate record.

12-7 Model Simulations of Glacial Maximum Climates

Glacial ice sheets are a critical boundary condition for simulations of glacial climate (Chapters 9, 11). Their central domes protruded upward as massive, icy plateaus, blocking and redirecting the flow of air. Climate model simulations suggest that a high-domed ice sheet over North America could have split the winter jet stream into two branches at the glacial maximum.

In modern winters, a single jet stream enters North America near the border between Canada and the United States. Storms associated with this jet bring wet winters to Oregon, Washington State, and British Columbia (Figure 12–11A). In contrast, during glacial times, the jet stream split into a northern branch located along the northern flank of the ice sheet and a southern branch over the American Southwest (Figure 12–11B).

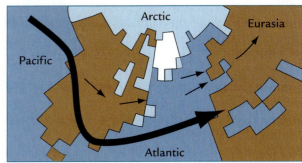

A Modern winters

B Glacial winters

Sea ice   Surface winds

Ice sheets   Jet stream

FIGURE 12-11 GCM simulation of climate near the northern ice sheets (A) Simulations run on climate models reproduce the modern path of the winter jet stream over North America. (B) For the last glacial maximum, a high-elevation ice sheet over North America splits the jet stream into two branches, one south and one north of the ice. At the surface, cold winds flow down off the North American and Scandinavian ice sheets and spiral in a clockwise pattern. (Adapted from COHMAP Project Members, "Climatic Changes of the Last 18,000 Years: Observations and Model Simulations," *Science* 241 [1988]: 1043–52.)

Ice sheets did not literally poke high enough into the atmosphere to block the flow of the jet stream and cause it to split in two. Ice sheets reach elevations of 2–3 km, whereas winter jet streams flow at altitudes of 10–15 km. But the ice did block the lower-level atmospheric flow, and the effect of this disruption was propagated higher into the atmosphere. These effects, along with the tendency of jets to flow above regions of strong temperature gradients at Earth's surface, caused the split jet. Model simulations using a high ice sheet (like that of the CLIMAP "maximum" reconstruction) split the jet to a much greater degree than do simulations based on lower-elevation ice.

Climate model simulations indicate that ice sheets caused other major changes in atmospheric circulation at Earth's surface (see Figure 12–11B). The models simulate a clockwise spiral of cold air moving down, off, and around the ice sheets in winter. Cold air flowing eastward along the northern flank of the North American ice sheet as part of this circulation blew southeastward over the western North Atlantic, chilling the ocean surface. A narrow layer of cold winds blew westward across the northern United States, reversing the west-to-east wind flow that dominates that region today. In Alaska, the clockwise pattern produced a south-to-north wind flow during the glacial maximum that may have prevented climate in the ice-free Alaskan interior from becoming even harsher than it is today.

A similar clockwise spiral of winds over the Scandinavian ice sheet brought cold, dry air southward into Europe (see Figure 12–11B). In addition, a strong upper-level jet stream crossed the Atlantic Ocean along latitudes between 45° and 50°N and entered Europe south of the ice sheet.

12-8 Climate Changes Near the Northern Ice Sheets

The most dramatic changes in climate at the glacial maximum were those in regions closest to and most directly influenced by the ice sheets. Most of the climate changes simulated by the models are consistent with independent geologic evidence.

The CLIMAP reconstruction based on the shells of planktic organisms shows the largest differences in estimated surface-ocean temperatures in the North Atlantic Ocean (see Figure 12–2). Frigid water and sea ice reached much farther south than they do today. The warm waters of the Gulf Stream and North Atlantic Drift flowed eastward toward Portugal instead of penetrating northeastward toward Scandinavia (Figure 12–12A). The flow of cold winds off the North American ice sheet was one important cause of this glacial cooling of the North Atlantic Ocean. Climate models that allow the ocean surface to react to the cold winds simulated changes in sea-surface temperature similar to those estimated by CLIMAP. In summer, the sea ice retreated to the north and the water warmed somewhat, but it remained well below modern temperatures. Later studies have indicated a larger summer retreat of sea ice and warmer temperatures than in the CLIMAP reconstruction.

Other large changes accompanied the North Atlantic cooling. Ice-rafted debris deposited in deep-ocean sediments across a broad band near 50°N latitude shows that icebergs broke off from continental ice sheets and drifted southward until encountering warm water and melting (Figure 12–12B).

Changes in North America An impressive example of agreement between observations and model simulations for the last glacial maximum occurs in the southwestern United States. Today this area is arid semidesert, except for deep winter snow pack on the mountains and small lakes maintained by meltwater runoff into the

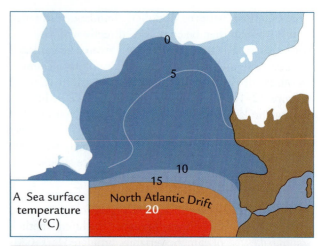

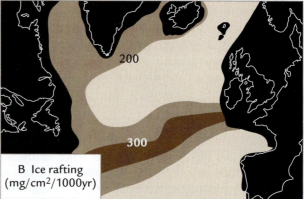

FIGURE 12-12 A cooler glacial North Atlantic Ocean (A) The region of largest ocean cooling in the CLIMAP reconstruction is the northern North Atlantic, which is surrounded by ice sheets. (B) Highest rates of deposition of ice-rafted debris occurred near 50°N, where southward-floating icebergs first encountered warm waters and melted. (A: Adapted from A. McIntyre et al., "Glacial North Atlantic 18,000 Years Ago: A CLIMAP Reconstruction," *Geological Society of America Memoir* 145 [1976]: 43–76. B: Adapted from W. F. Ruddiman, "North Atlantic Ice Rafting: A Major Change at 75,000 Years b.p.," *Science* 196 [1977]: 1208–11.)

basins. Most runoff is trapped in the basins and never reaches the ocean (Figure 12–13A).

At the last glacial maximum, this region was strikingly transformed, with hundreds of large new lakes where none exist today (Figure 12–13B). The most prominent of these, glacial Lake Bonneville near Salt Lake City, was ten times larger than today's Great Salt Lake. Dissolved salt that precipitated out of the brackish water into the lake muds created the Bonneville Salt Flats.

Climate model simulations of the last glacial maximum provide an explanation for this regionally wetter climate. The southern branch of the split jet stream entered North America over south-central California and produced two responses favorable to a moister climate than today's and a resulting expansion of lakes:

A

FIGURE 12-13 The glacial Southwest was wetter (A) Today most basins in the southwestern United States, such as Death Valley, are dry or are occupied only occasionally by temporary lakes. (B) At the last glacial maximum, lakes filled hundreds of basins because the southward displacement of the jet stream from Canada brought increased rain and cloud cover. (A: Peter Kresan. B: Adapted from G. I. Smith and F. A. Street-Perrott, "Pluvial Lakes of the Western United States," in *Late Quaternary Environments of the United States,* ed. S. C. Porter [Minneapolis: University of Minnesota Press, 1983].)

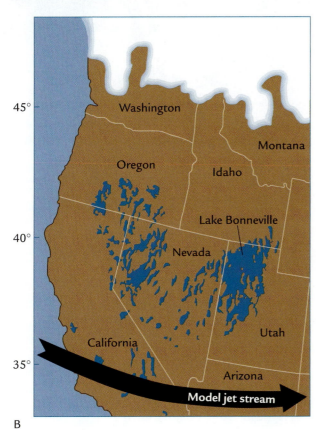

B

slopes of the Olympic Mountains in coastal Washington. At the last glacial maximum, this region was covered by grass and herb vegetation indicative of much drier conditions and less Pacific moisture.

The climate model simulations (see Figure 12–11B) suggest two reasons for this change. First, the shift of the winter jet stream to the southwestern United States displaced the main storm track and associated precipitation away from this region. In addition, the clockwise flow of cold, dry winds around the North American ice sheet produced more frequent low-level winds blowing westward from the dry mid-continent and replacing the flow of moist westerly winds from the Pacific.

The region with the most extensive coverage of lake cores and pollen data for testing climate models is eastern North America, today an area of temperate deciduous forests. In this region scientists can test the performance of climate models by checking for data-model agreement or disagreement about the magnitude, not just the direction, of climate changes.

East-central North America south of the ice sheet had a mixture of spruce trees, scattered deciduous trees, and grasses and herbs during the last glacial maximum. This mixture indicates a region of discontinuous tree cover interrupted by grassy openings. The more continuous forest cover south of 35°N was a mixture of pine and various deciduous trees.

Although the model-simulated pattern and the observed pattern of spruce in the northern United States at the glacial maximum match reasonably well (see Figure 12–10), the match does not hold up for several pollen types farther to the south. Pollen produced by deciduous trees such as oak and elm is much less abundant (or even nearly absent) in lake sediments from this region (Figure 12–14A) than the levels simulated by the climate models (Figure 12–14B).

more precipitation caused by winter storms following the path of the jet stream, and reduced evaporation caused by greater cloud cover and cooler temperatures.

In contrast to the wetter Southwest, the climate of the Pacific Northwest was colder and drier during the glacial maximum. In this region today, frequent winter storms from the Pacific Ocean bring moisture that sustains lush forests, including rain forests on the western

Distribution of elm pollen

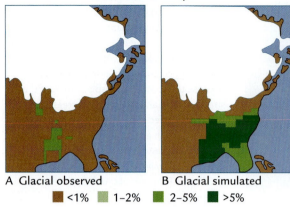

A Glacial observed   B Glacial simulated

■ <1%   ■ 1–2%   ■ 2–5%   ■ >5%

FIGURE 12-14 Data-model mismatch in the southeastern United States (A) Observed abundances of warm-adapted deciduous pollen such as elm in the southeastern United States during the glacial maximum are smaller than (B) the amounts simulated by climate models. (Adapted from T. Webb III et al., "Late Quaternary Climate Change in Eastern North America: A Comparison of Pollen-Derived Estimates with Climate Model Results," *Quaternary Science Reviews* 17 [1998]: 587–606.)

This mismatch suggests that the model-simulated cooling for the southeastern United States underrepresents the cooling that actually occurred by permitting too many warm-adapted trees. One cause of this mismatch may have been an unusual geographic configuration in which cold meltwater from the southern margin of the great Laurentide ice sheet flowed down the Mississippi River and emptied directly into the Gulf of Mexico at subtropical and tropical latitudes. If the sea-surface boundary conditions used in the model had incorporated this cold inflow, the model might have simulated cooler temperatures across a broad region of the southeastern United States influenced by air masses from the nearby Gulf. Disagreements like these in initial data-model comparisons point the way toward future improvements in model simulations and data interpretation.

Changes in Eurasia Europe was completely transformed at the glacial maximum. The conifer and deciduous forests typical of today's interglacial climate (Figure 12–15A) were absent from most of Europe south of the Scandinavian ice sheet. In their place, grass-covered steppes and herb-covered tundra vegetation covered much of the continent, with bits of forest scattered in the south (Figure 12–15B). The moderate maritime climate of today was preceded by a far harsher continental climate, more like that of modern northern Asia. These differences in vegetation agree with the dry, windy conditions indicated by the greater prevalence of windblown loess (see Figure 12–4).

Biome models simulate glacial vegetation in Europe similar to that observed, with Arctic tundra instead of

forest in the north near the ice sheets and grassy steppe vegetation prevailing farther to the south and east. One reason for this harsh glacial climate was the clockwise outflow of cold winds from the Scandinavian ice sheet (see Figure 12–11B). A second reason was the large chilling of the North Atlantic Ocean, which removed its moderating influence on winters in Europe (see Figure 11–3).

One of the most striking features of the last glacial maximum was the vast extent of steppe and tundra that covered much of northern Asia (Figure 12–16). A region covered today by forests of larch, birch, and alder trees was at that time a treeless expanse of grasses and herbs.

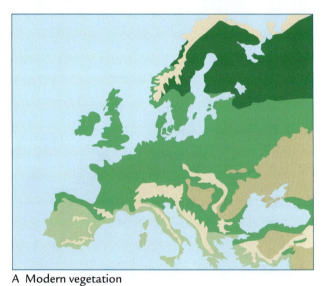

A Modern vegetation

□ Ice   ■ Boreal forest   ■ Mediterranean scrub
■ Tundra and   ■ Deciduous   ■ Prairie-steppe
   mountain   and conifer forest

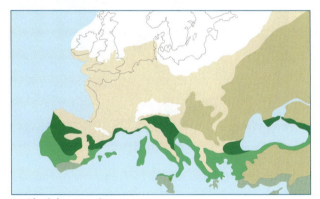

B Glacial vegetation

FIGURE 12-15 Glacial north-central Europe was treeless (A) Vegetation in modern Europe is dominated by forest, with conifers in the north and deciduous trees to the south. (B) At the glacial maximum, Arctic tundra covered a large area south of the ice sheet, with grassy steppe farther south and east and patchy forests near the Mediterranean coasts. (Adapted from R. F. Flint, *Glacial and Quaternary Geology* [New York: Wiley, 1971].)

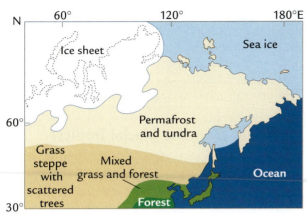

A Glacial maximum (observed)

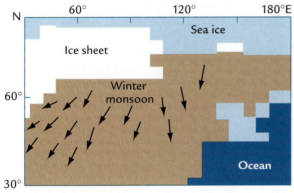

B Glacial maximum (model)

FIGURE 12-16 Glacial northern Asia was treeless (A) At the last glacial maximum, Asia was covered by permafrost and tundra in the north and steppe in the south, with little forest. (B) Climate models indicate that this distribution of vegetation resulted from a much stronger and colder winter high-pressure cell in northern Asia, with stronger cold winds blowing to the south. (A: Adapted from V. P. Grichuk, "Late Pleistocene Vegetation History," in *Late Quaternary Environments of the Soviet Union,* ed. A. A. Velichko et al. [Minneapolis: University of Minnesota Press, 1984]. B: Adapted from J. E. Kutzbach et al., "Simulated Climatic Changes: Results of the COHMAP Climate Model Experiments," in *Global Climates Since the Last Glacial Maximum,* ed. H. E. Wright et al. [Minneapolis: University of Minnesota Press, 1993].)

Forests were completely absent from the entire northern part of Asia, an indication that this region had an even harsher continental climate than it does today.

The harsh winter cold and sparse snow cover caused the ground to freeze tens of meters deep, forming **permafrost**, but in most regions the surface layer thawed in summer and allowed grasses and herbs to proliferate. South of the year-round and seasonal permafrost were extensive grassy steppes (Figure 12–16A). The greater area covered by tundra and steppe (rather than modern forest) made the ground surface much more reflective in the snowy season and further cooled this region.

Climate model simulations suggest that the main reason for this colder, drier glacial climate in Asia was a much stronger high-pressure cell in Siberia during winter that produced an increased outflow of cold, dry air southward (Figure 12–16B). In addition, icy conditions in the North Atlantic cut off much of the moisture source for the Asian interior.

The effects of these harsh winters in northern Asia extended into southeastern Asia. Today the influence of the warm, moist summer monsoon allows forests to grow along the Pacific coasts of China and Japan. At the glacial maximum, the stronger and more persistent winter monsoon sent cold air southward from Siberia and pushed the northern forest limit to the south (see Figure 12–16A). The North Pacific was also colder than it is today, with much more extensive winter sea ice along the coast of Asia and in the Bering Sea because of cold Siberian air blowing out over the ocean.

12-9 Climate Changes Far from the Northern Ice Sheets

Farther from the direct influence of the great northern hemisphere ice sheets, climate changes were less dramatic. In these regions the lower glacial concentrations of CO_2 and methane were probably the major cause of the climate changes (Chapter 11).

Large changes occurred in the Antarctic, where CLIMAP estimated that the winter limit of sea ice expanded northward by several degrees of latitude in the far-southern Atlantic and Indian oceans (Figure 12–17). Later analyses have reduced these limits only very slightly. Associated with this shift in sea ice was a northward displacement of the region of strongest upwelling and highest surface-water productivity, but productivity decreased in regions nearer Antarctica where the cover of sea ice persisted longer during summer.

The record of climate change on arid southern hemisphere continents remains sparse. Expanded desert dunes in Australia (see Figure 12–5) suggest an even more arid climate and an intensification of the modern counterclockwise wind flow. Lake levels and pollen data also provide supporting evidence for a drier interval near the glacial maximum. One factor that contributed to greater glacial aridity in northern Australia was the withdrawal of the ocean from a vast area just to its north (see Figure 12–2). Another factor was the lower concentrations of greenhouse gases in the atmosphere. A southward shift in the circum-Antarctic storm belt might also have led to drying of southern Australia.

Climate in much of South America is heavily influenced by winds from nearby oceans. Most ocean moisture is dropped in the Amazon rain forest and along the eastern flanks of the Andes. Scattered records from the Amazon Basin hint at the possibility that the rain forest

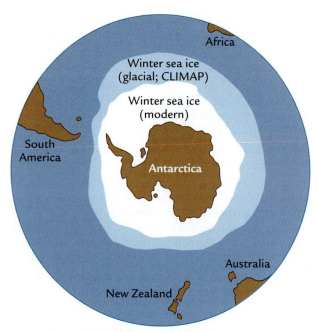

FIGURE 12-17 Glacial Antarctica was surrounded by more sea ice The CLIMAP glacial maximum reconstruction indicated that the seasonal maximum limit of sea ice in late winter and early spring expanded northward around Antarctica. (Adapted from J. D. Hays, "A Review of the Late Quaternary History of Antarctic Seas," in *Antarctic Glacial History and World Paleoenvironments*, ed. E. M. Van Zinderen Bakker [Rotterdam: Balkema, 1978], and from L. H. Burckle et al., "Diatoms in Antarctic Ice Cores: Some Implications for the Glacial History of Antarctica," *Geology* 16 [1988]: 326–29.)

may have been fragmented into smaller pieces than the massive area forested today.

Along the Andes, where most lake-sediment records have been found, pollen data generally indicate drier conditions at the glacial maximum. This drying is probably the combined result of less extraction of water vapor from the cooler oceans, the lowering of sea level by 110–125 m, and the cooler land temperatures resulting from lower CO_2 and methane levels in the atmosphere. Pollen data from far-southern latitudes indicate glacial climates wetter than today's west of the Andes but drier to the east. Climate model simulations show a southward shift of the axis of westerly winds and moisture-bearing storms, in agreement with the pollen evidence.

Because most of the tropics were more arid at the last glacial maximum, rain forest vegetation in both South America and Africa was probably less extensive than it is today. Yet despite this drier climate, total tropical biomass might actually have been greater. The large drop in global sea level exposed vast expanses of new land across continental shelves of Southeast Asia (see Figure 12–2). Because this region lay within the moist intertropical convergence zone, it would have supported tropical rain

forest vegetation. The increase in rain forest in this region may have offset the loss of biomass elsewhere in the tropics, but it could not offset the enormous decrease in forest biomass at high northern latitudes. As a result, the total glacial biomass on Earth's continents was about 25% lower than it is today (Chapter 10).

How Cold Were the Glacial Tropics?

For almost two decades climate scientists have argued about the amount of temperature change in the tropics and subtropics. Tropical sea-surface temperatures reconstructed by CLIMAP on the basis of fossil shells of ocean plankton averaged just 1°–2°C cooler than they are today, and in some regions such as the subtropical Pacific the ocean was estimated to have been more than 1°C warmer.

In contrast, other evidence suggests that temperatures over tropical landmasses and parts of the tropical ocean may have been 4°–6°C cooler than at present, far cooler than in the CLIMAP reconstruction. This discrepancy in estimates not only bothers scientists but also has much larger ramifications about an issue of modern (and future) importance—Earth's fundamental sensitivity to changes in atmospheric CO_2 and other greenhouse gases.

The tropics lie too far from the immediate thermal impact of the ice sheets to have been cooled by changes in atmospheric circulation (Chapter 11). In addition, solar insolation values at the last glacial maximum were close enough to those today that they could not have been a major factor in the glacial cooling of the tropics. What does explain the cooling in the tropics? By a process of elimination, the main cause must have been the 30% lower (190 ppm) levels of CO_2, along with the 50% drop in methane (Figure 12–18). When greenhouse-gas concentrations are low, less outgoing back radiation from Earth's surface is trapped in the atmosphere and the temperature falls. As a result, the amount of glacial cooling in the tropics should be a measure of the sensitivity of this part of the climate system to changes in atmospheric CO_2 and methane. With half of Earth's surface area lying between 30°N and 30°S, this cooling should give us a measure of the fundamental sensitivity of the climate system.

In Chapter 5 we examined Earth's response to higher levels of CO_2 in a greenhouse world. The last glacial maximum now provides a complementary perspective on the same relationship: Earth's response to lower levels of CO_2 in a full "icehouse" world. This analysis is directly relevant to future climate change, because the warming we face in the future will be caused by human-induced increases in atmospheric CO_2 and methane, and we need to know how large this warming will be.

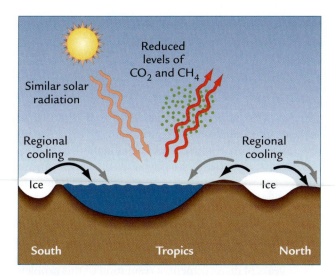

FIGURE 12-18 Lower CO$_2$ and CH$_4$ levels cooled the glacial tropics The tropics were too distant from the glacial ice sheets to feel their direct influence, and insolation values in summer and winter were close to those today. Lower levels of atmospheric CO$_2$ and CH$_4$ were the main cause of tropical cooling.

12-10 Evidence for a Small Tropical Cooling

The evidence for a small tropical cooling in the CLIMAP reconstruction was based on the small changes in planktic fauna and flora in low-latitude oceans. CLIMAP's technique for reconstructing sea-surface temperatures used the assumption that the distribution of species and assemblages of plankton is mainly determined by the temperature of the water in which they live. At higher northern latitudes during the glacial maximum, cold-adapted species moved into areas where warm-adapted species prevail today, indicating a large cooling in these regions. Across most low-latitude regions, however, the species that existed at the glacial maximum were not much different from the warm-adapted forms found there today (Figure 12–19). This lack of change in tropical plankton led CLIMAP to conclude that ocean temperatures in the tropics cooled by an average of only 1.5°C at the glacial maximum.

Evidence obtained from the biochemical composition of plankton shells supports the CLIMAP estimates in some regions. One technique is based on the relative abundance of complex organic molecules called

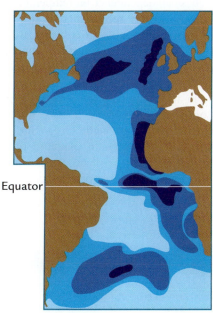

A Atlantic Ocean

Glacial plankton vs. plankton today
(Percent difference)

■ > 50   ■ 25–50   ■ 10–25   ■ < 10

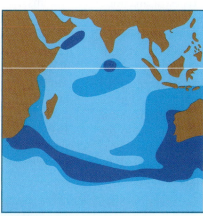

B Indian Ocean

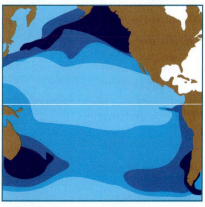

C Pacific Ocean

FIGURE 12-19 Planktic fauna of the glacial maximum vs. that of today The CLIMAP method of reconstructing glacial maximum ocean temperatures was based on temperature-sensitive plankton assemblages. Planktic assemblages in most low-latitude regions of the (A) Atlantic, (B) Indian, and (C) Pacific oceans differed only slightly from those of today, indicating little glacial cooling. (Adapted from T. C. Moore et al., "The Biological Record of the Ice-Age Ocean," *Palaeogeography, Palaeoclimatology, Palaeoecology* 35 [1981]: 357–70.)

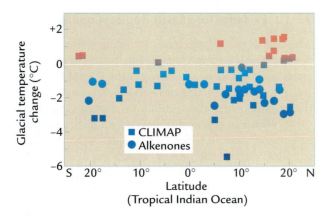

FIGURE 12-20 Confirmation of small Indian Ocean cooling A biochemical method of estimating past sea surface temperatures indicates a small cooling of the tropical Indian Ocean, similar to the values found by CLIMAP. (Adapted from E. Bard et al., "Interhemispheric Synchrony of the Last Deglaciation Inferred from Alkenone Paleothermometry," *Nature* 385 [1997]: 707-10.)

alkenones that constitute small fractions of tiny plant plankton (coccolithophores). The past abundances of these molecules can be measured in small $CaCO_3$ plates (coccoliths) deposited in ocean sediments (see Figure 2–14). The relative amounts of two types of alkenone molecules are sensitive to temperature in the modern ocean and can be used to reconstruct past temperatures.

In a north-south transect of cores across the western Indian Ocean, the cooling indicated by both methods is generally less than 2°C, with a larger cooling registered by the alkenone method only above 15°N (Figure 12–20). Temperature estimates based on magnesium/calcium ratios (Chapter 6) have also been made for many other regions. These methods indicate sea-surface temperatures cooler than those of CLIMAP by as much as 1°–2°C in some regions, but the agreement is closer in other regions.

12-11 Evidence for a Large Tropical Cooling

A different view emerges from other indicators, most of which come from continental records. The most compelling evidence is the descent of the lower limit of mountain glaciers by 600–1000 m throughout the tropics and middle latitudes (Figure 12–21). This drop in the elevation of the ice line has been interpreted as requiring a cooling of 4°–6°C over tropical mountains.

The lower limit of mountain glaciers today is determined mainly by temperature and secondarily by factors such as the amount of precipitation and the degree to which local mountain topography shelters the glaciers from direct sunlight. Glaciers exist today on tropical mountains higher than 5 km because the atmosphere cools by 6.5°C or more per kilometer of elevation,

resulting in subfreezing temperatures at higher elevations (companion Web site, pp. 15, 27–28). This relationship has been used to estimate a cooling of 4°–6°C to explain the lowering of tropical mountain glaciers by 600–1000 m during the glacial maximum.

Additional evidence for larger glacial cooling comes from the descent of the upper tree limit and other kinds of vegetation high on tropical mountains. In the harsh conditions on the upper flanks of mountains, temperature limits the growth of many kinds of vegetation, and the vertical drop in high-altitude vegetation limits during the last glaciation equals or even exceeds that of the mountain glaciers (see Figure 12–21).

12-12 Actual Cooling Was Medium-Small

After years of disagreement, a resolution of this problem seems to be emerging—the tropical cooling was neither as small as CLIMAP claimed nor as large as the critics initially thought but near the middle of the two estimates. One reason the CLIMAP estimates were too small is that plankton are less sensitive at low latitudes to changes in temperature than to changes in the availability of food. The low-latitude surface ocean is depleted in nutrients, and plankton are forced to adopt strategies for surviving where food is scarce. This requirement overwhelms the relatively small dependence on temperature.

In addition, the CLIMAP reconstruction for the Pacific Ocean was based on samples in which $CaCO_3$ had been extensively dissolved on the seafloor, thereby altering the assemblages of foraminifera and coccoliths. Similarly, the remnants of siliceous organisms (radiolaria and diatoms) left in the sediments are very different from those that originally lived in the surface waters. Estimates of temperatures at the Pacific Ocean surface derived from these different types of plankton often disagree, an indication that some or all are unreliable.

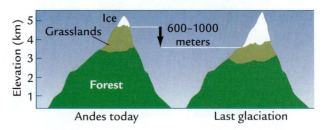

FIGURE 12-21 Descent of tropical mountain glaciers and forests The limits of mountain glaciers in the Andes were 600–1000 m lower during the last glaciation than they are today, and the upper limits of forests were similarly lower. These major shifts indicate a tropical cooling of at least 5°C, much larger than the 1°–2°C suggested by CLIMAP. (Adapted from T. van der Hammen, "The Pleistocene Changes of Vegetation and Climate in Tropical South America," *Journal of Biogeography* 1 [1974]: 3-26.)

The large-cooling view also has problems. The drier glacial climate in most of the tropics would have steepened the lapse rate from its present 6.5°C/km toward the 9.8°C/km rate typical of very dry air. A steeper lapse rate in the drier glacial tropics could account for part of the discrepancy between the ocean and land evidence. In addition, the evidence from mountain glaciers is poorly dated. Only a handful of regions have ^{14}C dates that closely constrain the glacial lowering of 600–1000 m to the exact glacial maximum time. Some glacial moraines initially thought to date from the glacial maximum turned out to have formed before 30,000 years ago during a time of cooler but also wetter climates. Lower glacier limits at such times could have been caused at least in part by greater snowfall.

Other factors contribute to the apparent discrepancy between land and ocean temperature changes. One factor is a result of sea level lowering. A sea level drop of 110–125 m increases the "height" of the mountains (relative to the new sea level) by that amount. For a lapse rate of 6.5°C/1000 m, this effect would make high mountain elevations cooler by ~0.75°C without any actual change in climate. Another factor is the greater responsiveness of the land than the ocean to climatic forcing. Climate model simulations of land-surface reactions to seasonal changes tend to exceed those of the ocean, which integrates year-round forcing. Still another factor is the fact that the very low glacial CO_2 values may fail to provide trees with enough CO_2 for photosynthesis and thereby cause the upper tree line to drop even without changes in temperature.

IN SUMMARY, the most likely resolution to the controversy over the glacial tropical cooling is that the cooling of some ocean regions (especially the Pacific Ocean) was larger than the CLIMAP estimate, but the cooling in many tropical land areas was not as large as the evidence from the mountains suggests. The CLIMAP estimates are also likely to be in error along some ocean margins and in nearly enclosed seas where CLIMAP had little or no core coverage or where glacial maximum planktic assemblages had unusual combinations of species that make it risky to apply techniques based largely on assemblages from the open ocean.

What is the implication for Earth's sensitivity to CO_2 and other greenhouse gases? We can estimate that the actual tropical cooling was about 3°C, roughly midway between the CLIMAP estimate (1.5°C) and the initial land-based estimate (5°C). And we know from measurements of air bubbles in ice cores that the glacial atmospheric CO_2 concentration was 90 ppm lower than the typical interglacial value of ~280 ppm, while the glacial methane concentration was about half of typical interglacial values of ~700 ppb.

A 3°C drop in tropical temperatures is close to the amount of cooling expected from general circulation model simulations of the tropical response to this lowering of greenhouse-gas concentrations. The evidence from the last glacial maximum indicates that the models capture the effect of greenhouse gases on climate reasonably well. This match gives climate scientists confidence that these same models are useful in forecasting future climate changes caused by increases in CO_2 and other greenhouse gases.

Key Terms

CLIMAP (Climate Mapping and Prediction) Project (p. 210)

Laurentide ice sheet (p. 212)

Cordilleran ice sheet (p. 212)

Scandinavian ice sheet (p. 212)

Barents ice sheet (p. 212)

glacial outwash (p. 214)

COHMAP (Cooperative Holocene Mapping Project) (p. 216)

biome models (p. 218)

permafrost (p. 223)

alkenones (p. 226)

Review Questions

1. What is the major uncertainty about the size of ice sheets at the glacial maximum?

2. Earth's radius r is 6371 km, and its surface area $(4/3 \, \pi \, r^2)$ is 70% water. If sea level was lower by 120 m during the most recent glacial maximum and if the surface of ice sheets was larger (45 million km² as against the modern surface area of 15 million km²), what was the average thickness of the ice sheets?

3. In what ways did ice sheets make the glacial world a "dirty" place?

4. How does the composition of pollen in lake sediments tell us about climate?

5. How and why did the glacial climate of the southwestern United States differ from the climate there today? Did the changes there generally agree with those in other regions?

6. How and why did glacial climates of Europe and northern Asia differ from the climate there today?

7. What caused the cooling of the tropics during the last glacial period?

8. Explain why a large versus small tropical cooling is important for understanding our future.

Additional Resources

Basic Reading

CLIMAP Members. 1981. *Seasonal Reconstruction of the Earth's Surface at the Last Glacial Maximum.* Map and Chart Series MC–36. Boulder, CO: Geological Society of America.

COHMAP Members. 1988. "Climatic Changes of the Last 18,000 Years: Observations and Model Simulations." *Science* 241: 1043–62.

Imbrie, J., and K. P. Imbrie. 1979. *Ice Ages: Solving the Mystery.* Short Hills, NJ: Enslow.

MacDougall, D. 2004. *Frozen Earth.* Berkeley, CA: University of California Press.

Rind, D., and D. Peteet. 1985. "Terrestrial Conditions at the Last Glacial Maximum and CLIMAP Sea-Surface Temperature Estimates: Are They Consistent?" *Quaternary Research* 24: 1–22.

Advanced Reading

Clark, P. U., J. M. Licciardi, D. R. MacAyeal, and J. W. Jenson. 1996. "Numerical Reconstruction of a Soft-Bedded Laurentide Ice Sheet During the Last Glacial Maximum." *Geology* 24: 679–82.

Fairbanks, R. G., and P. H. Wiebe. 1980. "Foraminifera and Chlorophyll Maximum: Vertical Distribution, Seasonal Succession, and Paleoceanographic Significance." *Science* 209: 1524–26.

Moore, T. C., Jr., W. H. Hutson, N. Kipp, J. D. Hays, W. L. Prell, P. Thompson, and G. Boden. 1981. "The Biological Record of the Ice-Age Ocean." *Palaeogeography Palaeoclimatology Palaeoecology* 35: 357–70.

Peltier, W. R. 1994. "Ice Age Paleotopography." *Science* 265: 195–201.

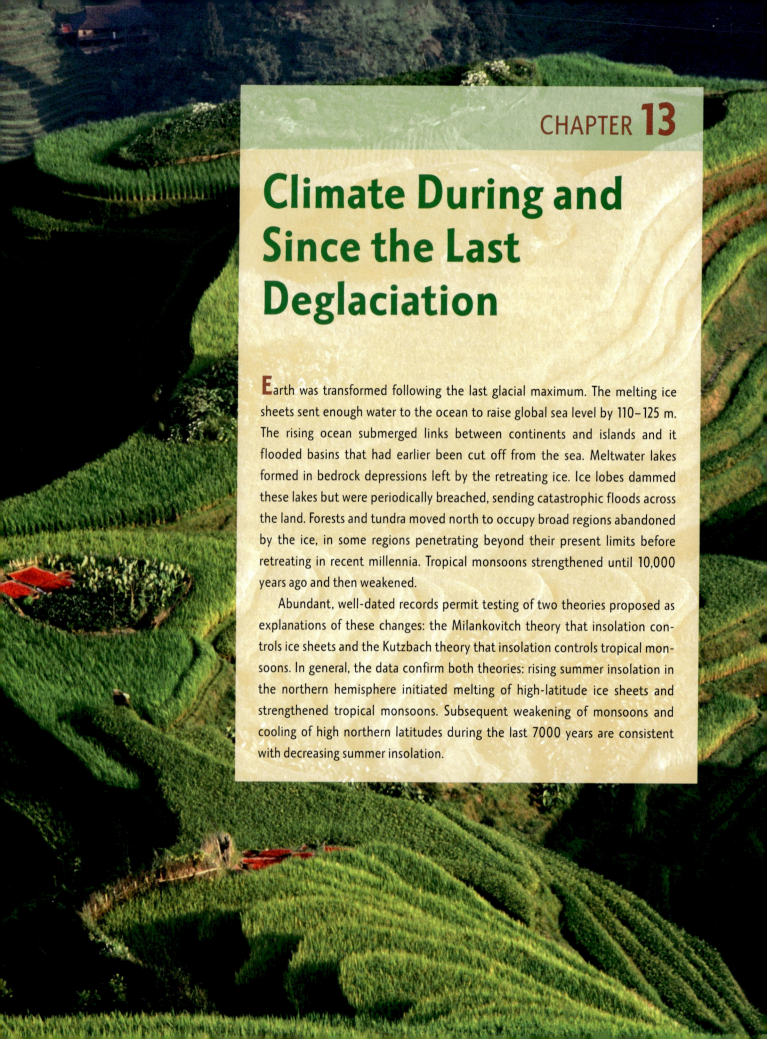

Climate During and Since the Last Deglaciation

Earth was transformed following the last glacial maximum. The melting ice sheets sent enough water to the ocean to raise global sea level by 110–125 m. The rising ocean submerged links between continents and islands and it flooded basins that had earlier been cut off from the sea. Meltwater lakes formed in bedrock depressions left by the retreating ice. Ice lobes dammed these lakes but were periodically breached, sending catastrophic floods across the land. Forests and tundra moved north to occupy broad regions abandoned by the ice, in some regions penetrating beyond their present limits before retreating in recent millennia. Tropical monsoons strengthened until 10,000 years ago and then weakened.

Abundant, well-dated records permit testing of two theories proposed as explanations of these changes: the Milankovitch theory that insolation controls ice sheets and the Kutzbach theory that insolation controls tropical monsoons. In general, the data confirm both theories: rising summer insolation in the northern hemisphere initiated melting of high-latitude ice sheets and strengthened tropical monsoons. Subsequent weakening of monsoons and cooling of high northern latitudes during the last 7000 years are consistent with decreasing summer insolation.

Fire and Ice: Shift in the Balance of Power

The main factors that explain why climate 21,000 years ago was different from climate today are the larger ice sheets and the lower atmospheric greenhouse-gas levels (primarily CO_2). During the subsequent deglaciation, a shift occurred in the balance of power among the factors that controlled global climate (Figure 13–1). Summer and winter insolation values that had been near modern levels during the last glacial maximum began to change. By 10,000 years ago, the angle of tilt of Earth's axis had reached a maximum at the same time that Earth's precessional motion moved it closest to the Sun on June 21. These orbital changes combined to produce a summer insolation maximum at all latitudes of the northern hemisphere.

The rise in summer insolation at higher northern latitudes triggered melting of the northern ice sheets. As the ice sheets melted, their influence on climate diminished, and the insolation anomalies (the departures from modern levels) became more important. The most recent deglaciation is mainly a story of this shift in the balance of power from ice (sheets) to fire (solar insolation). A second important change during this deglaciation was the increase in atmospheric CO_2 concentrations from 190 to 280 ppm, along with a doubling of methane levels. The increases in greenhouse gases coincided closely with ice melting.

13-1 When Did the Ice Sheets Melt?

Abundant evidence available from the recent deglaciation gives us an unusual opportunity to test explanations about how deglaciations occur. The Milankovitch theory (Chapter 9) predicts that the orbitally produced maximum in summer insolation near 10,000 years ago in the northern hemisphere should have caused significantly higher rates of ice melting.

It might seem that the way to quantify the rate of ice melting is to measure the gradual retreat of the ice sheet margins. Radiocarbon dating of material found in, under, or atop hundreds of moraines deposited by the ice shows that the retreat of the large ice sheet in North America began near 15,000 ^{14}C years ago, reached a midpoint near 10,000 ^{14}C years ago, and ended by 6000 ^{14}C years ago (Figure 13–2). The smaller Scandinavian ice sheet began to retreat at the same time as the one in North America, but it disappeared a few thousand years earlier. The timing of these retreats agrees with the Milankovitch theory.

Knowing the *area* covered by the retreating ice is a good start, but a complete analysis requires that these measurements be converted to ice *volume*. To make this conversion, we need to know the thickness of the ice as it retreated (thickness × area = volume). To complicate this analysis, the thickness of an ice sheet can be affected by the conditions in its basal layer. Portions of ice sheets that repeatedly slide on their bases are thin and relatively low in volume for a given area; portions that are frozen to their beds are thicker and larger in volume for the same area. Because of this uncertainty about thickness, records of changing ice area through time do not guarantee valid records of ice volume.

13-2 Coral Reefs and Rising Sea Level

The best record of ice sheet melting comes from tropical coral reefs far from the polar ice sheets (Chapter 9).

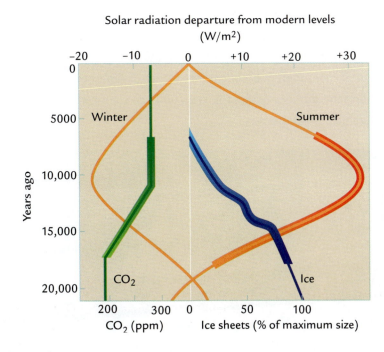

Solar radiation departure from modern levels (W/m²)

FIGURE 13-1 Causes of climate changes during deglaciation During the deglacial interval between 17,000 and 6000 years ago, climate changes were driven by rising summer insolation and by increased concentrations of CO_2 in the atmosphere; as the ice sheets shrank, their ability to influence climate diminished. (Adapted from J. E. Kutzbach et al., "Climate and Biome Simulations for the Past 21,000 Years," *Quaternary Science Reviews* 17 [1998]: 473–506.)

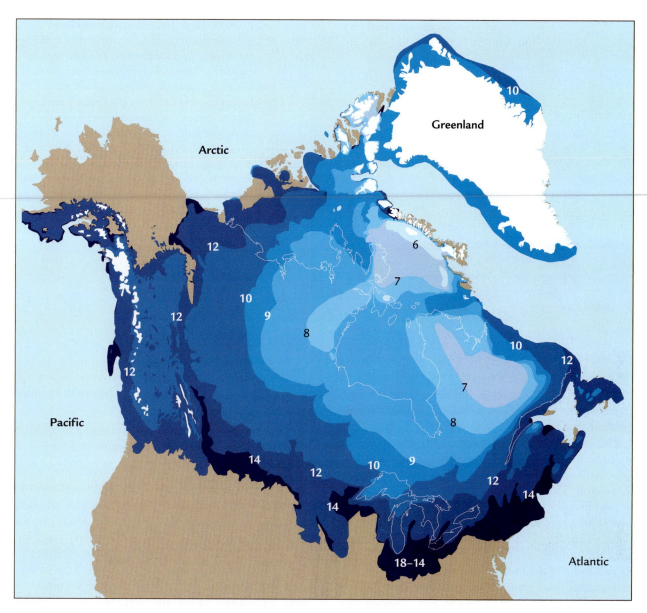

FIGURE 13-2 Retreat of the North American ice sheets Radiocarbon dating of organic remains shows that the margins of ice sheets in North America began to retreat near 14,000 ^{14}C years ago, and the ice disappeared completely shortly after 6000 years ago. The numbers indicate ^{14}C-dated ice limits in thousands of years. (Courtesy of Arthur Dyke, Geological Survey of Canada, Ottawa.)

Because several coral species grow just below sea level, the current elevation of older reefs built by corals can be used as a measure of past sea level, assuming that any tectonic movements of bedrock since the corals died can be removed.

Changes in sea level are directly related to changes in global ice volume because continental ice sheets are made of water taken from the sea. Coral reef measurements of lower sea level during the last glacial maximum and the subsequent deglaciation can be converted to a record of global ice volume, with each 1-m rise of sea level equivalent to 0.4 million km³ of ice.

In the late 1980s the marine geochemist Richard Fairbanks drilled and ^{14}C-dated a series of now-submerged coral reefs off Barbados, an island in the Caribbean. These reefs yielded a history of sea level rise from its low extreme at the last glacial maximum to its position during the modern interglaciation (Figure 13–3). Barbados is a region of slow tectonic uplift, and the present depth of each dated coral reef had to be adjusted by a few meters to remove this effect. The ^{14}C-dated deglacial sea level curve at Barbados supports the Milankovitch theory in a general way: the middle of the deglaciation occurred near the insolation maximum

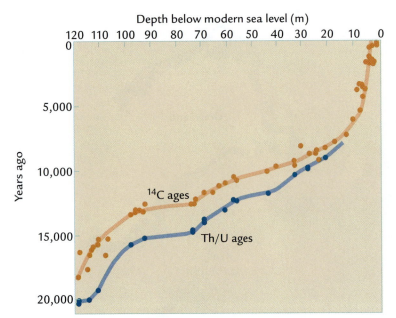

FIGURE 13-3 **Deglacial rise in sea level**
Submerged corals off Barbados in the Caribbean show the deglacial history of the rise in sea level caused by the return of meltwater from the ice sheets to the ocean. (Adapted from R. G. Fairbanks, "A 17,000-Year Glacio-eustatic Sea Level Record: Influence of Glacial Meltwater on the Younger Dryas Event and Deep-Ocean Circulation," *Nature* 349 [1989]: 637–42, and from E. Bard et al., "Calibration of the ^{14}C Time Scale over the Past 30,000 Years Using Mass-Spectrometric U-Th Ages from Barbados Corals," *Nature* 345 [1990]: 405–10.)

10,000 years ago, as expected. But the story is not that simple because the ^{14}C dates on the corals do not represent their true ages.

The evidence reviewed in Box 13–1 indicates that the dates of the rise in sea level from the thorium/uranium method are more accurate than those from the ^{14}C method. The Th/U chronology shifts the timing of the middle part of the deglaciation back in time by about 2000 years and the earlier parts of the deglaciation by as much as 3500 years (see Figure 13–3). In contrast, the timing of the summer insolation signal remains fixed by the independent (and highly accurate) astronomical time scale. As a result, the time of the major rise in sea level shifts back in time compared to the insolation curve.

Does this earlier timing for the deglaciation invalidate the Milankovitch theory? In a larger sense, it does not. Milankovitch chose summer as the critical season of insolation control of ice sheets, and the last deglaciation still occurred during a time when summer insolation was higher than it is now, although somewhat earlier in that interval than the Milankovitch theory predicts.

One factor that may contribute to this early response is simply the large amount of ice that was available to melt. When summer insolation rose to values high enough to melt ice prior to the insolation maximum 10,000 years ago, the ice sheets were still very large. In contrast, when declining summer insolation fell back to the same values just *after* the maximum, much less ice was left to melt, no matter how warm the climate had become. This bias would naturally tend to produce fastest rates of ice melting before the summer insolation maximum.

13-3 Glitches in the Deglaciation: Deglacial Two-Step

Another unexpected feature of the coral reef record is the observation that sea level did not rise smoothly through the deglaciation (see Figure 13–3). It rose quickly before 14,000 (Th/U or calendar) years ago, more slowly between 14,000 and 12,000 years ago, and then quickly again after 12,000 years ago. This pause in melting rates gives the deglaciation pattern a distinctive form sometimes called the *deglacial two-step* with a tempo of fast-slow-fast.

The pause in the rate of ice melting may have been larger than it seems from the sea level curve in Figure 13–3. Measured sea level values along this curve can be used to calculate the differences in sea level between successive intervals of time, and these differences provide a measure of the net rate of the flow of meltwater from the ice sheets back to the ocean during deglaciation. The meltwater influx signal calculated in this way (Figure 13–4) shows a slowing of melting rates during the middle of the deglaciation. Rates of ice melting were at least four to five times faster during the earlier and later intervals than during the pause in the middle.

The two-step deglaciation pattern tells us that the glacial ice sheets were not just giant ice cubes steadily melting in warmer air masses under a strengthening summer sun. Instead, the ice sheets exhibited more

BOX 13-1 TOOLS OF CLIMATE SCIENCE

Deglacial ^{14}C Dates Are Too Young

When the same Barbados corals that had been dated by the ^{14}C method were dated by the thorium/uranium technique, the Th/U ages turned out to be older than the ^{14}C ages by an amount that increased back in time. For samples with a ^{14}C age of 8000 years, the Th/U age was older by 1000 years, while for samples with a ^{14}C age of 18,000 years, the age difference increased to 3500 years.

With both sets of ages giving consistent-looking trends, scientists faced the problem of deciding which (if either) was correct. Fortunately, independent evidence was available from earlier work on the annual rings in long-lived trees. Individual rings in these trees had been dated both by the ^{14}C method and by counting backward year by year from the modern rings.

The ages derived from counting rings turned out to be older than those from the ^{14}C analyses. Near 8000 to 9000 years ago, the ages from the tree-ring counts were older than those from the ^{14}C analyses by 1000 years, the same as the offset of the Th/U coral ages from the ^{14}C ages. This agreement of two independent methods suggests that the ^{14}C ages are in error (younger than the actual ages).

The main reason the ^{14}C ages are too young is that the rate of production of ^{14}C atoms in Earth's atmosphere has varied in the past. The ^{14}C dating method is based on the assumption that ^{14}C has been produced in the atmosphere at a constant rate through time. Atoms of ^{14}C are produced when cosmic particles entering our atmosphere from other galaxies transform ^{14}N atoms into radioactive ^{14}C atoms, which then slowly decay away.

Because the rates of cosmic bombardment were higher during glacial times than they are now, more ^{14}C atoms were produced. Although many of those ^{14}C atoms have undergone radioactive decay, some of them still remain in the carbon-bearing material used for ^{14}C dating. This surviving excess of ^{14}C atoms makes it look as if less ^{14}C has decayed (and so less time has elapsed) than is actually the case.

Why would cosmic bombardment have been higher in the past? Earth is partly shielded from cosmic rays by its magnetic field. If this shield has at times been weaker than it is today, it would have provided less protection against cosmic bombardment. Two observations confirm this explanation. First, other elements known to be the product of cosmic bombardment were also more abundant

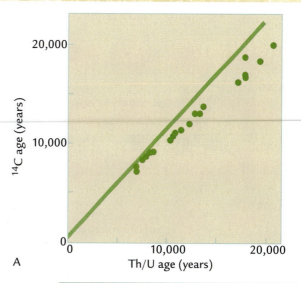

A

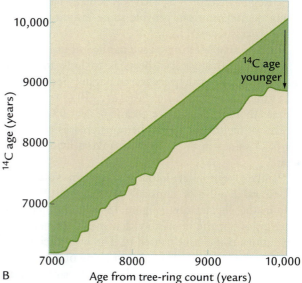

B

Offset of ^{14}C ages (A) Th/U dates from Barbados corals are older than ^{14}C ages by as much as 3500 years. (B) Ages derived by counting tree rings backward in time are offset from ^{14}C dates on the same layers by a similar amount. (A: Adapted from E. Bard et al., "Calibration of the ^{14}C Time Scale over the Last 30,000 Years Using Mass Spectrometric U-Th Ages from Barbados Corals," *Nature* 345 [1990]; 405–10. B: Adapted from M. Stuiver et al., "Climatic, Solar, Oceanic, and Geomagnetic Influences on Late-Glacial and Holocene Atmospheric ^{14}C/^{12}C Change," *Quaternary Research* 35 [1991]: 1–24.)

before 7000 years ago, confirming weaker magnetic shielding. Second, direct measurements of the past magnetic field in Earth's rocks and sediments show that it was weaker at that time.

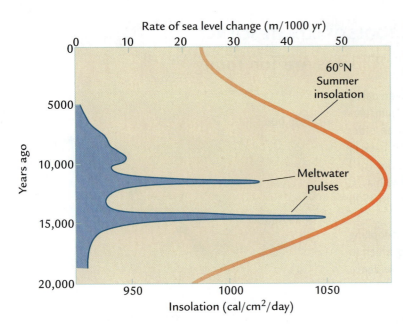

FIGURE 13-4 Influx of deglacial meltwater to the oceans The rate of the deglacial rise in sea level determined from submerged coral reefs can be used to calculate the rate at which water flowed into the oceans from melting ice sheets. The flow of meltwater slowed significantly during a pause in the deglaciation between 14,000 and 12,000 years ago. (Adapted from R. G. Fairbanks, "A 17,000-Year Glacio-eustatic Sea Level Record: Influence of Glacial Meltwater on the Younger Dryas Event and Deep-Ocean Circulation," *Nature* 349 [1989]: 637–42, and from E. Bard et al., "Calibration of the ^{14}C Time Scale over the Past 30,000 Years Using Mass-Spectrometric U-Th Ages from Barbados Corals," *Nature* 345 [1990]: 405–10.)

complex accelerations and decelerations in melting rates. Regional records of what was actually happening during deglaciation can provide some insight into the actual processes at work.

Rapid Early Melting One method of monitoring melting of individual ice sheets is to look for local pulses of meltwater delivery to the oceans. Because the δ^{18}O values of northern ice sheets are –30‰ to –35‰, whereas those in the surface ocean are near 0‰, major influxes of meltwater should be registered as pulses of low δ^{18}O values in the shells of plankton living in the ocean.

Planktic foraminifera in the northeastern Norwegian Sea record a pulse of unusually negative δ^{18}O values early in the deglaciation (Figure 13–5) and other evidence rules out the possibility that major temperature fluctuations could have caused it. The δ^{18}O oscillation is the result of an episode of early melting of the nearby Barents ice sheet, north of Scandinavia. Apparently this marine ice sheet, which had a base lying below sea level, was vulnerable to early destruction when summer insolation began to rise. A similar low-δ^{18}O pulse found in cores from the Gulf of Mexico indicates a short-term increase in the amount of meltwater flowing down the Mississippi River from the North American ice sheet.

In addition, ocean sediment cores taken southwest of Ireland contain a distinctive layer of sediment deposited 17,000 to 14,500 years ago that is rich in ice-rafted sand grains but nearly barren of the planktic foraminifera and coccoliths normally found in that region. This layer is evidence of a large influx of icebergs to the North Atlantic Ocean early in the deglaciation. The influx arrived during the first pulse of rapid sea level rise (see Figure 13–4). The evidence could mean that the major continental ice sheets lost a

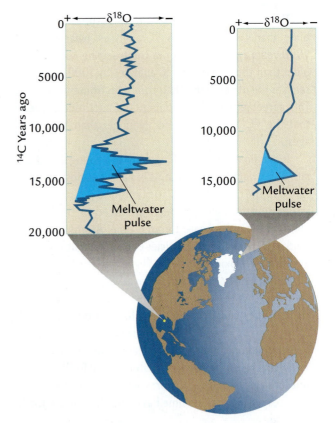

FIGURE 13-5 Local meltwater pulses $CaCO_3$ shells of ocean plankton from the Norwegian Sea and the Gulf of Mexico record pulses of low-δ^{18}O meltwater delivered from nearby ice sheets. (Top left: Adapted from A. Leventer et al., "Dynamics of the Laurentide Ice Sheet During the Last Deglaciation: Evidence from the Gulf of Mexico," *Earth Planetary Science Letters* 59 [1982]: 11–17. Top right: Adapted from G. Jones and L. D. Keigwin, "Evidence from Fram Strait (78 °N) for Early Deglaciation," *Nature* 336 [1988]: 56–59.)

substantial amount of their mass early in the deglaciation by calving icebergs to the ocean.

Mid-Deglacial Cooling: The Younger Dryas   The mid-deglacial pause in ice melting was accompanied by a brief climatic oscillation that is especially evident in records near the subpolar North Atlantic Ocean. Temperatures in this region had warmed part of the way toward interglacial levels, but this reversal brought back almost full glacial cold (Figure 13–6). The first evidence for this event came from pollen records in Europe.

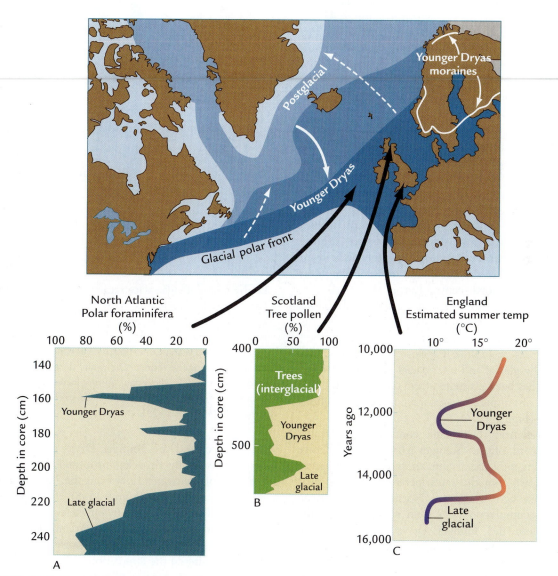

FIGURE 13-6 The Younger Dryas cold reversal Evidence of a cold episode that interrupted the general deglacial warming comes from (A) a southward readvance of polar water in the North Atlantic, (B) a reversal toward Arctic vegetation in Europe, and (C) a return to cooler temperatures indicated by fossil insect populations in Britain. (Top: Adapted from W. F. Ruddiman and A. McIntyre, "The North Atlantic Ocean During the Last Deglaciation," *Palaeogeography, Palaeoclimatology, Palaeoecology* 35 [1981]: 145–214. A: Adapted from W. F. Ruddiman, C. D. Sancetta, and A. McIntyre, "Glacial/Interglacial Response Rate of Subpolar North Atlantic Waters to Climatic Change: The Record in Oceanic Sediments," *Philosophical Transactions of the Royal Society of London* B 280 [1977]: 119–42. B: Adapted from G. R. Coope and G. Lemdahl, "Regional Differences in the Late Glacial Climate of Northern Europe," *Journal of Quaternary Science* 10 [1995]: 391–95. C: Adapted from T. C. Atkinson et al., "Seasonal Temperatures in Britain During the Past 20,000 Years, Reconstructed Using Beetle Remains," *Nature* 325 [1987]: 587–92.)

Because a distinctive Arctic plant called *Dryas* arrived in Europe during this episode, scientists call it the **Younger Dryas** event.

Later work on sediments in the North Atlantic Ocean also detected a clear Younger Dryas imprint: a rapid oscillation in the regional extent of icy polar water (Figure 13–6A). At the glacial maximum, polar water reached southward across the North Atlantic to 45°N. The southern margin of this cold water was defined by the **polar front,** a zone of rapid transition to the more temperate waters to the south. Early in the deglaciation, near 15,000 years ago, the polar front had shifted toward the northwest as if hinged at a point near eastern Canada. During this change, warm water began to flow northward along the European coast and to moderate climate enough to permit trees to advance northward from their full-glacial positions in far-southern Europe.

Near 13,000 years ago the polar front abruptly advanced back to the south, almost reaching its glacial position. At the same time, with the North Atlantic Ocean having again cooled, the Arctic vegetation (including *Dryas*) returned to northern Europe (Figure 13–6B). Later, near 11,700 years ago, the polar front abruptly retreated to the north, and forests began their final advance into north-central Europe.

The Younger Dryas readvance of the polar front represents a major reversal in circulation patterns. The estimated sea surface cooling in the Atlantic Ocean west of Ireland was at least 7°C, close to the difference between fully glacial and interglacial extremes. A similar cooling has been estimated from changes in the fossil remains of temperature-sensitive insect populations in England (Figure 13–6C).

Ice cores from Greenland contain a remarkably detailed record of the Younger Dryas event (Figure 13–7). During fully glacial climates, snow had been accumulating slowly, but the rates increased abruptly near 15,000 years ago when the North Atlantic Ocean warmed. Accumulation rates then slowed during the transition into the Younger Dryas event but again increased when it ended 11,700 years ago. Some of these transitions occurred in less than a century, with much of the change concentrated in a single decade. Similar changes occurred in ice-core concentrations of windblown dust, which peaked during the cold, dry, windy climate of the Younger Dryas and then abruptly decreased.

During the Younger Dryas event, the ice sheets in Scandinavia stopped retreating and in some regions readvanced a few hundred kilometers (see Figure 13–6). These pauses or small advances are thought to have been a response to the large regional cooling. The deglacial sea level curve derived from coral reefs (see Figure 13–2) shows that global ice volume continued to shrink during the Younger Dryas, although at a much slower rate than before or afterward.

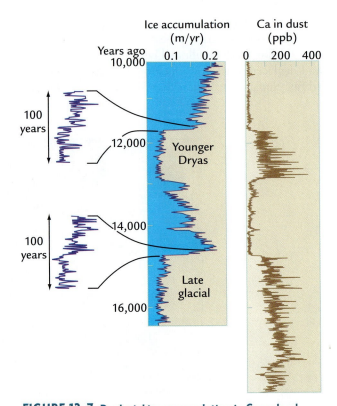

FIGURE 13-7 Deglacial ice accumulation in Greenland Rates of accumulation of ice in the Greenland ice sheet abruptly decreased during the Younger Dryas cold event and then increased when it ended, with the major changes occurring within 100 years. Concentrations of windblown dust increased during the Younger Dryas and decreased afterward. (Modified from R. B. Alley et al., "Abrupt Increase in Greenland Snow Accumulation at the End of the Younger Dryas Event," *Nature* 362 [1993]: 527–29).

What caused the Younger Dryas oscillation? The geochemist Wally Broecker proposed that changes in the path of meltwater flow from the North American ice sheet was the cause (Figure 13–8). He suggested that an abrupt diversion of the major meltwater route from the Gulf of Mexico to the North Atlantic Ocean during the Younger Dryas delivered a pulse of low-salinity water that altered the circulation of the North Atlantic by lowering the density of surface waters enough to prevent them from sinking and forming deep water. Because ocean surface waters give off heat when deep water forms in the North Atlantic Ocean, cutting off this process could have cooled climate in the North Atlantic and surrounding continents.

At first this explanation looked promising. Before the Younger Dryas event, the major pathway of meltwater had been down the Mississippi, into the Gulf of Mexico, and out into the Atlantic through the Gulf Stream. This early meltwater flow is recorded by the light-$\delta^{18}O$ signal in sediments from the Gulf of Mexico (see Figure 13–5). During the Younger Dryas event

FIGURE 13-8 Routes of meltwater flow During deglaciation, the direction of drainage of the North American ice sheet
changed, first southward to the Gulf of Mexico early in the
deglaciation, then east to the Atlantic Ocean (briefly) during
mid-deglaciation, and finally north into Hudson Bay and the
Arctic Ocean late in the deglaciation. (Adapted from J. Teller,
"Meltwater and Precipitation Runoff to the North Atlantic, Arctic,
and Gulf of Mexico from the Laurentide Ice Sheet and Adjacent
Regions During the Younger Dryas," *Paleoceanography* 5 [1990]:
897–905.)

(between 11,000 and 10,000 ^{14}C years ago), however,
the negative δ^{18}O pulse in the Gulf of Mexico weakened
because the meltwater was flowing eastward into the
Atlantic through the St. Lawrence region of eastern
Canada.

One criticism of this hypothesis is that the Younger
Dryas episode occurred at the same time that the rate of
global melting was slowing by a factor of 4 or 5 or more
(see Figure 13–4). With the overall flow of meltwater to
the oceans sharply diminished, it is hard to argue that a
diversion of flow would have greatly lowered the salinity of the North Atlantic. More recent investigations of
drainage patterns from lakes near the ice margin have
given contradictory evidence of fresh water outflow
directly to the Atlantic during the onset of the Younger
Dryas event.

Earlier suggestions that the Younger Dryas cooling
had a global expression have also proven to be incorrect. A small cooling evident in Antarctic ice cores that
was once interpreted as correlating with the Younger
Dryas event is not correlative. Instead, at least part
of the Antarctic was undergoing a slow warming
throughout the Younger Dryas interval. The origin
of the Younger Dryas cooling remains an enigma.

13-4 Positive Feedbacks to Deglacial Melting

Climate scientists basically agree that rising summer
insolation values caused by changes in Earth's orbital
tilt and precession set in motion the melting of the great
northern hemisphere ice sheets near 17,000 years ago.
But we still face an important question first explored in
Chapter 11: How did so small an insolation maximum
melt so much ice so quickly?

The answer to this question must be that positive
feedbacks accelerated the loss of ice. These feedbacks
must have been working most effectively when sea level
was rising (ice was melting) most rapidly (see Figure
13–4). The first of the two fast-melting phases (between
17,000 and 14,000 years ago) should have been a time of
especially active feedbacks, because the rapid rise in sea
level (loss of ice volume) at that time occurred well
before summer insolation had reached a peak.

Several of the bedrock interaction processes described
in Chapter 11 are in evidence during this interglaciation. The negative δ^{18}O pulse in Norwegian Sea cores
(see Figure 13–5) is evidence that a substantial part of
the marine ice sheet over the Barents Sea melted early
in the deglaciation. Because its base lay below sea level,
it may have been vulnerable to early destabilization.

In addition, two observations point to early thinning
of the Laurentide ice sheet on North America. The
major influx of icebergs to the North Atlantic Ocean
early in the deglacial sequence arrived at a time when
Laurentide ice had not melted back far from its glacial
position but when global ice volume was rapidly
decreasing. This observation suggests that ice streams
delivered large amounts of ice to the marine margins of
the ice sheet but with little loss of area.

In addition, the elevations of moraine deposits along
the southern ice lobes of the Laurentide ice sheet indicate thin profiles during the early-middle parts of the
deglaciation, consistent with the idea of ice that was sliding on a lubricated base in that region. Ice that flowed to
the southern margins of the ice sheets across land would
have melted relatively rapidly in the relative warmth of
the bedrock depression left behind by the once-thicker
ice. Again, the interior of the ice sheet could have been
thinned without any major retreat of the margins.

The mid-deglacial melting pause and the coincident
Younger Dryas cooling may represent an interval when
positive feedback processes slowed their impacts on the
climate system. Because summer insolation was still rising, melting continued but more slowly. The second
step of rapid ice melting and sea level rise after 11,500
years ago presumably reflects the return of the various
positive feedbacks to a more active role.

Another important feedback process was the rising
concentrations of greenhouse gases. As the ice sheets
melted, the CO_2 and methane levels rose in near

lockstep (see Figure 11–17). This very close timing suggests that the greenhouse-gas concentrations were largely controlled by the ice sheets and that the gases provided positive feedback to the process of ice melting.

Some climate scientists contest this north-centered view of the deglaciation. They note that evidence for even earlier deglacial warming responses can be found in the tropics and near Antarctica. Such evidence can be interpreted to indicate that the tropics or south polar region act as sensitive early triggers of melting in the north.

13-5 Deglacial Lakes, Floods, and Sea Level Rise

As the ice sheets melted back, the land in front of them remained depressed for some time rather than immediately rebounding to its former (ice-free) level (Chapter 9). Into these depressions poured meltwater from the retreating ice sheets, forming **proglacial lakes.** Because of the large volumes of meltwater arriving each summer, the lakes frequently cut new channels and overflowed into other lakes and then into rivers that carried water to the ocean.

The proglacial lakes existed in a highly dynamic landscape (Figure 13–9). As deglaciation proceeded, the ice margins fluctuated, with lobes of ice retreating and sliding forward but gradually shrinking farther and farther back over time. Each time the ice lobes retreated,

A Total area covered by deglacial lakes

B Lakes during deglaciation

FIGURE 13-10 Glacial Lake Agassiz During its several thousand years of existence, glacial Lake Agassiz flooded a total of more than 500,000 km² in western Canada, but much smaller areas were flooded at any one time. (A: Adapted from J. Teller, "Lake Agassiz and Its Contribution to Flow Through the Ottawa-St. Lawrence System," Geological Association of Canada Special Paper 35 [1987]: 281–89. B: adapted from J. Teller, "Glacial Lake Agassiz and Its Influence on the Great Lakes," Geological Association of Canada Special Paper 30 [1985]: 1–16.)

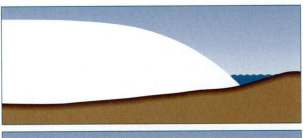

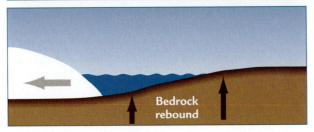

FIGURE 13-9 Proglacial lakes moving north Proglacial lakes develop in bedrock depressions left by melting ice sheets. Over time the lakes move north behind the ice sheets, while the land farther south rebounds toward its undepressed elevation.

they left behind new bedrock holes that formed the deepest parts of proglacial lakes. All the while, the slow rebound of bedrock caused the parts of the lakes farther south of the ice margins to become shallower and eventually disappear. Through time, the locations of the proglacial lakes moved north across the landscape, following the wave of depressed bedrock left south of the retreating ice.

The largest of the proglacial lakes in North America was Lake Agassiz in western Canada. Over its entire existence, it flooded an area greater than 500,000 km² (Figure 13–10A), but it flooded smaller areas at specific intervals during the ice retreat sequence (Figure 13–10B). At its

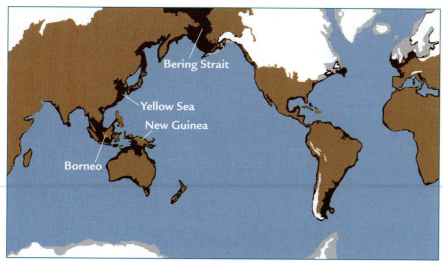

FIGURE 13-11 Deglacial flooding of coastlines The regions shown in dark brown were exposed at the last glacial maximum but flooded by the rise in sea level when the ice melted. (Adapted from CLIMAP Project Members, *Seasonal Reconstruction of the Earth's Surface at the Last Glacial Maximum,* Map and Chart Series MC–36 [Boulder, CO: Geological Society of America, 1981].)

◼ Land exposed by drop in sea level
◼ Land created under ice by drop in sea level

maximum size, it covered more than 200,000 km² to a depth of 100 m or more, forming a reservoir of 20,000 km³. Even this amount represented only a tiny fraction of the tens of millions of cubic kilometers of water stored in the glacial maximum ice sheets. Still, the water impounded in proglacial lakes and then released transformed parts of the landscape (Box 13–2).

The deglacial rise in sea level altered Earth's surface on a very large scale. Many regions of the world's continental shelves had been exposed during the low sea level at the glacial maximum, and many continents or ocean islands had been linked by land connections (Figure 13–11). One particularly large area was the expanse of dry land that joined Australia with New Guinea to the north. Land connections also linked

today's southeast Asian mainland with islands as far south as Borneo and joined northeastern Asia (Siberia) and westernmost Alaska across the present Bering Strait. England and Scotland were linked to the European mainland during the glacial maximum just south of the ice sheet. The return of some 44 million km³ of meltwater to the oceans during deglaciation submerged all these land corridors.

The lower level of the glacial ocean had also transformed smaller seas around the margins of the oceans, especially in the western Pacific. Today's Yellow Sea was dry land, and other seas in the western Pacific were more isolated from the open ocean because sea level was lower. Rising sea level flooded these seas and rejoined them to the open ocean.

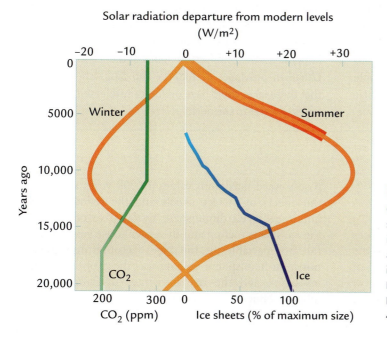

FIGURE 13-12 Causes of climate changes since deglaciation During the last 6000 years, with the ice sheets melted and CO_2 levels stabilized at or near interglacial levels, the main orbital-scale factor affecting climate was the gradual change in solar insolation toward today's values. (Adapted from J. E. Kutzbach et al., "Climate and Biome Simulations for the Past 21,000 Years," *Quaternary Science Reviews* 17 [1998]: 473–506.)

BOX 13-2 CLIMATE INTERACTIONS AND FEEDBACKS
Giant Deglacial Floods

In an unusual landscape called the **channeled scablands** in Idaho and east-central Washington, the bedrock consists of thick sequences of basalt deposited by lava flows during a time of heightened volcanic activity some 15 Myr ago. As the North American ice sheets were melting during the most recent deglaciation, the surface of these ancient lava flows was eroded into shapes suggesting the violent action of water on an immense scale. In the scabland region, deep canyons with nearly vertical walls were gouged into bedrock. At some locations, these now-dry channels abruptly plunge over steep cliffs into larger channels with depressions like those found at the base of modern waterfalls (but much larger). Huge boulders and displaced gravel and sand lie in the channels, but upland areas nearby have a thin cover of windblown loess typical of much of the rest of the Pacific Northwest outside the scabland region.

The geologist Harlen J. Bretz, working in the 1920s and 1930s, came to the conclusion that these erosional features must have resulted from a flood of immense proportions, one that within a few days carried a volume of water equivalent to all of Earth's rivers today. He suggested that the water in this flood ran wildly across the landscape, gouging and eroding the lower terrain but leaving the higher areas untouched before eventually flowing down the Columbia River into the Pacific Ocean. Bretz inferred that the source of all this water was a rapid melting event on the southern margin of the Cordilleran ice sheet that covered western Canada and extended southward into the northwest United States. He speculated that a volcano erupting beneath the ice margin had caused rapid melting and the sudden release of an enormous torrent of water.

For decades Bretz's ideas were rejected by geologists. At that time, most geologists took too literally the principle of uniformitarianism, the concept that slow geologic processes working today have, over immense spans of time, shaped and molded all of the features we see on Earth. Overenthusiastic application of this otherwise useful concept left little room for infrequent catastrophic phenomena; these events were rejected because they had not occurred during historical time. The problem with this view is that the human life span, and indeed all of recorded human history, is extremely short in relation to the age of the Earth, and our perspective on "normal" processes is narrow. With much longer life spans, we would naturally take a much broader view of what is normal.

In the 1950s, aerial photography revealed that the scablands were covered by giant ripple marks and gravel bars more than 20 ft high and spaced at intervals of 400 ft. Working on foot, Bretz had not recognized these enormous features because they were masked by scrubby vegetation. This new evidence convinced most geologists that the scablands had indeed been flooded, and further studies suggested that a discharge of some 25,000 to 30,000 m³ (750,000 ft³) per second flowing at 80 km/h (50 mi/h) was required to carve such a landscape. All the features Bretz had noted were indeed the result of rushing water on an unimaginably large scale.

A likely source of the water was glacial Lake Missoula, a proglacial lake ponded against the side of the ice sheet in Idaho. Although Bretz proposed only a single flood, the multiple layers of sediment left by the waters implied dozens of floods. One possibility is that each time a lobe of Cordilleran ice advanced far enough south to act as a dam, Lake Missoula filled up and released water in catastrophic bursts when the blocking ice lobe melted back. Geologists have also found evidence that large lakes hundreds of kilometers wide and tens of meters deep existed underneath the thin western lobes of the ice sheet in the Canadian provinces of Alberta and Saskatchewan, and that these lakes could have periodically released large volumes of water to the north.

Other Climate Changes During and After Deglaciation

Scientists have investigated two important changes during the late-deglacial and postglacial interval—the strength of north tropical monsoons and the warmth of summers in north polar latitudes. Monsoons grew stronger and summers warmer as summer insolation rose toward maximum values 10,000 years ago and as the effects of the ice sheets and the reduced greenhouse-gas levels diminished. By 6000 years ago, with the ice sheets melted and the greenhouse gases close to full interglacial levels, the major factor left to influence northern hemisphere climate was the drop in summer insolation and the rise in winter insolation toward modern values.

BOX 13-2 CLIMATE INTERACTIONS AND FEEDBACKS

CONTINUED

A

B

C

Channeled scablands Massive erosion by giant deglacial floods affected a large region in eastern Washington State and Idaho. (A) One source of meltwater for the floods may have been glacial Lake Missoula, ponded against a retreating lobe of the ice sheet in western Canada. (B, C) Evidence of erosion in the channeled scablands by floods of meltwater includes huge boulders, gravel, and sandbars on an immense scale and channels that lead to the edges of cliffs. (A: Adapted from G. A. Smith, "Missoula Flood Dynamics and Magnitudes Inferred from Sedimentology of Slackwater Deposits on the Columbia Plateau, Washington," Geological Society of America Bulletin 105 [1993]: 77–100. B and C: Victor Baker, University of Arizona.)

13-6 Stronger, Then Weaker Monsoons

Monsoons were strong near 10,000 years ago because of Earth's orbital configuration (Chapter 8). Summer insolation values over tropical and subtropical landmasses of the northern hemisphere at that time were 8% higher than those today (Figure 13–12). According to Kutzbach's orbital monsoon theory, increased insolation should

have driven a stronger summer monsoon circulation and produced higher tropical lake levels near 10,000 years ago than those today.

COHMAP (see Chapter 12) compared climate model simulations with geological data from this interval. Their simulations used summer insolation values from 9000 years ago, when northern hemisphere ice sheets were greatly reduced in size. The model simulated stronger monsoons across the entire north-tropical region of North Africa, southern Arabia, and southern Asia (Figure 13–13A).

A Model simulation

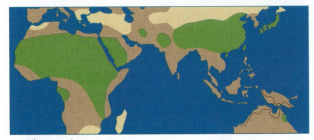

B Observations
Effective moisture (9000 years ago versus today)

■ Greater   ■ Less   ■ Same

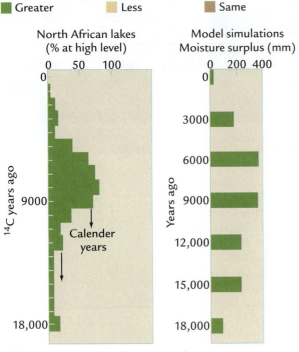

C Data-model comparison versus time

FIGURE 13-13 Tropical monsoon maximum Climate model simulations of stronger summer monsoons in the north tropics near 9,000 years ago agree with evidence in the climate record, such as higher lake levels. (A and B: Adapted from COHMAP Members, "Climatic Changes of the Last 18,000 Years: Observations and Model Simulation," *Science* 241 [1988]: 1043–52. C: Adapted from J. E. Kutzbach and F. A. Street-Perrott, "Milankovitch Forcing of Fluctuations in the Level of Tropical Lakes," *Nature* 317 [1985]: 130–34.)

Geologic evidence supports this simulation (Figure 13–13B, C). Between 10,000 and 7500 ^{14}C years ago, lake levels determined by ^{14}C dating of lake muds were substantially higher than they are today across most of North Africa between 15° and 30°N, in the southern half of Arabia, and over southeastern Asia. Lake Chad, in northern Africa, expanded to 300,000 km², an area comparable to the modern Caspian Sea.

Evidence in sediment cores from the Arabian Sea indicates that stronger monsoon winds blew along the coast of Somalia in eastern Africa, and along the southeast coast of Arabia 9000 years ago (Figure 13–14). These intensified winds drove water offshore and caused upwelling, which was recorded by key species of planktic foraminifera.

Scientists have also discovered evidence that ancient rivers flowed across regions that today are hyperarid desert. In central Arabia, a river flowed more than 500 km northeastward through the modern Arabian desert. Many parts of North Africa and Arabia that are now extremely dry were once grassy river valleys dotted with freshwater lakes and occupied by hippopotamuses, crocodiles, turtles, rhinoceroses, giraffes, and buffaloes (see Figure 8–7).

Although climate models and evidence from the geologic record confirm Kutzbach's theory of stronger north-tropical monsoons 10,000 years ago, a closer look reveals a mismatch in amplitude. The increase in rainfall in the models was small compared to geologic evidence from lake levels and pollen assemblages. Subsequent modeling efforts have explored positive feedback processes that could have amplified the small response simulated by early models.

One important feedback is the increase in recycling of water vapor provided by evapotranspiration of water vapor by vegetation (companion Web site, p. 34). The initial increase in monsoon rains in the early models causes trees to advance northward into grasslands, and grasses to move northward into deserts. In the later models, the advancing vegetation draws more moisture out of the soil and transfers it to the atmosphere through evapotranspiration. With more water vapor in the atmosphere, more rain falls, especially farther to the north (Figure 13–15). This positive moisture feedback from vegetation results in model simulations with wetter landscapes and higher lakes that are in closer (but still not full) agreement with the geologic evidence. One feedback not yet incorporated in models is increased recycling of moisture from small lakes and low swampy regions near rivers that are too small to be represented in climate-model grid boxes.

After reaching a peak near 10,000 years ago, summer insolation values at lower latitudes of the northern hemisphere have fallen continuously (Figure 13–16A). This decrease has occurred because Earth's precessional motion has carried it from a June 21 position close to the Sun 10,000 years ago to a June 21 position far from the Sun today (Chapter 7).

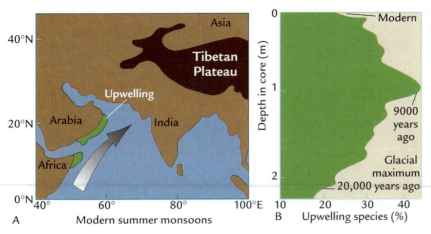

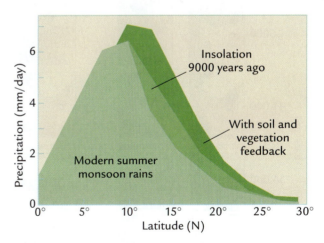

FIGURE 13-14 Upwelling in the Arabian Sea Climate model simulations of stronger summer monsoons over India 9000 years ago are supported by evidence from ocean cores indicating greater upwelling along the coasts of East Africa and Arabia in response to strong winds pushing surface waters offshore. (A: Adapted from W. L. Prell, "Monsoonal Climate of the Arabian Sea During the Late Quaternary: A Response to Changing Solar Radiation," in *Milankovitch and Climate,* ed. A. Berger et al. [Dordrecht: Reidel, 1984]. B: W. L. Prell, "Variation of Monsoonal Upwelling: A Response to Changing Solar Radiation," in *Climate Processes and Climate Sensitivity,* ed. J. E. Hansen and T. Takahashi [Washington, DC: American Geophysical Union, 1984].)

By 6000 years ago, summer insolation values in the northern tropics were still about 5% higher than the modern levels but were falling toward modern values. This slow decrease should have produced a corresponding decline in the strength of the tropical monsoons. Direct observations and ^{14}C dates of lakes across North Africa confirm a major drop in water levels during the last 9000 years (Figure 13–16B). Today, lakes are lower than they were between 9000 and 6000 years ago, and many have completely dried out.

Examined individually, most lake level histories in North Africa and India show large and abrupt changes during the overall transition to lower levels (Figure 13–16C). As was the case for the irregular rates of melting of northern ice sheets, these short-term changes in lake levels represent a type of climate response that cannot be directly attributed to the smooth, gradual forcing provided by changes in summer insolation.

13-7 Warmer, Then Cooler North Polar Summers

At the glacial maximum, the main controls on climate at high northern latitudes had been the regional cooling effects of the ice sheets and the global cooling caused by low CO_2 (and methane) values. As deglaciation proceeded, rising summer insolation values increasingly warmed land areas located far from the ice sheets and in time overcame the cooling effects of the shrinking ice sheets. Summer insolation values reached a peak 10,000 years ago, with the ice sheets much smaller but still present.

Northward Shifts in Vegetation The last deglaciation dramatically transformed the vegetation of the northern hemisphere continents. In North America, cold-tolerant spruce trees retreated from their glacial position in the central United States to their modern position in northeastern Canada (Figure 13–17A). Warm-tolerant trees such as oak moved a smaller dis-

tance from their glacial location in the far southeastern United States to their modern concentrations in mid-Atlantic states (Figure 13–17B).

The climate changes that occurred midway through the deglaciation produced unusual mixtures of plants called **no-analog vegetation** because no similar combination exists today. For example, spruce trees grew with hardwood deciduous trees (such as ash) in the northern Midwest of the United States early in the deglaciation, even though ash and other deciduous trees are rare today in regions where spruce trees grow.

FIGURE 13-15 Vegetation-moisture feedback Climate model simulations indicate that higher summer insolation 9000 years ago caused stronger summer monsoons and greater northward penetration of moisture into Africa. Additional model experiments show that positive moisture feedback from wetter soils and increased vegetation caused even greater penetration of moisture into the continental interior. (Modified from J. E. Kutzbach et al., "Vegetation and Soil Feedbacks on the Response of the African Monsoon to Orbital Forcing in the Early to Middle Holocene," *Nature* 384 [1986]: 623–26.)

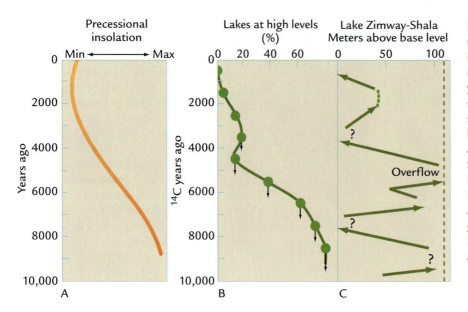

FIGURE 13-16 Weakening monsoons (A) Low-latitude summer insolation has slowly decreased since reaching a maximum value 10,000 years ago. (B, C) The decrease in summer insolation has weakened the summer monsoons and caused lake levels in North Africa to fall. (B: Adapted from J. E. Kutzbach and F. A. Street-Perrott, "Milankovitch Forcing of Fluctuations in the Level of Tropical Lakes," *Nature* 317 [1985]: 130–34. C: Adapted from R. Gillespie et al., "Postglacial Arid Episodes in Ethiopia Have Implications for Climate Prediction," *Nature* 306 [1983]: 680–83.)

No-analog mixtures developed because each vegetation type responded to a different combination of environmental variables from those that controlled the other types. These individualistic responses make it impossible to analyze past vegetation changes by lumping pollen together into larger communities or assemblages; each vegetation type has to be analyzed on its own.

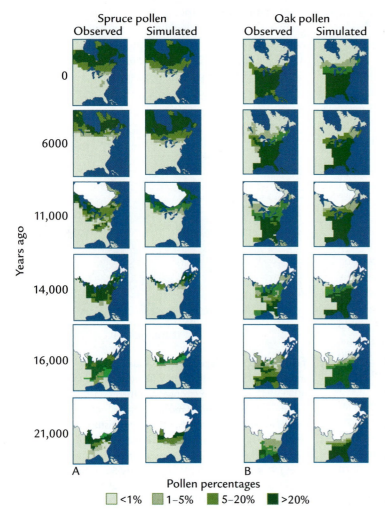

FIGURE 13-17 Data-model vegetation comparisons Pollen in lake sediments indicates large-scale changes in the distribution of spruce and oak pollen during the last deglaciation. Model simulations of climate and vegetation reproduce many but not all aspects of these observed patterns. (Adapted from T. Webb III et al., "Late Quaternary Climate Change in Eastern North America: A Comparison of Pollen-Derived Estimates with Climate Model Results," *Quaternary Science Reviews* 17 [1998]: 587–606.)

The distributions of spruce and oak pollen simulated by climate models are also shown in Figure 13–17 for comparison with the observed distributions. Both sets of maps show the same large-scale northward relocation of spruce, and they agree on the existence of a mid-deglacial interval when spruce became rare throughout eastern North America. Both sets of maps also show a similar northward expansion of oak, but the model simulates more oak in the southeast during deglaciation than the pollen data show, a mismatch similar to those noted for the glacial maximum (Chapter 12).

Peak Warmth Once CO_2 values had risen to full interglacial levels and only remnants of the ice sheets were left, summer insolation values became the main control on climate responses, particularly for vegetation. The high summer insolation values that had triggered ice melting remained greater than their current levels but had begun the decrease toward modern levels (see Figure 13–12). As a result, the warmest temperatures of the last several thousand years were registered immediately after the regional chilling effect of the ice sheets was removed, but before the renewed cooling effect caused by falling insolation levels. Some climate scientists refer to this warmer-than-modern interval as the *hypsithermal*, but the time of greatest warmth actually varies widely from region to region, depending on when the nearby ice melted and its regional cooling effect was removed.

With summer insolation values 6000 years ago still 5% higher than those today, the northern limit of boreal (spruce and larch) forest in Siberia and west-central Canada had moved as much as 300 km north of its modern position, narrowing the fringe of tundra bordering the Arctic Ocean. This expansion of forest beyond its modern limits confirms that summer temperatures on the northern continents were warmer than they are today. Winter insolation values lower than those today may have produced cooler winter temperatures, but the northern limit of boreal forest is mainly sensitive to temperature during the summer growing season.

Climate models have been used to assess this high-latitude summer warming 6000 years ago. With summer insolation values 5% above today's values, the models simulated a summer warming in northern Canada of as much as 2° to 3°C but a lesser warming in central Asia (Figure 13–18A). Additional simulations were run to incorporate the positive feedback effect of vegetation (companion Web site, p. 34). In regions where boreal forest advanced northward beyond its modern limits, the low albedo (25%) typical of these dark-green trees replaced the high albedo (60%) typical of scrubby tundra vegetation under partial snow cover. The low-albedo trees absorbed more sunlight than the underlying snow and further warmed the climate. Because far northern regions are snow-covered for

much of the year, this additional warming extended through most of the year and affected broad areas of northern Canada and Asia (Figure 13–18B). The net

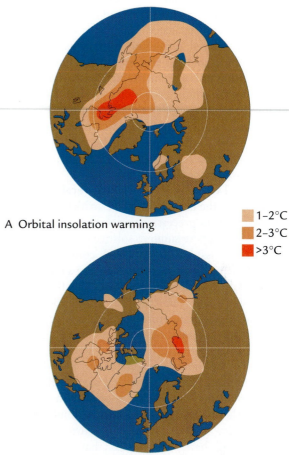

A Orbital insolation warming

1–2°C
2–3°C
>3°C

B Vegetation feedback warming

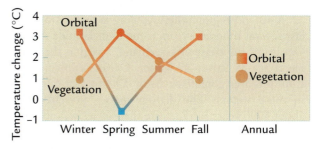

C Seasonal and annual averages

FIGURE 13-18 Peak deglacial warmth (A) Climate model simulations indicate that higher insolation 6000 years ago warmed high latitudes, especially in Canada in the summer. (B, C) Additional simulations show that positive feedback caused by northward expansion of low-albedo spruce forest into high-albedo tundra may have almost doubled this regional warming, with the largest changes in Asia during the spring. (Adapted from TEMPO (J. E. Kutzbach et al.), "Potential Role of Vegetation in the Climatic Sensitivity of High-Latitude Regions: A Case Study at 6000 Years b.p.," *Global Biogeochemical Cycles* 6 [1996]: 727–36.)

effect of this vegetation-albedo feedback almost doubled the initial insolation warming of high northern latitudes (Figure 13–18C).

Sea ice also contributed to these far-northern climate changes. High summer insolation caused the sea ice margin in the model to thin and retreat northward, and this change propagated into the rest of the yearly cycle, with delayed refreezing of seasonal sea ice in autumn, thinner and less extensive sea ice in winter, and earlier melting of sea ice in spring. As a result, despite the fact that the winter insolation values were considerably lower than those today, the reduced sea ice cover in winter and larger areas of open water near the coasts moderated the winter cooling of the continents. As a result, the simulated annual average change in this region showed a considerable warming.

Renewed Cooling in the Last Several Thousand Years During the last 6000 years, Earth's tilt has slowly decreased and its precessional motion has moved the northern hemisphere summer solstice toward the aphelion (distant-pass) position. These combined orbital changes have produced a 5% decrease in summer insolation and a 5% increase in winter insolation at high latitudes since 6000 years ago (see Figure 13–12). As a result, summer temperatures have fallen significantly during the last several thousand years in several regions at high northern latitudes (Figure 13–19).

Evidence of cooler summers comes from ice cores taken from small ice caps in several parts of the Arctic. Ice from the tiny Agassiz ice cap on Ellesmere Island, in far northern Canada, shows that summer melting episodes were far more frequent before 5000 years ago than they have been since that time (Figure 13–19A). This evidence supports a trend toward cooler summers.

A second region where cooling is evident over the last several thousand years is the high-latitude Atlantic Ocean off the coast of Greenland, a region that today has a sea-ice cover in winter. Ocean sediment cores from this area contain shells of diatoms that once lived in these waters. The diatom species present before about 5000 years ago indicate that sea ice was absent or scarce in this region (Figure 13–19B).

A third piece of evidence that indicates cooling in recent millennia is the increase in size of small glaciers on Arctic islands. Glacier margins on Arctic islands were located well back from their modern positions between 8000 and 3500 years ago (Figure 13–19C), and some of the ice caps melted entirely. These glaciers have reappeared or grown since 3500 years ago, oscillating toward progressively larger sizes, consistent with cooler summer temperatures and less melting.

A fourth indication of cooling is changes in the abundance of temperature-sensitive diatom species off the southwest coast of Norway. The estimated changes in sea surface temperature reconstructed from these assemblages show a gradual cooling since 6000 years ago (Figure 13–19D).

The boundary between tundra to the north and boreal forest to the south is still another climatic indicator in Asia and North America. This boundary in northern Canada was well north of its present limit 6000 years ago but has since advanced southward by up to 300 km (Figure 13–19E). This shift from forest to tundra vegetation suggests cooler summers and a shorter growing season.

Still another indication of summer cooling in the last several thousand years (not shown) comes from mountain glaciers. Like the much larger ice sheets, mountain glaciers at high latitudes will melt if summer insolation increases, but mountain glaciers respond to climate changes within just a few decades. Before 5000 years ago, mountain glaciers were small, but since that time their size has increased in most regions, consistent with the evidence of progressive cooling driven by a long-term decrease in summer insolation.

Current and Future Orbital-Scale Climatic Change

Astronomy tells us not only the changes in Earth's orbit that have already occurred but those that will occur in the future. This knowledge of the future gives us a good basis for predicting the course climate would follow if it responds only to orbital forcing (changes in precession and tilt). [Note that this analysis refers only to natural changes occurring at orbital scales. Other factors that will determine our near-term climatic future will be examined in Part V.]

Today June 21 occurs near the July 4 aphelion (distant-pass) position in Earth's eccentric orbit around the Sun. In another 10,000 years, Earth will have returned to the opposite configuration. June 21 will occur at perihelion, when Earth is closest to the Sun, just as it did 10,000 years ago (Figure 13–20 left). Because the 23,000-year cycle of orbital precession controls changes in the tropical monsoons, this shift in Earth's orbit will increase the amount of summer insolation across the northern tropics and drive a stronger monsoon over North Africa and southern Asia.

Predicting the effect on higher latitudes is more difficult. In the next 10,000 years, the tilt of Earth's axis will have fallen to the next minimum (Figure 13–20 center), and the resulting decrease in summer insolation will tend to cool climate. But the insolation decrease caused by the change in tilt will be opposed by the insolation increase caused by the change in precession. The combined signal (Figure 13–20 right) shows a larger effect of tilt than of precession.

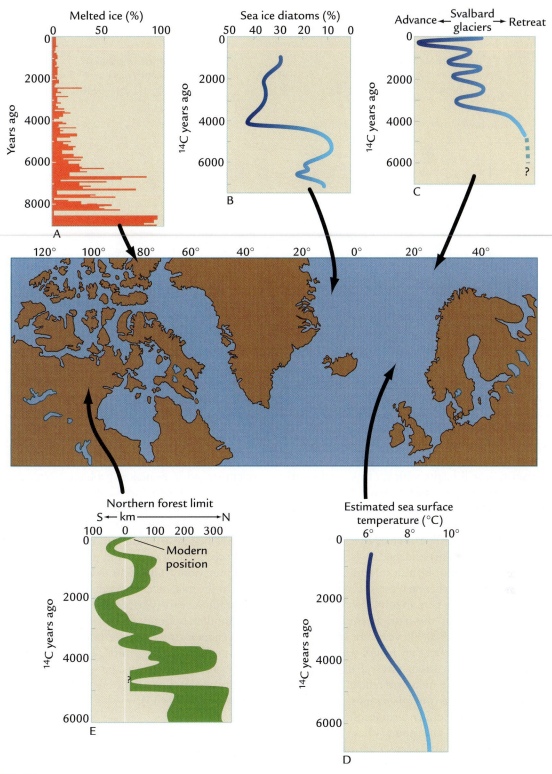

FIGURE 13-19 Cooling toward the present Evidence of a cooling of high northern latitudes during the last several thousand years includes (A) less frequent summer melting episodes in ice caps on Arctic islands; (B) more frequent sea ice off Greenland; (C) advances of ice caps on Arctic islands north of Europe; (D) lower temperatures in the Atlantic Ocean west of southern Norway; (E) a southward shift of the boundary between tundra and spruce forest in northern Canada. (A: Adapted from R. M. Koerner and D. A. Fisher, "A Record of Holocene Summer Climate from a Canadian High-Arctic Ice Core," *Nature* 343 [1990]: 630–31; B and D: adapted from N. Koc et al., "Paleoceanographic Reconstructions of Surface Ocean Conditions in the Greenland, Iceland, and Norwegian Seas Through the Last 14 Ka Based on Diatoms," *Quaternary Science Reviews* 12 [1992]: 115–40. C: Adapted from J. Lubinski, S. L. Forman, and G. H. Miller, "Holocene Glacier and Climate Fluctuations on Franz Joseph Land, Arctic Russia," *Quaternary Science Reviews* 18 [1999]: 87–109. E: Adapted from H. Nichols, "Palynological and Paleoclimatic Study of the Late Quaternary Displacement of the Boreal Forest-Tundra Ecotone in Keewatin and MacKenzie, N.W.T.," Institute of Arctic and Alpine Research Occasional Paper 15 [Boulder, CO, 1975].)

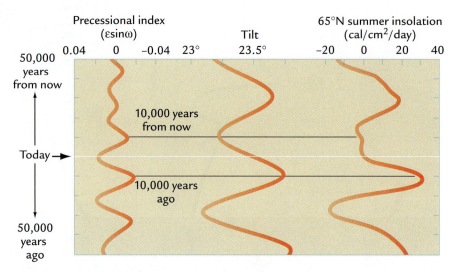

FIGURE 13-20 Future summer insolation trends During the next 10,000 years, precession-dominated insolation at low latitudes of the northern hemisphere will return to another maximum (left), while tilt-dominated insolation at very high northern latitudes will continue to fall toward the next minimum (center). The combined effects of tilt and precession will increase summer insolation, which will not fall to modern levels for another 50,000 years. (Adapted from A. Berger and M.-F. Loutre, "Modeling the Climatic Response to Astronomical and CO$_2$ Forcings," *C. R. Acad. Sci. Paris* 323 [1996]: 1–16.)

Climate scientists want to be able to predict when the next glaciation will begin, with ice caps starting to grow on North America, Eurasia, or both. Over the last million years, ice sheets have been present on northern hemisphere continents for about 90% of the time (see Figure 9–13), and times without any ice have only lasted for 10,000 years or less. This evidence suggests that the current interglaciation must be near its natural end.

One way to estimate when new ice sheets will grow at high northern latitudes is to examine what has happened at similar times in the past. Records of δ^{18}O changes covering the last 900,000 years are an approximate index of ice volume, but they have a sizeable temperature overprint. The current interval of low δ^{18}O values (minimal ice volume and warm temperatures) has lasted as long as previous interglaciations, and some 2-D models that simulate longer-term ice sheets predict that we are at or near the point when ice sheets should start growing again because of low summer insolation.

Other models predict that glaciation is not imminent and may not occur for the next 50,000 years, because summer insolation is now at a minimum and will not be significantly lower for the next 50,000 years (Figure 13–20 right). It seems that the ice-age cycles may have "skipped a beat." If new ice sheets have not begun to form during the present insolation minimum, why would they do so at the higher insolation levels of the next 50,000 years?

An ice-free interval 50,000 years in length presents scientists with a major problem. No such ice-free interval has occurred in the entire 2.75 million years of northern hemisphere glaciation, so why would one occur now? Has the natural operation of the climate system begun to shift in recent millennia toward unprecedented warmth and an ice-free state? Such an interpretation would seem to be in complete contradiction to the evidence for gradual cooling during the last few million years (Chapter 9).

Another option will be considered in Chapter 15. The climatic trend of the last several thousand years may not have been natural, because a new factor has arisen and prevented a new glaciation that should be underway by now. The new factor is agricultural humans (Chapter 15).

Key Terms

Younger Dryas (p. 236)
polar front (p. 236)
proglacial lakes
  (p. 238)

channeled scablands
  (p. 240)
no-analog vegetation
  (p. 243)

Review Questions

1. What is the best method of measuring the melting of ice sheets over the last 17,000 years?

2. To what degree does the timing of ice sheet melting support the Milankovitch theory that orbital insolation controls the size of ice sheets?

3. To what degree do changes in intensity of summer monsoons in the last 17,000 years support the Kutzbach theory that orbital insolation controls the intensity of monsoons?

4. What evidence suggests that variations in orbital insolation were not the only cause of climate changes during the last 17,000 years?

5. Describe how and why proglacial lakes travel slowly across the landscape behind melting ice sheets.

6. Why were summer temperatures at high northern latitudes warmer 6000 years ago than they are today? Would they also have been warmer at high southern latitudes?

7. What do orbital trends imply about future changes in monsoons and northern ice sheets?

Additional Resources

Basic Reading

COHMAP Members. 1988. "Climatic Changes of the Last 18,000 Years: Observations and Model Simulations." *Science* 241: 1043–62.

Kutzbach, J. E., and F. A. Street-Perrott. 1985. "Milankovitch Forcing of Fluctuations in the Level of Tropical Lakes from 18 to 0 Kyr B.P." *Nature* 317: 130–34.

Roberts, N. 1998. *The Holocene.* Oxford: Blackwell.

www.classzone.com/book/earth_science/terc/content/ visualization (Chapter15).

Advanced Reading

Bard, E., B. Hamelin, R. G. Fairbanks, and A. Zindler. 1990. "Calibration of the [14]C Time Scale over the Last 30,000 Years Using Mass Spectrometric U-Th Ages from Barbados Corals." *Nature* 345: 405–10.

Dyke, A. S., and V. K. Prest. 1987. "Late Wisconsinan and Holocene History of the Laurentide Ice Sheet." *Géographie Physique et Quaternaire* 41: 237–63.

Fairbanks, R. G. 1989. "A 17,000-Year Glacio-eustatic Sea Level Record: Influence of Glacial Melting on the Younger Dryas Event and Deep-Ocean Circulation." *Nature* 342: 637–42.

Teller, J. T. 1987. "Proglacial Lakes and the Southern Margin of the Laurentide Ice Sheet." In *North America and Adjacent Oceans during the Last Deglaciation,* ed. W. F. Ruddiman and H. E. Wright. Geology of North America, K–3. Boulder, Colo.: Geological Society of America.

Webb, T., III. 1998. "Late Quaternary Climates: Data Synthesis and Model Experiments." *Quaternary Science Reviews* 17: 587–606.

Webb, T., III, S. Howe, R. H. W. Bradshaw, and K. M. Heide. 1981. "Estimating Plant Abundances from Pollen Percentages: The Use of Regression Analysis." *Review of Paleobotany and Palynology* 34: 269–300.

Wright, H. E., Jr., J. E. Kutzbach, T. Webb III, W. F. Ruddiman, F. A. Street-Perrott, and P. J. Bartlein. 1993. *Global Climates since the Last Glacial Maximum.* Minneapolis: University of Minnesota Press.

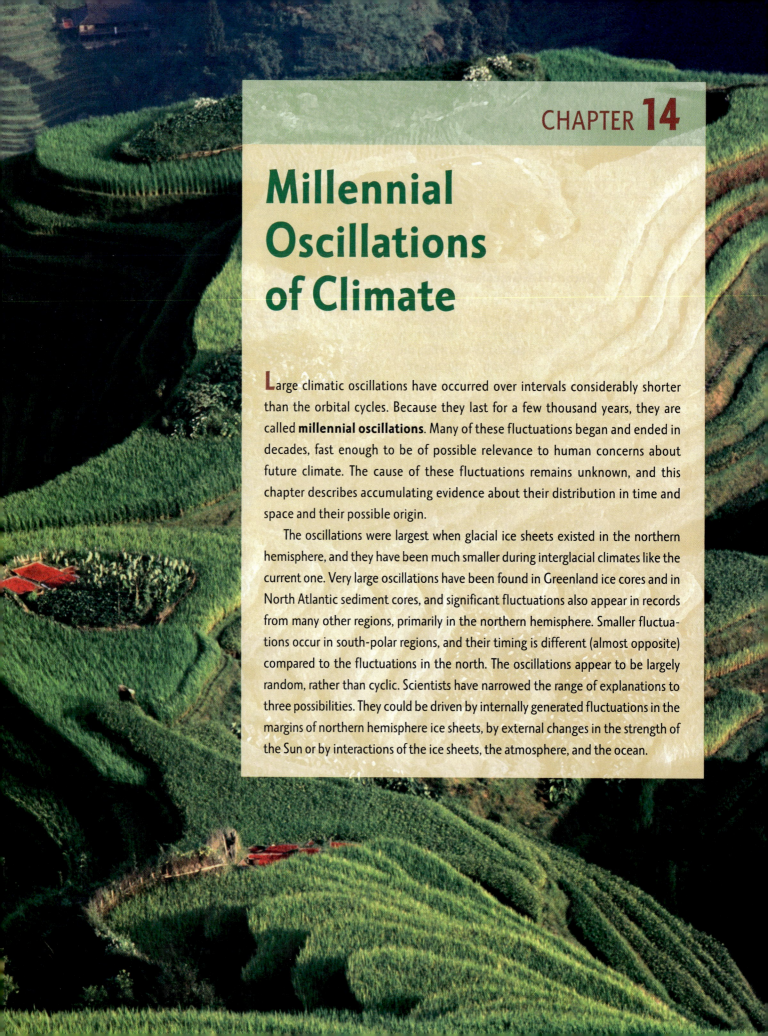

CHAPTER 14

Millennial Oscillations of Climate

Large climatic oscillations have occurred over intervals considerably shorter than the orbital cycles. Because they last for a few thousand years, they are called **millennial oscillations**. Many of these fluctuations began and ended in decades, fast enough to be of possible relevance to human concerns about future climate. The cause of these fluctuations remains unknown, and this chapter describes accumulating evidence about their distribution in time and space and their possible origin.

The oscillations were largest when glacial ice sheets existed in the northern hemisphere, and they have been much smaller during interglacial climates like the current one. Very large oscillations have been found in Greenland ice cores and in North Atlantic sediment cores, and significant fluctuations also appear in records from many other regions, primarily in the northern hemisphere. Smaller fluctuations occur in south-polar regions, and their timing is different (almost opposite) compared to the fluctuations in the north. The oscillations appear to be largely random, rather than cyclic. Scientists have narrowed the range of explanations to three possibilities. They could be driven by internally generated fluctuations in the margins of northern hemisphere ice sheets, by external changes in the strength of the Sun or by interactions of the ice sheets, the atmosphere, and the ocean.

Millennial Oscillations During Glaciations

The first critical clue that the climate system is capable of large changes over short intervals came from studies of the deglacial Younger Dryas event, which lasted less than 1500 years and began and ended very abruptly (Chapter 13). More recently, evidence has emerged that an ongoing series of similar short-term oscillations is superimposed on orbital-scale climatic cycles. These short-term fluctuations are largest and best defined during glacial intervals.

14-1 Oscillations Recorded in Greenland Ice Cores

Long ice cores taken on Greenland in the 1970s recovered records spanning much of the last interglacial-glacial cycle (Figure 14–1). The upper portions of these records were dated by counting annual layers, while the age of the lower section was estimated by using theoretical models of the flow of ice deeper in the ice sheets.

Two signals from this ice record were particularly important—the $\delta^{18}O$ composition of the ice and the concentration of the dust in the ice. Signals of $\delta^{18}O$ in ice cores record changes in the composition of the water vapor that falls as snow and consolidates into ice (Appendix 1). Of the several processes that can affect the $\delta^{18}O$ composition of ice (Table 14–1), local air temperature is the primary control. Chemical analysis of the dust has shown that the main source region was northern Asia. The transport path may have followed the northern branch of the split jet stream that moved across the Canadian margin of the North American ice sheet (see Figure 12–11B).

Both records show two distinctive features. One trend is the slow, underlying change from low dust concentrations and relatively positive (less negative) $\delta^{18}O$ values in the top of the section to higher dust concentrations and more negative $\delta^{18}O$ values in the middle part and then the return to positive $\delta^{18}O$ values and little dust in the bottom part. Imprecise dating at that time showed that the upper section is the current interglaciation, the middle part is the last glacial interval, and the bottom section is part of the previous interglaciation. The two interglacial intervals were warmer (with more positive $\delta^{18}O$ values) and relatively free of dust compared with the cold, dusty glacial interval in the middle.

These slower orbital-scale changes are difficult to see in this record because they are masked by a more prominent characteristic—the rapid oscillations over much shorter intervals between high and low dust concentrations and between negative and positive $\delta^{18}O$ values. The $\delta^{18}O$ fluctuations of 4‰ to 6‰ represent a large fraction of the difference between the full-glacial and full-interglacial values. Temperatures over Greenland during

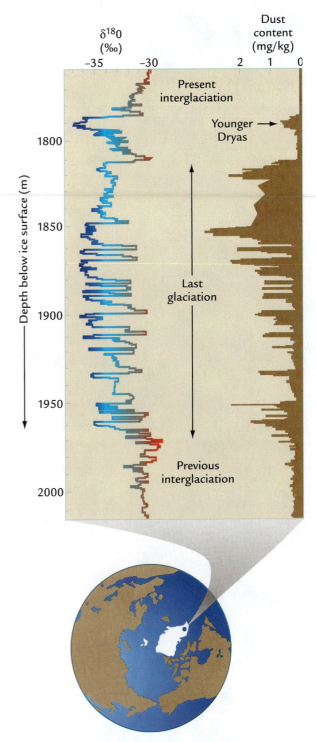

FIGURE 14-1 **Millennial oscillations in ice cores** An ice core drilled through the Greenland ice sheet in the 1970s contained records of $\delta^{18}O$ and dust concentrations. Large oscillations occur in the glacial portion of both signals, but not in the present interglaciation, or in the previous one. (Adapted from W. Dansgaard et al., "North Atlantic Climatic Oscillations Recorded by Deep Greenland Ice Cores," in *Climate Processes and Climate Sensitivity,* ed. J. E. Hansen and T. Takahashi [Washington, DC: American Geophysical Union, 1984].)

TABLE 14-1 Causes of δ¹⁸O Changes Recorded in Ice Cores

Negative ←	Change in δ¹⁸O values	→ Positive
Colder	Air temperature over ice	Warmer
Distant	Proximity of source region	Close
Low δ¹⁸O	δ¹⁸O composition of source	High δ¹⁸O
High	Elevation of ice	Low
Winter	Primary season of precipitation	Summer

glacial intervals oscillated rapidly between extremely cold intervals called *stadials* and relatively mild intervals called *interstadials*. These oscillations are often referred to as **Dansgaard-Oeschger oscillations** in honor of the geochemists Willi Dansgaard and Hans Oeschger, who first found and studied them. Their work suggested that the oscillations were spaced at intervals that ranged from as little as 1000 years to almost 9,000 years in length.

Each fluctuation toward more negative (glacial) δ¹⁸O values is matched by an abrupt increase in dust concentrations in the ice. Again, the range of variation in dust concentrations in these oscillations is a large fraction of the total difference between glacial and interglacial values. Geochemical analysis of the dust shows that most of it comes from distant source regions in Asia, not from nearby North America. The size of the dust particles is larger in the cold intervals than in the milder ones, indicating that strong winds lifted and transported the dust when climate was very cold. The colder intervals also contain larger amounts of sea salt (Na^{+1} and Cl^{-2} ions) plucked from salty sea spray above the turbulent ocean during cold and windy intervals and carried to the ice.

In the late 1980s, two long sequences were drilled on the summit of the Greenland ice sheet at sites named GISP and GRIP. These sites were carefully positioned over areas of smooth underlying bedrock to minimize the impact of changes in ice flow that can disturb deeper ice layers, and they were drilled less than 30 km apart to see whether or not they would reveal similar climate histories. The annual layering in these new cores extended to a greater depth and was used to date them further back in time, although stretching and thinning still introduced dating uncertainties deeper in the ice. The use of several annually deposited signals (dust, δ¹⁸O, and others) lessened the chance of miscounting

layers. The estimated counting errors increase from a few decades for ice 10,000 years old to several thousand years for ice 50,000 years old and to considerably more for older layers.

Both ice core sequences yielded nearly identical long-term climate records to within 200 m of the underlying bedrock. A portion of one of the ice core records is shown in Figure 14–2 (right). Because the two cores recorded nearly identical climatic signals over that entire interval, scientists had no doubt that both were reliable records of climate.

14-2 Oscillations Recorded in North Atlantic Sediments

During the 1980s and 1990s, millennial oscillations were also being discovered in North Atlantic sediments. Ocean sediments are normally not a promising archive for monitoring short-term climatic fluctuations because deposition rates are usually no greater than 1 or 2 cm/1000 yr, and small burrowing animals stir and mix the sediments to depths of 5–10 cm (Chapter 2). As a result, mixing usually obliterates climate oscillations shorter than 2500 to 5000 years.

Fortunately, places exist in the North Atlantic Ocean where deposition rates can be as high as 10–20 cm per 1000 years. Bottom currents carry fine sediments away from locations that are subject to swift flow and deposit them as large lens-shaped piles (called **sediment drifts**) in regions where the currents slow. The process is the same as the one that creates snowdrifts by scouring snow from exposed regions where the winds are strongest and piling it in regions where the wind speed slows. In this case, the coarser sand-sized sediments such as foraminifera and ice-rafted debris are not so easily moved by bottom currents as are the silts and clays. They tend to stay in place as a reliable record of climate changes even while bottom currents are delivering fine sediments that rapidly bury and preserve their climatic information.

In the mid-1980s, studies of these rapidly deposited sediments in the North Atlantic Ocean first detected shorter climate oscillations. The marine geologist Hartmut Heinrich found episodes of unusually abundant ice rafting separated by as little as 5000 years to as much as 15,000 years or more. These episodes are often referred to as **Heinrich events.** Later, the geologist Gerard Bond discovered even shorter-term variations in two climatic indices: (1) the percentage of the single polar species of foraminifera compared with the total population and (2) the relative amounts of the shells of foraminifera compared with the sand-sized grains of ice-rafted sand. As in the case of orbital-scale changes, he used higher percentages of cold-water foraminifera and larger concentrations of ice-rafted debris as an

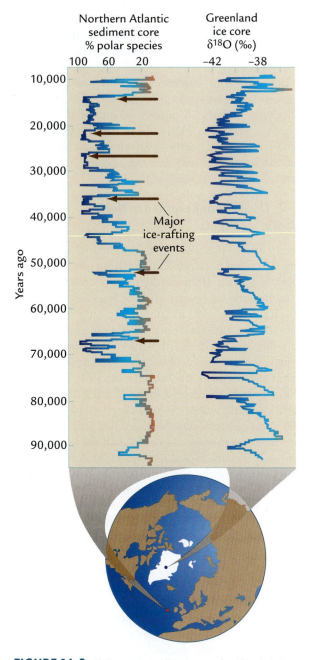

FIGURE 14-2 Millennial oscillations in the North Atlantic Ocean Millenial-scale fluctuations in the composition of North Atlantic foraminifera and in ice-rafting influxes (left) match δ¹⁸O changes in Greenland ice cores (right). (Modified from S. Stanley, *Earth System History*, ©1999 by W. H. Freeman and Company, after G. Bond et al., "Correlations Between Climatic Records from North Atlantic Sediments and Greenland Ice," *Nature* 365 [1993]: 143–47.)

indication of the presence of colder North Atlantic waters carrying larger numbers of icebergs.

Changes in the percentage of polar foraminifera in North Atlantic cores generally match δ¹⁸O changes in

Greenland ice (Figure 14–2). Times of colder air (more negative δ¹⁸O) over Greenland correlate with times of cold ocean temperatures (larger percentages of the polar species) in the North Atlantic Ocean. Dating of the youngest of the glacial ice core cycles by annual layer counts and of the ocean cores by the radiocarbon method (adjusted to calendar years) confirmed that the two sequences correlate closely for the part of the record younger than 30,000 years. The North Atlantic Ocean surface was cold when the air over Greenland was cold.

Both records show a similar pattern: repeated slow drifts toward colder, more glacial conditions followed by relatively abrupt shifts back to warmer conditions. The best dated of the ice-rafting events occurred at times when the climate had been cooling for several millennia, and each ice-rafting episode was followed by a rapid return to warmer temperatures. Some of the cooling sequences did not culminate in major ice-rafting episodes.

An initial question was whether the *relative* increases in the amount of ice-rafted debris compared to the foraminifera were caused by faster delivery of ice-rafted debris, slower deposition of foraminifera, or both. Radiocarbon dating (adjusted to calendar years) of the CaCO₃ shells of foraminifera contained in the younger ice-rafting layers indicated tenfold or larger increases in the rate of deposition of ice-rafted debris as well as smaller decreases (generally by less than half) in the rate of deposition of foraminifera.

Another major question was the source or sources of the ice-rafted debris. Although most of the ice sheets surrounding the North Atlantic contributed to the influxes, a large fraction of the grains deposited in the primary ice-rafting zone at 45°–50°N latitude came from the northeastern margin of the Laurentide ice sheet covering North America (Figure 14–3).

Initial investigations of limestone fragments during the major ice-rafting (Heinrich) events showed that source rocks in and north of Hudson Bay were the major source of this debris. Further evidence supporting this conclusion came from geochemical (isotopic) analysis of ice-rafted mineral grains that pointed to bedrock sources north and east of Hudson Bay. Detailed sampling of these large ice-rafting events showed, however, that the first debris deposited often came from smaller ice sheets around the North Atlantic Ocean. Only later did the distinctive limestone debris from North America arrive.

Much smaller amounts of ice-rafted debris were deposited during the smaller fluctuations. This debris came from a range of source regions, two of which left distinctive evidence (Figures 14–3 and 14–4). Fragments of clear and dark volcanic glass originated mainly from eruptions on Iceland. Iron-stained quartz grains came from several regions where outcrops of Pangaean-age sandstone contain quartz grains stained red by

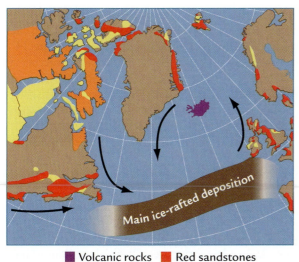

Volcanic rocks ■ Red sandstones
Limestones ■ Chemically distinctive rocks

FIGURE 14-3 Sources and deposition of ice-rafted debris
Highest rates of deposition of ice-rafted debris occur in the
North Atlantic Ocean between 45° and 50°N. During smaller
ice-rafting episodes, sources of debris include volcanic rocks
on Iceland and red sandstone rocks on several coastal
margins. During large ice-rafting events, massive amounts of
material come from eastern North America, including
limestone from Hudson Bay and fragments from other
regions with distinctive chemical signatures. (Adapted from
G. Bond et al., "Evidence of Massive Discharges of Icebergs into
the North Atlantic During the Last Glacial Period," *Nature* 360
[1992]: 245–49.)

iron oxidation during the ancient monsoon climates
(Chapter 4). Other regions around the Atlantic margins
also delivered debris during these smaller oscillations.

Long sequences of ocean sediments recovered by
the Ocean Drilling Program show that millennial fluctuations also occurred during previous glaciations, during both the 41,000-year cycles prior to 0.9 Myr ago
and the subsequent oscillations at ~100,000 years. The
largest millennial oscillations occurred during times
when ice sheets were large; fluctuations were negligible
during interglacial climates.

Because the millennial oscillations are so well developed in North Atlantic surface waters, scientists
searched for evidence that this signal had penetrated
into deep water formed in this region. The method
exploited was the same one used to measure similar
changes at orbital scales (Chapter 10): more negative
$\delta^{13}C$ values in the $CaCO_3$ shells of bottom-dwelling
(benthic) foraminifera mark times when deep water
from North Atlantic sources was replaced by bottom
water formed in the Southern Ocean. Because these
deep-water $\delta^{13}C$ signals are measured in the same cores
containing the planktic foraminifera used to monitor
changes in the surface waters, the relative timing of the
two kinds of changes can be determined, even without
accurate knowledge of absolute ages. Initial explorations
of these $\delta^{13}C$ trends detected millennial oscillations, but
the timing did not match the surface-ocean fluctuations
in a convincing way. We will return to this issue later.

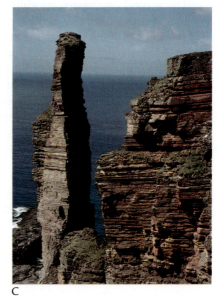

A   B   C

FIGURE 14-4 Sand-sized grains ice-rafted into the North Atlantic (A) Volcanic debris from
Iceland and (B) red-stained quartz grains from sandstone rocks around the Atlantic margins.
(C) Sources of red-stained quartz grains include red sandstones from the Orkney Islands, off
northern Scotland. (A and B: courtesy of G. Bond, Lamont-Doherty Earth Observatory of Columbia
University. C: John Forbes/PEP.)

IN SUMMARY, apparently synchronous millennial oscillations of very large amplitude are recorded in Greenland ice and North Atlantic sediments. These oscillations indicate coupled changes in several of the most important components of Earth's climate system: air and surface-ocean temperature, ice sheet margins, and ice rafting.

14-3 Detecting and Dating Oscillations in Other Regions

The verification of millennial oscillations in both Greenland ice and North Atlantic sediment set off a vigorous search for similar oscillations in other regions. Scientists who took part in this search (and are still doing so) faced two major problems: (1) Is the climatic archive that is being examined capable of recording such brief oscillations? (2) How accurately can the oscillations be dated?

Resolution of climate signals varies from archive to archive and from region to region. The ideal archive is one that allows resolution of annual changes and provides a record stretching well back into the last glaciation or preferably beyond. Unfortunately, few archives combine these characteristics, but many archives that reach far back in time can resolve climate changes lasting for tens or hundreds of years. This resolution is sufficient for detecting millennial oscillations, even if not at full amplitude.

The second (and more difficult) problem is determining whether or not the oscillations detected in other archives correlate with those found in Greenland ice and North Atlantic sediments. The ^{14}C method used to date most continental records (after correcting to calendar years) has analytical uncertainties of several thousand years for glacial-age material. Because these dating errors are comparable in size to the length of the oscillations, it is more often than not impossible to determine how the observed oscillations actually correlate.

Take the example of an oscillation that lasts for 2000 years and is perfectly dated in one region but has a dating uncertainty of +/− 1000 years in another (Figure 14–5). In this uncertainty, the two signals could be varying synchronously or changing with totally opposed tempos or varying with subtle leads and lags. At orbital time scales, leads and lags between climate signals provide critical clues to cause-and-effect relationships (Chapter 11). In the same sense, scientists who examine millennial-scale changes need accurate information about leads and lags, but dating uncertainties make it difficult and often impossible to obtain the required accuracy.

It is much easier to show that millennial-scale oscillations are present or absent in a given climate record, whatever their exact ages and correlations to other records. In a few cases, the millennial oscillations match the pattern of the changes in Greenland and the North Atlantic so closely that little doubt can exist that we are looking at the same millennial oscillations. But even in these cases, small leads or lags between the signals could exist.

14-4 Oscillations Elsewhere in the Northern Hemisphere

Other regions of the northern hemisphere also show millennial-scale oscillations, including records from Western Europe (Figure 14–6). Short-term oscillations appear in changes in the character of European soils. These soils were richer in organic material during warmer episodes but almost free of organic carbon during colder oscillations. In addition, European pollen records that showed orbital-scale changes from interglacial forests to glacial tundra also show short-term fluctuations within glacial intervals from full tundra to mixed grass steppe and forest vegetation. Although the correlation of the European records with the Greenland $\delta^{18}O$ fluctuations are not obvious, ^{14}C dating of the younger fluctuations indicates that cold-adapted vegetation occurred in Europe during times of colder air over

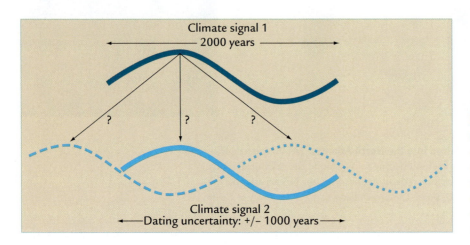

FIGURE 14-5 Uncertainties in dating millennial oscillations Typical dating uncertainties of 2000 years make it difficult to determine whether millennial oscillations in two separate climate records are synchronous, exactly opposite in timing, or offset slightly in timing.

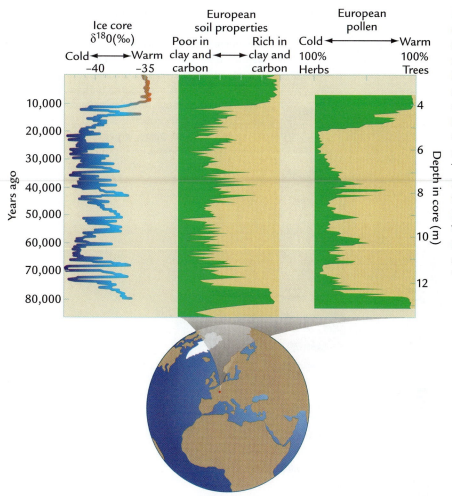

FIGURE 14-6 Millennial-scale climate changes in Europe Similar to $\delta^{18}O$ changes in Greenland ice (left), millennial-scale fluctuations occur in European soils (center) and pollen (right). (Left: Adapted from P. Grootes et al., "Comparison of Oxygen Isotope Records from the GISP and GRIP Greenland Ice Cores," *Nature* 366 [1993]: 552–54. Center: Adapted from N. Thouveny et al., "Climate Variations in Europe over the Past 140 Kyr Deduced from Rock Magnetism," *Nature* 371 [1994]: 503–6. Right: Adapted from G. M. Woillard and W. G. Mook, "Carbon–14 Dates at Grande Pile: Correlation of Land and Sea Chronologies," *Science* 215 [1982]: 159–61.)

Greenland and colder temperatures in the North Atlantic Ocean.

Short-term fluctuations have also been discovered in the glacial sections of windblown loess deposits in China. Coarse loess-rich layers that indicate physical weathering during very cold intervals alternate with finer, clay-rich soils that indicate greater chemical weathering during warm episodes. Although the Asian soil/loess sequences are not well dated, changes in that region may also match oscillations in and around the North Atlantic.

Other climate scientists have also found millennial oscillations in regions far from the North Atlantic Ocean. In the Santa Barbara Basin along the Pacific coast of North America at 35°N, several climatic signals match the $\delta^{18}O$ fluctuations in the Greenland ice sheet (Figure 14–7). Short-term oscillations in $\delta^{18}O$ values measured in the shells of planktic foraminifera indicate large (4°C or more) temperature changes in near-surface waters. In addition, the type of sediment deposited in the Santa Barbara Basin fluctuates between layers mixed by burrowing animals and intervals with varvelike layering still intact. Oxygen must have been

absent from the deep basin during warmer climates to prevent small creatures from burrowing and churning these delicate layers. Sediment mixing occurred during the oscillations toward colder climates, indicating that burrowing activity was vigorous. The obvious match in pattern between the records in the Santa Barbara Basin and those in Greenland ice makes it clear that very similar millennial oscillations affected both regions.

Other indications of millennial-scale oscillations in western North America come from fluctuations of glacial Lake Bonneville in Utah: the younger ^{14}C-dated lake level maxima appear to correlate with major ice-rafting events in the North Atlantic. Also, millennial-scale advances of mountain glaciers have been found in the Sierra Nevada of California, the Cascades of Oregon and Washington, and the Colorado Rockies. In the United States Midwest, far from the Atlantic margins of the Laurentide ice sheet, several ice lobes appear to have fluctuated in rough synchrony with the larger ice-rafting episodes, hinting at a possible link between the land and ocean margins. A pollen record from Florida also shows fluctuations during the last glaciation

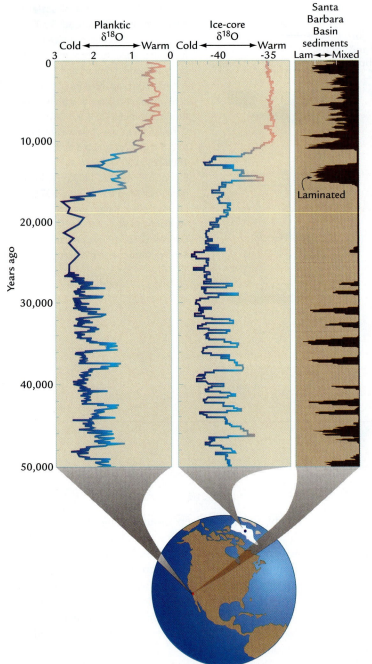

FIGURE 14-7 Millennial fluctuations in the Santa Barbara Basin The pattern of δ¹⁸O changes in planktic foraminifera from surface waters of California's Santa Barbara Basin (left) closely matches δ¹⁸O changes in Greenland ice (center). Fluctuations between intervals of varved sediments in the Santa Barbara Basin and intervals disturbed by burrowing animals show the same pattern (right). (Adapted from I. L. Hendy and J. P. Kennett, "Latest Quaternary North Pacific Surface-Water Responses Imply Atmosphere-Driven Climatic Instability," *Geology* 27 [1999]: 291–94.)

between pine pollen, indicative of wetter conditions, and grass and oak pollen, indicative of dry conditions. Fluctuations toward wetter conditions in Florida appear to correlate with major ice-rafting episodes in the North Atlantic. Evidence of millennial fluctuations has also been found in the northern tropics: fluctuations in the amount of dust delivered to the Arabian Sea, in the deposition of organic carbon off the coast of Pakistan; in the level of North African lakes, and in the abundance of plankton in the equatorial Atlantic.

14-5 Oscillations in Antarctica

The presence of millennial oscillations through much of the northern hemisphere raised the obvious question of whether or not they were global in extent, and ice cores from Antarctica are an obvious place to look for an answer. Antarctic ice cores do contain short-term δ¹⁸O oscillations, but they are smaller in size and do not have the same timing as those in Greenland ice. Even though the absolute dating accuracy is insufficient to

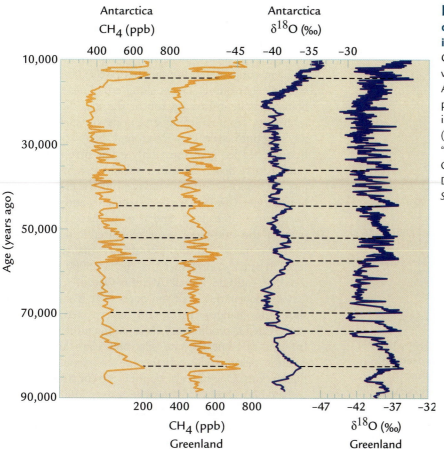

figure out the relative phasing of the changes in the north and the south, both records contain a common signal of atmospheric methane concentrations that can be used to correlate the two records very closely in a relative sense (Figure 14–8). The temperature oscillations over Antarctica turned out to be nearly opposite in timing to those in the north, but not quite. For example, the cold Younger Dryas episode in the Greenland/North Atlantic region occurred during a time when the Antarctic was gradually warming.

In general, slow warming trends in Antarctica occurred at times when Greenland had reached peak cold temperatures, and the fastest rates of warming in Greenland occurred when Antarctica had already reached maximum warmth. This pattern has subsequently been found to hold true for smaller millennial oscillations as well. The two regions are nearly but not precisely out of phase. As yet unanswered is the question of whether the entire Antarctic region followed the trend shown in Figure 14–8 (from the Byrd ice core in West Antarctica).

Millennial Oscillations During the Present Interglaciation

In contrast to the large millennial-scale fluctuations in many climate records during times when northern

hemisphere ice sheets were large, the fluctuations are muted or absent from the interglacial portions of the same records. The last two oscillations of significant size occurred during the late stages of melting of the northern ice sheets, not during full interglacial times.

The large-amplitude Younger Dryas episode may or may not have been linked to meltwater releases, along with icebergs shed into the Labrador Sea at this time (Chapter 13). A second oscillation occurred near 8200 years ago, by which time the Scandinavian ice sheet had completely melted and the Laurentide ice sheet in North America had shrunk to a small area (see Figure 13–2). This last oscillation occurred at the same time as a large release of meltwater from glacial lakes in the Hudson Bay region. The cold interval lasted only a few hundred years, but temperatures dropped markedly over Greenland and Europe.

Since 8100 years ago, millennial fluctuations have been much smaller in amplitude. They have also differed in pattern both from region to region and among different climatic indices within the same region and even within the same sediment or ice archive.

Millennial-scale $\delta^{18}O$ oscillations are not obvious in Greenland ice cores during the last 8000 years (see Figure 14–6), although small fluctuations do occur in the amount of sea salt (Na^{+1} and Cl^{-1} ions) from the

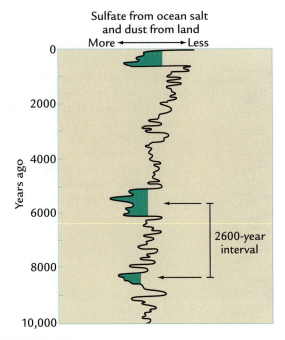

Sulfate from ocean salt
and dust from land
More ◀———————▶ Less

2600-year
interval

FIGURE 14-9 Changes in sea salt Over the last 10,000 years, ice cores from Greenland show nearly identical short-term fluctuations in the amount of sea salt (Na^{+1} and Cl^{-1} ions) and in dust from the continents, with a faint suggestion of a cycle near 2600 years. (Adapted from P. A. O'Brien et al., "Complexity of Holocene Climate as Reconstructed from a Greenland Ice Core," *Science* 270 [1995]: 1962-64.)

ocean (Figure 14–9). These oscillations have been interpreted as indicating changes in wind strength. Today salty spray from the sea surface is lifted by strong winter and spring winds and carried high onto the ice sheet, where it is deposited along with dust from continental sources.

Sediments from the North Atlantic Ocean indicate no major episodes of ice rafting during the last 8000 years, at least compared to those that occurred when ice sheets were present on North America and Eurasia. Carbon in the $CaCO_3$ shells of planktic organisms can be used for ^{14}C dating of these records. Several intervals record small increases in concentration of two kinds of mineral grains carried in by ice rafting: fragments of volcanic glass from Iceland and grains of iron-stained quartz from red sandstone rocks around the Atlantic margins (Figure 14–10). Although these influxes are 1000 to 100,000 times smaller in amplitude than those that occurred during full glacial intervals, they suggest that extremely small pulses of ice rafting may have occurred.

The kind of ice transport that could have delivered these grains is unclear. Although icebergs have been far less abundant during the last 8000 years, a small fraction of the icebergs shed by the Greenland ice sheet can

float as far south as 40°N (as the people aboard the *Titanic* found out). Icebergs from Greenland could have carried the red-stained quartz grains.

Sea ice is also capable of picking up and carrying small amounts of debris either because sediment freezes onto the bottom ice layers along coastlines or because material is deposited on top of the ice, such as glass fragments from volcanic eruptions or spring floods washing onto coastal sea ice around the Arctic margins. Because sea ice is at most only a few meters thick, it cannot carry debris far into a warm ocean before melting. Yet sea ice is common today along the east coasts of Spitsbergen (Svalbard) and Greenland and along the north coast of Iceland, and it could have transported many of the sand-sized grains measured in these cores.

For some reason, this ice-rafting signal is not registered in other records spanning the last 8000 years from the high-latitude North Atlantic Ocean. Changes in $CaCO_3$ concentrations from cores near Iceland do not show the changes that would be expected if surface water $CaCO_3$ productivity varied at the millennial scale. Influxes of noncarbonate silts and clays around the Atlantic margins also fail to register major millennial fluctuations. One index of bottom-current strength near Iceland does show oscillations at or near 1500 years.

A similar problem appears in compilations of advances and retreats of mountain glaciers, which have been superimposed on a slow drift toward slightly colder conditions during the last 8000 years. One compilation shows numerous short glacial advances (Figure 14–11A), but another reconstruction shows fewer, more widely spaced advances (Figure 14–11B). These differences may reflect different choices of the glaciers used in the compilations as well as the difficulty in obtaining reliable ^{14}C (or other) dates to constrain the upper and lower age limits of each ice advance.

Many other regions show what look like millennial-scale fluctuations during the last 8000 years. For example, North African lake levels have fluctuated markedly during that time (see Figure 13–16C), but these changes are difficult to date accurately. In general, the patterns of millennial-scale oscillations in high-resolution records covering the last 8000 years seem to disagree more than they agree. Part of this disagreement could result from the difficulty in detecting small climatic changes with measurement errors that are in some cases comparable in size to the signal being sought.

Another (more likely) explanation is that the separate parts of Earth's climate system simply went their own way during the last 8000 years. In the absence of any central driving force such as ice sheets, smaller-scale components of the climate system may have acted largely independently of each other during this interval. If so, the weaker millennial-scale oscillations found in particular regions during the current interglaciation

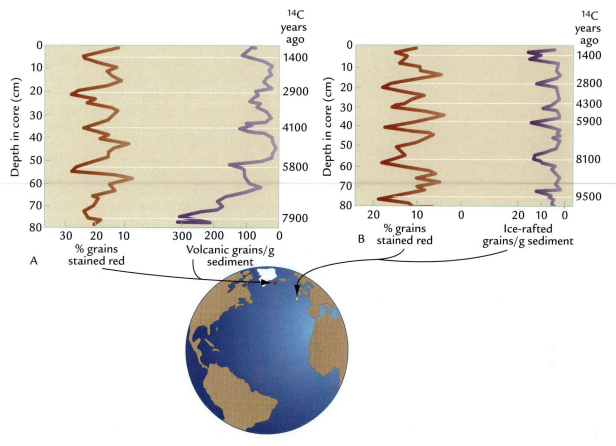

FIGURE 14-10 ~1500-year cycle in North Atlantic ice rafting? Detailed analysis of the last 10,000 years from two widely separated sediment cores shows small maxima in ice-rafted debris at intervals near 1500 years. Each ice-rafting peak contains volcanic glass fragments from Iceland, as well as iron-stained quartz and other grains from regions farther north. (Adapted from G. Bond et al., "A Pervasive Millennial-Scale Cycle in North Atlantic Holocene and Glacial Climates," *Science* 278 [1997]: 1257–66.)

may not even be the same phenomenon as the large fluctuations that occurred during times when ice sheets were present.

Causes of Millennial Oscillations

One important consideration is assessing the origin of millennial oscillations is whether or not they occurred in cycles. If the oscillations were cyclic, they should be useful for predicting the natural course of climate change in future centuries, and at first glance, many of the oscillations do look cyclic. The Dansgaard-Oeschger oscillations shown in Figure 14–2 have been referred to as *Dansgaard-Oeschger cycles*, and the longer-term variations in polar foraminifera and ice-rafted debris (which include the Heinrich events) are sometimes called *Bond cycles* after the geologist Gerard Bond who first detected them. In fact, however, the case for strong cyclic behavior is not convincing.

Some scientists claim to have found a ~1500-year cycle in both the ice sheet and ocean records, but this evidence remains questionable. The intervals separating major $\delta^{18}O$ oscillations in the Greenland ice core are irregular, ranging from approximately 1000 years in length to 9000 years or more. Time-series analysis of the Greenland ice core $\delta^{18}O$ record indicates that only a very small fraction of the observed variations falls within a band centered near 1500 years. Filtering of the $\delta^{18}O$ signal at this period reproduces a few of the shorter-term $\delta^{18}O$ variations, especially those during the interval near 35,000 years ago, but the cycle is not obvious in most of the rest of the record, where longer-duration oscillations dominate (Figure 14–12).

The ^{14}C-dated evidence for a 1500-year cycle in ice rafting in the North Atlantic Ocean during the last 10,000 years looks more convincing (see Figure 14–10), but the measured duration between ice-rafting maxima

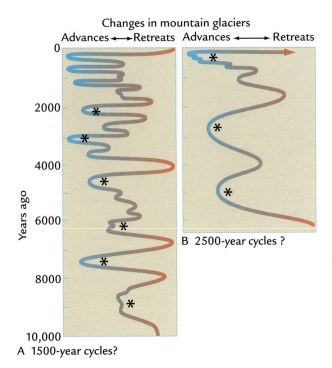

Changes in mountain glaciers
Advances ⟷ Retreats   Advances ⟷ Retreats

A 1500-year cycles?

B 2500-year cycles ?

FIGURE 14-11 Millennial-scale oscillations of mountain glaciers Two attempts to synthesize advances and retreats of mountain glaciers over the last several millennia have produced different interpretations. (A) One hints that advances may have occurred at intervals of 1500 years, but (B) the other indicates advances separated by about 2500 years. (A: Adapted from F. Rothlisberger, 10,000 Jahre Gletschergeschichte der Erde [Frankfurt: Sauerländer, 1997]. B: Adapted from G. H. Denton and S. C. Porter, "Neoglaciation," *Scientific American* 222 [1970]: 100-10.)

actually ranges between about 1000 and 2000 years. This signal is better described as "quasi-periodic" than cyclic. In any case, as noted earlier, oscillations at or near 1500 years are more often missing than present in other records of the current interglaciation from the high-latitude North Atlantic.

Alternatively, these oscillations may be **red noise.** The word *noise* means that the fluctuations are random and unpredictable, rather than cyclic and predictable. The term *red* refers to a characteristic behavior in which the longer-duration oscillations are larger in size than the shorter-term oscillations. As an analogy, consider a late spring afternoon with distinct clouds of many sizes drifting across the sky. Small clouds that block the Sun for a few minutes may cool temperatures at Earth's surface in a barely noticeable way. Larger clouds that block the sunlight for an hour or more cool the warm afternoon more obviously. A band of clouds that persists for most of the afternoon reduces the peak daily temperature even more. In this example of red noise, the passage of clouds in front of the Sun is random, and the

clouds alter local surface temperature in direct proportion to how long they block the Sun.

The glacial geologist Richard Alley has proposed that millennial oscillations fall somewhere between the extremes of cycles and red noise. His term for this behavior, **stochastic resonance,** is a composite of two concepts that seem in conflict with each other. The word *stochastic* means "random," whereas *resonance* implies cyclic behavior.

In Alley's view, resonance is evident in the fact that oscillations at a period near 1500 years do appear now and then in some records (Figure 14-13). At other times, however, the climate system skips past (fails to register) individual oscillations at 1500 years because of interference from the effects of random noise. When this happens, northern hemisphere climate drifts slowly toward colder conditions. Eventually, climate again responds to one of the multiples of the 1500-year cycle with an abrupt shift back to warmer climates. Because these abrupt transitions can occur over a wide range of intervals (after approximately 3000, 4500, 6000, 7500, or 9000

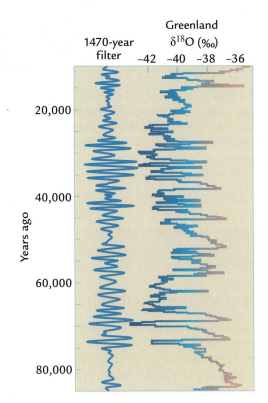

1470-year
filter

Greenland
δ^{18}O (‰)
−42 −40 −38 −36

FIGURE 14-12 A millennial cycle in Greenland ice? Filtering of the full δ^{18}O record from Greenland GISP2 ice covering the interval 85,000 to 10,000 years ago (right) reveals that a 1470-year cycle (left) is present at times but weak or absent during others. (Adapted from M. Stuiver, T. F. Brazunias, P. M. Grootes, and G. A. Zielinski, "Is There Evidence for Solar Forcing of Climate in the GISP2 Oxygen Isotope Record?" *Quaternary Research* 48 [1997]: 259-66.)

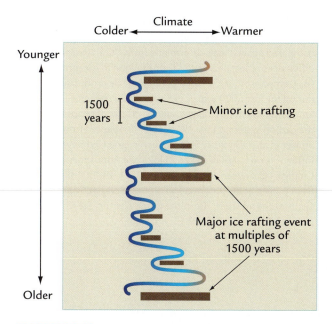

Climate
Colder ←――――→ Warmer

Younger

1500 years → Minor ice rafting

Major ice rafting event
at multiples of
1500 years

Older

FIGURE 14-13 Millennial-scale North Atlantic cycles?
One view of millennial-scale changes in the North Atlantic is
that short cooling cycles 1500 years in length gradually drift
toward colder conditions and occasionally culminate in major
ice-rafting episodes, followed by an abrupt return to warmer
conditions.

years), they appear to be random (stochastic) but are
actually an expression of highly irregular cyclic behavior.

> **IN SUMMARY,** the evidence that millennial oscillations
> occur at regular cycles is weak. The oscillations seem
> to be largely the product of random behavior (red
> noise) in the climate system. Several explanations for
> their origin are now under active consideration.

14-6 Solar Variability

One possibility is that millennial-scale oscillations are
caused by changes in the amount of solar radiation
arriving on Earth. No direct record exists of past
changes in the amount of incoming radiation at visible
or near-visible (infrared or ultraviolet) wavelengths.
This part of the spectrum contains most of the energy
that Earth receives from the Sun.

Instead, several available proxies provide scientists
with other measures of possible solar fluctuations. As
noted in Chapter 13, the difference between the ages
derived by counting (matching) tree rings and those
derived by ^{14}C dating of the same rings are one index.
The gradually increasing discrepancy between the ages
derived from these two methods (see Box 13–1)
indicates that Earth's magnetic field was weaker near
15,000 to 20,000 years ago and that the weaker solar

shielding permitted more bombardment by charged
cosmic particles (protons) and faster production of ^{14}C
atoms.

Shorter-term changes in age offsets are also apparent within the last 10,000 years (Figure 14–14). These
discrepancies may also reflect changes in the rate of
production of ^{14}C atoms in Earth's atmosphere,
although in this case the main cause is thought to be
changes in emissions from the Sun rather than the overprint of Earth's magnetic shielding. Particles streaming
from the Sun (called the *solar wind*) deflect some of the
incoming cosmic rays (protons) that would otherwise
enter Earth's atmosphere (Figure 14–15). Changes in
the amount of solar deflection over hundreds of years
could alter the ^{14}C production rate in the atmosphere
and explain the short-term differences in ages derived
from the two dating methods.

The major cycle of ^{14}C production in this 10,000-
year record was centered at 420 years, considerably
shorter than the millennial time scale explored in this
chapter. The only evidence of millennial changes was a
weak cycle in the band between 2000 and 2500 years
(~2100 years), and evidence for a cyclic link between
solar emissions and climate at this period is faint at best.
One reconstruction of the timing of mountain glacier
advances over the last 7000 years shows some indication
of a response near 2100 years (see Figure 14–11B), but
the other reconstruction does not (see Figure 14–11A).
A hint of changes at intervals near 2600 years occurs in
the sea salt signal in Greenland ice (see Figure 14–9),
but no such response occurs in the ice-rafting signals in
the nearby ocean (see Figure 14–10).

The absence of any obvious cycle of ^{14}C production
at a period of 1500 years weakens the case for a link
between solar forcing and climatic responses such as
North Atlantic ice rafting and Greenland ice $\delta^{18}O$. If
such a link exists, it would require enormous amplification from within the climate system (perhaps by
stochastic resonance). Still, the evidence against a cyclic
link between the Sun and Earth's climate does not
eliminate the possibility that a non-cyclic link exists.

Another proposed solar proxy is the ^{10}Be isotope,
which is produced by collisions with cosmic particles. As
in the case of ^{14}C, solar emissions can deflect the cosmic
particles and modulate production of ^{10}Be. Ice core
records show large millennial-scale changes in ^{10}Be concentration during the interval from 40,000 to 25,000
years ago, with an obvious correlation between ^{10}Be
maxima and $\delta^{18}O$ minima that indicate oscillations
toward cold climates. At first, this correlation seems to
point to a clear solar-climate link, but in fact the variations in ^{10}Be concentration are an artifact of climatic
changes. During cold oscillations, ice accumulation rates
dropped. With slower ice accumulation, the concentration of ^{10}Be in the ice would have increased even if its

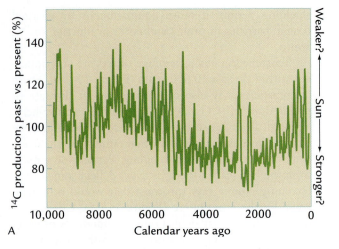

A

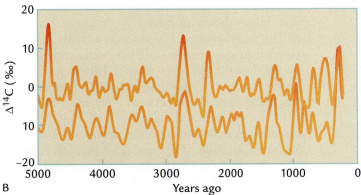

B

FIGURE 14-14 Millennial-scale changes in Sun strength?
(A) Ages determined by counting individual tree rings can be compared against ages determined from ^{14}C analyses. The age differences reflect changing ^{14}C production in the atmosphere. (B) Comparison of ^{14}C production and ^{10}Be production during the last 5000 years after smoothing to remove long-term magnetic changes and the effects of processes in the carbon system on the ^{14}C signal. (A: Adapted from M. Stuiver et al., "Climatic, Solar, Oceanic, and Geomagnetic Influences on Late-Glacial and Holocene Atmospheric ^{14}C/^{12}C Change," *Quaternary Research* 35 [1991]: 1–24. B: Adapted from J. Beer, et al. "Information on Past Solar Activity and Geomagnetism from ^{10}Be in the Camp Century Ice Core," *Nature* 331 [1988]: 675–79.)

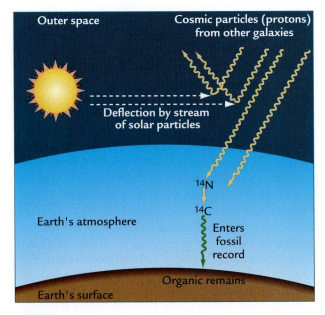

FIGURE 14-15 Changes in strength of solar emissions?
Changes in rates of ^{14}C production can be caused by changes in solar shielding of Earth's atmosphere from cosmic-ray bombardment.

rate of production in the atmosphere remained constant. Similarly, intervals of lower ^{10}Be concentrations during warm climates (δ^{18}O maxima) reflect dilution of the ^{10}Be signal by more rapid accumulation of ice. As a result, these older ^{10}Be fluctuations cannot be used to infer a connection to solar deflection of cosmic rays.

Changes in ^{10}Be concentration in ice cores have also occurred over the last 5000 years, and they have been compared with changes in ^{14}C production determined from the difference between tree ring counts and ^{14}C analyses. In making this comparison, allowance has to be made for differences in how the two signals are registered by the climate system. In effect, the ^{10}Be signal must be smoothed to mimic the more complicated processes that alter the ^{14}C record.

The very good match of the two signals shown in Figure 14–14B suggests that both the ^{10}Be and the ^{14}C signals may have varied in response to changes in solar deflection of the incoming cosmic rays. An equally possible explanation is that both oscillations are a response to changes in the internal dynamics of the climate system such as the circulation of the ocean. Because these changes occurred mainly over centuries and decades, we will return to this issue in Chapter 16.

IN SUMMARY, the evidence argues against a strong cyclical effect of solar variability on climate. A link between random (noncyclic) variations in Sun strength and climate at shorter (decadal to century) time scales remains a possibility.

14-7 Natural Instabilities in Ice Sheets

Another possible explanation of millennial oscillations in climate is that they resulted from natural internal variations in the behavior of northern hemisphere ice sheets. Considered as a whole, the great masses of ice lying on the continents have very slow response times of many thousands of years (Chapter 9). In contrast, the ocean margins of the ice sheets are capable of faster changes because they slide on soft sediments and release large amounts of ice (icebergs) to the ocean.

Along the marine margins of ice sheets, ice flows over bedrock with irregular bumps and depressions (Figure 14–16A). The bottom layers of ice scrape against higher-standing areas called **bedrock pinning points,** and the resulting friction slows the flow of ice. The bottom layers of ice can also freeze to the bedrock and slow the flow even more. Ocean water can produce the opposite effect: because the ice margins float in seawater, changes in sea level can lift the ice off its pinning points.

One idea is that the slow natural release of small amounts of heat from Earth's interior can melt the lower ice layers along ice margins (Figure 14–16B). The melting produces meltwater that trickles into the soft underlying sediments and makes them unstable, causing the ice margins to surge forward into the ocean. The surges release icebergs, which float away elsewhere and melt, and the thinned ice sheet margins retreat well inland from their previous positions. The new margins then slowly thicken and advance until the buildup of heat from below again destabilizes them.

A second idea focuses on a different kind of interaction between ice margins and the bedrock (Figure 14–16C). Over time, as the ice margins thicken, their weight depresses the underlying bedrock, which gradually sinks over thousands years. At some point, depression of the bedrock causes the ice to sink far enough with respect to sea level that it can be lifted and floated by ocean water. Because the ice margin is no longer anchored to the bedrock, it flows faster, and the ice streams release icebergs to the ocean. Once this outward flow of ice is exhausted, the ice stream retreats to another bedrock pinning point farther upstream (Figure 14–16D).

Both hypotheses are consistent with the observed episodes of accelerated ice rafting to the North Atlantic Ocean. First, ice-rafting pulses occurred only during

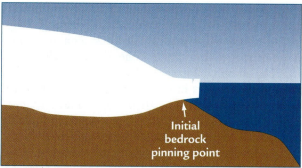

A Initial ice margin

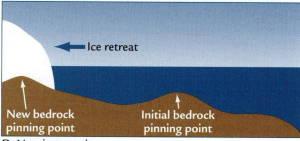

B Heat from below   C Depression of bedrock

D New ice margin

FIGURE 14-16 Natural oscillations of ice margins (A) Marine margins of ice sheets end in thin ice shelves flowing across upward-protruding bedrock knobs. This ice can be dislodged from these pinning points either by (B) Earth's heat escaping from below and melting ice or by (C) the gradual weighing down of bedrock under the heavy load of growing ice. Either way, (D) the ice margin retreats inland and stabilizes over another bedrock pinning point.

times when large ice sheets existed. In addition, the largest episodes occurred when the air and surface ocean were cold, rather than during the warmer intervals that might be expected to cause ice margins to collapse because of faster melting. This evidence argues against local warming as the driver of the iceberg pulses but allows for mechanisms based on internal ice sheet instabilities.

The ice instability hypotheses also have limitations. The fact that the composition of the debris deposited by the icebergs came from many distinct source regions (North America, Europe, Iceland) indicates that the margins of most of the ice sheets were involved in most ice-rafting events. But why would so many ice sheet margins be simultaneously involved?

One possible link is sea level (see Figure 14–16C). If one ice margin surged and sent icebergs into the ocean, the icebergs would melt and raise sea level. A rise in sea level could then destabilize other coastal ice margins by floating those ice shelves off bedrock pinning points and causing them to surge into the ocean as well.

A key question about this idea is whether or not the millennial-scale rises in sea level were large enough to link all the ice sheets. The amount of sea level change during these oscillations has been estimated by two methods. Variations in sea level based on coral reefs along the slowly uplifting coast of New Guinea indicate that sea level changes could have been as large as 10–15 m during these major oscillations. In addition, marine $\delta^{18}O$ signals recorded in the shells of benthic foraminifera from the deep tropical Pacific Ocean show millennial-scale variations of 0.1‰ or slightly larger during the same intervals. The coral reef and $\delta^{18}O$ evidence both point to sea level changes of 10 m or slightly more during the largest millennial oscillations (the Heinrich events). Smaller shorter-term variations in Pacific $\delta^{18}O$ values permit sea level changes of only a few meters during the smaller Dansgaard-Oeschger oscillations.

The massive North American ice sheet is the best candidate for causing sea level rises large enough to trigger reactions in the other ice sheets during major ice-rafting events. Yet the evidence in North Atlantic sediments shows that the first sand grains deposited during the major ice-rafting pulses came from the smaller ice sheets on Iceland and in coastal regions farther north. Only later did floods of debris arrive in icebergs originating in North America. This sequence appears to rule out the North American ice sheet as the initial sea level trigger during large ice-rafting events.

Changes in the sizes of the northern ice sheets during the shorter-term millennial oscillations were smaller and probably produced sea level fluctuations of just a few meters. It is more difficult to argue that these small changes could have provided the link between the northern ice sheet margins.

Another possibility is that the ice streams that delivered icebergs to the Atlantic Ocean pulled enough ice out of the interior of the North American ice sheet to alter atmospheric circulation. Climate changes over orbital time scales have been interpreted as a response to splitting of the jet stream because of changes in the elevation of the North American ice sheet (see Figure 12–13B). Some climate scientists have suggested that the same explanation might work for the shorter millennial-scale changes in response to ice volume changes of 1% to 10%. For this explanation to be viable, atmospheric circulation would have to have been *extremely* sensitive to small changes in the elevation of the North American ice sheet, perhaps because of the existence of a critical threshold.

IN SUMMARY, natural internal oscillations within ice sheets could plausibly have played a role in at least the larger millennial oscillations but probably not in the smaller ones.

14-8 Greenhouse-Gas Forcing

Because greenhouse gases play a major role in climatic changes at tectonic and orbital time scales, their behavior during millennial oscillations is worth considering. How large were the millennial-scale oscillations in CO_2 and methane, and do they indicate a forcing or feedback role for the gases?

Methane concentrations in ice cores show clear millennial oscillations (see Figure 14–8). Because these changes lag a few decades behind $\delta^{18}O$ (temperature) fluctuations, they appear to have been the result rather than the cause of the temperature oscillations. The changing methane concentrations would have acted as a positive feedback to millennial-scale temperature changes that had begun for other reasons. Methane production and release in the wetlands of northern Asia (Siberia) were particularly susceptible to changes in air temperature over millennial intervals.

Ice core records are less clear about millennial-scale CO_2 changes. Measurements in high-resolution ice cores from Greenland are suspect because the CO_2 in the air bubbles interacts chemically with $CaCO_3$ dust in the ice. In slowly deposited ice from Antarctica, millennial-scale events are too brief to be recorded in full amplitude because of the long time delay in sealing the air bubbles in the ice. The best records currently available suggest that some CO_2 oscillations may have been as large as 10 ppm or more, but the records are noisy and the patterns are not clearly developed.

IN SUMMARY, the role of CO_2 changes in millennial-scale oscillations remains unclear. Future drilling on the lower flanks of Antarctica should help to clarify the amplitude and timing of the CO_2 oscillations.

14-9 Other Natural Interactions in the Climate System

Another proposed explanation is that the millennial oscillations were produced by natural interactions within key components of the climate system in the Greenland/North America area, such as the ice sheets, the surface ocean, and the deep ocean. At orbital time scales, changes in this region influence temperature and precipitation across a broad area of the northern hemisphere (see Figures 11–2, 11–3, and 11–4).

An initial proposal was that a natural oscillation existed between the sizes of the (large) northern ice sheets, the amount of meltwater runoff, the salinity of the North Atlantic surface waters, and the rate of formation of deep water. Possible links among these responses are numerous: northward advection of warm surface water could promote ice melting, while a low-salinity meltwater lid on the surface ocean could stifle deep-water formation. While various aspects of this idea are still being explored, several problems are apparent. For example, the major iceberg releases occurred during times when the surface ocean was cold. In addition, the most intensively studied oscillation—the Younger Dryas cooling—may have occurred without any meltwater contribution from the ice sheets (Chapter 13).

The evidence that millennial temperature changes in the North Atlantic and Antarctic regions have opposed timing points toward another possible origin for the millennial oscillations (see Figure 14–8). This pattern, called the **bipolar seesaw,** has been interpreted as resulting from changes in the northward redistribution of heat by the Atlantic Ocean.

The typical pattern of ocean heat transport removes excess heat from the warm tropics and carries it toward the cold poles (companion Web site, pp. 22–24). The Indian and Pacific oceans both follow this pattern but the Atlantic Ocean does not. Instead, heat from the South Atlantic Ocean crosses the equator and moves into the high latitudes of the North Atlantic Ocean. The marine geologist Tom Crowley first proposed that changes in the northward transport of heat through the Atlantic Ocean accounted for the bipolar seesaw pattern (Figure 14–17). Large cross-equatorial transports of heat would leave the Southern Ocean cold while warming the North Atlantic Ocean. Weak transport would leave the Southern Ocean warm while cooling the North Atlantic region. Simulations by the climate modeler Tom Stocker and others have provided support for this idea.

The cause of this seesaw is not yet clear. The geochemist Wally Broecker proposed that changes in the amount of deep water formed in the North Atlantic Ocean controlled the amount of heat pulled northward into the Atlantic as part of a much larger-scale flow of heat through the world ocean termed the *conveyor belt* (companion Web site p. 23). Physical oceanographers have criticized this explanation by pointing out that the large-scale circulation of the surface ocean is driven not by deep-water formation but by winds.

Another problem with the conveyor belt hypothesis has come from high-resolution analysis of North Atlantic sediments. Changes in temperature of North Atlantic Ocean surface waters recorded by planktic foraminiferal assemblages fluctuate with the northern millennial-scale timing described earlier, but the bottom-water fluctuations measured in the same cores

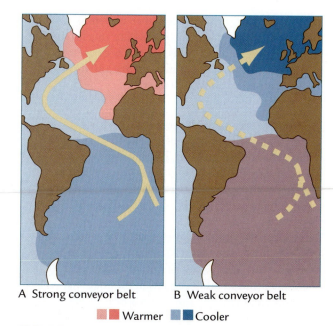

A Strong conveyor belt   B Weak conveyor belt

■■ Warmer   ■■ Cooler

FIGURE 14-17 Opposite hemispheric responses caused by ocean heat transport. (A) When cross-equatorial heat flow in the Atlantic is strong, it warms the North Atlantic but cools south-polar regions. (B) When cross-equatorial flow weakens, the temperature responses are reversed.

(in δ^{18}O values of benthic foraminifera) were unexpectedly found to have the south-polar timing. This finding indicates that deep-water fluctuations in the North Atlantic Ocean are not linked closely to processes operating in the north, as required by the conveyor belt hypothesis, but are controlled by changes in deep flow from the south.

Because the largest temperature changes are centered near Greenland and the North Atlantic Ocean, this area may still be the center of action of the millennial-scale oscillations. One possibility is that the atmospheric circulation in this region is unusually sensitive to small changes in surface climate of some kind (ice sheet elevation or some other feature). Because millennial-scale fluctuations in ice core δ^{18}O values tend to oscillate between similar extremes (see Figure 14–2), two relatively stable modes may exist in atmospheric circulation during glacial intervals, with abrupt switches back and forth between these modes at irregular (random) intervals.

Based on links already observed at orbital time scales (Chapter 11), large temperature changes over Greenland and the North Atlantic could alter temperature and precipitation patterns in Europe and parts of Asia, including the high-pressure cell in Siberia. These changes could disrupt the jet stream circulation in the upper atmosphere and propagate into more distant regions such as the Santa Barbara Basin. Evidence from a range of indicators suggests that millennial oscillations in the northern hemisphere were much larger in winter than in summer. This pattern is consistent with propagation of the millennial oscillation during the strong atmospheric flow of winter.

> **IN SUMMARY,** the origin of millennial climatic oscillations remains unknown, but considerable progress has been made in defining their time-and-space "footprint" in the climate system. The oscillations appear to be largely random rather than cyclic. Their amplitude is largest when ice sheets are present on North America and Eurasia, and the largest oscillations are centered near Greenland and the North Atlantic Ocean. Similar (but smaller) changes with nearly opposite timing occur over Antarctica and in other regions in the southern hemisphere, including the southeastern Pacific Ocean off the coast of Chile. The bipolar seesaw is a promising candidate for an internal oscillation of natural origin.

14-10 Implications for Future Climate

Because millennial oscillations occur much faster than orbital-scale changes, they have the potential to have a more immediate impact on our climatic future. Scientists and policy planners would like to know whether natural oscillations could cause climate to warm or to cool in future decades. Some scientists have speculated that a natural millennial-scale warming could be underway now. If this view is correct, it means that the observed warming of the last century or so could in part be the result of natural processes rather than human activities.

This claim is unjustifiable for several reasons. Because millennial oscillations are either completely random (red noise) or at best quasi-periodic, their present and future course cannot be predicted with any confidence. More critically, the largest oscillations occurred only during glacial climates, whereas the changes during the past 8000 years of warm interglacial climate have been small and local in scale. This observation argues against natural oscillations playing a major role in present and future climate change.

Of course, the Greenland and Antarctic ice sheets are still in place, and they remain susceptible to some degree of melting in the warmer climate of the future. Because ice sheets appear to have played a role in at least some of the glacial-age millennial oscillations, partial melting of today's ice sheets because of human activities could conceivably trigger changes in the climate system in the future, even if at a smaller scale. In this case, however, the cause of these changes will be human activities, not natural variations in the climate system.

Key Terms

millennial oscillations (p. 251)

Dansgaard-Oeschger oscillations (p. 253)

sediment drifts (p. 253)

Heinrich events (p. 253)

red noise (p. 262)

stochastic resonance (p. 262)

bedrock pinning points (p. 265)

bipolar seesaw (p. 267)

Review Questions

1. How do the processes that control $\delta^{18}O$ changes in ice sheets differ from those that control $\delta^{18}O$ fluctuations in ocean cores?

2. Why is it difficult to correlate millennial climatic oscillations in records from different regions?

3. What other regions show millennial oscillations like those in the North Atlantic and Greenland?

4. Are millennial oscillations true cycles?

5. How strong is the evidence that solar changes drive millennial oscillations?

6. What is the evidence for and against internal ice sheet processes causing millennial oscillations?

7. How could ocean flow cause opposite millennial oscillations north and south of the equator?

Additional Resources

Basic Reading

Alley, R. B. 2000. *The Two-Mile Time Machine.* Princeton, NJ: Princeton University Press.

Advanced Reading

Beer, J., et al. 1988. "Information on Past Solar Activity and Geomagnetism from [10]Be in the Camp Century Ice Core." *Nature* 331: 675–79.

Bond, G., W. S. Broecker, S. J. Johnsen, J. McManus, L. D. Labeyrie, J. Jouzel, and G. Bonani. 1993. "Correlations Between Climatic Records from North Atlantic Sediments and Greenland Ice." *Nature* 365:143–47.

Boyle, E. A. 2000. "Is Ocean Thermohaline Circulation Linked to Abrupt Stadial-Interstadial Transitions?" *Quaternary Science Reviews* 19: 255–72.

Brook, E. J., et al. 2005. "Timing of Millennial-Scale Climate Change at Siple Dome, West Antarctica,

During the Last Glacial Period." *Quaternary Science Reviews* 24: 1333–43.

Crowley, T. J. 1992. "North Atlantic Deep Water Cools the Southern Hemisphere." *Paleoceanography* 7: 489–97.

Denton, G. H., and W. Karlen. 1973. "Holocene Climatic Variations: Their Possible Causes." *Quaternary Research* 3: 155–205.

Hendy, I. L., and J. P. Kennett. 1999. "Latest Quaternary North Pacific Surface-Water Responses Imply Atmospheric-Driven Climatic Instability." *Geology* 27: 291–94.

MacAyeal, D. R. 1993. "Binge/Purge Oscillations of the Laurentide Ice Sheet as a Cause of North Atlantic's Heinrich Events." *Paleoceanography* 8: 775–84.

Shaffer, G., S. M. Olsen, and C. J. Bjerrum. 2004. "Ocean subsurface warming as a mechanism for coupling Dansgaard-Oeschger climate cycles and ice-rafting events." Geophysical Research Letters, 31, L24202, doi 10.1029/2004 GL 020968.

Stuiver, M., and T. F. Brazunias. 1993. "Sun, Ocean, Climate, and Atmospheric $^{14}CO_2$: An Evaluation of Causal and Spectral Relationships." *Holocene* 3: 289–305.

Sea ice: A major component of Earth's climate system Arctic sea ice had thinned and retreated significantly by the early twenty-first century, probably as a response to a global climate warning that will intensify in the future. (Johnny Johnson/Index Stock Imagery/Picture Quest.)

Historical and Future Climate Change

During the last 12,000 years, a new era in Earth history emerged. Prior to that time, scattered bands of Stone Age people lived a migratory hunter-gatherer-fisher existence. Then agriculture was discovered, and humans began to live in permanent dwellings. A new hypothesis suggests that early farmers began to alter climate thousands of years ago through increasing emissions of greenhouse gases. By 2000 years ago, something like modern life had emerged: iron tools, domesticated livestock, towns and cities, writing, and religions. By 100 years ago, still early in the industrial era, we had already moved more rock and soil than had all the water, ice, and wind in the climate system.

Climate changes over this interval are reconstructed from a wide array of archives and proxy indicators. The best archives for intervals spanning a few thousand years are annually layered ice cores and lake sediments. For recent centuries, annually layered tree rings and corals provide additional coverage, and historical observations are useful in a few regions. For the last 100 years, measurements made by instruments have become the major source of climatic histories.

Natural climate changes over the last 1000 years were much smaller than those over tectonic, orbital, and millennial time scales, amounting to 0.5°C at most on a global basis. Within the last 125 years, industrialization has had a rapidly accelerating effect on global climate, which has warmed by 0.7°C. Our growing overprint on climate seems destined to continue for centuries to come, and many parts of Earth's surface (particularly the Arctic) will be totally transformed by future warming.

To evaluate the extent of both natural and human causes of climate change during recent centuries and in the future, we explore these issues:

- When did humans begin to play a role in climatic change?
- What caused climate change during the last 1000 years?
- Was the warming since the late 1800s caused by humans or natural factors?
- What is the sensitivity of Earth's climate system to the by-products of the Industrial Revolution, including greenhouse gases (CO_2 and CH_4) and sulfur dioxide (SO_2)?
- What kinds of climate changes lie in Earth's future?

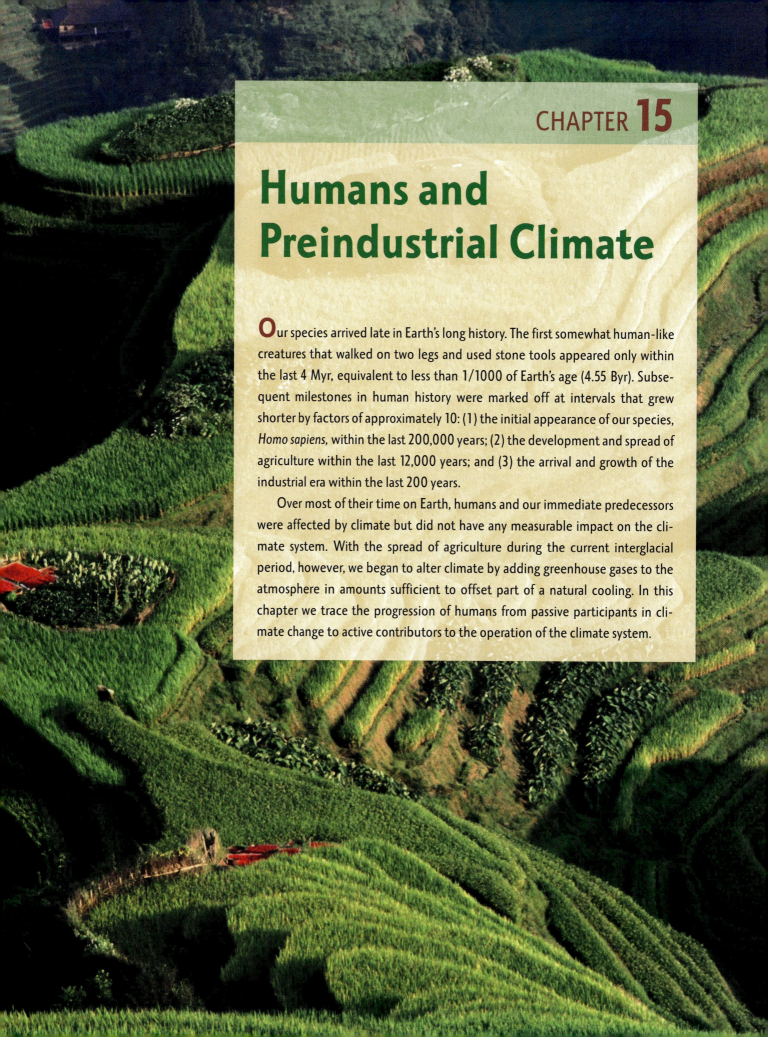

CHAPTER 15

Humans and Preindustrial Climate

Our species arrived late in Earth's long history. The first somewhat human-like creatures that walked on two legs and used stone tools appeared only within the last 4 Myr, equivalent to less than 1/1000 of Earth's age (4.55 Byr). Subsequent milestones in human history were marked off at intervals that grew shorter by factors of approximately 10: (1) the initial appearance of our species, *Homo sapiens*, within the last 200,000 years; (2) the development and spread of agriculture within the last 12,000 years; and (3) the arrival and growth of the industrial era within the last 200 years.

Over most of their time on Earth, humans and our immediate predecessors were affected by climate but did not have any measurable impact on the climate system. With the spread of agriculture during the current interglacial period, however, we began to alter climate by adding greenhouse gases to the atmosphere in amounts sufficient to offset part of a natural cooling. In this chapter we trace the progression of humans from passive participants in climate change to active contributors to the operation of the climate system.

Climate and Human Evolution

Anthropologists agree that humans evolved in Africa. Most of the earliest evidence for our ancient ancestors comes from plateaus along the eastern side of the continent (Figure 15–1). Volcanic activity in this region over millions of years deposited basalt layers that can be dated by radiometric methods. These dated basalts bracket the ages of intervening sediment layers that hold much of the record of human evolution.

15-1 Evidence of Human Evolution

Anthropologists focus either on distinctive events that break the continuous process of evolution into separate stages or on quantitative traits that can be measured as they gradually change. Human evolution is marked by five distinctive developments (Figure 15–2): (1) the initial branching off from primitive apes between 6 and 4 Myr ago; (2) the onset of bipedalism (a preference for moving upright on two legs) near 4 Myr ago; (3) the use of stone tools beginning near 2.5 Myr ago; (4) the branching of the prehuman line into the genus *Homo* and other forms by 2 Myr ago; and (5) the development of large brains since 2 Myr ago.

Appearance of Human Ancestors Human evolution can be traced back to small shrewlike mammals that evolved during the millions of years after the massive

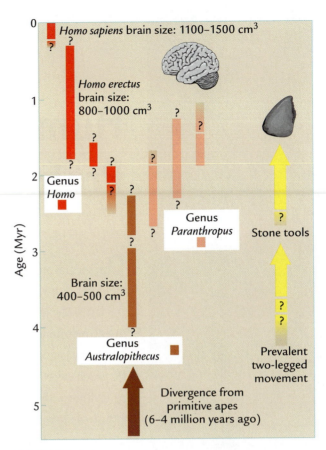

FIGURE 15-2 Human evolution The last 5 million years span the evolutionary line from primitive apes to the australopithecines ("southern apes"), our own genus, *Homo*, and finally our own species, *Homo sapiens* ("intelligent man"). The onset of two-legged walking appeared early in this evolutionary progression, followed by the first use of stone tools more than a million years later. (Adapted from P. B. deMenocal, "Plio-Pleistocene African Climate," *Science* 270 [1995]: 53–59.)

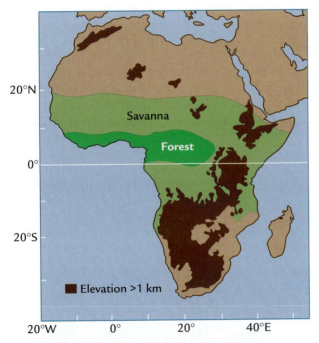

FIGURE 15-1 African topography and rain forests Eastern Africa is a region of broad plateaus at elevations not far above 1 km. Most rain forest vegetation occurs today in the wet intertropical convergence zone near the equator, encircled by a broad band of drier savanna.

extinctions at the time of the asteroid impact 65 million years ago (Chapter 5). These primitive mammals eventually evolved features we now associate with monkeys, such as grasping front paws and long prehensile tails, which led to the lemur family (a monkey-like tree-climbing animal). Modern lemurs are shown in Figure 15–3.

By 10 Myr ago, one such line had further evolved to primitive apes. Subsequently, a group of apes that included both our human ancestors and chimpanzees branched off from the primitive apes, with modern apes evolving from the other branch. Our prehuman ancestors are thought to have foraged for food in and near woodlands, moving at times on two legs. Radiometric dating of volcanic rocks in East Africa indicated that this branching occurred sometime between 10 and 5 Myr ago, but for a long time the timing was difficult to constrain more precisely.

FIGURE 15-3 Early mammals The line of early mammals from which humans evolved included creatures resembling these modern-day lemurs. (Frans Lanting/ Minden Pictures.)

A new source of evidence—molecular biology— reduced this uncertainty. Molecular biologists measure the composition of DNA molecules in the protein of living organisms. They postulate that DNA works like an evolutionary clock: the longer the time that has elapsed since two organisms branched off from a common ancestor, the more dissimilar their DNA will have become. If this DNA dissimilarity increases through time at a constant rate, the degree of dissimilarity can be used as a clock to measure elapsed time. Molecular biologists concluded that the line that led to humans diverged from the line that led to the modern great apes between 6 and 4 Myr ago (see Figure 15–2).

Walking Upright The evolutionary line that led to modern humans, called *hominins* (from the family *Hominidae*, meaning human-like), appeared by 4 Myr ago. Fossil remains of anklebones with a distinctive structure suggest that walking had become the primary means of movement by 4.3 Myr ago. These creatures were considerably smaller than modern humans and had chimpanzee-like faces with large, strong jaws. They

probably spent most of their time in trees, gathering fruits and nuts and avoiding predators, but they also moved on the ground when necessary.

A remarkable deposit dated to 3.6 Myr ago in Tanzania holds footprints of two creatures walking across a freshly fallen layer of volcanic ash that had cooled (Figure 15–4). The tracks show that one of the creatures turned, perhaps to look back at something, and then walked on. These human-like apes of the genus *Australopithecus* (known as *australopithecines*) walked upright to travel. Scientists argue whether these creatures developed the ability to walk in order to exploit food resources on the grassy savanna lying between stands of trees, or whether they developed upright postures in order to stand on the lower tree limbs (or on the ground) and reach up for fruits and nuts.

Use of Stone Tools The first firm evidence that hominins used stone tools dates to about 2.5 Myr ago. These early tools, used to butcher dead animals, produced marks on the animal bones that indicate crushing and scratching but are sometimes difficult to distinguish from similar marks made by the teeth of carnivores (lions, leopards, cheetahs).

One early hypothesis suggested that humans evolved mainly as "killer apes" because of the aggressive use of tools to kill their prey. But many anthropologists today believe that hominins simply made opportunistic use of the remains of animals previously killed by lions

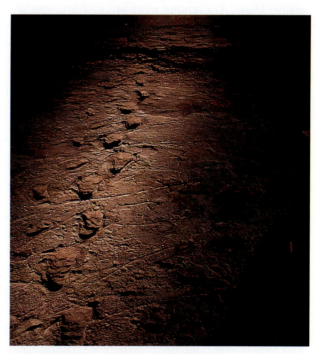

FIGURE 15-4 Footprints from 3.6 Myr ago Hominin (humanlike) creatures that walked across fresh volcanic ash 3.6 Myr ago in East Africa left their footprints, now fossilized. (Kenneth Garrett, National Geographic Society Image Collection.)

and leopards, and that they constantly had to contend with hyenas and other scavengers for this food. Brandishing tools as weapons may have helped them drive off competitors.

Use of tools for butchering implies an important change in diet, which must have previously consisted mainly of items collected from their environment: fruits, nuts, leaves, and small insects (mainly grasshoppers and termites). The use of crude cutting tools would have allowed hominins to take greater advantage not just of the meat but also the bone marrow and other internal animal parts. These protein-rich food sources would have more readily satisfied their energy needs. Stone tools could also have helped hominins dig out buried roots and tubers that were otherwise difficult to reach.

The available evidence suggests that tool making followed more than 1 million years after the ability to walk upright. Some scientists infer that tool making was a natural evolutionary development for creatures whose hands were freed for other uses when they began to walk on two legs.

Appearance of Homo Sometime after 2.5 Myr ago, the ancestral australopithecines evolved into several new forms (see Figure 15–2). One line led to the genus *Paranthropus*, stout creatures with large teeth and strong jaws used to crush protein-rich palm nuts and other hard food in their vegetarian diet. This group became extinct by 1 Myr ago.

The other major group carries the name of our own genus, *Homo*. These were more graceful (lean-bodied) creatures with larger heads and braincases. The earliest of these creatures is dated (with some uncertainty) to about 2.4 to 2.3 Myr ago, and our human ancestor *Homo erectus* (upright man) was definitely present by just after 2 Myr ago (see Figure 15–2). The wear on their teeth indicates a broader-based diet of meat, fruits, and vegetables.

The first appearance of these human-like creatures appears to follow closely after the earliest use of stone tools, but the precise timing is difficult to determine because the fossil record is fragmentary. The use of tools is likely to have favored those prehumans who were able to move easily across the landscape in pursuit of a large variety of seasonally changing food sources. Frequent movement may also have called on a greater use of intellect and imagination.

Brain Size Over time, our human ancestors developed larger brains, shown by the increasing size of preserved skulls that encased and protected the brain. In broad outline, the volume of the braincase tripled over the last 3 or 4 million years (Table 15–1).

The bipedal australopithecines had braincase volumes of 400–500 cm³. The *Homo erectus* ancestors who used stone tools had braincases twice as large, roughly 800–1000 cm³. Fully modern humans (*Homo sapiens*, or

TABLE 15-1 Growth in Size (Volume) of Hominid Braincases

Type of hominid	Age (Myr ago)	Braincase (cm³)
Homo sapiens	0.2–0	1100–1500
Homo erectus	2.4–1.8	800–1000
Australopithecus	4.1–3.1	400–500

intelligent man) first appeared between 200,000 and 100,000 years ago, with braincases ranging in size between 1100 and 1500 cm³. A tripling of brain size in 4 million years is unusually rapid compared to many evolutionary changes.

15-2 Did Climate Change Drive Human Evolution?

Many competing hypotheses have been proposed to explain human evolution. Some hypotheses focus on social factors, such as the need of early humans to gather food for their infants, who spent several years in a defenseless state. Others focus on the impact of new technology, such as the creation of tools to facilitate food gathering and to serve as weapons for defense and hunting. Other scientists propose that climate change has altered human evolution. Climatic hypotheses of human evolution fall broadly into two groups.

According to the **savanna hypothesis**, human evolution was driven by a long-term drying of African climate in which tropical rain forests became interspersed with semiarid grasslands, which gradually spread between groves of trees. Fragmentation of once continuous forest caused our ancestors to move on the ground for longer distances, requiring more rapid movement to cover longer distances and also greater resourcefulness. This drying trend began well before the divergence of humans and chimpanzees from great apes 6 to 4 Myr ago, and it has continued in subsequent times.

Remarkably little evidence from the earlier part of this interval has been found in North Africa, because aridity is so unfavorable to permanent deposition and preservation of sediment. Much of the information about the longer-term climate of Africa comes from ocean sediments that contain material blown out from the continent. Cores from the tropical eastern Atlantic Ocean show gradual increases in the rate of influx of continental dust (mostly quartz and clay) from North Africa after 4.5 Myr ago (Figure 15–5 left). This trend, also evident in sediments from the western Indian Ocean, has been interpreted as a sign of progressive drying that reduced the vegetation cover and exposed larger areas to erosion by winds. Dust transport may

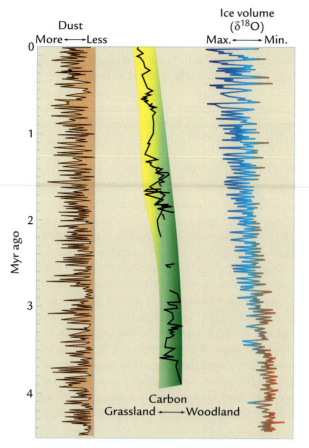

FIGURE 15-5 Long-term changes in African dust and vegetation Over the last 4.5 Myr, (left) increasing amounts of dust were blown from North Africa to the tropical Atlantic, and (center) vegetation cover in many regions gradually shifted away from trees and shrubs toward warm-season grasses. (Right) These changes occurred during a time of global cooling. (Adapted from P. B. deMenocal, "Plio-Pleistocene African Climate," *Science* 270 [1995]: 53–59.)

America, and North America. This change altered the carbon-isotopic composition of the carbon left behind in soils and in the teeth of grazing animals on all four continents. The trend toward heavier $\delta^{13}C$ values in North Africa (Figure 15–5 center) indicates a change from C3 vegetation (trees and shrubs) toward C4 vegetation (warm-season grasses). Greenhouse experiments suggest that such a shift could have happened because CO_2 concentrations fell past a critical threshold over several million years. The fact that a gradual change from C3 to C4 carbon occurred on all four continents suggests falling atmospheric CO_2 concentrations as a common explanation. As this trend developed, global climate was cooling, and ice sheets were growing (Figure 15–5 right).

Climate modeling points to several factors that may have driven a long-term drying in Africa. One factor was construction of volcanic plateaus in east Africa over the last 30 million years. Model simulations show that prior to plateau construction, moisture-bearing winds blew from the Indian Ocean westward into East Africa (Figure 15–6). As the plateaus took form, this influx of moisture was choked off, and the vegetation shifted from forests to grasslands across much of the region where most hominin remains are found but not across the rest of Africa. Another factor that may have contributed to this regional drying was a cooling of the

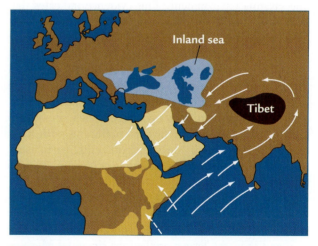

FIGURE 15-6 Tectonic effects on African climate Over the last 20 Myr, East African plateaus became drier because of local uplift in eastern Africa and cooling of the western Indian Ocean (orange), uplift of the Tibetan Plateau (yellow), and shrinkage of an inland sea in west-central Asia (blue). (Adapted from P. Sepulchre et al., "East African Aridification and Late Neogene Uplift," *Science* 313 [2007]: 1419–23, W. F. Ruddiman and J. E. Kutzbach, "Plateau Uplift and Climate Change," *Scientific American* 264 [1991]: 66–75, and G. Ramstein et al., "Effect of Orogeny, Plate Motion, and Land-Sea Distribution on Eurasian Climate over the Past 30 Million Years," *Nature* 386 [1997]: 786–95.)

also have increased because the spring winds that lift and carry the fine debris became stronger.

A long-term drying trend in North Africa is also evident in a sediment sequence from the Atlantic Ocean just offshore of the Sahara Desert that contains a history of pollen changes over portions of the last 3.7 Myr. Gradually decreasing amounts of forest pollen and increasing amounts of savanna and desert scrub pollen indicate a progressive trend toward dry-adapted vegetation. Other shorter-term pollen records from the East African plateau also reveal a long-term trend toward more open vegetation.

A long-term decrease in atmospheric CO_2 could also have been a factor in the change in vegetation. Over this interval, a major vegetation shift was underway not just in East Africa but also in parts of southern Asia, South

western Indian Ocean caused by the partial closure of a seaway near Indonesia and the resulting loss of warm tropical water from the western Pacific Ocean.

Tectonic changes in southern Asia may have also played a role. Continuing uplift of the northern and eastern parts of the Tibetan Plateau not only strengthened the Asian summer monsoon but also created a strong counterclockwise spiral of winds that drove hot, dry air out of the interior of Asia across Arabia and into northeastern Africa (Figure 15–6). These winds produced much drier conditions in summer across a huge arc extending from west-central Asia across the Arabian peninsula and into northeastern Africa. This drying trend may have been aided by the retreat of a vast interior seaway that once occupied west-central Asia and moderated its temperatures.

> **IN SUMMARY,** both the available data and model simulations confirms a gradual trend toward more open vegetation in Africa throughout the time of human evolution. This evidence seems to support the savanna hypothesis.

The actual record of hominin remains, however, suggests that the picture is more complicated. Hominins occupied environments ranging from woodlands and grasslands to river margins throughout their history, with no obvious trend toward greater occupation of the savanna environment. This evidence indicates that hominins have long been resourceful in making use of a range of environments and it has led to a different hypothesis of the effects of climate on human evolution.

The basic premise of the **variability selection hypothesis** is that rapid evolution occurred because alteration of habitat put new demands on our ancestors, some of whom then took advantage of new opportunities. Changes in climate favored hominims with traits useful for survival in new environments, particularly clans or tribes that had larger numbers of members with such traits.

Climatic variability has greatly increased at high latitudes during the last several million years, particularly with the onset of small glaciation cycles near 2.75 Myr ago and then the larger oscillations since 0.9 Myr ago (Chapter 9). In addition, the earliest appearance of our genus *Homo* is dated to between 2.4 and 2.0 Myr ago, not long after the first glacial cycles. Perhaps these early glaciations created cycles of cooling and drying in Africa that accelerated the pace of evolution by favoring individuals with greater adaptability.

Several lines of evidence suggest that northern hemisphere glaciation could have affected climate and hominin habitat in Africa. Pollen records from high terrain in East Africa show that the upper tree line began to retreat up mountainsides and give way to more

open vegetation at or near the onset of the glacial cycles. Subsequently, forest vegetation repeatedly fluctuated up and down the sides of the mountains at intervals of tens of thousands of years. These oscillations appear to be connected to the early northern hemisphere glacial cycles.

In addition, records from the eastern and southern plateaus of Africa also show a widespread change from woodland-adapted browsing (leaf-eating or grass-eating) animals to grazing animals strictly adapted to grasslands near and after 2.5 Myr ago. Many browsing animals became extinct, and new grazing animals appeared both by evolving and by emigrating from Asia. The evidence suggests a relatively rapid proliferation of open grasslands at the expense of closed forests. This marked change from browsers to grazers (and from forest to grassland habitat) has been attributed to the cooling and drying produced by northern hemisphere glacial cycles.

One problem with this proposed link to the northern ice sheets is that GCM simulations suggest that the ice sheets would have had relatively small direct effects on temperature and precipitation in North Africa. Even the large ice sheets that existed at the last glacial maximum, 20,000 years ago, cooled northern Africa by no more than $1°-4°C$, with smaller changes farther south and little change anywhere during the summer (monsoon) season. The smaller ice sheets that grew and melted during the glacial cycles that started 2.75 Myr ago would probably have had even smaller impacts on African climate.

It is possible that these early glaciations had some effect on vegetation and habitat in Africa because they were accompanied by drops in atmospheric CO_2 levels, like those that have occurred during the last several hundred thousand years (see Chapter 10). Lower CO_2 levels during these earlier glaciations could have caused some replacement of C3 trees and shrubs by C4 grasses during each glacial cycle, with fragmentation of forest habitat.

Another problem with the variability selection hypothesis is that the summer monsoons that control annual rainfall and vegetation types across the savannas south of the Sahara Desert and across most of the plateaus of East Africa appear to have continued for many millions of years with no strong trend toward larger or smaller oscillations. It appears that the monsoons did not contribute to a long-term increase in climatic variability.

Unfortunately, the fossil record of human remains is too sparse to determine any correlation between the onset of glacial cycles and hominid evolution.

> **IN SUMMARY,** the record of human remains is too sparse to test the variability selection hypothesis.

15-3 Testing Climatic Hypotheses with Fragmentary Records

Again and again we bump up against the problem of the fragmentary nature of the record of hominin remains. As noted earlier, preservation of fossils in Africa, Arabia, and southern Asia is sparse in part because of prevailing aridity. And for hominin bones made of easily dissolved calcium phosphate (Ca_3PO_4), preservation is even worse in the acid-rich soils of the rain forests. Because of these problems, the total record of human evolution over 5 million years is based on a few dozen fragments of skeletons, enough to reveal some of the broad outline of human evolution but few of the details.

We have already seen that sampling records of orbital-scale climate change even at intervals as close as a few thousand years can lead to gross misrepresentations of the shape of the actual climate signals (Chapter 7). This aliasing problem becomes all the more formidable in records with just one sample every 100,000 years or so.

Aliasing can produce erroneous indications of the time of first acquisition of new physical or technical skills (such as walking or the use of tools and fire) or of the first or last appearance of a new hominin species. With only a few samples, the actual first or last appearances are likely to be missed, and instead we see a much-reduced range (Figure 15–7). As a result, the true ranges of most hominins are probably longer than those shown in Figure 15–2. Similarly, aliasing also complicates attempts to define the relative timing between climate changes and the first use of new skills.

A second undersampling problem pertains to quantitative measurements of the evolution of physical traits. In this case, the fundamental problem is that a broad

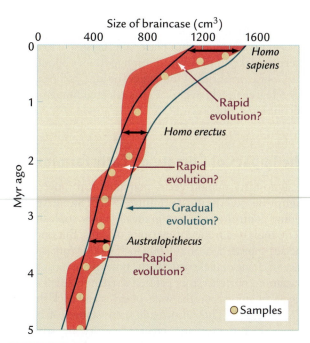

FIGURE 15-8 Undersampling of a measurable characteristic If a slowly evolving characteristic, such as the size of the human brain, varies widely at all points in time (black double arrows), scattered sampling (yellow circles) will permit an interpretation of either gradual or rapid evolution.

range of natural variation occurs within human (or pre-human) populations. For example, your classmates have heads that vary widely in size around the average value for the entire class. The size of a human head is closely tied to the size of the braincase (the part of the skull that cradles the brain), a key trait in the fossil record of human evolution.

The actual range of braincase sizes present at any one time in the past is fairly large in comparison with the evolution of the mean value (Figure 15–8). If the fossil record provides only one or two well-preserved specimens every hundred thousand years or so, a good chance exists that some of these specimens will not be representative of the brain size of the entire population living at that time but will fall above or below the mean. Depending on the specific samples collected and analyzed, an inaccurate picture of the long-term trend could emerge.

Even if the available fossil record is sparse, it should be possible to obtain a general sense of the direction in which the trend is moving, especially if the net amount of evolution far exceeds the natural range of variation at any one time. But the limitations of sparse data make it impossible to define the true rates of change. Figure 15–8 shows that sparse data on brain size could be interpreted equally well as either a slow, gradual trend or as rapid bursts of change.

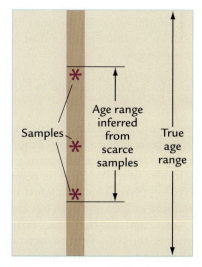

FIGURE 15-7 Undersampling of a fossil record If few samples of the fossil record of an organism are available, the true first and last appearances of the organism will be poorly estimated.

IN SUMMARY, the overall outline of human evolution has gradually fallen into place, but the sparse fossil record makes it difficult to test hypotheses about the cause or causes of human evolution. As a result, the degree to which climate has affected human evolution is unknown.

The Impact of Climate on Early Farming

In almost 2.5 million years, hominins had moved only slightly beyond the most primitive level of Stone Age life, adding control of fire and gradually more sophisticated stone tools to a meager repertoire of skills. But once our species appeared, near 200,000 to 150,000 years ago, the pace of change quickened. By the time of the most recent glaciation, our ancestors painted amazingly lifelike portrayals of animals on the walls of caves and rock shelters (Figure 15–9). They also made small statues of human and animal figures and created jewelry by stringing together shells, and they buried their dead with food and possessions for use in a future life. In these changes, we can recognize the early origins of a true human "culture."

More sophisticated stone tools designed for specific functions also appeared, and, for the first time, people began to use bone, a much more "workable" substance than stone, yet hard enough for many uses: as needles, awls (hole-punchers), and engraving tools. Needles made possible sewn clothing that fit closely, rather than loosely draped animal hides. With greater protection from the elements, people pushed northward into higher

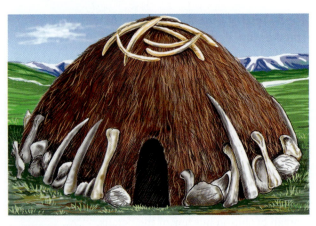

FIGURE 15-10 Early human buildings Humans living in northern Asia during the last glaciation constructed domed dwellings of hides draped over mammoth bones. Other bones served as anchors.

and colder near-Arctic latitudes. There they built dome-shaped houses with large mastodon bones supporting the superstructure and animal hides draping the roof for protection from rain and snow (Figure 15–10).

People also learned how to make rope from naturally available fibers and used the rope to make snares and lines and nets to catch small animals, birds, and fish. The hunting-gathering life became a hunting-gathering-fishing life. People shaped bone into spear throwers that held stone spear points and used rope to help bind the spear points to the shafts. This new technology produced a lethal and revolutionary new way of hunting that combined a weapon that could kill efficiently with a hand that could grasp it and an arm with the natural range of motion to throw it. Hunters could now bring down larger game—even mammoths—from a safer distance, as cave paintings show. By the start of the present interglaciation, an explosive alteration of basic human existence was underway.

15-4 Did Deglacial Warming Lead to Early Agriculture?

The first evidence of agriculture dates to just over 12,000 calendar years ago in a region of the Middle East called the **Fertile Crescent,** encompassing present-day Syria, Iraq, Jordan, and Turkey. The people living in this region (called Natufians) abandoned the hunting-and-gathering way of life and began to cultivate wheat, rye, barley, peas, and lentils rather than harvesting grains growing in the wild. Because agriculture eliminated the need for seasonal migrations to search for food, these people took up residence in permanent dwellings. Within 1000 years, the dwellings began to cluster into permanent village settlements.

FIGURE 15-9 Cave painting of the glacial era Despite the harsh glacial climate, our ancestors left beautiful, almost modern-looking paintings on the walls of caves in southern Europe. (Ferrero/Labat/Auscape International.)

Evidence of cultivation in the Fertile Crescent is derived from preserved remains of grains found in regions where the grains did not naturally grow and where their presence must have been aided by human efforts. Evidence of permanent occupation of villages comes from the dental remains of animals from the settlements. Layering in the teeth of these animals indicates the season when they died. Because the animals were killed in all seasons, the people must have stayed in the same place throughout the year. By 10,000 years ago, people had begun to domesticate cattle and other livestock in the Near East. Near the same time, people also began to grow barley and other crops in northern China.

Because of the close association in time between the later stages of the deglaciation and the origin of agriculture, several cause-and-effect links have been proposed. One seemingly plausible link is the possibility that the change from the harsh (colder and drier) glacial climate to the more accommodating (warmer and wetter) climate provided conditions more favorable for humans to begin the grand experiment of growing crops.

On the other hand, climatic data have been used as the basis for a totally different hypothesis that centers on the Younger Dryas climatic reversal between 13,000 and 11,700 years ago (Chapter 13). According to this idea, the Younger Dryas episode intensified the already dry conditions across the eastern Mediterranean region and forced people to retreat to dependable water sources. In these more closely clustered conditions, people who harvested and ate wild grains may have accidentally scattered some grains near their threshing sites, with the discarded grains sprouting in succeeding years as a form of primitive farming. Some evidence places the time of the earliest domestication of crops during the Younger Dryas.

Neither of these directly opposed hypotheses is easy to test. One problem is that agriculture may have begun earlier than the record indicates because the record is still incomplete (see Figure 15–7). Another problem is that the beginnings of agriculture in each region on Earth were one-of-a-kind events. Many such events, each related to a similar change in climate, would be required for a cause-and-effect relationship to be really conclusive.

15-5 Impacts of Climate on Early Civilizations

Climate change has been hypothesized as the cause of or at least a major factor contributing to the deterioration or collapse of early civilizations. One hypothesis focuses on the role of an early flood (Box 15–1), but drought is a more commonly invoked factor. In low-latitude regions where water was scarce, civilizations were more susceptible to drought than to changes in temperature.

The first advanced civilizations of the early Egyptian dynasties developed between 6000 and 5000 years ago, when the monsoon was still considerably stronger than it is today. Then and now, Egyptian life centered on the river Nile, fed by monsoon rains in the Ethiopian highlands and flowing northward through hyperarid desert (Chapter 8). When the Nile ran strong, large floods provided fertile soils and moisture for farming along the floodplain.

Climate in sub-Saharan North Africa turned much drier after 5000 years ago as the summer monsoon weakened. This drying trend affected the civilizations that had come into existence and grown in size during the wetter monsoon climates in the preceding millennia. The weakening of the summer monsoon after 5000 years ago greatly reduced the extent of summer flooding of the Nile. This change put greater stress on populations that had expanded in response to the stable food supply from large crop yields in a monsoonal climate.

The Akkadian empire, centered in what is now Syria, was the dominant civilization in Mesopotamia until 4200 years ago. Evidence from archeological investiga-

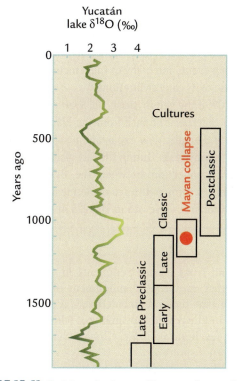

FIGURE 15-11 Did drought destroy Mayan civilization? Lake sediments indicate periods of prolonged drought during the time that Mayan civilization disappeared. (Adapted from J. H. Curtis, D. A. Hodell, and M. Brenner, "Climatic Variability on the Yucatán Peninsula (Mexico) During the Past 3500 Years, and Implications for Maya Cultural Evolution," *Quaternary Research* 46 [1996]: 37–47).

BOX 15-1   **CLIMATE DEBATE**

Sea Level Rise and Flood Legends

Many early cultures have a flood legend, a story about a great flood that swept away earlier people. Many of these legends share several features: a deluge sent by a higher authority as punishment for the sins of the people, a warning that a flood was coming, and preflood advice to gather up every kind of animal on a large vessel to preserve all forms of life in a postflood world. The story of Noah in the Hebrew Old Testament is the most widely known flood legend, but much the same story is found in the older Babylonian tale of Gilgamesh and in legends of other early cultures in the Old World. Storytellers passed these legends down over many generations.

In the eighteenth and nineteenth centuries, scientists attempted to reconcile the biblical story of the flood with their scientific knowledge by advocating the **diluvial hypothesis**. This hypothesis called on a great worldwide flood to explain the widespread deposits of unsorted debris (everything from clay to boulders) strewn across the northern continents. Today we recognize these deposits as moraines left by the retreating ice sheets. Scientists who now interpret the biblical flood legend less literally have continued to search for evidence of a major regional-scale flooding event, and many of them focus on the Near East because of its early civilizations.

In 1998 the geophysicists Bill Ryan and Walter Pitman pulled together evidence from the Black Sea region and posed a dramatic new explanation called the **Black Sea flood hypothesis**. During the last glacial maximum, the present connection between the Black Sea and the Aegean Sea through the Sea of Marmara in Turkey did not exist. Global sea level stood 110–125 m lower than it does

today, and the level of the Aegean, linked to the global ocean as part of the Mediterranean Sea, lay well below the threshold needed to connect the two bodies of water. A freshwater lake rimmed by reedy swamps covered a much smaller part of the basin than the modern Black Sea.

Later, as the very last portions of the great northern hemisphere ice sheets melted, the meltwater they returned to the ocean pushed the rising Aegean Sea into the Sea of Marmara, which then spilled into the freshwater lake (modern Black Sea). Radiocarbon dates in sediment cores from the Black Sea suggested that an abrupt transition occurred near 7600 calendar years ago. Freshwater mollusks that lived in a lake were replaced by mollusks and plankton that lived in salty ocean water. Within a short interval, it seemed, incoming seawater transformed this freshwater lake into a salty inland sea.

Ryan and Pitman also found a huge gorge lying buried underneath a thin layer of sediments at the bottom of the strait known as the Bosporus, between the Sea of Marmara and the Black Sea. Cut into bedrock, this gorge is clear evidence that an enormous flow of water poured into the Black Sea in the past. The coarse sediments along the floor of the gorge (sand, pebbles, cobbles, and even boulders) are arrayed in great dunelike shapes tilted toward the north, consistent with a flow from the Aegean Sea into the Black Sea. At the place where the Bosporus meets the Black Sea, a deep pool is cut into the underlying rock, apparently carved by an immense waterfall created by a torrent of incoming water.

Ryan and Pitman concluded that the rise of floodwaters caused by the torrent of seawater entering the

tions and from marine sediments in the Persian Gulf shows that an abrupt abandonment of major northern cities in that region coincided with a period of intense aridity and increased movement of windblown dust that lasted a few hundred years. Climate change is thought to have been responsible.

Another example is the sudden collapse of the Mayan civilization on the Yucatán Peninsula in Mexico in A.D. 860. Evidence from lake sediments indicates that this dislocation occurred during an interval of severe drought (Figure 15–11). A civilization that had built cities and massive stone temples, carved monumental statues out of stone, and developed a system of writing suddenly abandoned its inland cities and scat-

tered. Some of the people moved northward to coastal regions, perhaps because shallow groundwater was still accessible in such regions. Even these coastal populations dwindled.

Near 1300, the Anasazi people, who had created and occupied beautiful cave dwellings cut into the sides of cliffs in the southern Colorado Plateau, abruptly abandoned the entire region. Evidence from tree ring studies indicates that their sudden departure occurred during an interval of drought.

Depletion of resources is also a plausible competing explanation for some of these cultural changes. For example, although the Anasazi did abandon their cliff dwellings in the American Southwest during a drought,

BOX 15-1   CLIMATE DEBATE

CONTINUED

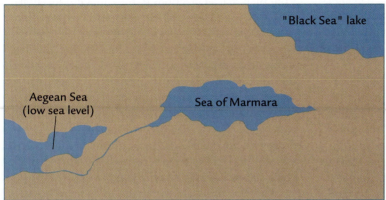

A Last glaciation

Black Sea flood Slowly melting ice sheets caused the Aegean Sea to rise until it flooded into a freshwater lake that was located in the region of the modern Black Sea. (Adapted from W. Ryan and W. Pitman, *Noah's Flood* [New York: Simon & Schuster, 1998].)

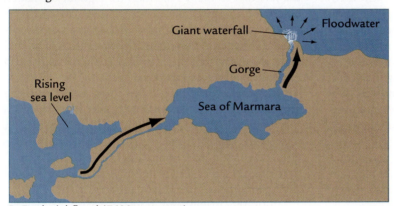

B Deglacial flood (7600 years ago)

Black Sea 7600 years ago would have inundated settlements on the shores of the freshwater lake within a single year, displacing thousands of people. Subsequent investigations of this region have suggested that some of the flooding of the modern Black Sea region occurred 7600 years ago but that some of the flooding occurred earlier, near 9400 years ago, perhaps because of runoff from the meltwater lakes to the north. The ancient flood legends may have their origin in the sea level rise at the end of the last deglaciation.

previous dry intervals of nearly comparable intensity had not driven them from the region. Abandonment also coincided with the near disappearance of tree pollen from climate records, indicating that the Anasazi had cut down the juniper and piñon trees previously used as fuel for cooking and for winter warmth on the high, cold plateau. If so, depletion of this crucial resource may also have been a major factor in the abandonment.

Similarly, the Mayas of Central America may have altered their own local environment. They may have contributed to regional drought by cutting trees and reducing the positive moisture feedback from evapotranspiration. Alternatively, their farming methods may have exhausted the limited supply of nutrient-rich soils in a region of limestone bedrock, making agriculture difficult or impossible. Wars with neighboring cultures may also have been a factor, along with disease. Isolating climate change as the sole cause of changes in early civilizations is often difficult.

Early Impacts of Humans on Climate

Eventually, the relationship between humans and their environment began to change. Instead of being passive players, humans began to actively alter their environment and Earth's climate.

15-6 Did Humans Cause Megafaunal Extinctions?

Populations of large mammals, called **megafauna**, decreased drastically late in the most recent glacial oscillation. Prior to 50,000 years ago, more than 150 genera of mammals larger than 45 kg (~100 pounds) existed. By 10,000 years ago, 50 or fewer genera were left. This interval of extinctions was unprecedented compared to millions of years of prior history.

During the most recent glaciation, near 50,000 years ago (within the uncertainties of the dating methods), many of the larger marsupials in Australia became extinct within a few millennia, including various kinds of kangaroos and wombats and a lion, as well as non-marsupials such as giant tortoises and flightless birds. Just before this time, humans had first entered Australia from southeast Asia, helped by the exposure of land masses by lowered sea level. These people, who used "fire sticks" to burn grasslands and drive game, have been proposed as the cause of these extinctions.

The megafaunal population of the Americas was very diverse until the late stages of the most recent deglaciation, with a rich array of mammals in North America (Figure 15–12). Then, within an interval of a few thousand years centered on 12,500 years ago, over half of the large mammal species living in both North and South America became extinct. The list in North America includes giant mammoths and mastodons (larger than modern elephants), horses the size of modern Clydesdales, camels, giant ground sloths, saber-toothed tigers, and beaver as large as modern bears.

One explanation for this rapid pulse of extinction is that major climate changes at the end of the glacial maximum created new environmental combinations to which many mammals were unable to adapt. These conditions included strong summer warming and drying of the land south of the ice sheets by high summer insolation, reduction of habitat in the cooler north by the slow retreat of the ice sheet, and unusual (no-analog) mixtures of vegetation that developed as forests and grasslands shifted from their glacial positions to their modern locations.

Critics of this climatic hypothesis note that no comparable pulse of extinction occurred in any of the 50 or so earlier deglaciations. In fact, the number of species that went extinct near 12,500 years ago exceeds the total during all of the previous 2.75 million years. The same basic combination of climatic conditions—high summer insolation, rising CO_2 levels, and rapidly melting ice sheets—had occurred during earlier deglaciations without causing pulses of extinction. So the critics ask why the extinctions occurred only during this one deglaciation and not the others.

Another vulnerability of the climatic hypothesis is the fact that so many mammals throughout the Americas, living in environments that ranged from semiarid grasslands to rain forests, suffered the same fate. Because these different environments followed different climatic paths during deglaciation, climate change cannot possibly explain all the extinctions.

A second explanation, called the **overkill hypothesis**, put forward by the paleoecologist Paul Martin, is that human hunting caused this extinction pulse. The immediate cause of the extinctions could have been either the first arrival of humans in the Americas or the first appearance of a new hunting technology or strategy among the people already present.

Both the origin and time of arrival of the first humans in the Americas were once thought to have been resolved. They supposedly came by land from Asia near 12,500 years ago, during the late stages of the last deglaciation. They crossed into Alaska over a land bridge in the Bering Strait exposed by the lower glacial

FIGURE 15–12 Mammals of the glacial maximum A rich array of large mammals lived on the North American plains prior to the most recent deglaciation, including several forms that became extinct: woolly mammoths, saber-toothed tigers, and giant ground sloths. (Courtesy of Smithsonian Institution, Washington, DC; painted by Jay Matternes.)

sea level and moved down the interior of North America east of the Rockies. They passed through the ice-free corridor opened by early melting and separation of the Laurentide ice sheet to the east and the smaller Cordilleran ice sheet over the Canadian Rockies (see Figure 13–2).

This view is now in dispute. Scattered but still disputed evidence hints at the arrival of humans 30,000 or 40,000 years ago, although most undisputed ^{14}C dates still support a much later arrival. Another challenge is that these people could have arrived by water, either traveling along northeastern Pacific coastlines or crossing at lower latitudes from eastern Asia.

Whatever the date of the first humans in the Americas, a new hunting technology appeared at the same time that the extinctions occurred (12,500 years ago). Many archeological sites that date near 12,500 calendar years ago contain spears fitted with a new and elegant kind of point fashioned by humans (Figure 15–13). This new technological development could have helped people hunt large mammals more effectively.

FIGURE 15-13 Pulse of mammal extinctions Woolly mammoths and other large mammals abruptly became extinct in North America near 12,500 years ago. Distinctive grooved spear points ("Folsom points," named for the site in New Mexico where they were first found, shown here with bones) suddenly appeared during this interval of widespread extinction. (Denver Museum of Natural History.)

One criticism of the hunting hypothesis is that people were too few in number to have caused so many extinctions, but studies from population models refute this criticism. Because reproduction (gestation) times for large mammals are long, hunters only need to cull an extra 1–2% of a species per year to drive them extinct within a single millennium. In addition, these people worked in groups to drive animals to their death over steep cliffs at the end of narrow bluffs. So many animals were killed that only a fraction were used for food and clothing.

Another criticism of the hunting hypothesis is that some creatures that do not seem likely to have been hunted also went extinct, including large meat-eating mammals that may have preyed on humans rather than becoming their prey. A plausible response is that carnivores that depended on the carcasses of large mammals for food may have gone extinct because much of their natural prey had gone extinct at the hands of humans. A still-unanswered criticism of the hunting hypothesis is that it does not explain why some large mammals that would seem to have been likely targets for hunters (moose, musk ox, and one species of bison) survived.

IN SUMMARY, both explanations of the pulse of extinctions have their critics, but the absence of any extinction pulse during all the previous ice-age cycles is a powerful argument against the hypothesis that climate was responsible. The cause appears to have been humans.

15-7 Did Early Farmers Alter Climate?

Introducing a controversial **early anthropogenic hypothesis**, the marine geologist William Ruddiman claimed that early agriculture had a substantial impact on greenhouse gases and on global climate thousands of years ago, much earlier than previously thought. He based this claim on the fact that concentrations of CO_2 and methane had fallen during the initial stages of the last four interglaciations, but they instead rose during the later part of the current interglaciation (Figure 15–14). Because the rest of this book will take an increasingly historical approach, the axes of all time plots from this point on have been rotated to the typical historical perspective: younger to the right (and warmer upward).

Deforestation is the proposed explanation of the anomalous rise in CO_2 that began near 8000 years ago. Stone Age humans with flint axes began to cut the forests of Europe, China, and India to create clearings for growing crops. The first appearance of cereal grains and other crop remains in hundreds of ^{14}C-dated lake

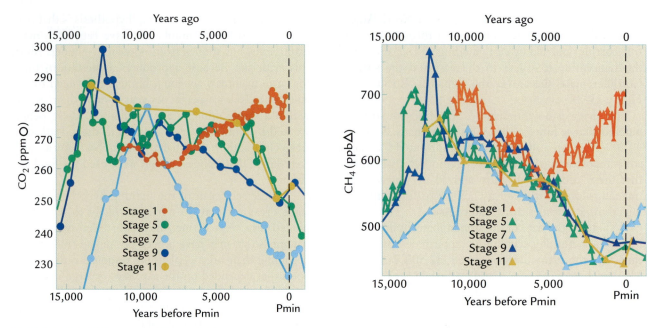

FIGURE 15-14 Wrong-way CO₂ and CH₄ trends Atmospheric concentrations of CO$_2$ and CH$_4$, which fell during previous interglaciations, instead rose during the current one. (Adapted from W. F. Ruddiman, "The Anthropogenic Greenhouse Era Began Thousands of Years Ago," *Climatic Change* 61 [2003]: 261–93.) Interglaciations aligned on insolation minima (Pmin) like the one today.

cores show the spread of agriculture from its place of origin in the Fertile Crescent to adjacent regions (Figure 15–15). Clearance of forests in southeast Europe began just after 8000 years ago, the same time that the CO$_2$ curve began its anomalous rise. By the start of the Bronze Age, 5500 years ago, agriculture had spread into virtually every part of Europe where farming is practiced today. Deforestation was also underway in China and India during this same interval. Whether the trees were burned or left to rot, CO$_2$ was delivered to the atmosphere.

Initially, farming occurred in small clearings, but by 2000 years ago, large parts of China, Southern Europe, and India had already been deforested. A survey of England in 1086 (the Domesday Book) found that 85% of the arable land was in pasture or crops, with only 15% still in forest. This gradual clearance of Eurasian forests contributed to the slow rise in CO$_2$ concentrations over the last 8000 years.

The methane concentration began its anomalous rise somewhat later, near 5000 years ago (see Figure 15–14). This increase could not have come from tropical

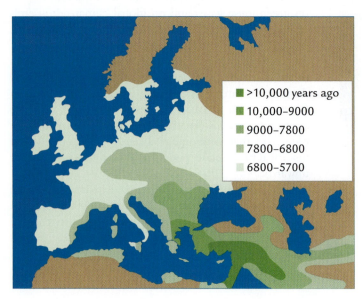

FIGURE 15-15 The spread of agriculture The practice of agriculture originated in the region north and east of the Mediterranean and gradually spread into Europe, North Africa, and other parts of Asia. (Adapted from N. Roberts, *The Holocene,* 2nd ed. [Oxford: Blackwell, 1998].)

wetlands because they were shrinking as the monsoon weakened (see Figures 13–13 and 13–16). Other studies indicate that the boreal wetlands could not have been the source of the extra methane, in part because Arctic summers were growing colder (see Figure 13–19), and in part because the wetlands in that region were slowly changing to a type that emitted less methane. Neither tropical nor boreal wetlands make sense as the cause of the anomalous methane rise.

Humans are the other possibility. Several human activities generate methane, including biomass burning, tending of livestock, and production of human waste, but the abrupt reversal in the methane trend seems to require a more dramatic explanation. Rice irrigation began in southeast Asia between 7000 and 6000 years ago and expanded rapidly by 5000 years ago, the same time the methane concentration began to rise. Rice and weedy plants grow in standing water and then die and rot. As the carbon in the plant remains is oxidized, the water loses its oxygen and begins to emit methane (with reduced carbon) into the atmosphere.

Irrigation had spread across all of Southeast Asia from China to India by 3000 years ago, as the CH_4 trend continued to rise. By 2000 to 1000 years ago, the Asian people were beginning to construct small rice paddies even on steep hillsides (Figure 15–16). The fact that so much effort was being put into adding so little cultivated area suggests that many flat valley floors were already in irrigation by this time. Rice is a highly nutritious food, and the success of Asians in growing this crop must have led to rapid increases in population. By 2000 years ago, census data show that 50 million people already lived in China. Emissions from livestock, biomass burning, and human waste must have risen accordingly.

Critics of the early anthropogenic hypothesis question whether human activities can explain the large anomalies in CO_2 (35–40 ppm) and methane (~225–250 ppb) that had developed prior to the 1800s (see Figure 15–14). Underlying this criticism is the reaction that too few humans were present on Earth to have taken control of atmospheric greenhouse-gas trends so many millennia ago. This criticism appears to have particular merit in the case of the large CO_2 anomaly, which cannot be accounted for even if all the forests of southern Eurasia had been cut by a few centuries ago. And yet, the anomalies remain, so they still require an explanation. One possibility now being explored is that positive feedbacks in the climate system have amplified the CO_2 emissions by humans during the last several thousand years.

Attempts at natural explanations face a seemingly more daunting problem. If natural processes in the climate system caused gas concentrations to fall early in the last four interglaciations (all of which occurred prior to agriculture), how can the same natural processes be

FIGURE 15-16 Hillside rice paddies By 2000 years ago, Asians were constructing rice paddies on steep hillsides like these in Guizhou Province, China. (Georg Gerster/Photo Researchers, Inc.)

called on to explain the rise in gas concentrations during this interglaciation? If the natural trend is down, then an upward trend cannot be natural.

Another implication of the early anthropogenic hypothesis is that the anomalous rises in CO_2 and CH_4 kept the atmosphere warmer than it would have been if nature had remained in control. Throughout the last several millennia, the decrease in summer insolation across the northern hemisphere cooled the high northern latitudes (see Figure 13–19). If the greenhouse-gas concentrations had fallen as they had done during the previous interglaciations, climate would have cooled even more, perhaps to the point of allowing new ice caps to form.

Sensitivity tests with general circulation models support this idea. If greenhouse-gas concentrations are reduced to the "natural" level predicted by the hypothesis (~240 ppm for CO_2 and ~450 ppb for CH_4), some

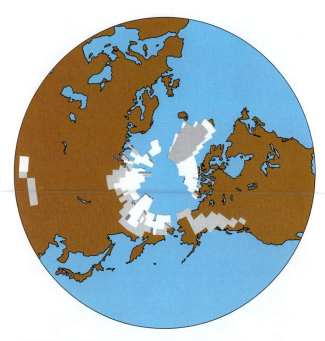

FIGURE 15-17 Is a glaciation overdue? Model simulations with present insolation but with CO_2 and CH_4 concentrations lowered to the natural levels they reached during previous interglaciations produce year-round snow cover (incipient glaciation) in northern Eurasia and North America. (Adapted from S. J. Vavrus, W. F. Ruddiman, and J. E. Kutzbach, "Climate Model Test of the Early Anthropogenic Hypothesis Using a Coupled Atmosphere-Slab Ocean Model," *Quaternary Science Reviews*, in review.)

regions at high northern latitudes would retain snow throughout the year (Figure 15–17). Because snow that persists through the summer and remains in autumn forms a base that can be added to during the following winter, simulating year-round snow cover is equivalent to simulating a state of incipient glaciation. Different models put year-round snow in different regions, including the Arctic margins of Canada and Russia, Baffin Island northeast of the Canadian mainland, and the northern Rockies. These model-to-model differences are the result of differences in the feedback processes included, particularly vegetation/albedo feedback and ocean dynamics.

Despite the differences, all the simulations agree that a new glaciation would be underway now if humans had not interfered with the natural operation of the climate system. Each of the new ice caps scattered across the north would probably be relatively small in size, but their aggregate area might be considerable, perhaps equivalent to that of modern Greenland ice sheet.

Key Terms

savanna hypothesis (p. 276)

variability selection hypothesis (p. 278)

Fertile Crescent (p. 280)

diluvial hypothesis (p. 282)

Black Sea flood hypothesis (p. 282)

overkill hypothesis (p. 284)

early anthropogenic hypothesis (p. 285)

Review Questions

1. What are the problems with the savanna hypothesis of human evolution?

2. Why is it difficult to determine whether or not climate affected human evolution?

3. Why is the climatic explanation of the large-mammal extinctions 12,500 years ago suspect?

4. Why are natural explanations for the CO_2 and CH_4 increases in recent millennia suspect?

Additional Resources

Faces of Earth (DVD set, 2007). Discovery Communications and American Geological Institute.

Basic Reading

Diamond, J. 1997. *Guns, Germs and Steel*. New York: Norton.

Roberts, N. 1998. *The Holocene*. Oxford: Blackwell.

Ruddiman, W. F. 2005. *Plows, Plagues and Petroleum*. Princeton, NJ: Princeton University Press.

Ryan, W., and W. Pitman. 1998. *Noah's Flood*. New York: Simon & Schuster.

Advanced Reading

Curtis, J. H., D. A. Hodell, and M. Brenner. 1996. "Climatic Variability on the Yucatán Peninsula (Mexico) During the Past 3500 Years, and Implications for Maya Cultural Evolution." *Quaternary Research* 46: 37–47.

Martin, P. S., and R. G. Klein. 1984. *Quaternary Extinctions*. Tucson: University of Arizona Press.

Potts, R. 1997. *Humanity's Descent: The Consequences of Ecological Instability*. New York: Avon.

Ruddiman, W. F. 2003. "The Anthropogenic Greenhouse Era began thousands of Years Ago." *Climatic Change* 61: 261–93.

CHAPTER 16

Climate Changes During the Last 1000 Years

Detailed records of climate change during the last 1000 years come from proxy indicators stored in archives such as mountain glaciers, tree rings, and corals. Historical observations systematically recorded by humans over several centuries are also available from a few regions, but instrumental measurements began only in recent centuries.

Climate changes over the last thousand years were small and variable in pattern from region to region until the large warming of the last century. In high northern latitudes surrounding the North Atlantic Ocean, a small cooling occurred between 1000 years ago and the time known as the **Little Ice Age** (1400–1900 years ago). The scope of this cooling elsewhere on Earth remains uncertain, as does the cause.

Little Ice Age

As the last millennium began, scattered evidence from Europe and high latitudes surrounding the North Atlantic indicates a time of relatively warm climate known as the **medieval warm period**. During this interval (1000–1300), Nordic people settled southwestern Greenland along the fringes of the ice sheet and managed to grow wheat. Sea ice, common during the past century around the Greenland coasts, is rarely mentioned in chronicles from this era.

The subsequent cooling into the Little Ice Age (1400–1900) affected people in northern Europe. With colder winters and a shorter growing season, grain crops and grape harvests repeatedly failed in far northern regions where they had been successfully grown during the medieval warm period, resulting in localized famine. Lakes, rivers, and ports froze throughout northern Europe during severe winters. Near the onset of the Little Ice Age, the settlements in Greenland were abandoned, in part because the marginal climate had become inhospitable but also perhaps because of conflicts with native Arctic peoples.

The mountains of Europe hold dramatic evidence that climate was cooler than it is today. People who lived in the Alps of Switzerland and Austria and the mountains of Norway in the fourteenth and fifteenth centuries experienced large-scale advances of glaciers firsthand. They wrote about the effects of these episodes on their lives and sketched the advancing ice. Many of the ice advances spread over alpine meadows where livestock had once grazed, and they destroyed farmhouses and even small villages where people had once lived. These historical expansions also prompted the glacial geologist Louis Agassiz to propose a theory that much larger ice ages had occurred tens of thousands of years earlier.

Other evidence confirms that temperatures were cooler in Europe and the nearby North Atlantic during the Little Ice Age. One documentary record shows the frequency of sea ice along the north and west coasts of Iceland (Figure 16–1). Because fishing was vital to the food supply on this isolated island, records were kept of times when coastal sea ice made it impossible for ships to go to sea. This record shows the number of weeks per year in which sea ice reached and blocked the northern coast of Iceland.

Sea ice appears to have been infrequent until 1600, although records kept before that date may be less reliable than later ones. It increased in frequency and reached a maximum in the nineteenth century but then all but disappeared from the coasts during the twentieth century except in occasional extreme years. This record indicates a major contrast near Iceland between the cold conditions of the Little Ice Age and the relative warmth of the twentieth century.

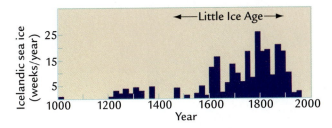

FIGURE 16-1 Sea ice on the coast of Iceland The frequency of sea ice along the coast of Iceland increased into the nineteenth century and then declined rapidly during the twentieth century. (Adapted from H. H. Lamb, *Climate—Past, Present, and Future*, vol. 2 [London: Methuen, 1977].)

Because major ice sheets did not actually form, the Little Ice Age was not a true ice age, but in some regions it was at least a small step in that direction. Much of the Canadian Arctic is a forbidding place with long, cold winters and short, chilly, mosquito-infested summers. In a few locations, small ice caps that persist at or near sea level are melting rapidly in the warmth of the modern climate. Rock outcrops in these regions are covered by **lichen,** a primitive, mosslike form of vegetation that can live on bare rock surfaces even under inhospitable conditions (Figure 16–2A). These organisms secrete acids that attack bedrock and break it down into mineral grains that provide nutrients. Small lichen growing in the Canadian Arctic today can be dated by their size (Figure 16–2B). They begin life as small specks and then expand into round blobs at predictable rates.

Scattered across portions of Baffin Island northeast of the Canadian mainland are broad halos of dead lichen surrounding small, modern ice caps and also occurring in high areas that now lack ice caps (Figure 16–2C). These halos of dead lichen are a product of the Little Ice Age. The size of the lichen indicates that they must have developed over warm intervals of several hundred years in the relatively recent past and subsequently been killed.

Because lichen are tolerant of extreme cold, it is unlikely that frigid temperatures killed them. A more likely explanation is burial beneath snowfields that blocked sunlight through the summer growing season. If summer melting failed to remove the previous winter's snow for many years in succession, the lichen would have died for lack of photosynthesis. The lichen halos are an indication that growing snowfields covered larger areas of high terrain on Baffin Island in the fairly recent past.

When did the lichen die off? The small, young lichen now growing on top of the dead ones indicate that they have been present only during the last century or so—since the end of the Little Ice Age. Expanded

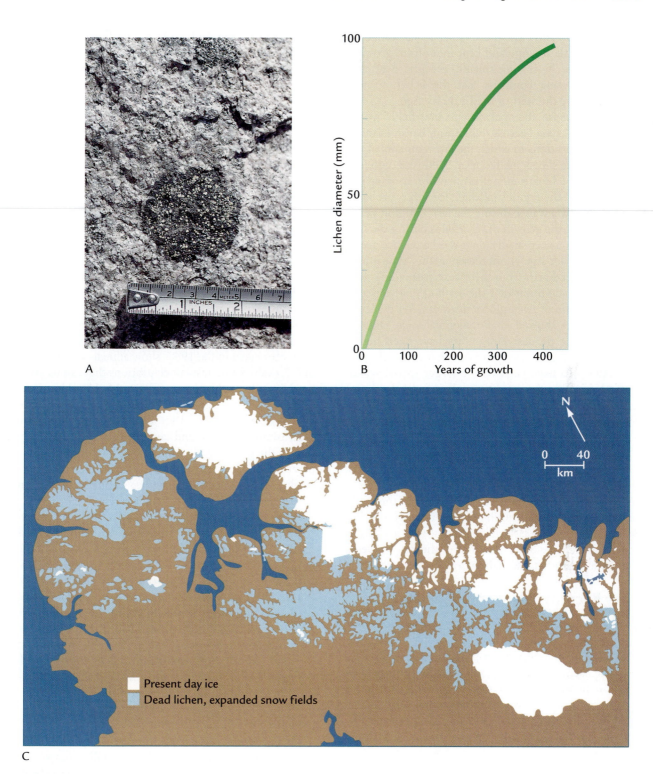

FIGURE 16-2 Lichen halos: a Little Ice Age snowfield? Lichen grow (A) on rock surfaces in the Arctic at (B) known rates. (C) Halo-like areas of dead lichen in the Canadian Arctic record an interval of expanded snowfields late in the Little Ice Age. (A: Courtesy of G. Falconer, Sidney, BC. B: Adapted from G. H. Denton and W. Karlen, "Holocene Climatic Variations: Their Pattern and Possible Cause," *Quaternary Research* 3 [1973]: 155–205. C: Adapted from J. T. Andrews et al., "The Laurentide Ice Sheet: Problems of the Mode and Inception," wmo/imap Symposium Proceedings 421 [1975].)

snowfields apparently killed the older lichen sometime during the Little Ice Age and kept them from growing back until late in the nineteenth century. Scattered radiocarbon dating confirms that the lichen formed until just before the start of the Little Ice Age.

It makes sense that permanent snowfields would have begun to expand across this region. Baffin Island is one of three locations in which the last remnants of the great glacial ice sheets melted near 6000 years ago (see Figure 13–2). Climate model simulations also suggest that it is one of the regions in which ice sheets probably began each new advance during the last 3 million years. In this context, the growth of snowfields during the Little Ice Age was a very small step toward a real ice age.

The Little Ice Age is often used as the "type example" of a cooler Earth from the recent past, along with the implicit assumption that the cooling that caused it was hemispheric or even global in extent. Yet evidence in Chapter 14 showed that millennial-scale climatic changes during the last 8000 years have been highly variable from region to region and that no single pattern appears to have persisted on larger spatial scales. From this point of view, the cooling in Iceland and nearby areas of the North Atlantic Ocean might just as easily have been a local phenomenon restricted to just this one region.

The next section examines several methods that provide climate records spanning the last 1000 years. The purpose of this exploration is to see what these methods indicate about the regional extent of the Little Ice Age cooling. By examining climate records covering major portions of the last millennium, we can also begin to address the issue of whether or not the last 100 years of climate change fit into natural longer-term trends.

Proxy Records of Historical Climate

Before the nineteenth century, people rarely kept records of temperature or precipitation. As a result, climate scientists have to rely mainly on proxy records in archives like those used for earlier intervals of Earth's history, such as lake sediments and ice cores. In addition, sources of proxy records such as tree rings and corals can also be used for this interval. These various archives provide limited (but steadily improving) coverage of climate changes across Earth's surface during the last few centuries.

16-1 Ice Cores from Mountain Glaciers

Like the huge ice sheets on Antarctica and Greenland, glaciers in mountain valleys and small ice caps covering mountain summits are also excellent climate archives (Figure 16–3). Some mountain ice dates back many thousands of years into the last glaciation, while other glaciers span only a few hundred years of climate history. Layers deposited at the surface usually contain

annual signals (Figure 16–3A), but ice flow may degrade the resolution deeper in the ice.

Retrieving ice cores from mountains is a formidable task. Heavy equipment (including solar-powered ice drills) must be hauled up to subfreezing mountain summits (Figure 16–3B). Mountain ice caps are between 100 and 200 m thick, and most expeditions drill and sample the entire thickness of ice at several locations on each ice cap (Figure 16–3C and D). Lack of oxygen at these altitudes quickly causes exhaustion and other problems. The ice cores that are extracted must be lugged down to lower elevations and kept from melting in the warmer air.

Only a few mountain glaciers have been cored. By far the most extensive efforts have been those by the intrepid glacial geologist Lonnie Thompson. His expeditions have retrieved ice cores at elevations of 5670 m (about 18,500 ft) above sea level from the Quelccaya ice cap in the Peruvian Andes. By counting annual layers and matching volcanic ash layers with historically documented eruptions, he found that these records extend back to 1500 years ago.

Cores taken in the 1980s show annual-scale changes in $\delta^{18}O$ values and dust concentrations that can be averaged over decadal intervals (Figure 16–4A). The variations in $\delta^{18}O$ values reflect the same processes as those affecting continent-sized ice sheets: changes in source area, transport paths, and amount of water vapor carried to the glacier, and particularly changes in the temperature at which the snow condenses above the ice. Higher dust concentrations indicate some combination of drier source areas and stronger winds.

Early cores from the Quelccaya glacier record registered a shift toward more positive $\delta^{18}O$ values and less dust near 1900, implying that a change toward some combination of warmer temperatures, weaker winds, and different source areas occurred at that time (see Figure 16–4A). This $\delta^{18}O$ record also resembles the Little Ice Age pattern shown in Figure 16–1, with more positive (warmer?) values from 1000 to 1400 and then more negative (cooler?) values after 1500. In contrast, the dust record does not show the expected match before 1600; if anything, the lowest dust concentrations occur within the early part of the Little Ice Age (1400–1600), with higher concentrations during the medieval warm period (1000–1300).

In a return expedition to Quelccaya in 1991, Thompson encountered something totally unexpected. During the 1970s and 1980s, previous coring expeditions at the top of the ice cap (Figure 16–4B) had found annual layering extending from the most recent ice all the way down to the deepest layers deposited 1500 years ago. In the new cores taken in 1991, however, meltwater percolating down from the surface had begun to destroy the annual layers (Figure 16–4C). This dramatic finding means that a tropical ice cap that had been continuously recording intact annual layering for 1500

FIGURE 16-3 Coring mountain glaciers Drilling ice cores from (A) annually layered mountain ice requires (B) hauling equipment and supplies up and down the sides of mountains and ice caps and then (C) drilling at elevations approaching 20,000 feet, sometimes with (D) towering summer storm clouds rising from lower elevations and looming in the background. (Courtesy of L. G. Thompson, Byrd Polar Research Institute, Columbus, OH.)

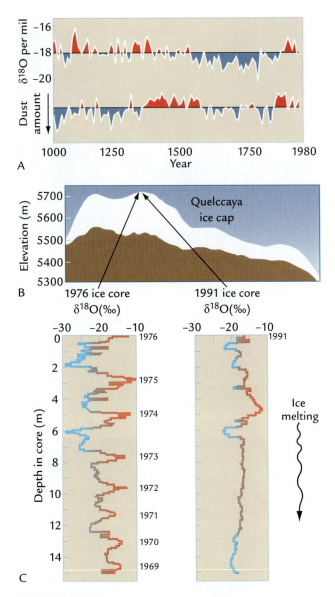

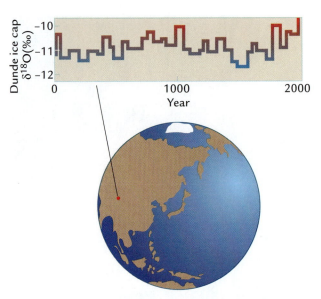

FIGURE 16-5 Dunde ice cap in Tibet An ice core from Dunde glacier in eastern Tibet shows lighter δ¹⁸O values during the last 50 years than in any such interval of the previous 2000 years. (Adapted from L. G. Thompson et al., "Holocene–Late Pleistocene Climatic Ice Core Records from Qinghai-Tibetan Plateau," *Science* 246 [1989]: 361–64.)

FIGURE 16-4 Quelccaya ice cap in Peru (A) An ice core taken in 1980 from the Quelccaya ice cap in the Peruvian Andes shows more negative δ¹⁸O values and higher dust concentrations from 1600 to 1900 than in the twentieth century. (B, C) A return expedition that cored the same location on the ice cap summit in 1993 found that the annual signal at the surface was being destroyed by melting, the first such melting event in the last 1500 years. (A: Adapted from L. G. Thompson et al., "The Little Ice Age as Recorded in the Stratigraphy of the Tropical Quelccaya Ice Cap," *Science* 234 [1986]: 361–64. C: Adapted from L. G. Thompson et al., "Recent Warming: Ice Core Evidence from Tropical Ice Cores, with Emphasis on Central Asia," *Global and Planetary Change* 7 [1993]: 145–56.)

years had suddenly begun to melt. In this location, the warming of the late twentieth century was obviously unprecedented for the last millennium and a half.

Thompson has also studied ice caps on subtropical mountains in Asia, including a much longer δ¹⁸O record

from Dunde ice cap in northern Tibet. Measurements averaged across 50-year intervals show some similarity to those in Figure 16–1, with more positive (warmer?) δ¹⁸O values before 1500 and especially around 1000, and more negative (colder?) values during much of the Little Ice Age interval (Figure 16–5). In this record, however, the transition to more positive (warmer?) δ¹⁸O values began near 1700, well before the Little Ice Age ended.

The average δ¹⁸O concentration at the Dunde site during the most recent 50-year interval (1937–1987) is more positive than in any other interval of the entire 12,000 years of record. Either the temperature of snow precipitation was uniquely warm during the mid-twentieth century or a major change has occurred in the sources or transport paths of the incoming water vapor.

IN SUMMARY, results from the Quelccaya and Dunde ice caps suggest that climate on low-latitude mountains may have been colder during the Little Ice Age and warmer (perhaps even uniquely so) during the twentieth century. On the other hand, some ice cores from the Antarctic and Greenland ice sheets do not provide distinct evidence of this pattern (Figure 16–6). These disagreements indicate that variations in climate recorded by δ¹⁸O changes in ice cores during the last 1000 years have in some cases been regional rather than global.

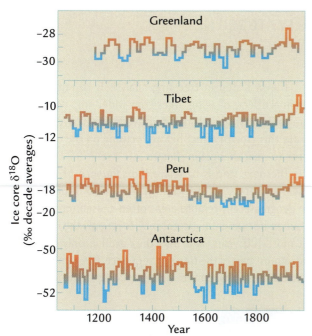

FIGURE 16-6 δ¹⁸O signals from ice cores in four regions
Cores taken from ice sheets and mountain glaciers in widely separated regions yield widely varying δ¹⁸O signals, and some do not show uniquely positive (warm?) δ¹⁸O values during the twentieth century. (Adapted from L. G. Thompson et al., "Recent Warming: Ice Core Evidence from Tropical Ice Cores, with Emphasis on Central Asia," *Global and Planetary Change* 7 [1993]: 145–56.)

16-2 Tree Rings

The use of tree rings to reconstruct climate change over the last several hundred years or more is called **dendroclimatology.** In regions of large seasonal changes, trees produce annual rings of varying clarity depending on species. The rings shift from lighter, low-density "early wood" of spring and early summer to darker, denser bands of "late wood" at the end of the growing season (Box 16–1).

Warmth and abundant rainfall during the growing season are favorable to tree growth, while cold and drought inhibit growth. The strategy is to search out regions where trees are most sensitive to climatic stress, usually at the limit of their natural temperature or precipitation ranges. In such regions, trees often grow alone or in isolated clusters. Years of unfavorable climate (low temperature or precipitation) are stressful to the trees and their growth slows, producing unusually narrow rings. Changes between favorable and unfavorable growth years produce distinctive variations in the widths and other properties of tree rings.

Tree ring studies have been carried out in many areas, many of which show some degree of distinctiveness due to local climate responses. Across the Arctic, tree ring studies primarily use trees such as spruce, larch, and Scotch pine (Figure 16–7A). An integrated signal

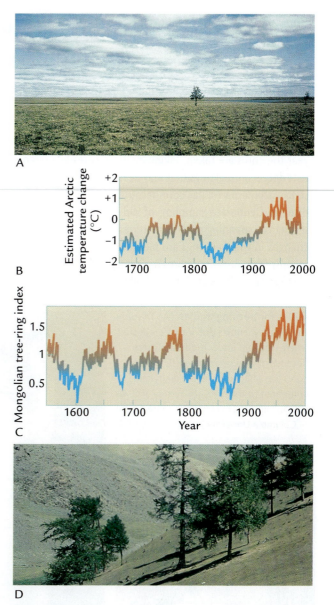

FIGURE 16-7 Arctic and Asian tree ring signals Signals from trees on northern continents, such as (A) Siberian larch, are combined to create (B) average circum-Arctic temperature changes over the last several centuries. (C) Similar-looking tree ring signals for Central Asia come from studies of (D) larch and pine in the mountains of Mongolia. Curve B shows departures from the 1951–1980 average. Curve C shows an index of changing tree ring width in Central Asia. (A and D: Courtesy of G. C. Jacoby, Lamont-Doherty Earth Observatory of Columbia University. B: Adapted from G. C. Jacoby and R. D. D'Arrigo, "Reconstructed Northern Hemisphere Annual Temperature Since 1671 Based on High-Latitude Tree-Ring Data from North America," *Climate Change* 14 [1989]: 39–49. C: Adapted from G. C. Jacoby et al., "Mongolian Tree Rings and 20th-Century Warming," *Science* 273 [1996]: 771–73.)

BOX 16-1   TOOLS OF CLIMATE SCIENCE
Analyzing Tree Rings

At chosen sites, a dozen or more trees are sampled by taking small radial cores (about 0.5 cm in diameter). The investigators date each tree by comparing the sequences of tree rings back in time, beginning from the year the cores are taken. Taking multiple cores lets scientists detect annual rings that may be missing or falsely present within a particular core because of local damage or disease. Deliberate sampling of old-age trees produces sequences spanning several centuries. Width and density changes in the tree rings are measured in each core, and records from dozens of cores are averaged to create a single representative signal for each site.

The first step in tree ring analysis is to remove the gradual effects of aging. Trees grow wider rings when they are young than at maturity, and investigators have to eliminate this growth effect before focusing on the effects of year-to-year changes in climate.

The next step is to relate the sequence of tree ring measurements to nearby instrumental records of climate. In geographically remote areas often chosen for tree ring studies, instrumental records may cover only the last 50 to at most 100 years, and thus they overlap only with the lattermost part of the tree ring sequence. By examining the correlations between the width or density of the tree rings

Coring for tree ring studies (A) To study tree rings, scientists drill into trees at sites where trees are under moderate stress because of cold or dryness. (B, C) The cores extracted are small in diameter compared to the trees. (Courtesy of G. C. Jacoby, Lamont-Doherty Earth Observatory of Columbia University.)

A

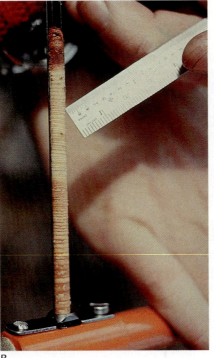

B

C

developed from trees distributed over the entire circum-Arctic region (Figure 16–7B) shows cool conditions in the late seventeenth and early eighteenth centuries, some warming in the middle and late eighteenth century, a deeper cooling in the early to mid-nineteenth century, a slow but substantial warming after 1850, a brief cooling between 1950 and 1970, and a small warming since 1980.

The 320-year trend in Figure 16–7B covers the middle and end of the Little Ice Age and the instrumentally recorded warming of the twentieth century. Estimated regional temperature variations larger than 1°C

occurred around the Arctic during the Little Ice Age, which was clearly not a time of unrelieved cold. The warming of the Arctic in the mid-twentieth century reached values unique for the 320 years of record, but temperatures in the late twentieth century were similar to those in the 1700s.

Another important region for tree ring studies is Central Asia. Because Asia is the largest continent, its climate is less subject to the moderating effects of ocean water than other regions. Climate signals derived from tree rings in Mongolia (Figure 16–7C) show a trend

BOX **16–1   TOOLS OF CLIMATE SCIENCE**

CONTINUED

and the instrumental record of climate change, scientists determine which aspects of climate control tree growth.

The final step is to use correlations defined from the calibration period of tree ring and climate data to model the tree ring/climate relationship and to estimate past climate using the tree ring data as predictors. Projections into the past assume that the climatic controls on tree ring properties found over the calibration interval were also in operation during the earlier interval.

In cold regions, temperature early in the growing season is often the strongest influence on tree growth, but a response to temperatures in other seasons is also present. In relatively arid regions, precipitation is more important. In some cases, precipitation in one year can affect growth in the next by providing moisture to the soil, which favors growth the following spring. On the other hand, deep winter snows can slow growth in the spring in colder areas.

Many records from the circum-Arctic margins show an unexpected decrease in tree ring growth in the last several decades, despite the fact that temperatures have been warming and CO_2 (fertilization) levels rising. This seemingly anomalous trend is not well understood.

The period of instrumental observations has also been a time of steadily rising atmospheric CO_2 levels. Controlled experiments with vegetation in greenhouses show that tree growth is enhanced by higher levels of CO_2 and photosynthesis. Some scientists speculate that correlations between tree ring properties and climate (temperature and precipitation) during the **calibration interval** (the interval for which tree ring data can be compared against climatic measurements) have ignored the "fertilization" effect of rising CO_2 on plants. CO_2 fertilization could masquerade as a climate signal, falsely indicating

gradual warming in cold regions or a trend toward wetter climates in dry regions.

In arid regions, some evidence suggests that rising CO_2 levels and faster fertilization of leaf pores have reduced the exposure of vegetation to the dry air that normally constrains growth in these regions. If this evidence is valid, then CO_2 fertilization may have been a factor in faster tree growth in dry regions, regardless of other climate changes.

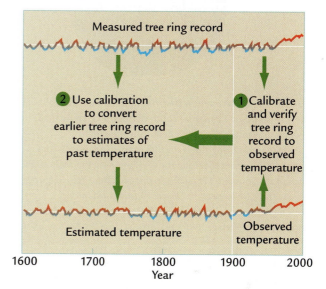

Calibrating tree ring signals Tree ring analysis is based on correlating the width (or density) of individual tree rings with monthly changes in temperature and precipitation recorded in the last half of the instrument record of climate (the last few decades). After these relationships are tested against the first half of the instrumental record, they are used to convert older tree-ring characteristics into estimates of past temperature and precipitation using the entire instrumental record for calibration.

similar to that of the circum-Arctic region. Intervals of warmth occurred during the Little Ice Age, both in the mid-eighteenth century and earlier, and colder temperatures occurred in the late sixteenth, late seventeenth, and mid- to late nineteenth centuries. The warming of the middle and late twentieth century appears unprecedented within the 450 years of record. These signals come from larch and pine trees growing on the high flanks of mountains in Central Asia (Figure 16–7D).

Far fewer data are available from trees at high latitudes of the southern hemisphere, in part because most

of that region is ocean. Records from pines on the island of Tasmania, south of Australia, extend back more than 2000 years, and the warmth of the last several decades matches any levels reached during the 2000-year interval (Figure 16–8). The interval of the Little Ice Age after 1500 is cooler than the late twentieth century but does not stand out distinctively against earlier centuries.

Other tree ring signals from the southern hemisphere vary considerably in character. Records from New Zealand, Chile, and Argentina extending back a few hundred years show unique warmth in the twentieth century.

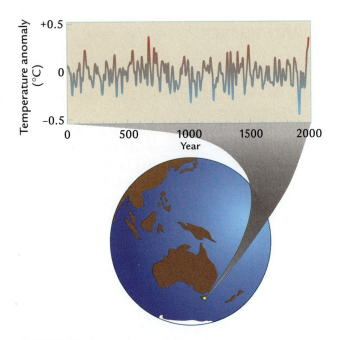

FIGURE 16-8 Tasmanian tree rings Tree ring records from the island of Tasmania, south of Australia, show that temperatures in the twentieth century are nearly unprecedented during the last 2000 years. (Adapted from E. Cook et al., "Interdecadal Climatic Oscillations in the Tasmanian Sector of the Southern Hemisphere: Evidence from Tree Rings over the Past Three Millennia," in *Climatic Variations and Forcing Mechanisms of the Last 2000 Years,* ed. P. D. Jones and R. S. Bradley [Berlin: Springer-Verlag, 1996].)

IN SUMMARY, tree rings, like ice cores, tell us that climate has varied from region to region over the last several hundred years so that no one record fully describes the trends in all areas. Viewed in their entirety, tree ring signals tell us that climate varied significantly within the Little Ice Age, in some regions even warming at times to levels comparable to those observed during part of the twentieth century. Many records show substantial and in some cases unprecedented warmth in the 1900s.

16-3 Corals and Tropical Ocean Temperatures

Observations of climate changes at annual or decadal resolution are not widely available from the oceans because of slow deposition and mixing of sediments by burrowing organisms. In recent years, climate scientists have begun to exploit corals as climate archives, using annual bands in their $CaCO_3$ structures. Because most corals grow in warm tropical or subtropical oceans, the information they provide complements ice core and tree ring studies from higher latitudes and altitudes. Most coral studies come from the tropical Pacific, which is dotted with volcanic islands surrounded by corals.

The most widely used climatic index in corals is $\delta^{18}O$ measurements at seasonal or better resolution. The two major controls on $\delta^{18}O$ variations are temperature (warmer waters produce lighter $\delta^{18}O$ values) and salinity (heavier rainfall produces lighter $\delta^{18}O$ values). The temperature effect on coral $\delta^{18}O$ values is analogous to the changes recorded in the shells of planktic foraminifera (Appendix 1).

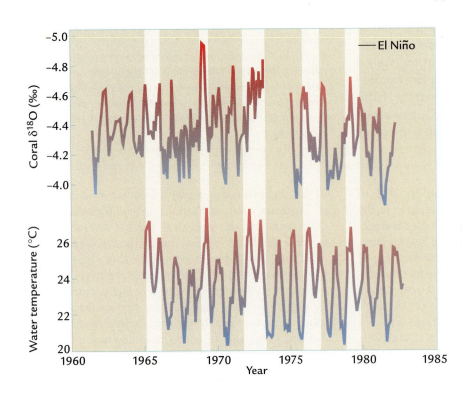

FIGURE 16-9 Coral $\delta^{18}O$ records: the eastern tropical Pacific Corals from the Galápagos Islands in the eastern Pacific tend to record low (negative) $\delta^{18}O$ values during warm El Niño years. (Adapted from R. B. Dunbar et al., "Eastern Pacific Sea Surface Temperatures Since 1600 A.D.: The $\delta^{18}O$ Record of Climate Variability," *Paleoceanography* 9 [1994]: 291–315.)

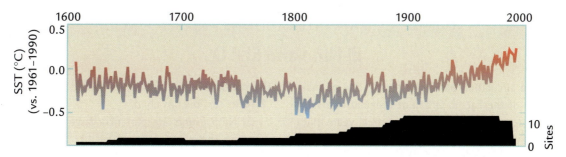

FIGURE 16-10 Stacked δ¹⁸O records from corals A record produced by stacking individual δ¹⁸O records from corals in the topical Pacific and Indian Oceans shows lighter (warmer, wetter?) values in the 1700s, heavier values in the 1800s, and very light values in the 1900s. (Adapted from R. Wilson et al., "Two-Hundred-Fifty Years of Reconstructed and Modeled Tropical Temperatures," *Journal of Geophysical Research* 111 [2006]: C10007, doi:10.1029/2005JC003188.)

Longer-term δ¹⁸O trends in the Pacific Ocean are overprinted by large year-to-year fluctuations in the El Niño and ENSO system (Box 16–2). These shorter-term δ¹⁸O variations largely reflect temperature changes in the eastern tropical Pacific Ocean and combined temperature-salinity changes in the central and western Pacific Ocean.

Modern corals deposited on the coast of the Galápagos Islands in the eastern Pacific record seasonal temperature changes similar to those measured directly by thermometers in surface waters (Figure 16–9). The match between the δ¹⁸O record and ocean temperature is not perfect because salinity changes also affect the δ¹⁸O values. Prominent El Niño years appear as δ¹⁸O minima that indicate warm temperatures.

Corals spanning several hundred years have recently become more widely available from the tropical oceans. A regional average signal created by combining records from the Pacific and Indian Oceans shows that the late twentieth century is the warmest period for at least the last 400 years (Figure 16–10). An interval of heavier (colder or drier) δ¹⁸O values during the 1800s was preceded by one of greater warmth during the 1700s, although the earliest part of the record is based on just a handful of sites.

16-4 Other Historical Observations

As human civilizations developed, people in some regions began to keep records of climatic phenomena for reasons unique to the cultures in which they lived. Climate scientists who attempt to use these early historical records to reconstruct past climate have to weigh their reliability carefully. Were the people recording the phenomena continuously present or relying on second-hand information? Was the task of observing passed to dependable observers through the generations? The phenomena recorded also vary widely from place to place. They include: the frequency and timing of first and last frosts and of droughts and floods, the timing of autumn lake or river freeze-up and spring ice breakup, the first flowering of shrubs and trees (such as cherry blossoms), and the dates of harvests.

Immediately after conquering the Inca Empire, the Spanish began to make environmental observations along the coast of Peru. Ships' logs are the major source of information from this region, supplemented by records kept by missionaries and others. These observations include responses now understood to result from El Niño events, including sea-surface temperatures warmer than normal, reduced catches of anchovy and other fish, departure of sea birds from coasts and islands, unusually heavy rains and floods, and outbreaks of cholera and malaria. The records start in 1525 and continue through most of the twentieth century (Figure 16–11).

El Niño events during this interval have been ranked by historians on a qualitative scale ranging from none to very severe. With 115 events in 465 years, the time between successive El Niño events averages 4 years, but the timing varies widely around this number. El Niño events tend to cluster within certain intervals (the late nineteenth century) but are rare in others (the mid-seventeenth century).

For several reasons, histories of climatic phenomena such as the El Niño record in Figure 16–11 and the Icelandic sea ice record in Figure 16–1 are difficult to use to compare large-scale climate changes. The records come from widely scattered locations that do not even provide regional, much less global, coverage. Also, different indices are sensitive to climate changes during different seasons of the year. The extent of sea ice is sensitive to cold winter temperatures and low ocean salinity, the freezing of lakes to prolonged autumn cold, the blooming of cherry blossoms to warmth in spring, and the length of the growing season to unusual overnight frosts in spring and autumn. As a result, these indices record changes in parts of the climate system with widely varying response times. For this reason, historical observa-

BOX 16–2   CLIMATE INTERACTIONS AND FEEDBACKS

El Niño and ENSO

The **El Niño** circulation pattern interrupts the normal circulation of the Pacific Ocean at irregular intervals ranging from 2 to 7 years. During the normal years that occur between El Niños, surface temperatures along the coasts of Peru and Ecuador and in the eastern equatorial Pacific are near 18°C (50°F) in winter—far cooler than typical tropical temperatures (25°C, 77°F) and in fact the coolest tropical surface water on Earth.

The lower non–El Niño temperatures result from upwelling driven by strong winds in southern hemisphere winter (August). The winds from the south drive warm surface waters westward away from the coast of South America, and cooler water wells up from below. The winds then turn to the west near the equator and drive warm surface water toward the southwest, causing cool water to well up near the equator. Upwelling water brings nutrients to the surface, supplying food to an ecosystem ranging from plant plankton to fish (anchovies and tuna), sea birds, and marine mammals (seals and sea lions). Non–El Niño years are also dry along the coast of South America because cool, upwelling waters are a poor source of water vapor for the atmosphere. As a result, the coastal deserts of Peru and Chile are normally among the driest regions on Earth.

El Niño years change all this. During El Niño winters, strong southerly winds fail to blow in the eastern and tropical Pacific, upwelling does not occur, and the surface waters along the South American coast warm by 2°–5°C. Without upwelling, the plankton populations crash, and most fish die or move away. Without fish, sea birds on tropical islands cannot feed their young, and they abandon their nests to fly elsewhere in search of food. In severe El Niño years, a significant fraction of the year's population of young sea birds and mammals dies. Ocean warming near the coastal South American deserts produces a large source of moisture, and rain falls in cloudbursts that produce flash floods in regions with little or no natural vegetation cover to absorb the water. The warm rains also favor the breeding and spread among humans of tropical diseases such as malaria and cholera.

Even though the El Niño circulation pattern reaches its height during southern hemisphere winter in August, the first hint of unusual warming of the surface ocean is often detected during the previous summer, near Christmas. For this reason, Peruvian fishermen named this phenomenon "El Niño," or "the boy child."

El Niño events are part of a larger-scale circulation spanning the entire tropical Pacific. In the 1920s the atmospheric scientist Gilbert Walker found matching changes in atmospheric pressure between the western Pacific (northern Australia and Indonesia) and the south-central Pacific island of Tahiti. High pressures over Australia correlate with low pressures in the south-central Pacific and vice versa. Low atmospheric pressures are associated with rising air motion and rainfall, while high surface pressure is associated with sinking motion and dry conditions (see the companion Web site, pp. 15–18). The opposing pressure trends through time across the tropical Pacific are part of an enormous circulation cell called the **Southern Oscillation.** Sinking and rising motions occur at opposite times over northern Australia and Indonesia in the west and across the south-central Pacific in the east.

El Niño and the Southern Oscillation are linked. El Niño years, with warm ocean temperatures and heavy rains in Peru, are times of high pressure and drought over northern Australia and of low pressures and high rainfall in the south-central Pacific. Non–El Niño years, with cool ocean temperatures near South America, are times of low pressure and increased rainfall in northern Australia and of higher pressures and reduced rainfall in the south-central Pacific. This linked circulation is known as **ENSO** (El Niño–Southern Oscillation).

The physical link between these two systems occurs in the lower atmosphere and the upper ocean. Strong trade winds that cause upwelling in the eastern Pacific during non–El Niño years also drive warm surface water westward across the tropical Pacific. Warm water piles up in the western Pacific at a height several tens of centimeters above the level of the eastern Pacific and forms a natural source of moisture for evaporation and precipitation in northern Australia and Indonesia. Some of the rising air flows eastward at high elevations and sinks in the east-central Pacific, contributing to the normally cooler and drier conditions near South America.

During El Niño years, without strong trade winds pushing water westward, some of the pool of warm water in the western Pacific flows back eastward and becomes a source

BOX **16–2** **CLIMATE INTERACTIONS AND FEEDBACKS**

CONTINUED

of latent heat and moisture for local rains. When the flow reaches the Americas, it deflects northward and southward along the coast, bringing warmer and wetter conditions north to California and south to Peru, along with reduced upwelling. In the western Pacific, slightly cooler conditions during El Niño years result in the drying in

northern Australia and Indonesia. Eventually El Niño conditions subside and the tropical ocean reverts to its normal state. On occasion it overshoots its normal state and produces abnormally cool sea-surface temperatures in the eastern Pacific. This overshoot is called **La Niña,** or "the girl child."

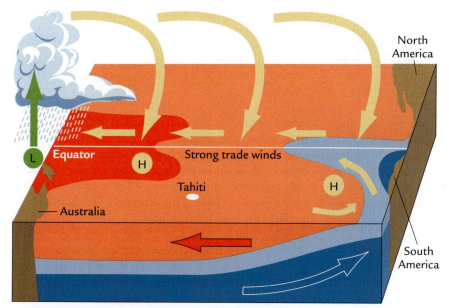

A Non El Niño year

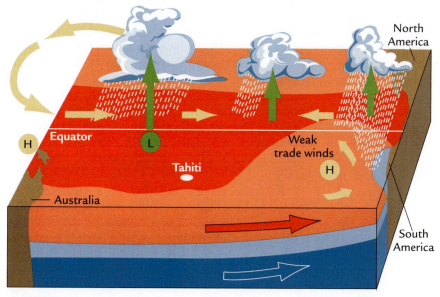

B El Niño year

Circulation in El Niño and non–El Niño years El Niño events change atmospheric and ocean circulation across the entire Pacific Ocean, from Australia and Indonesia to the west coast of South America.

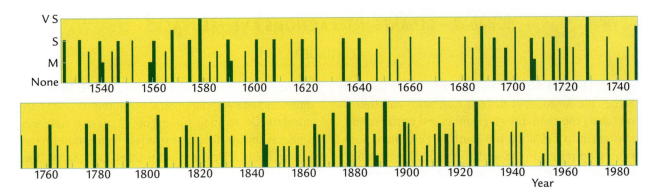

FIGURE 16-11 **Historical records of El Niño** Historical chronicles of unusual phenomena along the South American coast reveal El Niño events since the early sixteenth century. (VS = very strong, S = strong, M = moderate.) (Adapted from W. H. Quinn and V. T. Neal, "The Historical Record of El Niño Events," in *Climate Since A.D. 1500,* ed. R. S. Bradley and P. D. Jones [London: Routledge, 1992].)

tions recorded before the era of weather instruments give us only anecdotal information about climate changes during preceding centuries.

In summary, exploration of a variety of proxy climate indicators (ice cores, tree rings, and tropical corals) and of a few historical records has improved our knowledge of climate changes that occurred during the past 1000 years. Despite these efforts, coverage still remains well short of global.

Reconstructing Hemispheric Temperature Trends

Several attempts have been made to synthesize high-resolution records from ice cores, tree rings, and corals into a single estimate of northern hemisphere temperature changes during the last millennium. These efforts face daunting problems. Proxy records are all linked in some way to temperature, but they are also affected in complex ways by other climatic, biological, and ecological factors. Because the links between the proxy variations and climatic variables are complicated, scientists are forced to rely on best-fit statistical approximations that simplify the true underlying relationships. Another problem is that the proxy indicators may be sensitive primarily to temperature changes in summer or winter or spring, yet the reconstructions combine them into a single "common" temperature trend. Still another problem is the extreme scarcity of proxy data. More than 100 proxy records are available for the last century or two, but only about two dozen are available for the year 1500 and only half a dozen for the year 1000.

Despite these complications, the reconstructed signals (Figure 16–12) all show a gradual but erratic temperature decline for almost 900 years, ending with a dramatic warming that began just before 1900. The amplitude of the changes that occurred prior to the instrumental

record of the last 125 years varies by a factor of two or more among the reconstructions. Reconstructions based on records that are more heavily weighted toward higher latitudes show the largest range of variations, while those with a greater representation of lower latitudes tend to have a smaller range. This difference is consistent with the fact that high latitudes are more climatically reactive than low latitudes because of the amplifying effects of albedo feedback from reflective surfaces of bright snow or sea ice (companion Web site, p. 8; also see Chapter 3).

Because of the sparse coverage of sites, all the reconstructions inevitably have large uncertainties. The shaded region in Figure 16–12 shows one attempt to estimate the uncertainty in the temperature reconstruction through time. The estimated uncertainty is large even for recent centuries, and it reaches almost ten times the size of the estimated temperature variations prior to 1500.

The better-constrained portions of all reconstructions show a cool interval equivalent to the peak of the Little Ice Age in the 1600s, 1700s, and 1800s. In comparison, the warming since the 1880s stands out as highly unusual both in the abrupt rate of change and in the level of warmth attained. This unprecedented warming will be examined more closely in Chapters 17 and 18.

Evidence for the existence of a medieval warm period and a subsequent cooling into the Little Ice Age is more ambiguous, in part because of the very large uncertainties inherent in the sparse records from those intervals. In the reconstructions that attempt to balance the latitudinal distribution of sites, the net century-scale cooling between 1000–1200 and 1400–1900 averages about 0.2°C, with larger fluctuations during particular decades. Different reconstructions place the main steps in the cooling at different times.

This cooling trend may or may not persist when a reasonably complete "global" reconstruction becomes

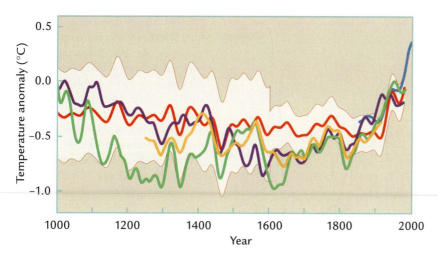

FIGURE 16-12 Northern hemisphere temperatures during the last millennium
A synthesis of high-resolution climate records spanning all or part of the last millennium shows a small gradual cooling for 900 years followed by a large and abrupt warming in the twentieth century. Light shading indicates uncertainty in estimated temperature. (Adapted from P. D. Jones and M. E. Mann, "Climate over Past Millennia," *Reviews of Geophysics* 42 [2004]: 2003RG000143.)

available. Attempts to reconstruct meaningful temperature trends for the entire southern hemisphere are not yet possible because so few records are available. One very preliminary attempt at such a reconstruction shows no obvious cooling trend from 1000–1200 to 1400–1900. When this estimate was combined with one from the northern hemisphere to calculate a global average, the century-scale global cooling from the interval 1000–1200 to the interval 1400–1900 was reduced to less than 0.1°C. If later reconstructions confirm this result, the small cooling between 1000–1200 and 1400–1900 could turn out to be just a local oscillation in and around the North Atlantic Ocean.

IN SUMMARY, reconstructions of northern hemisphere temperature during the last 1000 years show that temperatures between 1400 and 1900 were considerably cooler than during the last century. Evidence for an earlier interval of moderate warmth between 1000 and 1200 is far less convincing because of fewer records and larger uncertainties. The notion of a medieval warm period, followed by a cooler Little Ice Age, is a valid description of trends that occurred from eastern Canada across Greenland and Iceland and east into northern Europe, but it may or may not characterize the changes across the remaining 90–95% of Earth's surface.

Proposed Causes of Climate Change from 1000 to 1850

The northern hemisphere temperature reconstructions in Figure 16–12 show a small cooling that began between 1000 and 1200 and developed slowly until 1600–1800. This cooling has several possible explanations: orbital-scale forcing, variability linked to the millennial bipolar seesaw, solar variability, volcanic eruptions, and anthropogenic factors.

16-5 Orbital Forcing

Orbital forcing may have contributed to the part of the cooling that occurred in higher northern latitudes. Over the last 6000 years, portions of the circum-Arctic have cooled by 1°–2°C because of decreasing summer insolation at both the tilt and precession cycles (see Figures 13–18 and 13–19). If the rate of cooling was uniform over 6000 years, the fraction of the cooling that would have occurred within the last 1000 years would have amounted to 0.16°–0.33°C. Allowing for the fact that temperature changes are amplified near the poles, the mean hemisphere-scale cooling would probably have been a factor of 2 or 3 smaller, or about 0.1°C, equivalent to about half the amount observed in the reconstruction for the northern hemisphere (Figure 16–12).

16-6 Millennial Bipolar Seesaw

Millennial oscillations have been proposed as another explanation for the cooling during the last millennium, despite the fact that longer-term evidence from the current interglaciation generally suggests that climatic trends have been highly irregular and local in scope (see Chapter 14). If the north polar cooling into the Little Ice Age was part of a small oscillation linked to the bipolar seesaw, it should have been accompanied by an even smaller warming in the Antarctic region. This pattern was typical of large glacial-age oscillations (see Figure 14–18). At this point, proxy coverage of the last 1000 years in the southern hemisphere is insufficient to test this idea.

16-7 Solar Variability

The ^{14}C and ^{10}Be evidence examined in Chapter 14 suggested that solar variability does not cause changes in climate over millennial time scales, but the possibility was left open that changes in solar output could still play a significant role in changes at decadal and century scales.

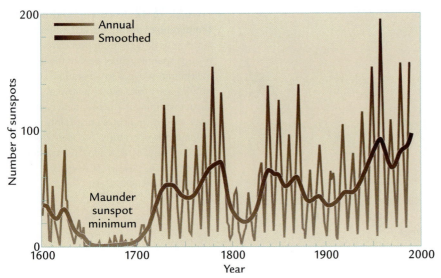

FIGURE 16-13 Sunspot history from telescopes Measurements made with telescopes over the last several hundred years show that the number of sunspots increase and decrease at an 11-year cycle. During the interval known as the Maunder minimum, sunspots were scarce for several decades. (Adapted from C.-D. Schonwiese et al., "Solar Signals in Global Climatic Change," *Climatic Change* 27 [1994]: 259–81.)

One possible solar-terrestrial link for the last several centuries is tied to the 11-year **sunspot cycle.** Continuous and accurate observations of the number of dark sunspots visible on the Sun began with the invention of the telescope more than 400 years ago. The most obvious feature of the long-term record is the 11-year cycle (Figure 16–13).

Intuitively, it would seem likely that the presence of dark spots on a bright surface would reduce the total amount of emitted solar radiation, but the relationship actually has the opposite sense. Most of the radiation emitted by the Sun streams out from its polar regions and from bright rings around the sunspots called **faculae** (Figure 16–14). During years when sunspots are abundant, the amount of radiation emitted in solar flares is at a maximum, because mechanisms operating within the Sun simultaneously regulate both sunspots and net solar emissions. As a result, the amount of solar radiation arriving at Earth during sunspot maxima is at a maximum

rather than a minimum. Satellite measurements in the past two decades indicate that the range of variation in solar irradiance is about 0.11% (see Chapter 18).

A longer-term trend is also apparent in the sunspot record from telescopes, with larger sunspot numbers (and larger variations) in the last three centuries compared to the late 1600s. An interval between 1645 and 1715 that had almost no sunspots is called the **Maunder sunspot minimum,** after the astronomer who discovered it. A similar interval between 1460 and 1550 is called the **Sporer sunspot minimum.** Despite the relatively crude observational techniques available at those times, the existence of these two intervals is certain.

Some scientists initially proposed that variations in solar output during the last several centuries were larger than the small range observed during the brief satellite era. They suggested that solar output might have been as much as 0.25–0.4% weaker than today during the Maunder and Sporer minima, with a gradual rise in

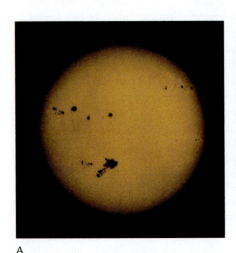

FIGURE 16-14 Sunspots and solar emissions Intervals when (A) sunspots were abundant were also times when (B) strong solar emissions from the bright margins of the sunspots sent increased levels of radiation to Earth. (National Optical Astronomy Observatory.)

irradiance during the last few centuries. In addition, because these intervals of low sunspot activity persisted for decades, the slow-responding parts of the climate system would have had time to respond to them more fully compared to the low values during the 11-year cycle. This link seemed particularly appealing because the sunspot minima occurred at times when northern hemisphere temperatures were considerably cooler than they are today. As a result, some scientists suggested that changes in solar irradiance accounted for 50–75% of the 0.6°C increase in temperature during the 1900s.

Later astronomical observations failed to support this claim. Archival images of Sunlike stars from several observatories failed to show variations comparable to those proposed for the Sun. No evidence for variations greater than those at the 11-year cycle has been detected. Recent estimates place the contribution from solar irradiance changes since 1880 at less than 0.07°C, or about 10% of the amount of the 0.7°C warming shown in Figure 16–12.

This finding has not ended the debate about possible solar effects on climate during this interval. Although most of the Sun's emissions arrive as visible or near-visible (ultraviolet and infrared) radiation, the Sun also sends out a plasma or ionized gas called the "solar wind," which interacts with Earth's stratosphere as it is deflected by Earth's magnetic field. One possibility under consideration is that the solar wind affects the formation of ozone, which in turn alters the formation of clouds in the troposphere and thereby affects climate at Earth's surface.

The similarity of ^{14}C and ^{10}Be trends during the last 5000 years seems to point to a common origin from solar changes (see Chapter 14). Scientists have looked for a correlation between these isotopic trends and climatic proxies during recent millennia with only mixed success. Temperature-sensitive changes in $\delta^{18}O$ within the last millennium show a substantial correlation with ^{10}Be variations, and the existence of this link would seem to imply a solar role. On the other hand, the fact that no such relationship is evident in previous millennia of the Holocene greatly weakens this case. The possibility also remains that the similar changes in ^{14}C and ^{10}Be are both a response to changes occurring within the climate system (changes in ocean circulation, rate of ice accumulation, etc).

IN SUMMARY, changes in solar irradiance featured prominently in several initial attempts to explain climatic trends during the last millennium, but recent evidence suggests that irradiance changes were small enough to have had little impact on climate during this interval. Still, the possibility exists that even very small changes in irradiance might "excite" internal oscillations in naturally varying parts of the climate system like ENSO or the North Atlantic Oscillation (see Chapter 17).

16-8 Volcanic Explosions

Explosive eruptions of volcanoes cool climate over intervals of a few years. Volcanoes erupt sulfur dioxide (SO_2) gas, which mixes with water vapor in the air and forms droplets and particles of sulfuric acid called **sulfate aerosols.** Highly explosive eruptions can reach 20 to 30 km into the stratosphere, where the aerosols block some incoming solar radiation and keep it from reaching the ground. With solar radiation reduced, Earth's surface cools.

Latitude determines the geographic extent of the impact of volcanic eruptions. Volcanoes that erupt poleward of about 25° produce particles that stay within the hemisphere in which the eruption occurs, and the cooling impact is limited to that hemisphere. Explosions that occur in the tropics are redistributed by Earth's atmosphere to both hemispheres and have a global impact on climate.

Ocean-island volcanoes with iron- and magnesium-rich compositions tend not to cause explosive eruptions but instead emit lava that flows easily across the land. Volcanic particles sent into the air by these eruptions rarely reach the stratosphere but stay within the troposphere. Within a few days, the particles are brought back down to Earth by rain. With so brief a stay in the atmosphere, the particles cannot be widely enough distributed around the planet to produce large-scale effects on climate.

In contrast, volcanoes along converging plate margins are fed by magmas richer in silica and other elements found in continental crust (Chapter 4). Their eruptions are more explosive because the natural resistance of this kind of molten magma to flow causes internal pressures to build up to the point where volcanic particles can be injected into the stratosphere, well above the level where precipitation can wash them out. Because of the pull of gravity, slow settling of these fine particles takes years, long enough for the particles to be distributed within a hemisphere or across the entire planet.

Sulfate aerosol concentrations in the stratosphere reach their maximum regional distribution within months and then begin to decrease as gravity removes the particles. The decrease follows an exponential trend: each year about half of the remaining particles settle out, and within two to three years aerosol concentrations are much reduced (Figure 16–15 top). The effect of the aerosols on temperature follows the same trend, with a maximum initial cooling that soon fades away (Figure 16–15 bottom). If several explosions follow within an interval of a few years, their impact on climate may be sustained for a decade or more.

Climate scientists face difficulties in trying to reconstruct the effects of older volcanic explosions on climate. They may get some idea about the magnitude of ancient eruptions from the sizes of the craters left by the explosion or from the volume of volcanic ash

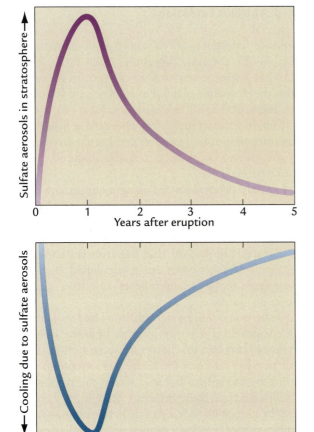

FIGURE 16-15 Volcanic explosions and cooling Large volcanic eruptions (top) launch sulfate aerosols into the stratosphere and (bottom) cool climate for a few years. (Adapted from R. S. Bradley, "The Explosive Volcanic Eruption Signal in Northern Hemisphere Continental Temperature Records," *Climate Change* 12 [1988]: 221–43.)

deposited nearby. The geographic area over which the ash is distributed may also provide clues about the power of the eruption and whether the volcanic particles might have reached the stratosphere. But sulfur forms only a small and variable fraction of the total volume of erupted material, and the volume of ash cannot directly be used to estimate the amount of sulfur erupted. As a result, it is difficult to estimate the climatic effects of ancient eruptions reliably.

Despite these obstacles, several attempts have been made to estimate the effect of volcanic eruptions on the northern hemisphere temperature trends plotted in Figure 16–12, including the example shown in Figure 16–16. Comparison of the volcanic and temperature histories indicates that sequences of large eruptions played a role in decadal-scale cooling.

IN SUMMARY, the more frequent clusters of eruptions after 1300 appear to have contributed to the small cooling trend into the Little Ice Age interval.

16-9 Greenhouse-Gas Effects on Climate

Several high-resolution CO_2 records from Antarctic ice cores have been accurately dated based on the presence of numerous layers of volcanic ash of known age. CO_2 concentrations were relatively high (~283–284 ppm) near 1000–1200 but had fallen to ~276–277 ppm by 1600–1800 (Figure 16–17). Some scientists have suggested that solar-volcanic changes caused this 7–8 ppm CO_2 drop by cooling the surface ocean (which increased CO_2 solubility) and by reducing the CO_2 emitted from the oxidation of litter on land. Carbon cycle models suggest, however, that these natural explanations can account for only a small part (~2–3 ppm) of the CO_2 decrease without violating the small size of the ocean cooling.

Another proposed explanation of the CO_2 drop is that some or all of it was anthropogenic in origin. The

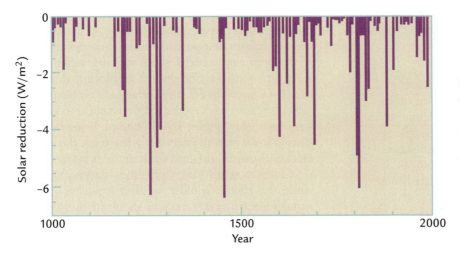

FIGURE 16-16 Volcanic explosions and solar radiation An estimate of the history of volcanic explosions during the last millennium shows their decadal-scale (and longer) effect in reducing the intensity of incoming solar radiation. (Adapted from T. Crowley, et al., "Modeling Ocean Heat Content Changes During the Last Millennium," *Geophysical Research Letters* 30 [2003]: GL017801).

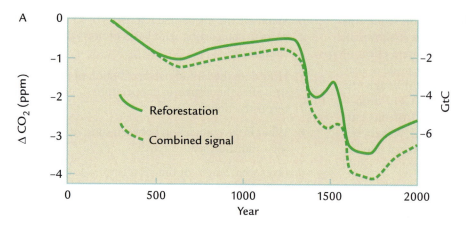

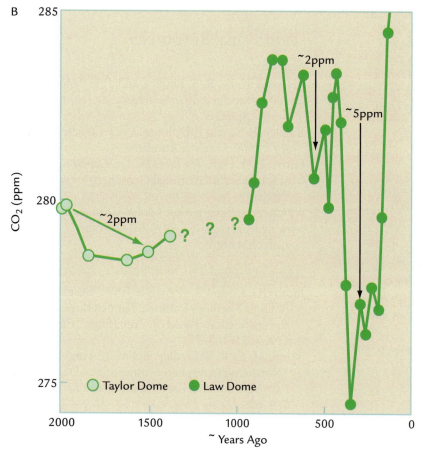

FIGURE 16-17 **Effect of pandemics on atmospheric CO_2** Mortality caused by pandemics led to farm abandonment, reforestation, and storage of carbon in growing trees. The two parts of the figure show a simulation of carbon sequestered by reforestation and related effects from the American pandemic (1500–1700), indicating a reduction of atmospheric CO_2 half as large as that measured in ice cores (Law and Taylor Domes; Adapted from W. F. Ruddiman, "The Early Anthropogenic Hypothesis: Challenges and Responses," Reviews of Geophysics 45, RG 4001, doi 10.1029/2006 RG 000207, 2007).

"early anthropogenic hypothesis" summarized in Chapter 15 was grounded in the assumption that deforestation by humans, aided by feedbacks from the climate system, caused the slow increase in the amount of atmospheric CO_2 during the last 8000 years (see Figure 15–14). From this point of view, any subsequent drop in the rising CO_2 trend could have had an origin connected in some way to humans.

Two great pandemics occurred during the interval of falling CO_2 values: (1) the "Black Death," an outbreak of bubonic plague between 1347 and 1352 that killed 25 million Europeans, ~33% of the population at that time;

(2) the American pandemic, a host of diseases carried by Europeans that killed some 50 million Native Americans between 1492 and 1700, or 85–90% of the previous population. At the times these pandemics occurred, many Europeans and most Native Americans farmed areas in clearings cut into natural forests. When the pandemics decimated the populations, the forests grew back in fields left untended, and the carbon required by these growing trees for photosynthesis was taken from the CO_2 in the atmosphere (as well as the surface ocean and the biosphere elsewhere on Earth). Much of the reforestation took place within 50 to 100 years.

A simulation of the effect of this carbon sequestration on atmospheric CO_2 concentrations is shown in Figure 16–17. By this estimate, the combined effect of the two pandemic-driven reforestation episodes may explain a CO_2 drop of at least 4 ppm. Although the likely temperature effect of a 4-ppm CO_2 drop—0.05°C—seems trivial, it represents 25% of the estimated northern hemisphere cooling of ~0.2°C between 1000–1200 and 1600–1800. If the global mean cooling was only half as large (0.1°C), reforestation could account for half the total.

IN SUMMARY, the estimated cooling from 1000 years ago into the Little Ice Age is small, and any or all of several factors could have played an important causal role. Far greater geographic coverage is needed to define the global climatic response before meaningful cause-and-effect conclusions can be drawn. In contrast, no such ambiguity exists about the large, rapid and global warming since 1850.

Key Terms

Little Ice Age (p. 289)

medieval warm period (p. 290)

lichen (p. 290)

dendroclimatology (p. 295)

calibration interval (p. 297)

El Niño (p. 300)

Southern Oscillation (p. 300)

ENSO (p. 300)

La Niña (p. 301)

sunspot cycle (p. 304)

faculae (p. 304)

Maunder sunspot minimum (p. 304)

Sporer sunspot minimum (p. 304)

sulfate aerosols (p. 305)

Review Questions

1. What evidence indicates a cooler climate in Europe and nearby regions during the Little Ice Age?

2. What evidence from ice cores suggests that the warming during the twentieth century reached levels unprecedented over the last 1000 years?

3. Why are the rings of environmentally stressed trees ideal for detecting climate signals?

4. How could rising CO_2 levels complicate interpretations of changes recorded in tree rings?

5. What factors influence $\delta^{18}O$ values recorded in corals and how?

6. What are the major characteristics of El Niño years in comparison with years of normal circulation?

7. How does the latitudinal distribution of proxy sites affect hemispheric temperature reconstructions?

8. What is the connection between sunspots and solar radiation sent to Earth?

9. Over what length of time can large volcanic explosions alter climate?

10. How could disease have affected climate during the last millennium?

Additional Resources

Basic Reading

Alverson, K. D., R. S. Bradley, and T. F. Pedersen. 2003. *Paleoclimate, Global Change, and the Future.* Berlin: Springer.

Bradley, R. S., and P. D. Jones. 1992. *Climate Since A.D. 1500.* London: Routledge.

Grove, J. M. 1988. *The Little Ice Age.* London: Methuen.

www.igbp.net (International Geosphere-Biosphere Progam)

www.pages-igbp.org (Past Global Changes Project)

www.ngdc.noaa.gov.paleo/global warming /paleolast.html

Advanced Reading

Jacoby, G. C., and R. D'Arrigo. 1993. "Secular Trends in High Northern Latitude Temperature Reconstructions Based on Tree Rings." *Climatic Change* 15: 163–77.

Mann, M. E., R. S. Bradley, and M. K. Hughes. 1999. "Northern Hemisphere Temperatures During the Past Millennium." *Geophysical Research Letters* 26: 759–62.

North, G. A., et al. 2006. *Surface Temperature Reconstructions for the Last 2000 Years.* Washington, DC: National Academy of Sciences.

Quinn, W. H., V. T. Neal, and S. E. Antunez de Mayolo. 1987. "El Niño Occurrences over the Past Four and a Half Centuries." *Journal of Geophysical Research* 92: 14449–61.

Thompson, L. G., E. Mosley-Thompson, M. E. Davis, P. N. Lin, T. Yao, M. Sdyurgerov, and M. Dai. 1993. "Recent Warming: Ice Core Evidence from Tropical Ice Cores, with Emphasis on Central Asia." *Global and Planetary Change* 7: 145–55.

Wilson, R., A. Tudhope, P. Brohan, K. Briffa, T. Osborn, and S. Tett. 2006. "Two-Hundred-Fifty Years of Reconstructed and Modeled Tropical Temperatures." *Journal of Geophysical Research* 111: C10007, doi:10.1029/2005JC003188.

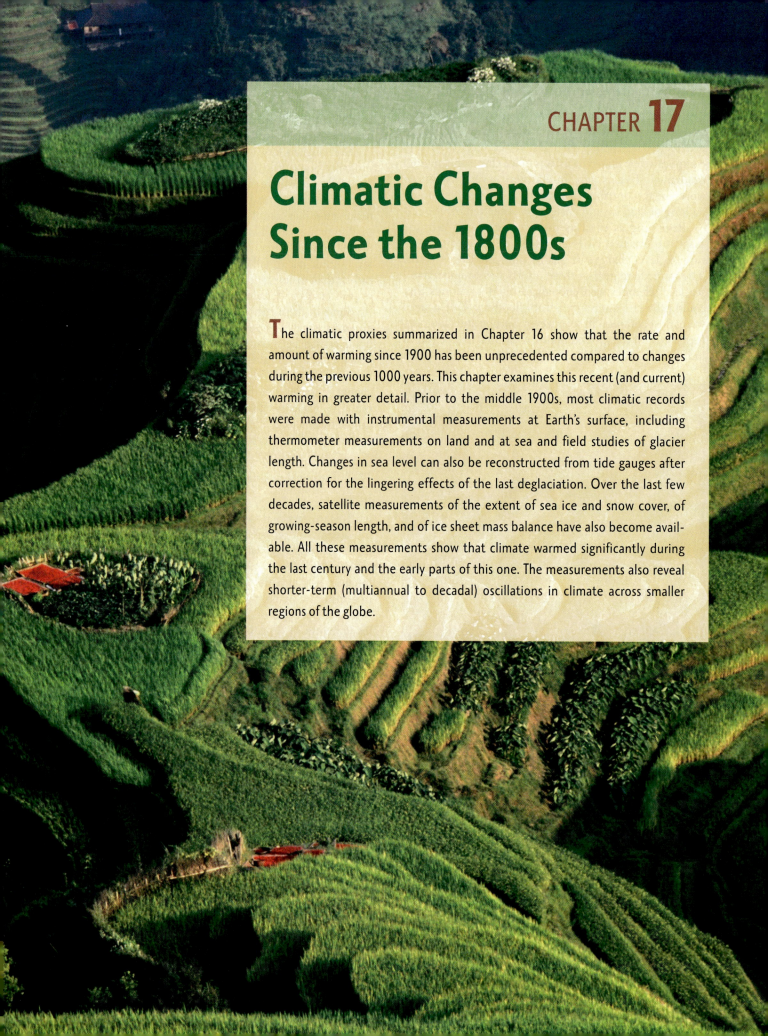

CHAPTER 17

Climatic Changes Since the 1800s

The climatic proxies summarized in Chapter 16 show that the rate and amount of warming since 1900 has been unprecedented compared to changes during the previous 1000 years. This chapter examines this recent (and current) warming in greater detail. Prior to the middle 1900s, most climatic records were made with instrumental measurements at Earth's surface, including thermometer measurements on land and at sea and field studies of glacier length. Changes in sea level can also be reconstructed from tide gauges after correction for the lingering effects of the last deglaciation. Over the last few decades, satellite measurements of the extent of sea ice and snow cover, of growing-season length, and of ice sheet mass balance have also become available. All these measurements show that climate warmed significantly during the last century and the early parts of this one. The measurements also reveal shorter-term (multiannual to decadal) oscillations in climate across smaller regions of the globe.

Reconstructing Changes in Sea Level

One of the key sources of information on global climatic trends over the last 150 years is the average level of the ocean. Reconstructing changes in sea level is greatly complicated by lingering effects of ice sheets that melted thousands of years ago. In addition, several factors at work today contribute to sea level change, including melting of land ice and changes in ocean temperature.

Beginning as early as the late eighteenth century, seaport towns and cities installed **tide gauges** to measure sea level changes caused by tides and large storms. The immediate goal of these efforts was to understand sea level changes well enough to build structures that could protect communities from flooding. Tide gauges were most common in seaports in Europe and along the East Coast of the United States, areas that had begun to industrialize early (Figure 17–1). Some 100 to 200 years

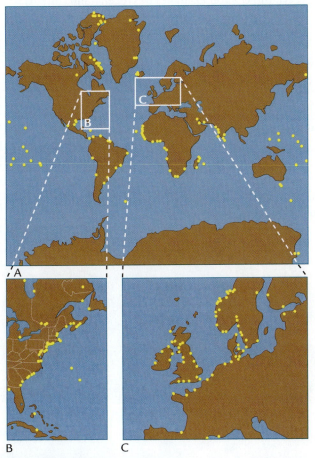

FIGURE 17-1 Tide gauge stations (A) Tide gauge records spanning several decades to as much as two centuries are available from hundreds of coastal locations on Earth's surface. (B, C) The largest concentrations are in eastern North America and northwestern Europe. (Adapted from A. M. Tushingham and W. R. Peltier, "Ice-3G: A New Global Model of Late Pleistocene Deglaciation Based upon Geophysical Predictions of Post-Glacial Relative Sea level Changes," *Journal of Geophysical Research* 96 [1991]: 4497–4523.)

later, these tide gauge records show not only short-term changes caused by tides and storms but also longer-term histories of sea level change in the decades and centuries since their installation.

Deriving sea level trends from tide gauge records is difficult. Some tide gauge records indicate rapid sea level falls, but others show a slow sea level rise, and still others indicate faster rises. At first, it doesn't seem to make sense that completely different trends could occur in different areas. The world ocean is one interconnected body of water, and it would seem that global sea level should rise or fall by the same amount everywhere, rather than rising in one place and falling in another over intervals of decades or centuries.

Some of these regional differences result from modern processes that vary regionally. Areas of active tectonic uplift caused by mountain building or of active subsidence caused by recently added sediment loads (such as river deltas) have to be avoided in determining the actual change in the level of the ocean. In addition, adjustments must be made for human effects such as subsidence caused by pumping of groundwater and impoundment of rainfall runoff in reservoirs behind dams.

17-1 Fading Memories of Melted Ice Sheets

By far the greatest problem in reconstructing past sea levels is the fact that the bedrock of the land and under the oceans still retains a memory of the ice sheets from the most recent glaciation. Even though the last remnants of the glacial maximum ice sheets finished melting near 10,000 years ago in Europe and 6000 years ago in northern Canada, the rock in Earth's upper mantle is still in the process of adjusting to the previous load of ice (Chapter 4). This bedrock "memory" causes different behaviors in today's movements of Earth's land and seafloor surfaces. The types of long-term change in relative sea level defined by tide gauges fall in three geographic groups (Figure 17–2).

One group of tide gauges shows rapid drops in relative sea level in recent centuries (Figure 17–3). These gauges are located in regions that were once directly beneath the ice sheets, such as the Hudson Bay region of Canada and the Baltic Sea region of Scandinavia. Relative sea level is now falling rapidly in these regions primarily because the bedrock is still rebounding from the removal of the ice sheet load thousands of years ago. A rise in bedrock means a fall in relative (but not global) sea level.

Bedrock at depths of 100 to 200 km has a slow viscous component of response, and it takes many thousands of years to recover fully from loads imposed on it or removed from it (Chapter 9). The enormous load of glacial ice thousands of years ago depressed bedrock surfaces beneath the central parts of the ice sheets by as much as 1 km, causing deep rock to flow slowly outward at great depths. Later, when the ice melted, rock slowly

FIGURE 17-2 Patterns of sea level change Relative sea level today is changing in several ways regionally because of bedrock movement. Bedrock is rapidly rising in areas formerly covered by thick ice and sinking in regions surrounding the former ice sheets. Farther from the ice sheets, ocean basins are sinking under the added weight of meltwater. (Adapted from A. M. Tushingham and W. R. Peltier, "Ice-3G: A New Global Model of Late Pleistocene Deglaciation Based upon Geophysical Predictions of Post-Glacial Relative Sea level Changes," *Journal of Geophysical Research* 96 [1991]: 4497–4523.)

flowed back into this region, causing the depressed bedrock to rise gradually toward its former elevation. Even now, thousands of years after ice melting, rates of relative sea level fall caused by ongoing bedrock rebound are as high as 10 mm per year.

As a result of this slow bedrock rebound, ancient beach ridges surround the lower-lying parts of Hudson Bay (Figure 17–4A). The modern beach is at sea level, and older beach ridges occur at successively higher elevations

away from the coast of Hudson Bay (Figure 17–4B). The highest-elevation ridge dates to 7000 ^{14}C years ago, just after the last ice melted from the area and ocean water flooded back into the depression in Hudson Bay caused by the ice. The stair-step series of beach ridges shows that the land beneath the former ice sheet has been rising for thousands of years, with the rates gradually slowing as the bedrock memory of the ice sheet load has weakened.

The second major group of tide gauge responses shows a relatively fast rate of sea level rise. This group is clustered in a halo pattern surrounding the former ice sheets but extending well beyond the ice margins (see Figure 17–2). In North America, these gauges are found along the east coast from southern New England south to Florida, and in Europe they are found in a narrower band across England, France, and northern Germany.

Today's rapid rise of sea level in these regions is also caused by a memory of the glacial maximum ice sheets, even though these areas were not located directly beneath the ice loads. During glacial times, the deep rock displaced from beneath the center of the ice sheet load had to go somewhere. Flowing outward beyond the margins of the ice sheets, it caused an increase in the elevation of the land, called a **peripheral forebulge** (Figure 17–5). As a rough analogy, the weight of a person sitting on a partly inflated air mattress in water will depress the center of the air mattress into the water, but the excess air pushed to the edges of the mattress will cause the edges to bulge up out of the water.

After the ice sheets melted, the rock displaced beyond the ice margins gradually flowed back into the region where the ice sheet had been. This return flow

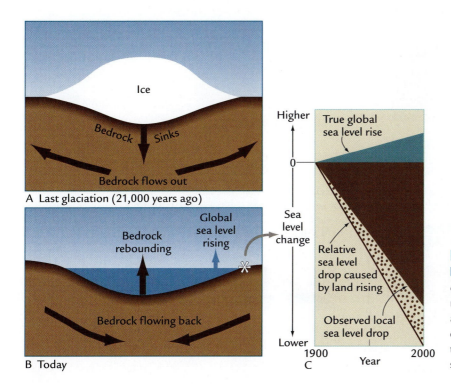

FIGURE 17-3 Bedrock rebound and sea level fall In the regions where ice sheets once were present, relative sea level is rapidly falling today. Bedrock in these areas is still rebounding in response to the earlier melting of ice, and the rebound of the land overwhelms the true global rise of sea level.

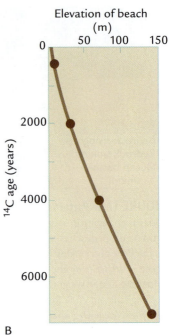

A

B

FIGURE 17-4 Old beach ridges (A) A series of old beach ridges surrounds Hudson Bay. (B) The beaches increase in age with elevation because the land has been slowly rising for 7000 years. (A: Courtesy of Claude Hillaire-Marcell, University of Quebec, Montreal. B: Adapted from W. R. Peltier and J. T. Andrews, "Glacial-Isostatic Adjustment. I. The Forward Problem," *Geophysical Journal of the Royal Astronomical Society* 46 [1976]: 605–46.)

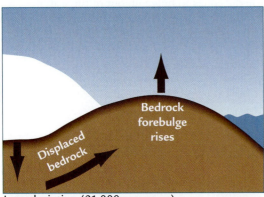

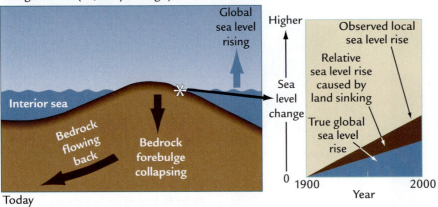

FIGURE 17-5 Bedrock sinking and sea level rise In regions surrounding glacial ice sheets, relative sea level is rising rapidly today. The continued flow of bedrock away from these areas and into the former ice sheet centers causes the land to sink and adds to the true global rise of sea level.

caused the peripheral forebulge to collapse and the land surface to sink (see Figure 17–5). The land in the region of the collapsing forebulges is still sinking, which adds to the true rise of global sea level and produces a fast rise of relative sea level in these regions. Dating of older (now submerged) beaches in these regions indicates that this pattern of unusually rapid rise in relative sea level has persisted for thousands of years.

The third group of tide gauge responses comes from coastlines located far from the northern hemisphere ice sheets (Figure 17–6). Relative sea level in these regions is rising at rates slightly less than the global rate of sea level rise. It might seem that regions located so far from the glacial ice sheets should be free of memory effects from the ice, but they are not. The return of glacial meltwater to the oceans has added an extra load on the bedrock beneath the ocean floor in these regions.

At maximum size, the ice sheets extracted a layer of water at least 110 m thick from the world ocean. With this water load removed, the average level of the crust in the ocean basins rose over 30 m compared to the level of the nearby continents that were not directly affected by the weight of the ice sheets or the weight of the ocean water. When this layer of water returned to the oceans during ice melting, it loaded down the ocean crust and caused it to sink (see Figure 17–6). This slow

sinking of the ocean crust is still going on today (because of the viscous memory effect), and it counteracts a small part of the true rise of global sea level, producing a slightly reduced rise in relative sea level in regions far from the ice sheets.

While the great glacial ice sheets are long gone, they are not forgotten, at least not by the bedrock and the shorelines. Bedrock memories of these shifting loads of ice and water are the major obstacles to determining the true rate of global sea level rise during the past century. Unfortunately, many of the longest and most reliable tide gauge records happen to have been located in just those regions of Europe and North America where the lingering overprints from the ice sheets are largest.

Attempts to remove all these complications from the melted ice sheets indicate that sea level rose by about 17 cm during the 1900s (Figure 17–7). Humans have been adjusting to this slow rise of the ocean for years. In coastal plain regions with very low slopes, a 17-cm rise in sea level within a century can translate into an advance of 1700 cm (17 m, or almost 60 ft) across the land. Many lighthouses built along coastal land a century or more ago now sit marooned in the ocean, protected at least in the near future by constructed boulder walls. Some, such as Cape Hatteras Light in North Carolina, have been moved inland to better-protected positions.

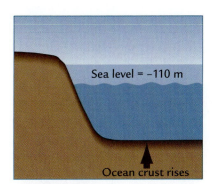

Sea level = –110 m

Ocean crust rises

Last glaciation
(21,000 years ago)

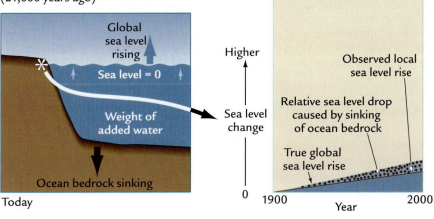

Global sea level rising

Sea level = 0

Weight of added water

Ocean bedrock sinking

Today

Higher

Sea level change

0

Observed local sea level rise

Relative sea level drop caused by sinking of ocean bedrock

True global sea level rise

1900   Year   2000

FIGURE 17-6 Ocean bedrock sinking and sea level rise In coastal regions far from glacial ice sheets, relative sea level is not rising as fast as the true global average. The continued sinking of ocean bedrock under the weight of 110 m of meltwater counteracts a small fraction of the true global rise in sea level.

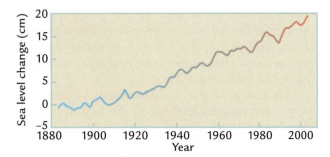

FIGURE 17-7 Global rise in sea level during the twentieth century Sea level rose almost 20 cm during the twentieth century because land ice has melted and seawater has warmed and expanded. (Adapted from B. C. Douglas, "Global Sea Rise: A Redetermination," *Surveys in Geophysics,* 28 [1997]: 279–92.)

Because this rise in sea level has been tied to several aspects of the warming that has occurred during the last century or more, the next two sections explore the instrumental and satellite evidence that document the recent warming. At the end of the chapter, we return to the issue of the cause of sea level rise.

Other Instrumental Records

The "modern" era of instrumental measurements of climate can be divided into two parts: (1) a recent era extending back four decades or less in which changes have been remotely sensed from satellites in space and (2) an earlier era in which measurements of climate were made using instruments invented during the era of scientific exploration that began centuries ago.

17-2 Thermometers: Surface Temperatures

Thermometers have been used to measure air temperature at a few locations in Eurasia and North America for over 200 years (Figure 17–8). At the same time, the surface temperature of the ocean has been measured along heavily traveled shipping routes and at ocean islands, but large gaps in coverage exist at middle and high latitudes of the southern hemisphere because of frequent storms and extensive sea ice. Only since the late 1800s have enough stations been recording temperature to permit reasonable estimates of the surface temperature of the entire planet.

Measurements both on land and at sea have been difficult. Ocean temperatures were once measured by scooping up seawater in a canvas bucket and inserting a thermometer. If a few minutes elapsed between collecting the water and measuring its temperature, evaporation could cool the water by several tenths of a degree centigrade. More reliable measurements came later from thermometers embedded in the outer parts of the

intake valves that drew in seawater to cool the ship's engines.

On land, the largest complication has been the growth in populations around many land stations. Near large towns and cities, the spread of asphalt surfaces and the loss of vegetation has led to increased absorption of solar radiation during the day and greater back radiation of heat at night. The result has been a significant extra warming effect at these stations. Although this warming reflects real temperature changes at these stations, it is not characteristic of changes across much more extensive rural areas. Care must be taken not to project this bias, called the **urban heat island** effect, into rural regions. Stations in regions of little or no population growth provide a check on the heat island effect. Adjusting for this bias reduced initial estimates of warming trends during the last century by about 30%.

Reconstructions of global temperature during the last 150 years by different groups are very similar,

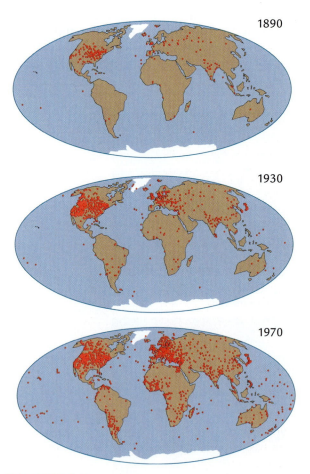

FIGURE 17-8 Temperature stations Coverage of land-based stations that measure surface temperature expanded significantly during the twentieth century. (National Climate Data Center, NOAA, Asheville, NC.)

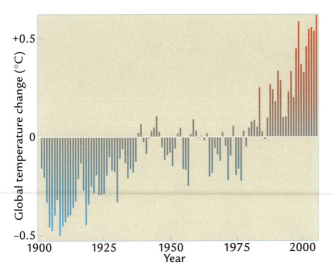

FIGURE 17-9 Change in surface temperature since 1900
Reconstructions of global surface temperature based on surface station thermometer measurements show a warming trend of 0.7 °C since 1900, interrupted by a small cooling from the late 1940s to the mid–1970s. (National Climate Data Center, NOAA, Asheville, NC.)

disagreeing mainly in the early 1900s when station coverage was still sparse. Temperatures have warmed by about 0.7°C over the last 110 years (Figure 17–9). Temperatures were considerably cooler before the early 1900s, rose quickly during the 1920s to early 1940s, stabilized or fell slightly from the late 1940s through the late 1970s, and have again risen abruptly since 1980.

17-3 Subsurface Ocean Temperatures

The ocean has the capacity to store enormous amounts of heat for long intervals of time (companion Web site, pp. 9–11). Changes in surface climate do not easily penetrate below the upper 100 m that are mixed by winds, but important information on deeper ocean trends during the last half-century has come from a painstaking examination of millions of subsurface temperature profiles by the climatologist Sid Levitus and his colleagues.

From these observations a detailed picture has emerged of the slow penetration of heat from the atmosphere into the subsurface layers below 100 m. Slow, downward molecule-by-molecule diffusion has transferred some of the surface heat to depths of a few hundreds of meters, and near-horizontal movement of heat has transferred even more heat into the subsurface ocean. This heat enters the ocean at higher middle latitudes and moves toward the equator along layers of equal density. Still slower penetration to depths of 1000 m or greater has also occurred in areas of deep overturning like the subpolar North Atlantic Ocean.

The integrated temperature increase for the global ocean down to 3000 m for the last half of the 1900s is 0.06°C (Figure 17–10). Although this warming is much smaller than the 0.7°C rise in surface air temperature, the amount of heat generated during the late part of the industrial era and stored in the deep ocean exceeds that stored in the atmosphere by more than a factor of 10. This heat storage in the ocean is direct evidence of a marked change in the heat balance of the entire climate system compared to earlier decades.

17-4 Mountain Glaciers

Today, mountain glaciers cover 680 km² of Earth's land surface and represent about 4% of the total surface area of land ice on Earth today. Mountain glaciers at middle and high latitudes respond to local climate, primarily changes in summer temperature and also variations in winter snowfall. At lower latitudes, solar radiation and precipitation are also important. Because of these differences in sensitivity, mountain glaciers in different regions can show varying behavior. Individual glaciers

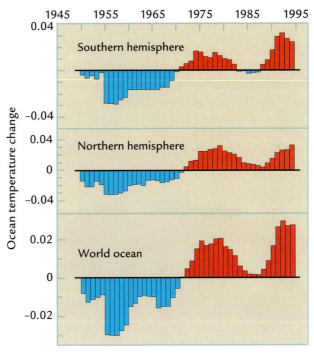

FIGURE 17-10 Subsurface ocean warming The mean ocean heat content of the upper 3000 m of the world ocean has increased in both the northern and the southern hemisphere and in the global average ocean by an average of 0.06 °C. Despite the small size of this warming, the amount of heat stored in the ocean exceeds all the other reservoirs combined. (Adapted from S. Levitus, J. I. Antonov, T. B. Boyer, and C. Stevens, "Warming of the World Ocean," *Science* 287 [2000]: 2225-9.)

may respond to climate at rates that range from about a decade to as much as several hundred years. The response times of most mountain glaciers fall in the range of 10–40 years.

Despite this wide range of possible responses, historical observations of the lower limits of glaciers between 1860 and 1900 show that 35 of 36 glaciers examined were already in retreat. More recent studies of glacier lengths have been supplemented in some cases by analyses of glacier thickness that permit calculations of full glacier volume (Figure 17–11). Between 1900 and 1980, 142 of 144 glaciers analyzed retreated (Figure 17–12). The average retreat of all glaciers between 1850 and 2000 was ~1750 m, or just over a mile. The energy used to melt these ice sheets (and polar sea ice) also used up a fraction of the excess heat generated during the industrial era, but it used far less than was stored in the deep ocean.

Exceptions to this general pattern of retreat exist. Some glaciers in the mountains of Norway advanced during the 1960s and 1970s during an interval of cooling in the Norwegian Sea, but the prevailing trend during the twentieth century has been melting. In recent decades, the rate of melting has accelerated for many glaciers. All tropical mountain glaciers studied are in retreat, and some have disappeared entirely.

This pervasive, near-global retreat cannot be explained by reduced precipitation. This explanation would require an average drop in precipitation of 25% in many sites across the globe. Instrumental evidence indicates that precipitation changes in most regions are much smaller than 25%, with increases in some glacier areas and reductions in others. In contrast, the temperature increases observed during the late 1800s and the

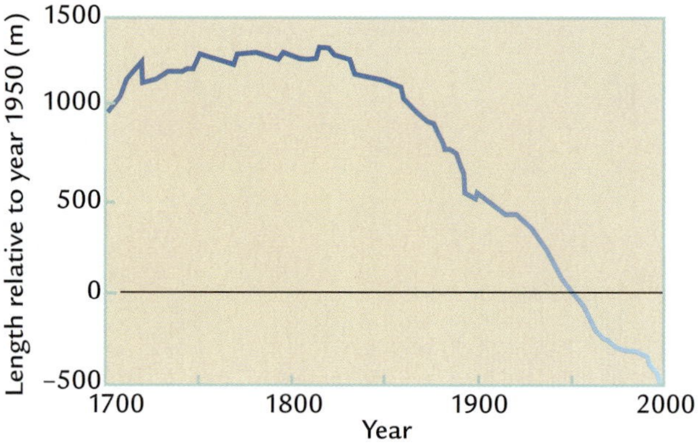

FIGURE 17-12 Retreat of mountain glaciers since the 1800s Mountain glaciers around the world have retreated by an average length of more than 1.5 km since the 1800s. (Adapted from J. Oerlemans, "Extracting a Climate Signal from 169 Glacier Records," *Science* 308 [2005]: 675-77.)

1900s have been both global in scale and of the right magnitude to explain the glacial melting.

17-5 Ground Temperature

Heat probes inserted into soils or bedrock can measure past changes in temperature that have slowly penetrated from the atmosphere and ground surface into subsurface layers. These profiles are sensitive to longer-term (century-scale) temperature changes at the surface but much less so to shorter, decadal-scale variations. Subsurface temperature records have been taken at hundreds of stations in both hemispheres (Figure 17–13A), and most profiles show warmer temperatures in the near-surface layers than a few tens of meters below. The measurements indicate that a warming has occurred at the surface in the last century or two and that it is in the process of penetrating to deeper layers. This warming of subsurface continental areas also used up a small fraction of the excess heat generated during the industrial era.

Models that simulate the penetration of the temperature anomalies beneath the surface indicate that the warming during the last two centuries lies at the upper end of the range of surface temperature reconstructions based on climatic proxies (Figure 17–13B). This match in part reflects the fact that both the ground temperature profiles and the proxy reconstructions showing larger variations tend to be based on sites in extratropical latitudes, where temperature responses are larger than the global average. Other complications with ground temperature profiles include depth of snow, which may shield the ground from extreme temperature changes in winter, and clearance of forests, which can cool local temperatures because of the higher albedo of open land surfaces.

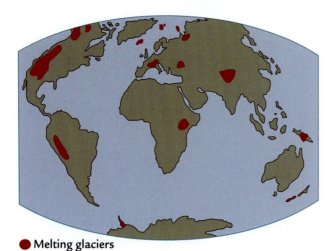

● Melting glaciers

FIGURE 17-11 Locations of retreating mountain glaciers Mountain glaciers in many regions retreated during the 1900s (Adapted from M. F. Meier, "Contribution of Small Glaciers to Sea Level," *Science* 226 [1984]: 1418-21.)

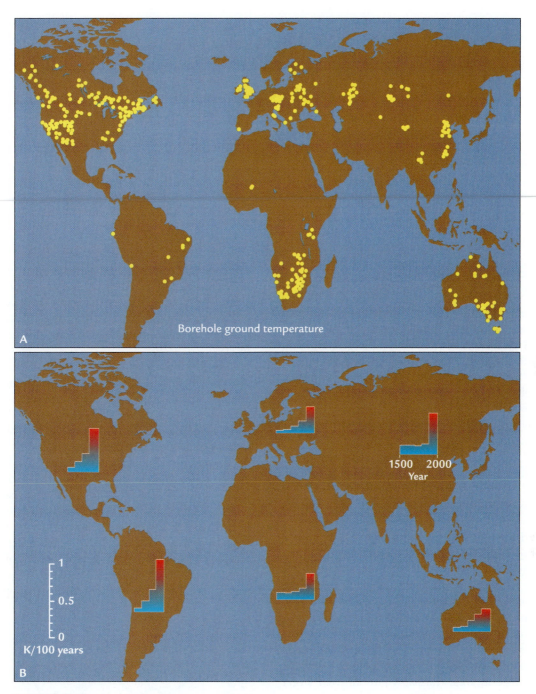

FIGURE 17-13 Ground temperature (A) Ground temperature profiles have been measured at hundreds of stations at middle and high latitudes. (B) Model simulations based on these profiles show a warming on all continents from the 1500s (far-left bar) to the 1900s (far-right bar). (Adapted from S. Huang et al., "Temperature Trends over the Past Five Centuries Reconstructed from Borehole Temperature," *Nature* 403 [2000]: 756–8.)

IN SUMMARY, instrumental measurements of temperature at the surface, in the subsurface ocean, and in the ground, along with measurements of mass balance of mountain glaciers, provide consistent evidence that a large warming has occurred during the last 125 years.

Satellite Observations

The range of measurements of Earth's climate increased markedly with the advent of satellite sensors in the late 1900s. Different satellite sensors have come on line at different times, including a few within the last decade.

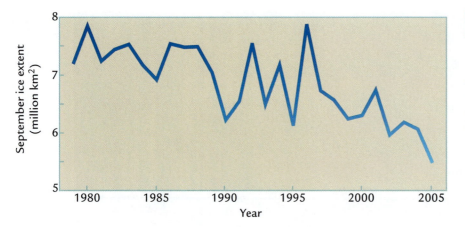

FIGURE 17-14 Decrease in Arctic sea ice cover Satellite measurements show a 6% decrease in the extent of sea ice cover in the Arctic Ocean since the early 1970s. (Intergovernmental Panel on Climate Change, "Climate Change 2007: The Physical Science Basis" [Geneva: World Meterological Association, 2007].)

17-6 Circum-Arctic Warming

Perhaps the most dramatic climatic responses during recent decades are those in the Arctic Ocean and over nearby continents. The region north of 60°N has warmed by an average of 1°C, with the largest warming over North America, but a small cooling over parts of Russia. This warming trend has been accompanied by large changes in a number of climatically sensitive indices.

The average annual extent of sea ice in the Arctic Ocean has decreased markedly since the late 1970s (Figure 17–14). The minimum extent of sea ice occurs in September at the end of the summer melt season. In the three decades since the 1970s, the extent of summer sea ice has decreased by ~25%, and losses accelerated in the early 2000s. If this trend continues, summer sea ice will not exist in the Arctic by the middle of this century.

Until recently, the extent of winter sea ice had not changed much, but significant decreases in winter sea ice began to occur in the early 2000s. In addition, measurements made by acoustic soundings from submarines show that the thickness of multiyear ice near the center of the Arctic Ocean thinned by 40% between the 1950s and mid–1990s, shrinking from just over 3 m to less than 2 m. This 40% thinning, combined with the 25% shrinkage in area, translates into an even larger loss in the total volume of sea ice.

The satellite records of sea-ice extent are consistent with the long-term warming indicated by surface temperature measurements and other observations. Yet even high-quality satellite records spanning several decades need to be interpreted with some caution. One complication is that shorter-term regional climatic changes that persist for a decade or longer may be hard to distinguish from a long-term global warming trend.

For example, sparse observations indicate that sea-ice limits were also reduced during the 1920s and 1930s. These observations agree with ground station

evidence that those decades were relatively warm in the Atlantic sector of the Arctic, at least in part because this interval was part of the early stage of the global warming trend. The extent of sea-ice retreat in recent years has now exceeded the losses during this earlier warm interval.

Satellite measurements from high latitudes of the northern hemisphere also show other trends that are consistent with major regional warming. Satellite measurements show decreases of northern hemisphere snow cover (Figure 17–15), with earlier melting of snow in spring and later initiation of snow cover in autumn.

An additional indication of warming at high northern latitudes comes from satellite and surface station measurements of the length of the growing season. Surface measurements in central Alaska indicate an irregular increase in the growing season by two weeks over the past 50 years (Figure 17–16). Satellite sensing of the chlorophyll produced by vegetation north of 45°N has also shown that by the mid–1990s the growing season was beginning a week earlier in spring than it had in the

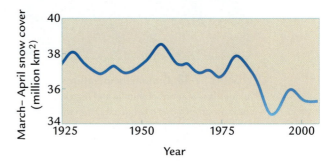

FIGURE 17-15 Decrease in snow cover over the northern hemisphere In the last several decades, satellite measurements show a decrease of snow cover in the northern hemisphere. (Adapted from Intergovernmental Panel on Climate Change, "Climate Change 2007: The Physical Science Basis" [Geneva: World Meterological Association, 2007].)

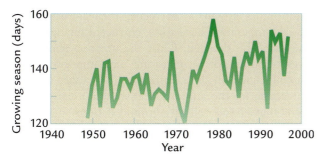

FIGURE 17-16 The growing season lengthens in Alaska
Surface temperature measurements indicate that the length of
the growing season increased in Alaska during the last half of
the twentieth century. (Adapted from S. W. Running et al.,
"Radar Remote Sensing Proposed for Monitoring Freeze-Thaw
Transitions in Boreal Regions," *EOS* 80 [1999]: 213–21.)

early 1980s and that it was ending half a week later in
autumn.

17-7 Ice Sheets

Precise measurements of the volume of ice sheets first
became possible near the start of the twenty-first century
(Figure 17–17). Satellites flying above the ice measure the
elevation of the ice surface with sufficient accuracy to
detect changes occurring within just a few years. Multiple
passes across the ice sheets combined with computer
analysis of the radar images received have made it possible to obtain accurate measurements of the entire surface
of the ice sheets.

Because the elevation of bedrock under the ice can
also change through time, ice thickness and ice volume
cannot be determined solely from measurements of sur-
face-ice elevation. Until recently, the elevation of bedrock

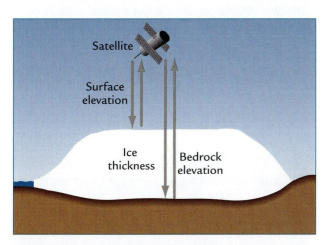

FIGURE 17-17 Changes in ice volume Changes in the
thickness and volume of ice sheets are being monitored from
satellites using radar and gravity measurements.

beneath the ice was measured by labor-intensive efforts in
which recording stations moving across the ice surface
sent out radar waves that bounced off the bedrock and
returned to the station. The travel time of the radar pulses
is a direct measure of ice thickness. This time-consuming
technique could be used only to measure selected lines
across small portions of the ice sheets. More recently,
radar measurements of bedrock elevation have been made
quickly and inexpensively from satellites, providing full
coverage of these continent-sized masses of ice.

In addition, surface stations installed in critical locations with global positioning receivers linked to satellites
measure both the elevation of the ice and the gravity
field (the strength of the pull of Earth's gravity on the
satellite receiver). Because rock is almost three times
denser than ice, the gravity field at each station primarily
measures changes in the elevation of the underlying
bedrock in response to ice melting or growth. Thinning
of the ice allows the crust to rebound quickly because of
the elastic part of the response of Earth's mantle to
unloading, while thickening of the ice depresses the
crust for the same reason (see Chapter 9). This information from the gravity data complements satellite and
radar measurements of ice elevation and thickness.

The evidence from almost a decade of wide-ranging
measurements of the Greenland ice sheet is that the central portion of the ice sheet at higher, colder elevations
above 2 km has been relatively stable, but the lower
coastal margins have been rapidly thinning. Overall, the
annual loss of ice on Greenland now exceeds 200 km³
per year, enough to raise global sea level by more than
5 mm/year (5 cm/century).

The annual rate of thinning along many coastal out-
let glaciers in the warmer southern half of Greenland is
1 m/year, and many of these thinner ice margins are
retreating by hundreds of meters each year. With the
ice retreating from the coasts, some areas that were
mapped as peninsulas until just a few years ago have
now been discovered to be islands, and geographers are
redrawing their maps. Much of the northern margin of
Greenland ice is also ablating, but at much slower rates
than farther south.

Changes in Antarctic ice have been less dramatic
and more variable from region to region. The part of
the Antarctic ice sheet along the peninsula that juts
northward toward the southern tip of South America
has been rapidly shedding ice during the last decade.
The huge central mass of the East Antarctic ice sheet
has areas that show both thickening and thinning, but
overall this ice sheet appears to be nearly stable. Satellite evidence has also shown that narrow ice streams
drain ice from the interior of the East Antarctic ice
sheet at rates as much as 100 times faster than the slow
flow across the rest of the ice sheet.

IN SUMMARY, a host of satellite measurements show that north polar regions have warmed dramatically. These trends add to the already overwhelming instrumental evidence that the planet has warmed over the last 125 years, particularly during the last 30–40 years.

For a time, however, one of the most important measurements made from satellites did not agree with this warming trend—measurements of the temperature of the troposphere (the atmosphere below elevations of ~10 km). Satellite data initially suggested no warming since 1980 and perhaps even a small cooling, in dramatic contradiction to the evidence from surface stations. Scientists skeptical about global warming pointed to these measurements as evidence that the temperature trend assembled from the surface stations must be in error. Their case was strengthened by the fact that temperature measurements made from **radiosondes** (metal devices suspended from balloons rising through the troposphere) agreed better with the satellite data than with the surface station data in some regions.

This mismatch has been resolved in recent years, and both the satellite estimates and some of the radiosonde estimates turned out to have been invalid. Satellite estimates of the temperature of the lower atmosphere are based on measurements of the *brightness* (energy emission level) of molecules of oxygen (O_2), which is the second largest constituent of Earth's atmosphere. This brightness parameter correlates with the temperature of the oxygen molecules in the air and of the air itself. The satellite sensors integrate energy emissions from oxygen molecules across the entire lower atmosphere (0–10 km) as well as the lower part of the stratosphere. If Earth's troposphere warms, the stratosphere will cool. As a result, changes in the warming of the lower troposphere from satellite measurements have to be derived as a residual after subtracting the cooling of the lower stratosphere.

Subsequent reexamination found that incorrect adjustments had been used to remove the changes in stratospheric temperatures to isolate the temperature changes in the troposphere. Additional problems were caused by the effect of friction on the satellite orbits: a gradual drift toward lower altitudes and a slightly later arrival over particular locations on Earth's surface. Still other problems arose when new satellites replaced older ones. When all these complications were taken into account, the satellite data came into close alignment with the surface station trend shown in Figure 17–9.

The problem with the radiosonde data turned out to be an overcorrection for solar heating of the metal shields. Nighttime radiosonde data agreed with the evidence from ground stations, but daytime data did not show the warming recorded at the surface. Because radiosondes are made of metal, their measurements of air temperature have to be corrected for heating of the metal by the Sun. Close examination showed an overcorrection for solar heating, thus imposing a false cooling signal on the measurements. A more accurate correction for the Sun's heating showed a temperature trend that agreed well with ground stations.

With this problem resolved, scientists who argued that global warming is not real lost one of their main supporting arguments. The full range of instrument and satellite data now indicates that a major warming has occurred in the last century or more.

Sources of the Recent Rise in Sea Level

We now have the information needed to return to the problem of the origin of the ~17-cm rise in average sea level during the last century. This increase is primarily the result of three factors (Table 17–1).

Because water expands slightly when heated above 4°C, warming of the ocean will cause sea level to rise. The subsurface ocean warming trend shown in Figure 17–10, along with less complete data from earlier decades, indicate that thermal expansion can explain 4 cm of sea level rise since 1960. In addition, even though mountain glaciers account for only about 1% of the ice present on land, 99% of them have been melting and retreating since the middle 1800s and adding water to the ocean (see Figure 17–12). These glaciers account for at least another ~5 cm of the rise in sea level during the last century. Current estimates are that about 2 cm of sea level rise during the 1900s came from melting of the Greenland and Antarctic ice sheets, although the net mass balance of the Antarctic ice sheet over most of the past century is not well constrained.

IN SUMMARY, the combination of thermal expansion, melting of mountain glaciers, and melting of ice on Greenland and Antarctica can account for 11 cm of the estimated sea level rise of ~17 cm during the twentieth century. This rise is yet another consistent part of the picture of global warming.

TABLE 17-1 Factors in the Rise of Sea Level in the Twentieth Century (in centimeters)

Ocean Thermal expansion	+4
Mountain glaciers	+5
Greenland and Antarctic ice	+2
All factors	+11
Observed sea level rise	+17

The rate of ice melting on Greenland has accelerated markedly since the 1990s. Satellite sensors deployed during the last decade show more rapid melting along the lower margins of the Greenland ice sheet, especially in the southern portions. Estimates of sea level change in future years will be much better constrained because of new instrumentation deployed in recent years. Satellites that have been measuring the altitude of the ocean surface since 1992 (Figure 17–18 top) indicate a mean sea level rise of 3 mm per year, or almost twice the 1.7 mm/year average for the 1900s. When satellites have been measuring the altitude of the sea surface for several more decades, it should be possible to see through shorter-term changes in sea level caused by weather phenomena and El Niño events and accurately quantify sea level rises caused by the melting of land ice and warming of ocean water.

Another technological innovation may help scientists find out what portion of the future sea level change is caused specifically by warming of the ocean. The velocity of sound waves moving through the ocean depends on water temperature: the velocity averages about 1500 m/sec but increases by 4.6 m/sec for each 1°C of warming of the water. The SOFAR (SOund Fixing And Ranging) channel at a water depth near 1 km is particularly favorable for transmission of sound waves (Figure 17–18 bottom). Sound waves moving through the overlying and underlying ocean layers are gradually bent into this channel from above and below because its temperature and density make the waves move slightly faster than in the surrounding layers. Scientists can use these far-traveled sound waves to measure the average sound velocity across the paths the waves follow and therefore the average temperature across large stretches of the subsurface ocean.

Shorter-Term Oscillations

Several of the trends toward greater warmth during the last century show oscillations or even small-scale reversals. For example, global temperature rose prior to the 1940s but fell during the 1950s through the 1970s, before beginning the rapid rise that continues today. In addition, satellite records of polar warmth span only the last four decades at most. The short length of these records leaves open the possibility that natural oscillations in climate over multiyear to decadal intervals could have contributed to some of the observed trends.

The large-scale ENSO fluctuations in the tropical Pacific Ocean are the most prominent short-term oscillation in the climate system (Chapter 16). During El Niño years, sea level pressure falls and ocean temperatures warm across the east-central tropical Pacific Ocean, while sea level pressure rises and precipitation decreases in the far western tropical Pacific Ocean and over New Guinea and northern Australia (Figure 17–19). El Niño events recur irregularly within a broad band of 2–7 years, and each fluctuation lasts for about a year.

In addition to the ENSO changes, other oscillations have been detected in smaller regions. These oscillations appear as changes in surface pressure, temperature, and winds that may persist for many years or even decades.

The **Pacific Decadal Oscillation** (PDO) has a spatial distribution similar to ENSO, but unlike ENSO it can persist for decades rather than just a single year. During "warm" (positive) PDO phases, sea-surface temperatures are warmer in the tropics and along the Pacific coast of western North America and cooler in the west-central North Pacific (Figure 17–20A). "Cool" PDO phases have the opposite spatial pattern. Based on a little more than a century of high-resolution instrumental observations, PDO patterns can persist for several decades (Figure 17–20B). A prominent change from a "cool" to a "warm" PDO pattern occurred in 1976, near the time that global temperature shifted from a few decades of stable or cooling climate to the rapid warming that continues today. The PDO pattern has been less persistent in the last 15 years, as global warming has continued. The century-length record is too short to determine whether the PDO has a cyclical signature of multidecadal or century length.

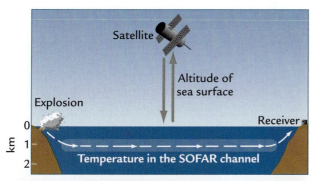

FIGURE 17-18 Changes in subsurface ocean temperature
Warming of the ocean is monitored by measuring increases in the height of the ocean caused by thermal expansion of seawater and increases in the velocity of sound traveling through subsurface layers.

A

FIGURE 17-19 El Niño and the Southern Oscillation (ENSO)
Temperatures are warmer off western South America and across the eastern tropical Pacific during El Niño episodes which occur every 2 to 7 years. (Adapted from E. M. Rasmussen, "El Niño and Variations in Climate," *American Scientist* 73 [1985]: 108–77.)

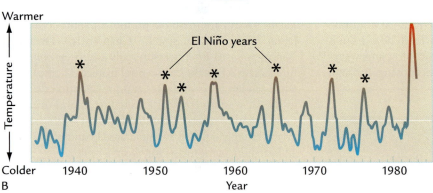

B

The **North Atlantic Oscillation** (NAO) is a fluctuation in atmospheric pressure between a subpolar low-pressure center near Iceland and a high-pressure center in the Azores-Gibraltar region. The NAO is best developed in winter. The "positive" NAO mode features lower pressure over Iceland, higher pressure over the Azores, and a strengthening of westerly winds across intervening latitudes of the Atlantic Ocean between these two pressure centers (Figure 17–21A). The subtropical Atlantic Ocean is warmer in a large region extending from the mid-Atlantic and southeast coast of the United States eastward to the Azores Islands. The warm, moisture-bearing winds arriving from this part of the Atlantic make Europe warmer and wetter than during years of negative NAO. Cooler temperatures occur off the west coast of Africa, where strong trade winds also send extra amounts of dust out across the ocean toward the Caribbean Sea.

The NAO has varied in strength over decadal scales, with persistent positive years in the early 1900s, numerous negative years in the 1960s and 1970s, and a large and persistent strengthening during the 1980s and early 1990s (Figure 17–21B). Some scientists have claimed that the rapid retreat of Arctic sea ice during the 1990s was affected by the North Atlantic oscillation. However, the North Atlantic oscillation weakened during the late 1990s, but the retreat of sea ice has not only continued but even intensified.

Still another possibility is that recent climate may have been influenced by solar forcing at the relatively weak and short-term cycles of 440 years or less that were identified by differences between tree ages derived by counting rings and by ¹⁴C dating (Chapter 14). Some high-resolution records show small tendencies toward cyclic behavior at these periods, but many of them also show similar tendencies at other periods that are not connected to known solar changes. In addition, many other records show no evidence of short-term solar forcing. At this point, the possibility of short-term cycles in response to solar forcing is not settled.

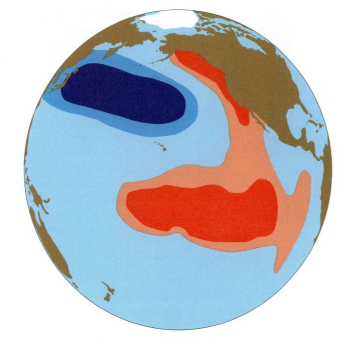

A

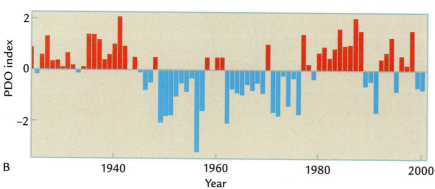

B

FIGURE 17-20 Pacific Decadal Oscillation (PDO) Like El Niños, positive PDO years have warm temperatures off the west coast of the United States but cool temperatures in the northwest Pacific ocean. PDO oscillations can persist for many years to decades. (Adapted from N. J. Mantua et al., "A Pacific Interdecadal Climatic Oscillation with Impacts on Salmon Production," *Bulletin of the American Meteorological Society* 78 [1997]: 1069–79, and from K. E. Trenberth and J. W. Hurrell, "Decadal Atmosphere-Ocean Variations in the Pacific," *Climate Dynamics* 9 [1994]: 303–19.)

IN SUMMARY, it is premature to dismiss the possibility that short-term oscillations have played some role in the global trend toward greater warmth since the late 1800s, but it is equally premature to conclude that they have. As the many climatic sensors now in place register year after year of record or near-record warmth, the current warming trend looks more and more like a long-term anthropogenic trend rather than a response to natural short-term oscillations.

Key Terms

tide gauges (p. 310)

peripheral forebulge (p. 311)

urban heat island (p. 314)

radiosondes (p. 320)

Pacific Decadal Oscillation (PDO) (p. 321)

North Atlantic Oscillation (NAO) (p. 322)

Review Questions

1. How do ice sheets that melted many thousands of years ago complicate efforts to determine the global sea level rise during the past century?

2. What is the urban heat island effect? How does it complicate attempts to synthesize trends of regional, hemispheric, or global temperature change?

3. Name four kinds of satellite evidence that support a gradual warming of high northern latitudes in the last two decades.

4. If Arctic sea ice has retreated by 25% and thinned by 40% in the last 50 years, what has been the percentage loss in its volume?

5. Why is it difficult to determine whether or not ice sheets are growing or shrinking?

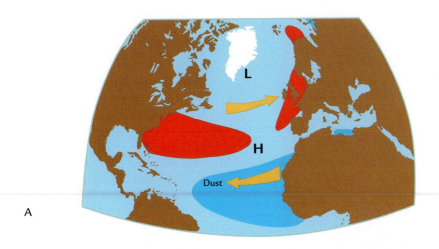

A

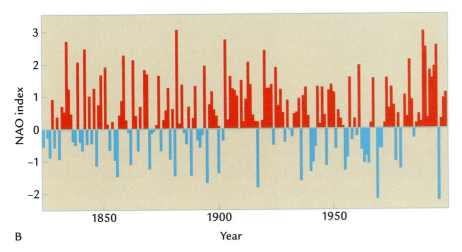

B

FIGURE 17-21 North Atlantic Oscillation (NAO) During positive NAO years, the western subtropical North Atlantic Ocean is warm, and strong winds (upper yellow arrow) blow this warmth and moisture into north-central Europe. NAO conditions (positive or negative) can persist for many years. (Adapted from J. W. Hurrell et al., "An Overview of the North Atlantic Oscillation," in *The North Atlantic Oscillation; Climate Significance and Environmental Impact*, ed. J. W. Hurrell et al., *Geophysical Monograph Series* 134 [2003]: 1–35.)

6. How does warming of the ocean affect sea level?

7. How do the North Atlantic Oscillation and Pacific Decadal Oscillation (NAO and PDO) complicate efforts to detect global warming?

Additional Resources

Basic Reading

www.igbp.net (International Geosphere-Biosphere Program)

http://wcrp.wmo.int/ (World Climate Research Program)

www.ncdc.noaa.gov/oa/climate/globalwarming.html (National Climate Data Center)

www.ipcc.ch (Intergovernmental Panel on Climate Change)

www.pages-igbp.org (Past Global Changes Project)

Advanced Reading

Chung, C. E., V. Ramanathan, and J. T. Kiehl. 2003. "Effects of the South Asian Absorbing Haze on the Northeast Monsoon and Surface-Air Heat Exchange." *Journal of Climate* 15: 2462–76.

Huang, S. H., N. Pollack, and P.-Y. Shen. 2000. "Temperature Trends over the Past Five Centuries Reconstructed from Borehole Temperature." *Nature* 403: 756–8.

Hurrell, J. W. 1995. "Decadal Trends in the North Atlantic Oscillation: Regional Temperatures and Precipitation." *Science* 269: 676–9.

Intergovernmental Panel on Climate Change. 2007. "Climate Change 2007: The Physical Science Basis." [Geneva: World Meterological Association, 2007].

Levitus, S., J. I. Antonov, T. B. Boyer, and C. Stevens. 2000. "Warming of the World Ocean." *Science* 287: 22285–93.

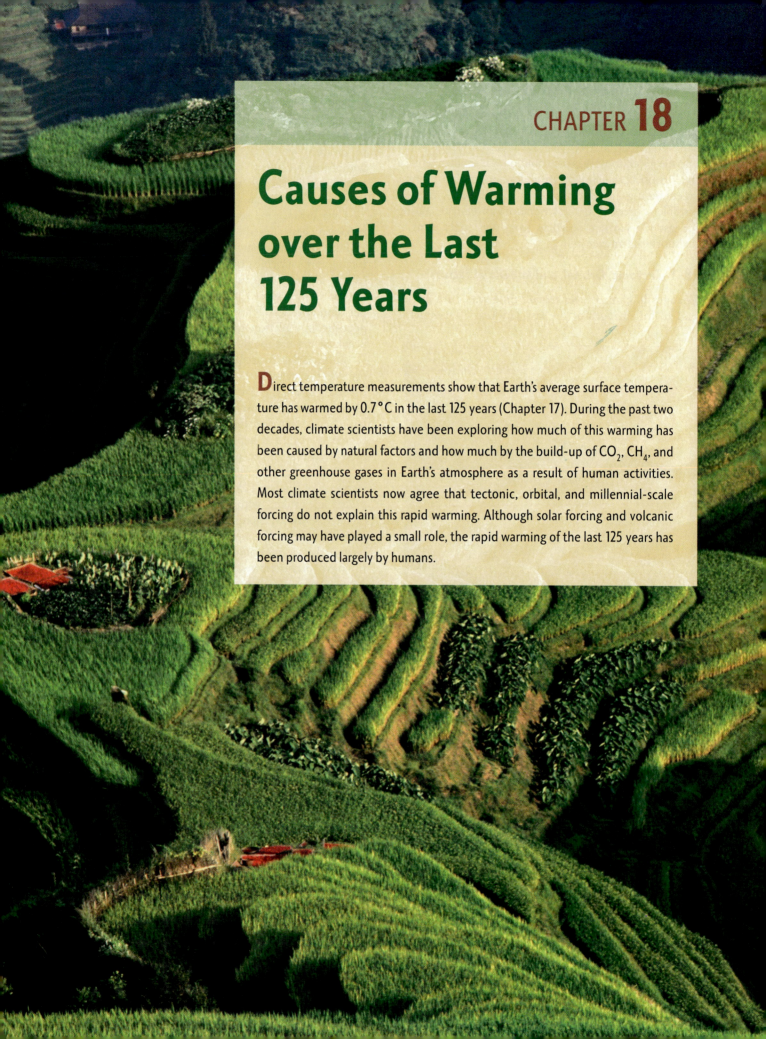

Causes of Warming over the Last 125 Years

Direct temperature measurements show that Earth's average surface temperature has warmed by 0.7 °C in the last 125 years (Chapter 17). During the past two decades, climate scientists have been exploring how much of this warming has been caused by natural factors and how much by the build-up of CO_2, CH_4, and other greenhouse gases in Earth's atmosphere as a result of human activities. Most climate scientists now agree that tectonic, orbital, and millennial-scale forcing do not explain this rapid warming. Although solar forcing and volcanic forcing may have played a small role, the rapid warming of the last 125 years has been produced largely by humans.

Natural Causes of Recent Warming

A key question in the 0.7°C global warming since the late 1800s is the role of natural changes in climate. A large contribution from natural changes would imply that the greenhouse-gas contribution was smaller, while a small contribution from natural forcing implies a larger greenhouse-gas role (Figure 18–1). Here we examine all possible sources of natural variations in climate during the last 125 years, proceeding from the longer-term to the shorter-term factors.

18-1 Tectonic, Orbital, and Millennial Factors

Changes in climate over *tectonic* time scales are irrelevant to the changes of the last 125 years. During the

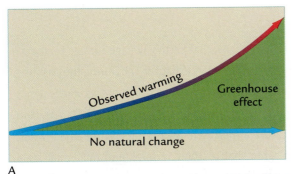

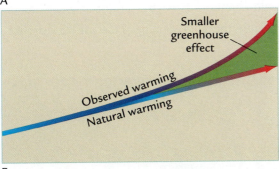

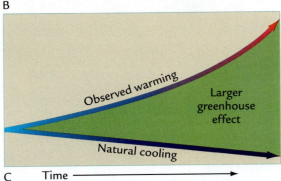

FIGURE 18-1 Natural warming and greenhouse effects
The fraction of the observed warming during the twentieth century that can be attributed to increased concentrations of greenhouse gases depends on the trend in natural climate, variously shown as (A) no change in temperature, (B) a natural warming, and (C) a natural cooling.

transition from greenhouse (ice-free) conditions to the current icehouse state, Earth's climate has cooled by at most 5°–10°C over 100 Myr. The average rate of cooling (~0.00001°C per century) has been much too slow to produce any detectable effect on Earth's climate within just a century or so. Shorter intervals of faster tectonic-scale change also fall well short of the rates needed to alter climate measurably in 125 years.

Over *orbital* time scales, changes in Earth's tilt and precession have altered the amount of insolation received at different latitudes and in different seasons, but orbital forcing is not a viable explanation of the recent warming. Global average temperature during the last 6000 years has cooled by at most 1°C, at an average rate of 0.016°C per century or less. The recent 0.7°C global warming is opposite in direction to this gradual cooling and has occurred at a rate roughly 35 times faster.

Millennial-scale oscillations were large when northern ice sheets existed, but they weakened as the ice melted. During the last 8000 years of the current interglaciation, climatic oscillations at the millennial scale have been weak and highly irregular in pattern from region to region (Chapter 14). In fact, it is not completely clear whether or not true "millennial oscillations" even existed during this interval. In contrast, the warming of the last 125 years has been global or near-global in extent, with ever-fewer regions trending counter to the warming pattern. This global (or near-global) response does not match the bipolar seesaw pattern of the glacial millennial oscillations, in which the timing of Antarctic responses is opposite those in the North Atlantic region. Millennial-scale oscillations do not appear to be a factor in the recent global warming.

18-2 Century- and Decadal-Scale Factors: Solar Forcing

Satellite measurements of the amount of radiation arriving from the Sun began in 1978 and have now documented changes over almost three 11-year cycles (Figure 18–2 top). During those cycles, solar radiation has varied by 0.15%, or 2 W/m², compared to the global average of 1370 W/m².

Climate models indicate that a change of 0.15% in the Sun's strength could alter global mean temperature by as much as 0.2°C if it persisted for many decades. Half an 11-year cycle, however, does not give the climate system time to register its full equilibrium response. As a result, Earth's mean temperature in the models warms and cools by less than 0.1°C in response to the 11-year variations in Sun strength. Temperature changes this small should be difficult to distinguish from the natural variability produced by all the other factors in Earth's climate system, and very few observational records show convincing evidence of an 11-year temperature signal.

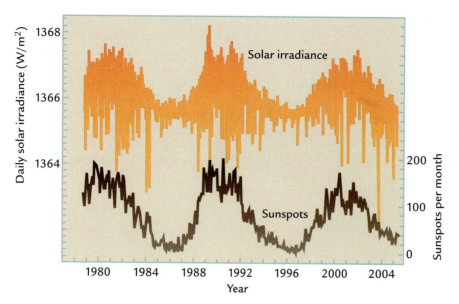

FIGURE 18-2 Solar radiation and sunspots Since 1978, satellites have measured changes in the solar radiation arriving at the top of the atmosphere (top), and these changes correlate with the observed numbers of sunspots (bottom). (Solar irradiance values adapted from www.pmodwrc.ch/, and sunspot data from sidc.oma.be/index.php3.)

As noted in Chapter 16, scientists estimated until a few years ago that the Sun may have been 0.25–0.4% weaker than it is now during several long sunspot minima that occurred prior to the industrial era. Such a change is about twice the range measured by satellites in recent decades. If these larger changes persisted for decades, the climate system would have had time to come much closer to its full equilibrium response, and model simulations indicate that temperatures could have fallen by as much as 0.3°–0.5°C below the temperatures of the late twentieth century. If so, half or more of the 0.7°C warming during the 1900s might have been accounted for by solar forcing.

This hypothesized Sun-climate link has not held up well under closer scrutiny because Sunlike stars do not vary over multidecadal intervals. Current estimates suggest that changes in solar irradiance have accounted for less than 0.1°C of the 0.7°C warming since the late 1800s. During the three decades of direct satellite observations, the solar contribution to the rapid warming since 1980 has been negligible. A proposed climate/solar link operating through the stratosphere remains a possibility, but it is a somewhat speculative one.

18-3 Annual-Scale Forcing: El Niños and Volcanic Eruptions

Two factors have had measurable effects on global climate over time scales of a year or two: major El Niño events and large volcanic explosions. In the last century, both factors have altered global temperature by less than 1°C, and their effects have disappeared into the background noise of the climate system within a year or two.

Large El Niño episodes can warm the eastern tropical Pacific sea surface by 2°–5°C and add 0.1°–0.2°C to the global mean temperature trend. The major El Niños that occurred in 1983 and 1998 produced one-year warm spikes in temperature (Figure 18-3), but their effects had disappeared by the following year.

Several large volcanic eruptions occurred during the era of instrumental temperature records, including Katmai (1912), Agung (1963), and El Chichón (1981). Although these eruptions probably cooled global climate by ~0.1°C for a few years (see Figure 18-3), the amount of cooling is difficult to determine because of uncertainties about the sulfur content of each eruption and the height in the atmosphere to which the SO_2 was ejected.

The large Mount Pinatubo eruption in the Philippines in 1991 was the first volcanic explosion measured in sufficient detail to assess its effect on global climate. Global climate cooled by 0.6°C during the summer after the eruption and by an average of ~0.3°C for the first full year after the eruption. After two years, the cooling effect of Pinatubo disappeared into the background noise of natural year-to-year temperature variability.

> **IN SUMMARY,** the gradual 0.7°C increase in global temperature over the last 125 years cannot be explained by natural forcing operating at tectonic, orbital, or millennial time scales, nor is it the result of short-term forcing from volcanic explosions or El Niño events. Up to 10% of the warming (0.07°C) could result from changes in solar irradiance.

Anthropogenic Causes of the Recent Warming

The bulk of the recent warming is anthropogenic in origin. We explore here several kinds of emissions from human activities that have altered climate: the greenhouse gases carbon dioxide and methane, chlorine-bearing

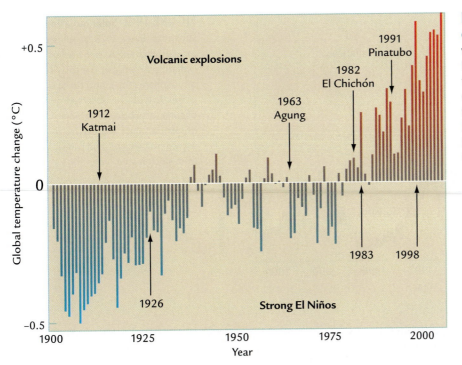

FIGURE 18-3 Brief volcanic cooling and El Niño warming Large volcanic explosions and major El Niño events cause short-term changes that can be large enough to be detected in the instrumental temperature record but leave the long-term baseline trend unaffected. (Adapted from National Climate Data Center, NOAA, Asheville, NC.)

chemicals of various kinds, and sulfate and carbon aerosols.

18-4 Carbon Dioxide (CO_2)

Bubbles of ancient air trapped in ice and instrumental measurements begun by the geochemist Charles Keeling in 1958 show an accelerating rise in the CO_2 concentration during the last two centuries. By the early 2000s the concentrations had passed 380 ppm, well above the 180–300 ppm range of natural (glacial-interglacial) variations (Figure 18–4A).

The additional carbon emitted from human activities has come mainly from two sources (Figure 18–5). Throughout the late 1700s and most of the 1800s, the main source of carbon was clearing of forests to meet the needs of an increasing human population: farmland for agriculture, wood for home heating, and charcoal to fuel the furnaces of the Industrial Revolution. Cutting and burning of forests (and resulting emission of CO_2) were particularly intensive in eastern North America during that interval.

After 1900, most of the extra carbon added to the atmosphere came from fossil fuel reservoirs buried

beneath Earth's surface. At first coal was the main fuel, but later on oil and natural gas became the major sources of energy. Gradually the carbon released by fossil fuels came to exceed the amount produced by land clearance by ever-larger amounts. Today industrial

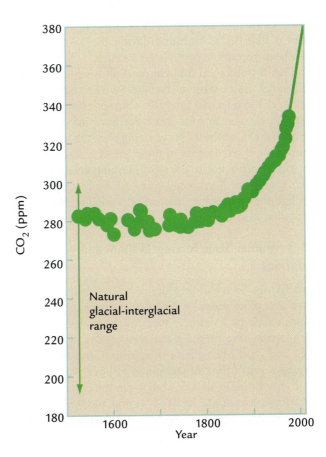

FIGURE 18-4 Preindustrial and anthropogenic CO_2 The combined atmospheric CO_2 record from bubbles in ice cores and from instrument measurements since 1958 shows an accelerating increase of CO_2 in the last 200 years above the natural baseline of 280 ppm. (Adapted from Intergovernmental Panel on Climate Change, "Climate Change 2007: The Physical Science Basis" [Geneva: World Meteorological Association, 2007].)

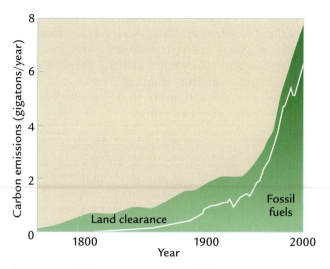

FIGURE 18-5 Human production of CO₂ Two factors account for the increase in atmospheric CO_2 caused by human activities in the last 250 years: (1) burning of carbon in trees to clear land for agriculture and (2) burning of carbon in fossil fuels—coal, oil, and gas. (Adapted from H. S. Kheshgi et al., "Accounting for the Missing Carbon Sink with the CO_2-Fertilization Effect," *Climate Change* 33 [1996]: 31-62, and from data in T. A. Boden et al., *Trends '91: A Compilation of Data on Global Change,* ornl/cdiac-46 [Oak Ridge, TN: Oak Ridge National Laboratory, 1991].)

phere, no one measurement in the ocean can provide a representative history of the average change in ocean CO_2 concentrations through time. Because the ocean is not as well mixed as the atmosphere, many measurements in many areas are needed to characterize its average change in CO_2 content.

The excess CO_2 from human activities has already been well mixed into the upper tens of meters of the ocean, which quickly exchange molecules of gas with the atmosphere. By comparison, the shallow subsurface ocean below 100 m is more out of touch with the atmosphere, and most of the deeper ocean below 1 km is even more isolated from the surface. As a result, smaller amounts of the excess CO_2 produced in the last two centuries have penetrated below 100 m, and very little has entered the deep ocean except in regions of active turnover like the subpolar North Atlantic Ocean.

Even in the surface layer of the ocean, the exchanges vary by region and by season. On an annual average, cold high-latitude ocean water acts as a net CO_2 sink and takes CO_2 from the atmosphere, while warm low-latitude ocean surfaces act as a CO_2 source and give some of it back (Figure 18–7). One reason for this pattern is that CO_2 gas is more easily dissolved in cold water than in warm water. Local air-sea exchanges are also governed by the relative concentration of CO_2 in the surface ocean versus the overlying atmosphere, and by other physical and biochemical processes that control carbon exchanges with subsurface waters.

For all these reasons, huge numbers of measurements made over vast regions of the ocean during all seasons are required to quantify the slow penetration of CO_2 into and beneath the ocean surface. Despite these problems, growing numbers of measurements confirm that the ocean takes up only about 25–30% of the total human carbon input.

carbon emissions (mostly in the northern hemisphere) account for most of the fossil fuel total, while cutting and burning of tropical rain forests are the largest land-clearance contribution.

In recent decades, 55% of the excess carbon produced each year has ended up in the atmosphere and another 25–30% has been added to the ocean (Figure 18–6). Unlike measurements of the well-mixed atmos-

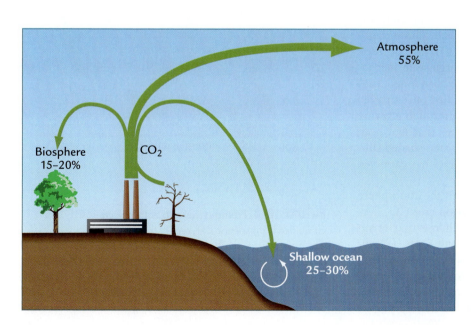

FIGURE 18-6 Where does the CO₂ produced by humans go? Of the carbon added to the climate system by humans, 55% ends up in the atmosphere, 25-30% enters the surface ocean, and the rest is stored in the biosphere (vegetation on land, litter, and organic carbon in estuaries).

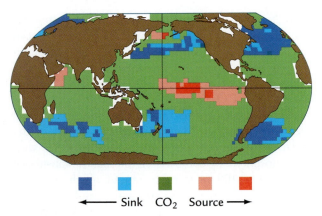

Sink ← CO₂ → Source

FIGURE 18-7 Ocean sources and sinks of CO₂ Annually averaged CO₂ concentrations in ocean surface waters are close to those in the overlying atmosphere, but the higher-latitude oceans act as net sinks that absorb carbon from the atmosphere, while the tropical oceans are net sources that give some of it back. (Adapted from T. Takahashi et al., "Global Air-Sea Flux of CO₂: An Estimate Based on Measurements of Sea-Air pCO₂ Differences," in *Carbon Dioxide and Climate Change* [Washington, DC: National Academy of Sciences, 1997].)

If 55% of the excess carbon ends up in the atmosphere and 25–30% in the oceans, where does the other 15–20% go? The only major reservoir left to take up the rest of the carbon is the biosphere, both live vegetation (trees and grasses) and dead organic litter in soils and coastal estuaries (see Figure 18–6).

Although burning of forests to clear land has been a major source of extra carbon for the atmosphere and the ocean for millennia (Chapter 15), in some areas carbon has recently been returning from the atmosphere to the vegetation. One way this can happen is by regrowth of forests in previously cleared regions.

In eastern North America, forests had been almost completely cut by the early 1900s for farming and for fuel. A century later, these regions look completely different. Forests now grow in areas where photographs from the early 1900s show landscapes devoid of trees. As the Midwest was opened up to large-scale mechanized farming, farms in the East were abandoned, particularly in New England. Rock walls that once marked the boundaries of open fields now run through growing forests. Near eastern cities, rural areas that had been cleared of trees gradually turned into tree-shaded suburbs. This widespread regrowth of trees has extracted CO₂ from the atmosphere.

A second way to remove CO₂ from the atmosphere is through CO₂ fertilization. Vegetation uses CO₂ during photosynthesis to create the cellulose that forms leaves, blades of grass, tree trunks, and roots. Greenhouse experiments show that most plants obtain carbon more easily from a CO₂-rich atmosphere and grow faster as a result. Scattered evidence suggests that the 35% rise in atmospheric CO₂ in the last 200 years has

increased this fertilization effect and taken more carbon from the atmosphere through several mechanisms. The vegetation grows faster; it becomes more varied in composition and grows more densely; the amount of woody material in tree branches, trunks, and roots increases; and trees and shrubs shed more fresh carbon litter into soils and coastal estuaries.

IN SUMMARY, ice cores and instrumental measurements show that atmospheric CO₂ levels have risen by ~35% in the last 200 years. This increase accounted for more than 60% of the total observed increase in the greenhouse-gas effect. Because greenhouse gases trap outgoing radiation from Earth's surface, the rising CO₂ levels have warmed the planet.

18-5 Methane (CH₄)

The concentration of the greenhouse gas methane in the atmosphere has also increased as a result of human activity. The influence of humans is again evident from trends measured both in ice core air bubbles and (since 1983) from instrumental observations (Figure 18–8). Since the 1800s the methane concentration has risen to over 1750 ppb, well above the natural range of 350–700 ppb during the previous 400,000 years.

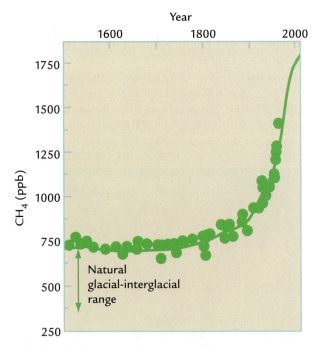

FIGURE 18-8 Preindustrial and anthropogenic CH₄ The combined atmospheric CH₄ record from bubbles in ice cores and from instrument measurements since the early 1980s shows an accelerating rise of CH₄ above the preindustrial baseline of 700–725 ppb. (Adapted from Intergovernmental Panel on Climate Change, "Climate Change 2007: The Physical Science Basis" [Geneva: World Meteorological Association, 2007].)

Methane added to the atmosphere comes from sources rich in organic carbon but lacking in oxygen: swampy bogs where plants decay, cattle and other grazing and browsing animals that digest vegetation, animal and human waste, and burning of grassy vegetation. In the absence of oxygen, bacteria break down the vegetation and extract the carbon, which combines with hydrogen to form methane gas.

Increased emissions of CH_4 during the last 200 years have resulted from the explosion of populations of humans and livestock. Today the amount of CH_4 produced by human activity is more than twice as large as that from natural sources (see Table 17–1). Ever-increasing areas of tropical land have been put into rice paddy cultivation in Southeast Asia, and these artificial wetlands produce methane. The growing numbers of cattle and other livestock have increased the amount of methane gas sent to the atmosphere.

Methane also acts as a greenhouse gas by trapping outgoing radiation from the Earth's surface. Although its concentration is much lower than that of CO_2, it is far more effective on a molecule-by-molecule basis in trapping radiation than CO_2. The enormous increase in methane concentration during the last two centuries has also caused the planet to warm. Based on the observed increases during the last 200 years, methane has accounted for ~16% of the total greenhouse-gas effect during the industrial era, although its rate of increase has slowed since the late 1990s.

18-6 Increases in Chlorofluorocarbons

Another group of chemical compounds that has increased in abundance in the atmosphere are the CFCs, or **chlorofluorocarbons**. These compounds contain chlorine (Cl), fluorine (Fl), and bromine (Br) and can be lifted high into the atmosphere by winds.

CFCs have for decades been produced for use as refrigerator and air conditioner coolants, chemical solvents, fire retardants, and foam insulation in buildings. Released at ground level, they slowly mixed upward through Earth's atmosphere. Because CFCs stay in the atmosphere for an average of 100 years, they eventually reached the stratosphere, where their concentrations increased during the late 1900s. CFCs act as a greenhouse gas, trapping outgoing back radiation from Earth's surface. Their increase in the 1900s explains about 10% of the total greenhouse-gas warming.

Several decades ago, the biologist James Lovelock, originator of the Gaia hypothesis, inferred that the rising production of the CFCs might be causing dangerous increases in their concentration in the stratosphere. In the 1970s, the atmospheric chemists Sherwood Roland and Mario Molina confirmed that the CFCs were indeed having a dangerous impact because of their interaction with stratospheric ozone.

Ozone (O_3) occurs naturally in the stratosphere, with the largest concentrations between 15 and 30 km. Incoming ultraviolet (UV) radiation from the Sun liberates individual O atoms from oxygen (O_2) and produces ozone:

$$\text{Radiation} + O_2 \rightarrow O + O$$
$$\text{(UV)}$$
$$O + O_2 \rightarrow O_3$$

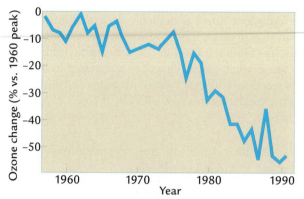

A Antarctic ozone concentration

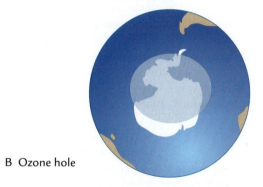

B Ozone hole

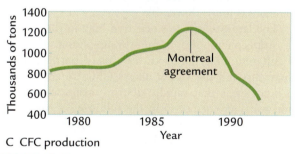

C CFC production

FIGURE 18-9 Anthropogenic CFC increases Increasing CFC concentrations in the atmosphere in recent decades have been the result of human production. (A, B) The total amount of ozone in a column of air over Antarctica in spring decreased by half between the 1960s and 1980s because of rising CFC levels. (C) The 1987 international Montreal agreement has begun to reduce CFC production. (A: Adapted from T. E. Graedel and P. J. Crutzen, *Atmosphere, Climate, and Change,* Scientific American Library, ©1997 by Lucent Technologies. B, C: Adapted from T. E. Graedel and P. J. Crutzen, *Atmosphere, Climate, and Change,* Scientific American Library, ©1997 by Lucent Technologies, after J. C. Farman et al., "Large Losses of Total Ozone in Antarctica Reveal Seasonal Interaction," *Nature* 315 [1985]: 201–10.)

Ozone is naturally converted back to oxygen (O_2) in the atmosphere by a similar process, but in this case the radiation source can be light in either ultraviolet or visible wavelengths:

$$O_3 + \text{Radiation} \rightarrow O_2 + O$$
$$\text{(UV or visible)}$$
$$O + O_3 \rightarrow 2O_2$$

With visible radiation far more abundant than ultraviolet radiation, ozone is naturally destroyed much faster than it is produced. As a result, ozone is a short-lived gas. In addition, the rate of conversion back to O_2 increases when certain chemicals are present to speed up the reaction. Chlorine reacts with ozone and destroys it, forming chlorine monoxide (ClO):

$$Cl + O_3 \rightarrow ClO + O_2$$

Chlorine then reacts with free oxygen molecules and is liberated from ClO:

$$ClO + O \rightarrow Cl + O_2$$

These liberated chlorine atoms then begin a new cycle of ozone destruction. This cycle is important to humans because ozone in the stratosphere forms a natural protective barrier that shields life forms from levels of ultraviolet radiation that would otherwise produce cell mutations including skin cancers.

In the last century, human activities have greatly accelerated the natural destruction of ozone by adding extra chlorine to the stratosphere. Measurements from 1960 to 1990 showed that the amount of ozone in a column of air over Antarctica had decreased considerably in the region where stratospheric chlorine is unusually abundant (Figure 18–9A, B).

The largest decreases occurred high in the Antarctic stratosphere during the spring season. Isolation of Antarctic polar air from the rest of Earth's atmosphere through the winter allows CFCs to accumulate to high levels that rapidly destroy ozone when solar radiation increases early in the following spring. The region over Antarctica in which stratospheric ozone is much less abundant than elsewhere is called the **ozone hole.**

This clear connection between CFCs and ozone depletion caused so much alarm that the world's nations signed a treaty in Montreal in 1987 to reduce and ultimately eliminate the use of CFCs. Production immediately began to decline (Figure 18–9C), and the concentrations of the type of CFCs that industries found easiest to replace stabilized and began a slow decline. Other CFCs that are still in widespread use have continued to increase, but at slower rates. Stratospheric ozone levels have stopped falling, but have not yet begun a significant recovery toward natural levels.

Ozone also occurs naturally in much greater abundance in the lower troposphere. It originates from both natural and anthropogenic processes, including biomass burning and oil production in refineries. At these lower levels in the atmosphere, ozone generally plays a positive environmental role by cleansing carbon monoxide (CO) and sulfur dioxide (SO_2) from the air.

At high concentrations, however, ozone is toxic to plants and an irritant to human eyes and lungs. In the lowermost atmosphere, ozone trends have moved in the opposite direction during the industrial era. Human activities have caused large ozone increases that have produced periodic smog alerts in many large cities. Slow-moving air masses settle over urban areas and allow concentrations of ozone and other pollutants to build to dangerous levels in summer. Tropospheric ozone acts as another greenhouse gas, and its buildup in the troposphere has added about 10% to the industrial era warming of Earth's surface.

IN SUMMARY, the increase in CFC concentrations during the late twentieth century has added to the greenhouse-gas warming, but the destruction of stratospheric ozone has cooled the planet slightly. Increases in tropospheric ozone have contributed to the warming.

18-7 Sulfate Aerosols

Industrial era smokestacks emit the gas sulfur dioxide (SO_2) as a by-product of smelting operations in furnaces and from burning coal. SO_2 reacts with water vapor and is transformed into sulfate particles, called **sulfate aerosols.** Because these aerosols stay within the lower several kilometers of the atmosphere, their primary impact on climate is regional in scale.

Until the 1950s, smokestacks in Europe and North America were small and most SO_2 emissions stayed close to ground level, producing thick industrial hazes and sulfur-rich acidic air around cites. Building facades and cemetery monuments made of limestone and marble were deeply etched by these acidic hazes. In the 1970s, taller smokestacks were built to disperse SO_2 emissions higher in the atmosphere (up to 3 km). This effort dramatically improved air quality in many cities, but it created a different problem in more distant areas.

The sulfate particles that are now being sent higher in the atmosphere are carried by fast-moving winds across broad areas. Although sulfates stay in the atmosphere for only a few days before rain removes them, they can be carried 500 or more kilometers downwind from source regions. Today large plumes of sulfate aerosols are carried far from sources in Eastern Europe, east-central North America, and China (Figure 18–10).

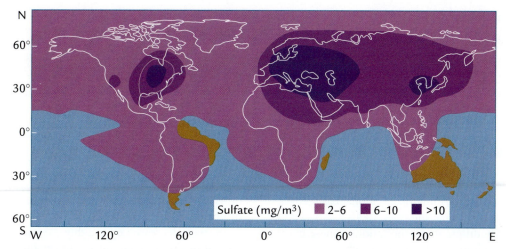

FIGURE 18-10 Sulfate aerosol plumes Sulfur dioxide emissions from industrial smokestacks in the North American Midwest, Eastern Europe, and China produce sulfate aerosols that prevailing winds carry eastward in large plumes. (Adapted from R. J. Charlson et al., "Perturbation of the Northern Hemisphere Radiative Balance by Backscattering from Aerosols," *Tellus* 43 [1991]: 152–63.)

Part of the history of SO_2 emissions is recorded in Greenland ice (Figure 18–11). Throughout the 1800s, industrial SO_2 emissions had been smaller than those from natural sources, but by the middle 1900s, industry emissions became the dominant SO_2 source. The sulfate concentration in Greenland ice rose sharply during World War II industrialization. Concentrations in Greenland ice began to drop sharply after 1980, when the United States (and then other nations) acted to limit SO_2 emissions.

Climate scientists have inferred that the large plumes of sulfate aerosols cause regional cooling downwind from smokestack sources. Like sulfate aerosols created by volcanic explosions, industrial sulfate particles reflect and scatter some of the incoming solar radiation back to space and keep it from reaching Earth's surface. The reduction in radiation cools climate regionally, and these effects show up in tabulations of global mean temperature change.

A second potential climatic effect of sulfate aerosols is less well understood. The tiny particles form natural centers (nuclei) around which water vapor can condense, forming droplets and then clouds. Clouds can have two opposing effects on climate: the surfaces of thick clouds can reflect more incoming solar radiation and cool climate, but thinner, higher clouds can also absorb more outgoing radiation from Earth's surface and increase the greenhouse effect. Because sulfate aerosols mainly affect low-level clouds, their indirect effects are thought to cool climate, but the amount is highly uncertain.

IN SUMMARY, the net global effect of increased sulfate aerosols is thought to be a cooling, but estimates of the size of the cooling vary widely.

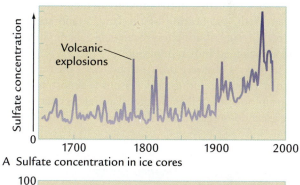

A Sulfate concentration in ice cores

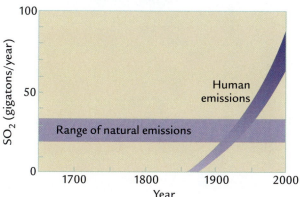

B Estimated global SO_2 emissions

FIGURE 18-11 Preindustrial and anthropogenic sulfates (A) Measured sulfate concentrations in Greenland ice reveal a significant regional increase since the nineteenth century. (B) Estimated global SO_2 emissions from smokestacks now exceed natural emissions. (A: Adapted from P. A. Mayewski et al., "An Ice-Core Record of Atmospheric Responses to Anthropogenic Sulphate and Nitrate," *Atmospheric Environment* 27 [1990]: 2915–19. B: Adapted from R. J. Charlson et al., "Climate Forcing by Anthropogenic Aerosols," *Science* 255 [1992]: 423–30.)

18-8 Brown Clouds

In the late 1980s and 1990s, scientists (and anyone traveling on airplanes) could see brown hazes hanging over many regions, particularly Eurasian megacities with millions of people. These layers of haze contain a wide range of chemical constituents, including carbon-rich organic aerosols generated by small stoves used for cooking and heating, by biomass burning, and by other activities.

Because dark carbon-rich aerosols absorb the Sun's radiation, scientists initially suspected that these hazes heated the lower atmosphere and added to the net amplitude of global warming on a regional basis. Investigations in the late 1990s and early 2000s led, however, to a surprisingly different interpretation of these hazes, called **brown clouds.** The carbon-rich aerosols do absorb radiation, but they heat the layer of air 2–3 km above the surface. As a result, the clouds block a portion of the incoming solar radiation and prevent it from reaching Earth's surface, which cools. One important consequence of this ongoing decrease in solar radiation has been a reduction of the intensity of the hydrological cycle. With less heating of the land, evaporation has decreased, contributing to greater sub-Saharan drought and a weakened Indian monsoon.

In regions of severe brown-cloud hazes, the reduction in solar radiation during peak seasons is almost an order of magnitude larger than the global average increase from the greenhouse-gas effect (Figure 18–12). The effects of the brown clouds have also been found to extend thousands of kilometers downwind from source regions.

> **IN SUMMARY,** these findings indicate that part of the true greenhouse-gas warming effect has been masked in recent decades in regions where brown clouds occur. The actual greenhouse warming effect in these areas (and for the planet as a whole) has been larger than previously thought.

During the late 1900s and early 2000s, satellite observations detected a related phenomenon called **global dimming.** The brightness of Earth's surface viewed from space decreased by roughly 7% over four decades. Part of this trend is connected to the buildup of carbon aerosols in brown clouds. With the brown clouds intercepting more solar radiation, the land received less solar radiation and became dimmer (less bright). Other factors include blocking of solar radiation by contrails emitted from jets and other emissions from urban areas.

In several older industrial regions where aerosol emissions have been reduced, such as the closing of polluting industries in the former Soviet Union, the long-term dimming trend has reversed since the early 1990s and the surface is brightening. The dimming trend continues over Southeast Asia, Africa, and other regions where emissions from urbanization and other sources are still growing.

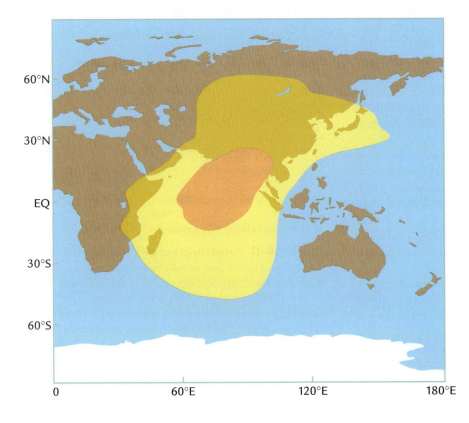

FIGURE 18-12 Brown clouds
Brown-cloud hazes in the lower atmosphere shroud regions downwind of large populations in India and China. (Adapted from C. E. Chung and V. Ramanathan, "South Asian Haze Forcing: Remote Impacts with Implications to ENSO and AO," *Journal of Climate* 16 [2003]: 1791–1806.)

18-9 Land Clearance

Human activities, primarily cutting of forests for agriculture, have altered much of the land surface of the planet. Forest clearance causes a net increase in albedo, as darker forests are replaced by brighter pastures and croplands. With more solar radiation reflected, the surface cools. At tropical and subtropical latitudes, forest clearance also reduces the amount of evapotranspiration. With reduced moisture availability, land surfaces dry out and bake in the intense summer Sun. On a global average basis, the net effect of land clearance has been a small cooling of the planet.

Earth's Sensitivity to Greenhouse Gases

Two independent sources of evidence put constraints on Earth's sensitivity to greenhouse gases: (1) numerical models of the climate system (GCMs and simpler models) and (2) climate reconstructions of intervals from Earth's history when greenhouse-gas concentrations differed from those today.

18-10 Sensitivity in Climate Models

GCMs and simpler climate models have long been used to simulate Earth's sensitivity to changes in greenhouse gases. For convenience, this sensitivity is quantified as the global average change in surface temperature caused by a doubling of CO_2 concentrations from the modern (preindustrial) level of 280 ppm. With a doubled CO_2 concentration used as an initial boundary condition, the models are run until the simulated temperature comes into equilibrium with this higher CO_2 level. The global average increase in simulated temperature is the **2 × CO_2 sensitivity** for that model.

The CO_2 concentration has increased by ~35% since the start of the industrial era. Climate scientists also calculate the combined heat-trapping effects of the other greenhouse gases by converting changes in the other gases into equivalent changes in CO_2. For example, the 150% increase in methane concentrations over the last 150 years has had an additional greenhouse effect equivalent to a 12% increase in CO_2. This is counted as a 12% increase in **equivalent CO_2**. The 35% increase in CO_2, combined with the 12% equivalent increase in methane and the 13% increase in all other greenhouse gases, has produced a combined increase equivalent to a CO_2 rise of ~60%.

Calculating the effect of greenhouse gases on Earth's climate involves two steps. Step 1 is to determine the **radiative forcing** provided by the gases, using the same W/m^2 units with which incoming solar radiation and Earth's back radiation are measured (Box 18–1). The radiative forcing excludes the complicating effect of feedbacks in the climate system. Because clouds are the largest uncertainty among these feedbacks, radiative forcing is sometimes called "clear sky forcing," meaning the amount of warming from the gases in the absence of any clouds produced as an indirect effect. Climate scientists widely agree that the direct radiative effect of doubling CO_2 (or equivalent amounts of other greenhouse gases) would increase global temperature by ~1.25°C.

The next step is to multiply the estimates of total radiative forcing by estimates of the sensitivity of the climate system to the forcing. GCM simulations over the last several decades have produced a wide range of estimates of Earth's temperature sensitivity to the 2 × CO_2 (or equivalent CO_2) level (Figure 18–13). These estimates average around 2.5°C but vary by 1°–1.5°C around that central estimate.

The reason for this wide range of estimates is uncertainties about feedbacks in the climate system. Positive feedbacks add to the warming produced by the baseline radiative forcing, while negative feedbacks counter some of the warming. The most prominent feedbacks come from changes in water vapor, the albedo of snow and ice, and clouds.

Water vapor, the major greenhouse gas in Earth's atmosphere today, provides positive feedback to the warming initiated by increases in greenhouse gases. In a clear, cloudless sky, the amount of water vapor that can be present in air increases rapidly at higher temperatures (companion Web site, pp. 13–14). This feedback helps make the tropics warmer than the poles and summers warmer than winters.

The same positive feedback should occur in response to temperature changes initiated by increases in CO_2 and other gases. The initial 1.25°C warming caused by doubling CO_2 levels should increase the amount of water vapor and cause an additional increase in global temperature of about 2.5°C, or twice the amount caused by CO_2 alone (Figure 18–14).

Another important positive feedback arises from reflection of solar radiation by ice and snow. If a warming initiated by greenhouse gases causes a retreat of snow and ice toward the poles, the reduced extent of these high-albedo surfaces will increase the absorption of solar radiation at high latitudes. The resulting positive feedback should increase the initial 2 × CO_2 warming by about 0.6°C.

By far the largest feedback-related uncertainty in models is clouds (see Figure 18–14). Different types of clouds vary in the amount of solar radiation they reflect (which cools climate) compared to the amount of back radiation from Earth's surface they absorb (which warms climate). High, wispy clouds tend to warm Earth's climate slightly, because they are composed of ice crystals that are better at absorbing outgoing radiation than at reflecting incoming radiation. Thicker, lower clouds

BOX 18-1   CLIMATE INTERACTIONS AND FEEDBACKS
Radiative Forcing of Recent Warming

The movement of heat through Earth's climate system can be traced from the incoming solar radiation through the subsequent reflection and absorption of that radiation, the trapping and redistribution of Earth's back radiation by greenhouse gases, and the outgoing radiation. Each of these processes redistributes radiative energy and each can be measured in units of W/m².

Of the 343 W of incoming solar radiation per square meter of Earth's surface, an average of 240 W/m² penetrates into the climate system. Prior to the industrial era, naturally occurring greenhouse gases such as water vapor, CO_2, and CH_4 trapped ~328 W/m² of back radiation from Earth's surface in a natural greenhouse effect. This trapping of energy and the internal feedbacks that resulted helped to make Earth 33 °C warmer than it would have been without greenhouse gases (and kept it from freezing).

Greenhouse gases produced by humans since 1800 have added to this natural greenhouse effect. The radiative forcing effect of these added gases is their impact on climate in the absence of feedback effects from clouds, water vapor, snow and ice albedo, and vegetation. Estimates place the radiative effect of the industrial era buildup of greenhouse gases at 2.7 W/m², or just over 1% of the total amount of incoming solar radiation (240 W/m²). Carbon dioxide has contributed 60% of the total increase in radiative forcing; CH_4, CFCs, and N_2O contribute the remaining 40%.

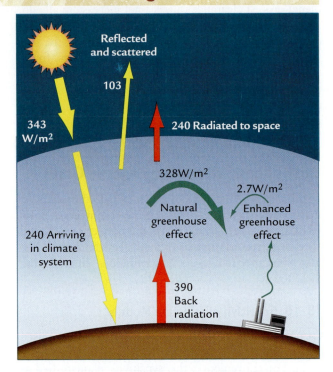

Effects of increases in greenhouse gases on radiation
Human activities since the start of the industrial era have increased greenhouse gas concentrations enough to enhance the natural greenhouse effect by 1.8%. (Adapted from IPCC Scientific Assessment Working Group, *Radiative Forcing of Climate Change,* ed. J. T. Houghton et al. [Cambridge: Cambridge University Press, 1994].)

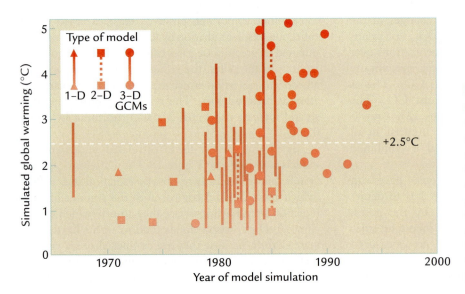

FIGURE 18-13 Model simulations of 2 × CO_2 sensitivity Simulations with several kinds of climate models in recent decades have yielded estimated sensitivities of global mean temperature to doubled concentrations of atmospheric greenhouse gases in the range of 0.5°–5°C. (Adapted in part from J. Adem and R. Garduno, "Feedback Effects of Atmospheric CO_2-Induced Warming," *Geofísica Internacional* 37 [1998]: 55–70.)

BOX 18-1   CLIMATE INTERACTIONS AND FEEDBACKS

CONTINUED

Because 2.7 W/m² represents a 1.8% addition to the natural greenhouse effect of 150 W/m², it is referred to as an **enhanced greenhouse effect.** In response to this anthropogenic enhancement, Earth's surface and lower atmosphere have warmed, and the atmosphere now radiates additional heat to space to compensate for the warming.

When the climate system has time to reach an equilibrium temperature response to an initial change in radiative forcing, the amount of warming depends both on the increase in radiative forcing and on the sensitivity of the climate system, which is only known within rather broad limits

(see Figure 18–13). For a mid-range estimate of 2.5°C for the climate system sensitivity to a CO_2 doubling, the industrial-era radiative forcing of 2.7 W/m² can be converted to an equilibrium temperature response by multiplying it by the climate-system sensitivity of 0.625°C/W/m²). According to this calculation, Earth's climate would have warmed by 1.7°C (2.7 × 0.625) during the industrial era if it had reached full equilibrium with the greenhouse-gas concentration in 2006. Because the climate system has a response time of several decades, however, it has not responded fully to greenhouse gases added during recent decades.

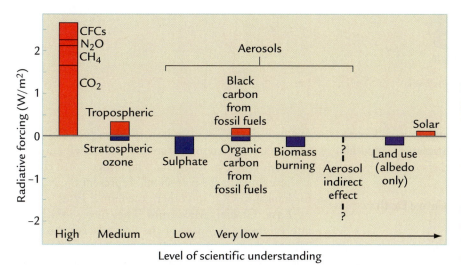

Radiative effects of greenhouse gases Several greenhouse gases have contributed a total of 2.7 W/m² to the greenhouse effect since 1850. The contributions of nongreenhouse factors (aerosols and other factors) are less certain. (Adapted from Intergovernmental Panel on Climate Change, "Climate Change 2007: The Physical Science Basis" [Geneva: World Meteorological Organization, 2007].)

cool climate because they are better at reflecting incoming solar radiation than at trapping outgoing radiation.

The problem facing scientists is assessing all the changes in the many types of clouds as Earth's climate warms. Even the best climate models have grid boxes much too large to simulate in a realistic way individual clouds and the processes that operate within them. These small-scale processes have to be estimated based on statistical probability. For example, if the air temperature across a specific region in a model simulation cools to a particular value, the grid boxes within that region are directed to produce a certain fraction of cloud cover that may deliver precipitation.

At present, it is not even possible to predict whether the total amount of cloud cover on Earth would increase or decrease as greenhouse gases increase. A warmer atmosphere will evaporate more water vapor from tropical oceans, thereby increasing the amount of

water vapor available to form clouds, but a warmer atmosphere also gains in its capacity to hold water vapor, which reduces the likelihood that vapor will condense into clouds. Which of these competing effects would win out in a warming world remains unclear. As a result of these uncertainties, the treatment of clouds in different climate models over the last several decades has yielded a net overall feedback effect ranging from slightly positive (a small warming) to highly negative (a large cooling).

The assessment of the 2007 **Intergovernmental Panel on Climate Change** (IPCC) is that the negative feedback from clouds cancels more than half of the positive feedback from water vapor and albedo (see Figure 18–14). The estimated effect of combining all the feedbacks is a 1.25°C warming, equivalent in size to the initial radiative warming caused by greenhouse gases alone. The IPCC estimates the net 2 × CO_2

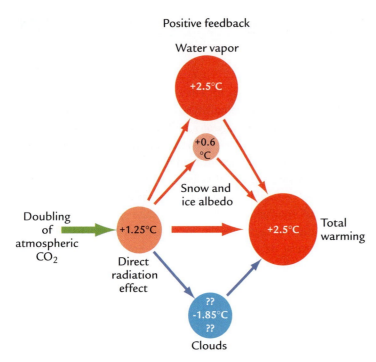

Positive feedback

FIGURE 18-14 Components of 2 × CO₂ warming
Higher concentrations of greenhouse gases in the atmosphere alter global temperature both by increasing the amount of heat trapped in a clear (cloud-free) atmosphere and by activating positive feedbacks (water vapor, snow and ice) and negative feedbacks (low-level clouds) that amplify or reduce the amount of temperature change.

sensitivity at +2.5°C or slightly higher (see Figure 18–13), but the range of possible sensitivities could lie anywhere between +1.8°C and +4°C.

18-11 Sensitivity to Greenhouse Gases: Earth's Climate History

Earth's climatic history provides additional evidence for Earth's sensitivity to CO₂ changes. The preindustrial temperature and CO₂ concentration represent just one point along a continuous curve of possible climatic states (Figure 18–15). For a given amount of change in CO₂, larger temperature responses occur toward the low-CO₂ end of the range than at the high-CO₂ end.

One reason for this varying sensitivity is the greater extent of snow and ice cover when CO₂ concentrations are lower. The increased areas of snow and sea ice in these colder climates provide a larger albedo-temperature feedback to the initial CO₂ changes. In contrast, at the high-CO₂ end of the range, snow and ice are reduced in extent, and little albedo feedback occurs. A second reason for the varying sensitivity is that the trapping of Earth's back radiation becomes less efficient at higher CO₂ concentrations as the atmosphere gradually becomes saturated with CO₂.

The model used to generate Figure 18–15 had a 3°C increase in global temperature for a CO₂ doubling from 280 to 560 ppm, just above the mid-range estimate of 2.5°C (see Figure 18–13). The range of uncertainty from other GCM models (shown by the dashed lines in Figure 18–13) can be tested by examining inter-

vals from Earth's climate history when past CO₂ levels and global mean temperature are either well known or reasonably well constrained. The temperature changes at these times in the past should represent the response to changes in CO₂ and other greenhouse gases.

Last Glacial Maximum The most informative interval from the past is the last glacial maximum, 21,000 years ago. At that time atmospheric CO₂ values were near 190 ppm, about 32% lower than the preindustrial level of 280 ppm, and methane values were 350 ppb, about 50% lower than their preindustrial value. The methane reduction translates into a 6% loss in CO₂, bringing the net drop in equivalent CO₂ concentration to 38%. As discussed in Chapter 12, greenhouse gases must have been the major factor affecting ocean temperatures at tropical and lower subtropical latitudes, because the ice sheets were too distant to have had large direct effects through changes in atmospheric circulation patterns, and because insolation values were close to those today.

Estimates of the amount of glacial cooling in the tropics range from as small as –1.5°C (CLIMAP) to as large as –4° or –5°C. This range of estimates indicates that Earth's sensitivity to CO₂ largely falls within the range indicated by climate models (see Figure 18–15). The smaller estimate of cooling from the CLIMAP reconstruction would put Earth's sensitivity at the lower end of the range simulated by the models, while the larger (4°–5°C) cooling indicated by expanded mountain glaciers would push Earth's sensitivity slightly beyond the high

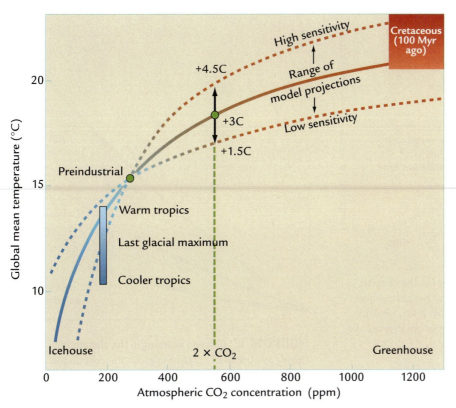

FIGURE 18-15 Estimates of 2 × CO₂ sensitivity from Earth history Estimated changes in global temperature and measured or estimated changes in atmospheric CO₂ concentrations for the same intervals of Earth's history fall within the range of sensitivities indicated by climate models. (Adapted in part from R. J. Oglesby and B. Saltzman, "Sensitivity of the Equilibrium Surface Temperature of a GCM to Changes in Atmospheric Carbon Dioxide," *Geophysical Research Letters* 17 [1990]: 1089–92.)

end of the range. The emerging consensus estimate of a ~3°C tropical cooling (Chapter 12) points to a sensitivity toward the higher end of the model range.

Cretaceous (100 Myr Ago) The Cretaceous world of 100 Myr ago can also be compared to the modern world. Estimates of greater warmth during this greenhouse interval range from +5° to +11°C, with recent estimates favoring the middle of this range (Chapter 5). Unfortunately, CO_2 values for the Cretaceous are not tightly constrained. Several techniques based on analysis of carbon isotopes in the remains of fossil plants and soils suggest values considerably higher than those today, but the estimates range from four to twelve times the preindustrial CO_2 level. The upper limits of possible changes in CO_2 concentrations lie well off the limits plotted in Figure 18–15. With Cretaceous temperatures warmer by substantially more than 5°C, Earth's sensitivity appears to fall toward the high end of the wide range of uncertainty indicated by the dashed lines in Figure 18–15.

IN SUMMARY, analyses of past intervals generally support the range of CO_2 sensitivity estimated from climate models. For the best-constrained case, the most recent glacial maximum, the sensitivity lies toward the higher end of the range of model estimates.

Why Has the Warming Since 1850 Been So Small?

The evidence summarized to this point indicates that natural factors have played a small role in the global warming trend since ~1880 and that greenhouse gases have been the dominant factor. Yet some climate skeptics continue to resist or reject this conclusion because of the small size of the observed global warming (0.7°C) compared to the large (35%) rise in CO_2 and the even larger (60%) rise in equivalent CO_2. They argue that this large an increase in greenhouse gases should have warmed Earth's surface by far more than the 0.7°C increase actually measured and that Earth's sensitivity to CO_2 and other greenhouse gases is therefore well below the lower end of the range that the climate models indicate. Mainstream climate scientists have responded that a direct comparison of this kind is invalid because it ignores two other factors that have also affected temperature changes since 1880.

18-12 Delayed Warming: Ocean Thermal Inertia

One factor affecting temperature change is the effect of thermal inertia in delaying the full response of the climate system to the higher levels of greenhouse gases. When external factors begin to alter climate, some parts of the climate system react more slowly

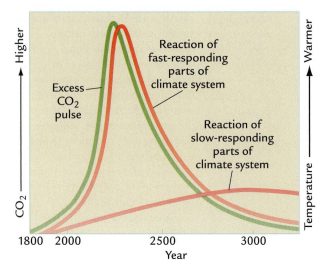

FIGURE 18-16 **Different response times in the climate system** Parts of the climate system such as the atmosphere can respond to imposed changes in days or weeks, while ice sheets require thousands of years. The upper ocean response occurs over decades.

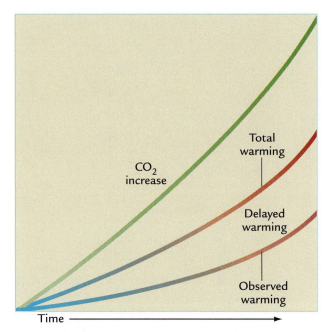

FIGURE 18-17 **Delayed warming in the climate system?** During an interval of rapidly rising CO_2 concentrations, the warming measured at any time is smaller than the warming that will be realized when Earth's delayed response is registered.

than others because of their greater thermal inertia (Figure 18–16).

The most important source of thermal inertia for the climate system as a whole is the ocean, which covers 70% of Earth's surface and stores enormous amounts of heat. The upper layer of the ocean (0–100 m) is stirred by lower-atmospheric winds and acts as a relatively fast-responding part of the climate system. Most of the volume of the ocean, however, lies below the wind-mixed layer and is only slowly affected by changes at the surface. The overall response time of the ocean to forcing from the atmosphere is measured in decades. As a result of this slow ocean response, the amount of increase in global surface temperature observed at any time within the last 125 years (including the present) represents only a part of the warming that is eventually going to occur, even with no further increase in gas concentrations (Figure 18–17).

The implication of this delayed warming is that estimates of Earth's true $2 \times CO_2$ sensitivity must be considerably larger than the value calculated by comparing the CO_2 increase between the 1800s and the early 2000s with the amount of warming that took place during that interval. Because the rise in all greenhouse-gas concentrations during the last half-century has been extremely rapid, some of the warming they will eventually cause is still "in the pipeline."

18-13 Cooling from Anthropogenic Aerosols

In the late 1970s the climate scientist Murray Mitchell proposed that the size of the greenhouse-gas warming was being significantly reduced by an offsetting cooling effect caused by smokestack emissions of SO_2, followed by production of sulfate particles in the atmosphere (Figure 18–18). SO_2 emissions in most areas were rising with an exponential trend similar to the CO_2 emissions, because both were by-products of the era of industrialization. Today, three major sulfate plumes lie over and

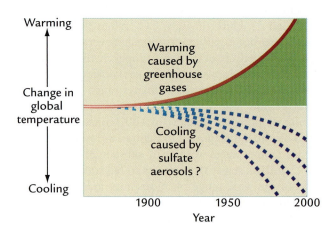

FIGURE 18-18 **Aerosol cooling counteracts parts of the greenhouse gas warming** The warming effect of greenhouse gases is partly canceled by the cooling effect of sulfates produced by SO_2 emitted in smokestacks. (Adapted from J. M. Mitchell, "The Natural Breakdown of the Present Interglacial and Its Possible Intervention by Human Activities," *Quaternary Research* 2 [1973]: 436–45.)

downwind of the eastern United States, Eastern Europe, and China (see Figure 18–10).

Current estimates are that sulfate aerosols have reduced incoming solar radiation by enough to offset about 15% of the radiative forcing (and global warming) caused by greenhouse gases (see Box 18–1). Ignoring this offsetting anthropogenic cooling effect makes the climate system appear less sensitive than it actually is.

The brown-cloud hazes have the same kind of effect on estimates of Earth's sensitivity as the sulfate aerosols. In regions of very strong hazes, the carbon-rich brown clouds are estimated to have offset as much as 50% of the local radiative forcing from greenhouse, with a smaller (and highly uncertain) effect on a global mean basis.

> **IN SUMMARY,** the combined effects of oceanic thermal inertia, sulfate aerosols, and carbon (brown-cloud) aerosols reconcile the seemingly small amount of global warming measured at present relative to the considerable greenhouse-gas buildup during the last century.

Global Warming: Summary

By 2006, global climate had warmed by 0.7°C above the level in the late 1800s. Climate scientists agree that the atmospheric concentrations of CO_2 and other greenhouse gases produced by human activities have increased markedly in the last century, and they agree that the gas increases have caused climate to warm. The major disagreement over global warming has focused on whether the greenhouse-gas increases explain some, all, or little of the observed warming.

As recently as the late 1990s (when the first edition of this book was written), it was still possible to make a case that the role of humans in this warming had been relatively small, but subsequent evidence has seriously undercut the central arguments that supported this position. Satellite-measured temperature trends once failed to show a warming since 1980, but later investigations found that proper adjustments for a range of complicating factors yielded a warming trend in line with the one recorded by surface stations. Earlier hypotheses that the variability of Sunlike stars might have accounted for half or more of the observed warming were undercut by observations that Sunlike stars do not show the large variations in irradiance proposed. Earlier speculation that carbon aerosols may have been falsely boosting the apparent size of the true greenhouse-gas warming were upended by the finding that the carbon aerosols in brown cloud hazes have instead kept Earth's surface cooler and have actually countered a significant fraction of the true greenhouse-gas warming.

It would be inaccurate to say that this debate is "over" because the scientific process constantly reexamines its facts and its assumptions. It would be accurate to conclude that an enormous range of evidence now supports the mainstream view that Earth's climate has warmed in the last 125 years mainly because of humans and to claim that no serious counterarguments to this view remain.

Meanwhile, Earth continues to register the many effects of global warming year after year. Most of the warmest years in the last 125 years have occurred in the 1990s and 2000s. Circum-Arctic snow and sea ice are shrinking to limits that are unprecedented during the satellite era. Almost every mountain glacier on Earth is melting, and most will disappear within 25 years at their current melting rates. The margins of the Greenland ice sheet are melting at increasing rates. Humanity's great experiment with the climate system is well underway, and the ongoing increases in greenhouse gases promise much more of the same in the future.

Key Terms

chlorofluorocarbons (CFCs) (p. 331)

ozone (p. 331)

ozone hole (p. 332)

sulfate aerosols (p. 332)

brown clouds (p. 334)

global dimming (p. 334)

$2 \times CO_2$ sensitivity (p. 335)

equivalent CO_2 (p. 335)

radiative forcing (p. 335)

enhanced greenhouse effect (p. 337)

Intergovernmental Panel on Climate Change (IPCC) (p. 337)

Review Questions

1. What human activities produce CO_2 and how have they changed in the last 200 years?

2. Where does the CO_2 produced by humans go?

3. How high in the atmosphere do sulfate aerosols from smokestacks reach?

4. Why do chlorofluorocarbons (CFCs) reach much higher in the atmosphere than sulfate aerosols?

5. What are the strongest positive and negative feedbacks on changes in Earth's temperature?

6. In a net sense, do feedbacks increase or decrease the direct radiative effects of greenhouse gases on global temperature?

7. What factors complicate attempts to estimate Earth's sensitivity to CO_2 by directly comparing the observed twentieth-century warming to the measured rise in greenhouse gases?

8. Some climate skeptics point out that temperatures were warmer in north polar regions 6000 years ago, and conclude that modern greenhouse-gas concentrations have not produced warmth that is unusual by natural standards. Evaluate the relevance of this conclusion based on what you have learned from this book.

Additional Resources

Basic Reading

Henson, R. 2006. "The Rough Guide to Climate Change." London: Rough Guides.

Karl, T. R., and K. E. Trenberth. 1999. "The Human Impact on Climate." *Scientific American* (December), 100–105.

Schneider, S. H. 1997. *Laboratory Earth: The Planetary Gamble We Can't Afford to Lose.* New York: Basic Books.

www.igbp.net (International Geosphere-Biosphere Program)

http://wcrp.wmo.int/ (World Climate research Program)

www.ncdc.noaa.gov/oa/climate/globalwarming.html (National Climate Data Center)

www.ipcc.ch (Intergovernmental Panel on Climate Change)

www.pages-igbp.org (Past Global Changes Project)

Advanced Reading

Charlson, R. J., S. E. Schwartz, J. M. Hales, R. D. Cess, J. A. Coakley, J. E. Hansen, and D. J. Hoffman. 1992. "Climate Forcing by Anthropogenic Aerosols." *Science* 225: 423–30.

Foukal, P., G. North, and T. Wigley. 2004. "A Stellar View on Solar Variations and Climate." *Science* 307: 68–9.

Hansen, J. E., A. Lacis, D. Rind, G. Russell, P. Stone, I. Fung, K. Ruedy, and J. Lerner. 1984. "Climate Sensitivity: Analysis of Feedback Mechanisms." *American Geophysical Union Geophysical Monograph Series* 29: 49–52.

Hoffert, M. I., and C. Covey. 1992. "Deriving Global Climate Sensitivity from Paleoclimate Reconstructions." *Nature* 360: 573–76.

Intergovernmental Panel on Climate Change. 2007. "Climate Change 2007: The Physical Science Basis." Geneva: World Meterological Organization.

Liepert, B. G. 2002. "Observed Reductions in Surface Solar Radiation in the United States and Worldwide from 1961 to 1990." *Geophysical Research Letters* 29, doi 10.1029/2002GL014910.

National Research Council. 2003. *Understanding Climate Change Feedbacks.* Washington, DC: National Academy of Sciences Press.

Ramanathan, V., et al. 2001. "The Indian Ocean Experiment: An Integrated Assessment of the Climate Forcing and Effects of the Great Indo-Asian Haze." *Journal of Geophysical Research* 106: 28371–99.

Wigley, T. M. L., P. J. Jaunman, B. D. Santer, and K. E. Taylor. 1998. "Relative Detectability of Greenhouse Gas and Aerosol Climate Change Signals." *Climate Dynamics* 14: 781–90.

CHAPTER 19

Future Climatic Change

The climatic effects of rising CO_2 levels will become increasingly significant in the future. Consumption of estimated reserves of fossil fuels (mainly coal) in the next two or three centuries will add far more CO_2 to the atmosphere than has accumulated so far, unless technology or extreme conservation efforts reduce this influx. Atmospheric CO_2 concentrations will increase to levels at least twice and possibly four or more times those that have existed during all of human existence. Concentrations will be comparable to levels last seen millions to tens of millions of years ago when a much warmer greenhouse world existed. This warming will completely overwhelm natural variations in climate and cause changes unprecedented in human experience. Atmospheric CO_2 levels will eventually decrease as the ocean slowly absorbs much of the excess carbon, but concentrations will still remain well above natural values for thousands of years into the future.

Future Human Impacts on Greenhouse Gases

Over the next few centuries, natural factors will continue to affect climate change, but their impacts will remain small (as in recent millennia). The largest driver of future climate will be emissions of greenhouse gases and aerosols from human activities.

19-1 Factors Affecting Future Carbon Emissions

In recent decades, atmospheric CO_2 concentrations have been rising at a rate of ~2 ppm/year, mainly because of burning of fossil fuels and secondarily because of deforestation. This rise will continue in the future, but the rate is impossible to predict. The uncertainties center mainly on the amount of carbon we will emit but also on the way the climate system distributes the additional CO_2 among its carbon reservoirs.

For as long as fossil fuels remain reasonably abundant, future carbon emissions can be approximated by multiplying three factors:

increase in carbon emissions = increase in population × change in emissions per person × changes in efficiency of carbon use

The number of humans is critical to these projections for obvious reasons: expanding numbers of humans require more fossil fuel for industry, transportation, and home heating, and they cut more forest to clear land for farming and urban/suburban growth. The number of humans has increased from 1.5 to 6 billion in just over 100 years as medical and agricultural advances have extended life expectancy.

Attempts to project future population increases are complicated by the tendency of birth rates to fall as per capita income rises, by centralized efforts by nations such as China to slow population growth, and by the efforts of some organizations to avoid constraints on reproduction. Global population is projected to rise rapidly from 6 billion during the year 2000 to more than 9 billion near 2050 but with a leveling off of the trend after mid-century (Figure 19–1). Slowing of the explosive growth that occurred between 1900 and 2000 will result from developing countries reaching higher levels of wealth and parents in those countries choosing to have fewer children. This trend has already been underway for decades in fully industrialized countries.

The change in carbon emissions per person is linked to the average standard of living and to the efficiency of use. In most nations, as the standard of living has increased over time, the process has required more carbon-based fuel for industrialization, transportation, and home heating/cooling. In the near term, the largest changes will occur in Southeast Asian nations rapidly

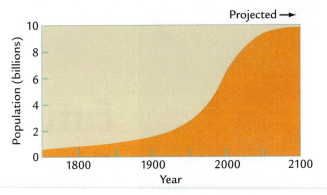

FIGURE 19-1 Future population A United Nations projection of human population during the twenty-first century at first continues the explosive increase of the twentieth century but later levels off at ~10 billion people. (Modified from F. Press and R. Siever, *Understanding Earth*, 2d ed., © 1998 by W. H. Freeman and Company.)

moving from semiindustrialized to industrialized status. Forecasting future living standards is difficult, but the Intergovernmental Panel on Climate Change (IPCC) has estimated a range of possible increases for the next few centuries. The efficiency with which carbon is used is especially hard to predict because it reflects many factors, including the kind of fossil fuel used and the technology used to burn it.

The next few decades will see peaks in annual production of oil and then later of natural gas, after which production will decline. Compared to coal, oil and gas are relatively "clean" fuels that emit smaller amounts of CO_2 per unit of energy produced. When oil and gas use declines, coal will by far be the dominant carbon-based fuel until the end of the fossil fuel era. Most accessible high-grade anthracite coal was burned in the early years of industrialization, leaving mostly lower-grade bituminous coal in the ground. Because bituminous coal produces far more CO_2 per unit of usable energy, global carbon emissions will increase. If the bulk of the vast carbon reservoirs stored in tar sands and oil shales becomes commercially worth exploiting, CO_2 emissions will go even higher.

To this point, the three factors in the future carbon emissions equation (population, quality of life, efficiency) seem likely to drive CO_2 emissions much higher in the near future. Because their effects are multiplied, the increase in future emissions seems likely to be very large.

The best hope for reducing (or limiting the increase of) future carbon emissions lies in technology, a part of the "efficiency" term in the equation. Humans are a uniquely innovative species, and our ingenuity will certainly make future breakthroughs in improving the efficiency with which we use fossil carbon sources. Signs of this kind of innovation have begun to appear during the

early 2000s as oil and gas have become more expensive. Promising areas as of the mid–2000s include solar and wind energy, biofuels (organically based fuels derived from agriculture), hydrogen-powered vehicles, and nuclear power plants that do not produce CO_2 or other greenhouse gases. Future increases in the cost of fossil fuels will help to accelerate the technology-based search for alternative energy.

19-2 Projected Carbon Emissions and CO_2 Concentrations

The IPCC has predicted a range of possible future carbon emission trends. The trends are uncertain even within this century and impossible to predict further in the future. The two projections of future CO_2 emissions shown in Figure 19–2 encompass the range of the IPCC estimates for the next century as well as far more uncertain projections for later centuries. The upper curve is based on the assumption of minimal efforts to curb emissions, with economic benefit continuing to be the major influence on decision making by nations and individuals. This trend shows emission rates reaching a level of almost 30 billion tons per year (four times the current rate of 7.5 billions tons per year) late in this century and continuing to rise before reaching a peak sometime between 2100 and 2200 and falling back to lower values in future centuries.

The second projection assumes that individuals, cities, and nations take strong action to curb emissions and that technology plays a role in reducing emissions. With these assumptions, the lower curve shows world-wide carbon emissions peaking at ~9 billion tons per year near 2040, falling below the modern rate of emissions late in this century, and continuing to drop slowly in future centuries.

Estimating the future path of CO_2 concentrations in the atmosphere is more difficult than estimating emission levels. In addition to the uncertainties in the emissions trend, questions exist about the way the climate system will redistribute the pulse of excess CO_2 among its carbon reservoirs. The easiest assumption to make is that the atmosphere will continue to receive just over half the total carbon emissions, as it has been doing for decades, with the rest entering the ocean and the biosphere. This assumption will remain valid only if the climate system continues to operate as it does today.

The part of the CO_2 emissions that will be taken up by the ocean is difficult to predict. Geochemists have gained some insight into this problem by tracking the gradual penetration of the products of recent human activities into the ocean. One example is the pulse of extra ^{14}C produced by nuclear testing in the middle 1900s. This bomb-produced ^{14}C has penetrated into the uppermost 100–1000 m of the ocean in most regions. As the surface ocean becomes warmer or lower in salinity in the future, formation of deep water at high latitudes may change, possibly resulting in less CO_2 absorbed compared to today's ocean.

Other climate scientists are investigating the role of vegetation in absorbing excess CO_2. In future decades, more CO_2 could be taken up by vegetation because of increased fertilization of plants, as well as expansion of trees into the now-frozen north and movement of scrub vegetation into arid regions. On the other hand, this trend may be countered by several factors. For example, studies in which high-CO_2 air is blown across plots of vegetation indicate that the carbon fertilization effect may decrease after only a few years when the vegetation has reached saturation with the extra CO_2. In addition, humans may continue to reduce the area of CO_2-absorbing forests, especially in the tropics.

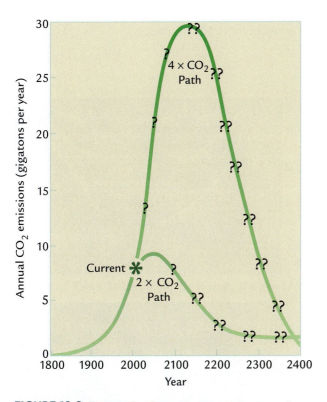

FIGURE 19-2 Projected carbon emissions Projections of future carbon emissions vary with uncertainties in future populations, living standards, and conservation efforts and technological innovations. An optimistic projection (the $2 \times CO_2$ scenario) shows an earlier decrease in emission rates, but a projection of present trends (the $4 \times CO_2$ scenario) shows a continuous rise. Both projections of carbon emissions eventually decline when reserves of fossil fuels are largely consumed. (Adapted from Intergovernmental Panel on Climate Change, "Climate Change 2007: The Physical Science Basis" [Geneva: World Meteorological Organization, 2007], and from H. S. Kheshgi et al., "Accounting for the Missing Carbon Sink with the CO_2-Fertilization Effect," *Climate Change* 33 [1996]: 31–62.)

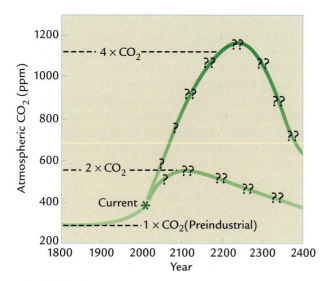

FIGURE 19-3 Projected CO$_2$ concentrations Concentrations of CO$_2$ in the atmosphere are projected to reach levels somewhere between twice (2 × CO$_2$) and four times (4 × CO$_2$) the preindustrial value of 280 ppm within the next two to three centuries. (Adapted from Intergovernmental Panel on Climate Change, "Climate Change 2007: The Physical Science Basis" [Geneva: World Meteorological Organization, 2007], and from H. S. Kheshgi et al., "Accounting for the Missing Carbon Sink with the CO$_2$-Fertilization Effect," *Climate Change* 33 [1996]: 31–62.)

Projections of future atmospheric CO$_2$ concentrations ultimately depend on the interplay between the ongoing emissions of excess (anthropogenic) CO$_2$ to the atmosphere and the slow removal of this excess by the ocean and the biosphere. The two projections of future CO$_2$ concentrations in the atmosphere shown in Figure 19–3 differ in the amplitude and timing of the CO$_2$ peaks attained. In the upper curve, the CO$_2$ concentration rises to more than 800 ppm by 2100 and then to more than 1200 ppm between 2200 and 2300, a level more than four times the preindustrial value of 280 ppm. The concentration then begins a slow decline that lasts for centuries (millennia). In the lower curve, the CO$_2$ concentration rises to a value twice the preindustrial level by the year 2100 and then begins a gradual decline.

These projections ignore the "equivalent CO$_2$" increases in other greenhouse gases that have amounted to 40% of the industrial era total to date. CFC emissions have begun to decline and will play a smaller role in the future, and methane emissions have also recently leveled off and could continue to decrease. As a result, the future trend will increasingly be dominated by CO$_2$.

Atmospheric CO$_2$ levels twice or four times the preindustrial value would be without precedent in the 100,000 to 200,000 years that our species has existed and also in the millions of years in which our human ancestors (hominins) evolved. The history recorded in

Antarctic ice cores shows that CO$_2$ values have not risen above 300 ppm for at least the last 760,000 years. By some estimates, concentrations as high as 550 ppm have not existed since 5 to 20 million years ago, and concentrations as high as 1200 ppm have not existed for tens of millions of years, possibly since the Cretaceous greenhouse world.

The projected trends of CO$_2$ concentration (see Figure 19–3) reach their peak values later than the projected CO$_2$ emissions trends (see Figure 19–2) and then remain at high levels for a long time. The reason for this delay is the slow rate of CO$_2$ removal by the ocean. Although individual CO$_2$ molecules move back and forth between the air and the surface ocean within a few years, it takes many centuries for the ocean to absorb most of the excess CO$_2$ emitted by humans (Figure 19–4A), and the remaining 10% will stay in the atmosphere for 10,000 years or more. As a result, new CO$_2$ emitted to the atmosphere adds to older CO$_2$ left from emissions in prior decades and centuries.

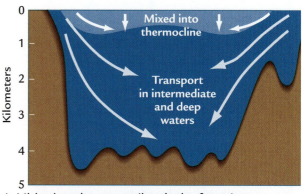

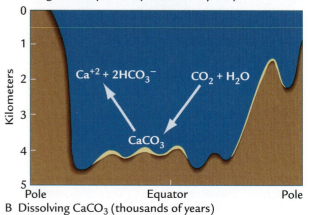

FIGURE 19-4 Fate of the CO$_2$ pulse created by humans (A) Over decades to centuries, most of the excess CO$_2$ in the atmosphere will be mixed into the subsurface ocean. (B) Over centuries to millennia, the excess CO$_2$ will make seawater more acidic and will dissolve seafloor CaCO$_3$.

Even after human CO_2 emissions begin to fall, the slowly decreasing yearly CO_2 emissions continue to push the atmospheric concentration to even higher levels. Not until about a century after the peak in CO_2 emissions does the rate of additional input by humans drop below the rate of removal by the ocean. At this point, the CO_2 concentration begins to decline.

Over even longer intervals, other considerations come into play. Over the course of hundreds to thousands of years, the acidity produced by the CO_2 absorbed in the deep ocean dissolves some of the $CaCO_3$ on the seafloor (Figure 19–4B). This process neutralizes CO_2 in the ocean. The ultimate fate of our excess CO_2 pulse will be a slow-acting chemistry experiment in the deep ocean.

19-3 Other Human Effects on the Atmosphere

Other emissions by humans may also be important. Methane production is likely to increase, but by a smaller amount than CO_2. Most of the land that can be used to grow rice is already in irrigation, and future increases in the extent of these CH_4-emitting "rice wetlands" are likely to be negligible. As the number of humans on Earth levels out after mid-century (Figure 19–1), the number of CH_4-emitting livestock that people tend should also stabilize.

Unlike CO_2, methane stays in the atmosphere for only a decade before being oxidized to other gases. One interesting question is whether a warmer future world will cause large Arctic reservoirs of now-frozen methane to melt. If that happens, unknown amounts of CH_4 gas could be added to the greenhouse effect as a positive feedback (Box 19–1).

Future additions of SO_2 and carbon aerosols to the atmosphere by humans are also difficult to predict. The positive impact of environmental cleanup efforts in some industrialized nations will be countered to an unknown extent by increased burning of sulfur-rich coal in nations still undergoing industrialization. A net increase in SO_2

BOX 19-1 CLIMATE INTERACTIONS AND FEEDBACKS
Will Frozen Methane Melt?

Methane exists as a gas in the atmosphere, but in Earth's colder regions it also occurs in a frozen form known as **methane clathrate**, a mixture of methane and slushy ice. Clathrates occur in deep-ocean sediments along continental margins, where the pressure produced by overlying water and sediments makes CH_4 stable at temperatures well above freezing (5°C or more). Clathrates also occur in the Arctic, both in shallow ocean sediments and below permafrost on land. The volume of CH_4 stored in these reservoirs is enormous, far exceeding the volume in all wetlands and livestock reservoirs combined.

Without major changes in climate, most methane clathrate will remain trapped in its present form, but with the large future warming projected for north polar regions, the question is whether or not it will remain trapped in the slushy ice. Permafrost is expected to continue to melt from the top down, but will this warming reach deep enough to tap the methane clathrates and liberate some of the trapped CH_4 to the atmosphere?

Because it will take centuries for the higher surface temperatures to penetrate far into permafrost and ocean sediments, most (but not all) scientists doubt that large amounts of CH_4 will be released. Still, if future greenhouse

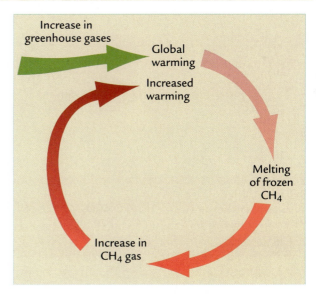

Methane clathrate feedback Future warming of the deep coastal ocean and melting of polar permafrost could release frozen methane cause additional global warming.

warming of the poles or the deeper ocean causes the release of even a small fraction of the vast mass of frozen methane, it could provide a significant positive feedback to the initial greenhouse-gas warming.

and carbon aerosol input to the atmosphere could produce a cooling effect that would counter part of the future CO_2 warming but by probably only a small fraction.

Future Climate Changes Caused by Increased CO_2

As atmospheric CO_2 levels rise in the future, Earth's climate will continue to warm. Attempts to estimate the amount of future warming are subject to the cumulative uncertainties from three factors: the amount of excess CO_2 emitted by humans, the levels of atmospheric CO_2 reached as the excess carbon is redistributed among various carbon reservoirs, and Earth's sensitivity to higher CO_2 concentrations. Emissions of other greenhouse gases are also a source of uncertainty.

The two projections of future CO_2 concentrations shown in Figure 19–3 span the possible range of increases during the next few centuries. The best estimate of Earth's sensitivity to a doubling of CO_2 is 2.5°C, a value that falls in the middle of the range indicated by climate models (Chapter 18). Two estimates of projected future temperature change based on this climate sensitivity are plotted in Figure 19–5. The lower projection shows the amount of warming for the emissions trend that drives atmospheric CO_2 concentrations

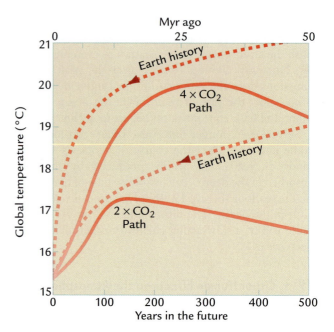

FIGURE 19-6 Temperature: Past and future Average global temperature in the next 200 years is projected to reach levels comparable to those that last occurred many millions of years ago.

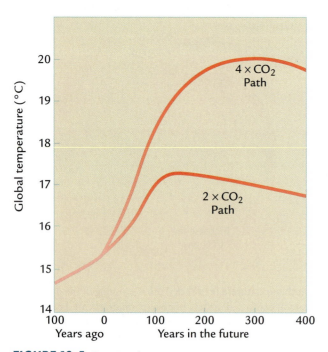

FIGURE 19-5 Projected temperature increases In the future, higher levels of CO_2 and other greenhouse gases are projected to cause global temperature to increase by at least 2.5 °C for the 2 × CO_2 scenario and perhaps 5 °C or more for the 4 × CO_2 scenario.

to twice their preindustrial value (the 2 × CO_2 case). The result is an additional 2°C warming by the year 2100, with further warming into the early 2100s. For this projection, the full warming (industrial era and future) by 2100 is 2.5°C.

The upper projection shows the additional warming for the emissions pathway that reaches four times the preindustrial CO_2 level (the 4 × CO_2 case). In this projection, an additional warming of 3.8°C occurs by 2100 and the peak warming is more than 5°C. A future warming of this amount would be more than 7 times the warming during the 1900s.

19-4 A World in Climatic Disequilibrium

As future CO_2 concentrations reach levels not seen for many millions of years, Earth's climate system will retrace in just a few centuries a journey that natural (tectonic-scale) forces required millions or tens of millions of years to produce (Figure 19–6). Although it might seem possible to use past intervals from Earth's climate history as direct analogs for future climate, this approach can at best be only partly successful. The problem is that the high-CO_2 pulse will arrive too quickly for all parts of the climate system to come into equilibrium with the warmer temperatures. The fast-responding parts of the climate system will react to the warming within just a few decades, but the slow-responding ice sheets on Greenland and Antarctica will not.

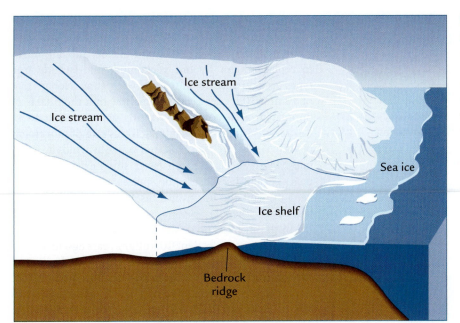

FIGURE 19-7 Vulnerable ice shelves Ice from the interior of Greenland and Antarctica flows in ice streams to the shelves along the margin. In a warmer world, ice shelves may be vulnerable to destruction by rising ocean temperatures, which may in turn accelerate flow in the ice streams. (Adapted from R. A. Bindschadler et al., "What Is Happening to the West Antarctic Ice Sheet?" *EOS* 79 [1998]: 256–65.)

The margins of ice sheets can fluctuate rapidly over intervals of a few centuries (Chapter 14), but the great mass of continent-sized ice sheets respond only over many thousands of years (Chapter 9). The greater warmth of the next few hundred years will greatly increase melting along the ice margins, but the central portions will melt more slowly. As a result, the main bulk of the ice will survive for much longer. The part that survives will be like a gigantic block of ice on a summer day, out of place in a warmer world.

Many climate scientists think that most of the Greenland ice sheet will survive the pulse of high CO_2 levels during the next few centuries and persist into the subsequent era of decreasing CO_2 concentrations (see Figure 19–3), but others disagree. They point out the ablation that has been occurring at surprisingly high elevations on the ice sheet in recent summers and the accelerated melting caused by rapid flow of ice in marginal ice streams (Figure 19–7). Current evidence is insufficient to tell which of these views is correct.

In any case, the world of the future will be a strange no-analog combination of the slow-responding ice sheets and deep ocean and the fast-responding atmosphere, land surface, vegetation, and upper ocean. Near the slow-responding ice sheets, which strongly influence regional climates by their high albedo and their effect on atmospheric winds, the lingering cold caused by the ice will suppress part of the response of the atmosphere and nearby surface ocean to the higher CO_2 levels. Farther from the ice, the fast-responding parts across most of the climate system will react strongly to the new warmth.

Because this kind of disequilibrium has not occurred in Earth's past, no exact analogs for future climates exist. Still, even the partial analogs that do exist provide general indications of future climate.

19-5 Partial Future Analogs: 2× and 4× Preindustrial CO_2 Concentrations

Future CO_2 concentrations will reach somewhere between twice and four times the preindustrial value of 280 ppm (see Figure 19–3).

2 × CO_2 World CO_2 levels are currently 35% higher than the preindustrial level of 280 ppm, and equivalent CO_2 levels are 60% higher. Rising rates of emissions in the early years of the twenty-first century make it highly unlikely that atmospheric greenhouse-gas concentrations will fail to reach the equivalent 2 × CO_2 level by the middle of the current century. If this happens, Earth's average temperatures will register a near-complete equilibrium response to this doubled CO_2 value by the end of the century. For the fast-responding parts of the climate system, this 2 × CO_2 world will be analogous to the world that existed about 10 million years ago.

One striking difference in the world of 10 million years ago was the much-reduced extent of ice in the Arctic Ocean. With less sea ice, the atmosphere extracted heat stored in the ocean each summer, and this transfer moderated air temperatures during Arctic winters. One consequence of the warmer winters was that the broad bands of permafrost and tundra that now surround the Arctic Ocean in Eurasia and North America (Figure 19–8) were absent, and conifer trees grew in their place.

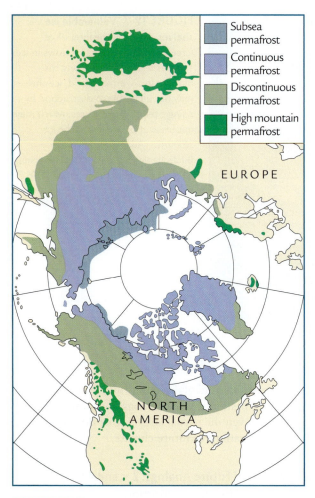

FIGURE 19-8 Melting permafrost Today's large ring of permafrost around the Arctic Ocean will become vulnerable to gradual melting in the warmth of a $2 \times CO_2$ world. (From F. Press and R. Siever, *Understanding Earth,* 2d ed., © 1998 by W. H. Freeman and Company.)

Because sea ice and vegetation are relatively fast-responding parts of the climate system, we can expect much larger transformations of polar sea ice, tundra, and northern forests as climate warms to the $2 \times CO_2$ level (Figure 19–9). The high climatic sensitivity of this region is obvious from the northward retreat of tundra and sea ice limits caused by summer insolation values ~5% higher than those today near 6000 years ago (Chapter 13). Trees will move north of the Arctic Circle, and sea ice will retreat from the coasts. As a hint of what the future holds, summer sea ice limits have shrunk by more than 20% in the last four decades. Melting of surface permafrost (already underway) will continue at a rapid rate, but much of the deeper subsurface permafrost will not be affected.

At northern mid-latitudes 10 million years ago, forests of deciduous trees grew much farther north than they do today. Coupled models of climate and vegetation simulate a northward shift of mid-latitude hardwood trees like maple and beech during the next century, with warm-adapted trees like oak and hickory moving north to replace them. By one reckoning, mid-latitude tree types already need to shift northward at an average rate of 10 m (>30 ft) per year to remain within their areas of optimal growth conditions during the current rate of warming. Most species can match this rate of movement by dispersal of pollen, seeds, and cones by winds and animals. With the larger and faster shifts expected in the next century, however, some species may not be able to keep up with the northward displacement of their optimal environment.

In the tropics and subtropics, scrub and tree vegetation were more prevalent 10 million years ago in several arid regions: the sub-Himalayan region of India and Pakistan, the western North American high plains, the South American pampas region of Argentina, and parts of sub-Saharan Africa and the East African highlands. Higher levels of CO_2 in the atmosphere allowed C3 vegetation (trees and shrubs) to live in arid regions, but then the gradual CO_2 lowering during the last 10 million years made C3 vegetation less competitive with C4 grasses. In the next century, we will pass through the same $2 \times CO_2$ threshold but heading in the opposite direction and at a much faster rate. As C3 shrubs and trees replace C4 vegetation, some arid and semiarid regions could become greener (see Figure 19–9).

The warming during the next century will also alter regional patterns of precipitation and evaporation in significant ways. Evaporation will increase worldwide because warmer temperatures will permit air to carry more water vapor. With more water vapor in the air, global average precipitation will also increase but in patterns that may vary from region to region. With evaporation increasing, areas that fail to receive more precipitation will become drier while those that receive more precipitation could become wetter. Unfortunately, because climate model simulations of regional precipitation often disagree, moisture trends are difficult to predict region by region.

No evidence of mountain glaciers has been found in North or South America, Africa, or Asia before about 7 million years ago. The only place where mountain glaciers may have existed was in far northern Scandinavia and on Greenland, where the combined effect of high latitudes and high altitudes may have made temperatures cold enough for ice to persist. Almost every mountain glacier on Earth has already been retreating for a century or more, and the rates are accelerating (see Chapter 17). At a prevailing lapse rate of 6.5°C/km, a 2.5°C warming in the future should cause a vertical retreat of glaciers up the sides of mountains by some 330 m (about 1000 ft). Because mountain glaciers can begin to respond to climate changes within just decades,

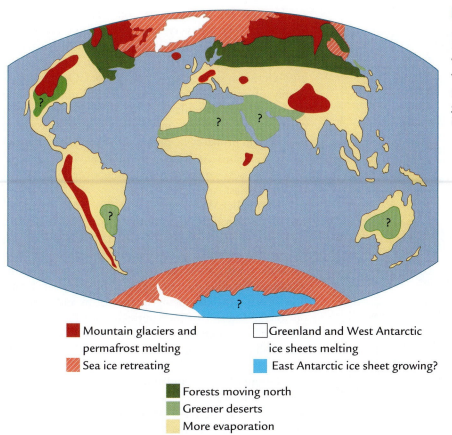

FIGURE 19-9 2 × CO₂ world The $2 \times CO_2$ world likely to exist by the year 2100 will in many ways be similar to the world that existed 10 million years ago, with less sea ice and permafrost in polar regions, far fewer mountain glaciers, and greener scrub vegetation in some semiarid desert regions.

■ Mountain glaciers and permafrost melting
▨ Sea ice retreating
□ Greenland and West Antarctic ice sheets melting
■ East Antarctic ice sheet growing?
■ Forests moving north
■ Greener deserts
■ More evaporation

most glaciers should completely disappear in a $2 \times CO_2$ world. The melting ice will contribute to a further global rise in sea level in the future.

The disappearance of mountain glaciers will have consequences for water supplies in some arid regions. In many arid areas, winter rains deliver water during a time when crops cannot be grown. Water during the early part of the growing season comes from melting of winter snows on nearby mountains. Later in the growing season, runoff from melting glaciers becomes a major source of water. In a $2 \times CO_2$ world, higher evaporation will put more stress on water supplies. Mountain snow packs will be thinner because more winter precipitation will instead fall as rain. In addition, runoff from mountain glaciers will end when the glaciers disappear. As a result, mountains will store and deliver less summer season water for irrigation and other human uses.

Fauna and flora on mountainsides will also be affected by warming. To remain in an optimal temperature regime, both will have to shift to higher elevations, and the transition will be easier for relatively mobile life forms than less mobile ones. In some cases, the warming may push the preferred environment "off the top" of the mountains and cause species extinctions.

The ocean (especially the deep ocean) is one of the slower-responding parts of the climate system, but it will be affected in future centuries and millennia by **ocean acidification.** The ocean has an "alkaline" pH that ranges from 7.8 to 8.5, compared to the 7.0 pH boundary between acid and alkaline conditions. Because the pH scale is logarithmic, a shift from 8.0 to 7.0 indicates a tenfold increase in relative acidity (or decrease in alkalinity).

Over the several thousand years of deforestation and the past few hundred years of other carbon emissions, humans have already added ~120 billion tons of carbon to the ocean, and the average pH has shifted by 0.1 unit. By 2100, the pH will have shifted by another 0.2 unit or more, in effect doubling the acidity level of the ocean. As this happens, snaillike organisms called pteropods with carbonate shells made of aragonite (a form of $CaCO_3$) will have increasing trouble producing shells, as will some species of corals. Coral reefs will also face more frequent episodes of "bleaching," with the unusual warmth during large El Niño years killing temperature-sensitive species.

The slowest-responding parts of the climate system are the ice sheets on Greenland and Antarctica. Prior to ~7 million years ago, no ice sheet existed on Greenland, apparently because temperatures were too warm to permit ice accumulation at near-polar latitudes. In the future, we face a different situation: we start off in a world with

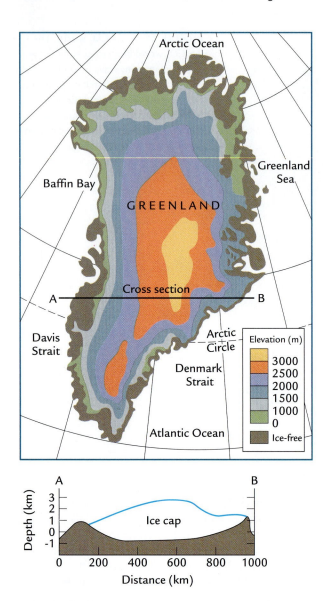

FIGURE 19-10 Greenland ice sheet Lower margins of the Greenland ice sheet will melt rapidly in the warmth of a $2 \times CO_2$ world, but the highest central surface of the ice sheet may be less affected. (From F. Press and R. Siever, *Understanding Earth,* 2d ed., © 1998 by W. H. Freeman and Company, and from R. F. Flint, *Glacial and Quaternary Geology,* © 1971 by Wiley.)

the Greenland ice sheet already in existence (Figure 19–10) and we need to predict how it will respond in a warmer world. Simulations with ice sheet models indicate that the 2.5°C global warming in a $2 \times CO_2$ world will produce widespread melting of the surface of the Greenland ice sheet.

As noted earlier, the extent of future melting of Greenland ice is highly uncertain. By some projections, only the margins will melt. Other scientists think that accelerated flow in marginal ice streams could lead to much greater losses of ice in future decades and possibly near-total loss in future centuries.

The enormous ice sheet on eastern Antarctica (Figure 19–11) was very large 10 million years ago. Today, this frigid ice sheet is starved for snow, with only a few centimeters per year falling across most of its high-elevation surface. In a warmer $2 \times CO_2$ world, the supply of snow should increase and allow faster and thicker annual accumulations of ice in the interior. This change toward positive mass balance in the ice sheet interior will probably be opposed by faster flow in marginal ice streams. At present, it is unclear which of these two processes would dominate the overall ice mass balance in East Antarctica.

The smaller West Antarctic ice sheet was less extensive 10 million years ago than it is now, if it existed at all. The extensive ice shelves that now fringe this ice sheet are vulnerable to destabilization because they are in contact with an ocean that could become slightly warmer. Destabilization of the ice shelves can accelerate flow in ice streams moving to the ocean and draw ice out of the interior of the continent (see Figure 19–7). This evidence suggests that the western Antarctic ice sheet will be vulnerable to greater melting in a $2 \times CO_2$ world.

Melting of Greenland and Antarctic ice will cause a future rise in sea level, as will expansion of ocean water as it slowly warms. The 2007 IPCC report projects that the rate of sea level rise is likely to double from the rate of 17 cm during the last century to about 30 cm (approximately 1 ft) during the current century. Sea level rises could be much higher than this estimate if the Greenland ice sheet proves to be as vulnerable as some scientists predict.

$4 \times CO_2$ World By one projection (see Figure 19–3), atmospheric CO_2 concentrations could reach values more than four times the preindustrial level between 2200 and 2300. Projections this far into the future are inherently speculative, especially because of the likelihood that technological innovations will avert this large a change. But if we do reach a $4 \times CO_2$ world, all the warming trends described for the $2 \times CO_2$ world will be amplified. These changes will move Earth's climate toward the world of 50 to 100 million years ago, the last time CO_2 levels were so high.

The climate of 50 million years ago was much warmer than today. The Arctic margins were surrounded by a mixed forest of hardwoods and evergreens adapted to relatively mild winters. No sea ice existed in the Arctic. Temperate beech (*Nothofagus*) forests existed on an ice-free Antarctic continent. Mountain glaciers probably did not exist anywhere on Earth. A 5°C warming would cause glacial ice to retreat 660 m (more than 2000 ft) up the sides of mountains, enough to eliminate today's mountain glaciers.

In a $4 \times CO_2$ world, north polar regions will warm enough to eliminate shallow permafrost, tundra, summer

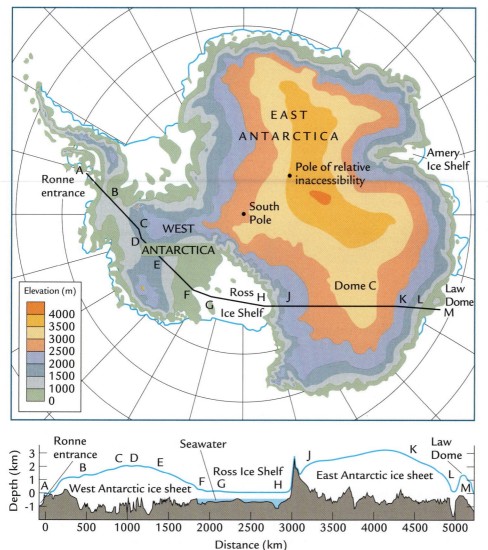

FIGURE 19-11 Antarctic ice sheet Fringing ice shelf margins of the small, low-altitude West Antarctic ice sheet may begin to melt in a 2 × CO₂ world, but the higher, colder East Antarctic ice sheet will be less vulnerable. (From F. Press and R. Siever, *Understanding Earth,* 2d ed., © 1998 by W. H. Freeman and Company, and after Uwe Radok, "The Antarctic Ice," *Scientific American,* August [1985]: 100, based on data from the International Glaciological Project.)

sea ice, and probably most winter sea ice (Figure 19–12). The belt of conifer forest will slowly move toward and reach the Arctic Ocean. Sea ice around Antarctica would also probably disappear. Changes at middle and lower latitudes will be similar to but larger in magnitude than those for the 2 × CO₂ world.

In trying to project ice sheet changes so far in the future, we once again start with the fact that the ice sheets now exist but will be completely out of equilibrium with the warmer climate all around them. The Greenland ice sheet will become even more vulnerable, but it is unclear whether or not irreversible melting would occur. West Antarctic ice will also be out of equilibrium with a warmer ocean, and the margins of East Antarctica will also melt at rates that are hard to predict. With increased ice sheet melting and thermal expansion of the ocean, sea level will rise faster and higher, reaching a level perhaps 1–2 m above that of the present, and possibly more.

19-6 Greenhouse Surprises?

Because the mixture of slow and fast responses to the large CO₂ pulse of the future will create a climatic disequilibrium unprecedented in Earth history, interactions within the climate system may produce unanticipated phenomena, or "greenhouse surprises."

One frequently mentioned possibility is that faster melting of Greenland ice could send enough freshwater to the North Atlantic Ocean to lower its salinity and thereby slow or stop the formation of deep water. A relatively small drop in salinity in the Labrador Sea beginning in the 1970s lowered the density of the surface waters enough to prevent them from sinking during winters over the next two decades. Melting of the margins of the Greenland ice sheet or altered precipitation patterns caused by greenhouse warming could conceivably add enough low-salinity water to the North Atlantic to slow or stop the formation of deeper water.

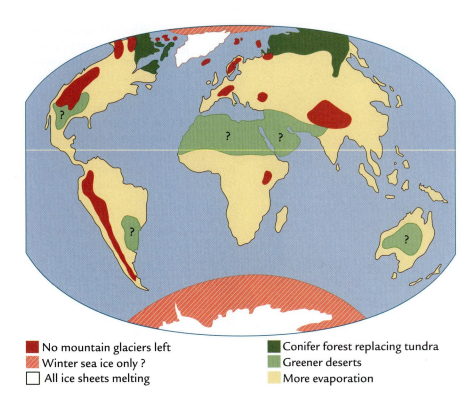

FIGURE 19-12 $4 \times CO_2$ **world** The $4 \times CO_2$ world that may come into existence centuries from now would be similar to the one that existed 50 or more million years ago, with little or no sea ice or mountain glaciers and forests growing at high Arctic latitudes.

- ■ No mountain glaciers left
- ■ Winter sea ice only ?
- □ All ice sheets melting
- ■ Conifer forest replacing tundra
- ■ Greener deserts
- ■ More evaporation

A likely consequence of such a circulation change would be colder temperatures in northern Europe. Today the heat extracted from North Atlantic surface waters during the formation of deep water is carried eastward and helps to keep Europe warmer in winter than North American or Siberian regions at the same latitude. Without this heat, Europe would become more like northern Canada, perhaps 5°C or more colder than it is now.

Recent reassessments suggest, however, that a complete cutoff of deep-water formation in the Atlantic is unlikely. The planetary wind system will continue to drive relatively warm and salty surface water northward from the tropical Atlantic Ocean, where relatively cold winter air masses will continue to extract heat. Partial reductions (or relocations) of deep-water formation could occur, but they are unlikely to offset the opposing effect of the ongoing warming in Europe.

19-7 How Will Greenhouse Warming Change Human Life?

The projected climatic transformations in the $2 \times CO_2$ and $4 \times CO_2$ worlds are dramatic, but they omit an everyday perspective on how climate will change for people living at middle latitudes, including the children and grandchildren of the students now reading this book.

Consider the $2 \times CO_2$ world with a 2.5°C global mean temperature increase. Poleward amplification of this temperature change by albedo feedback will increase the warming at middle latitudes (40°–60°N) where many humans in North America, Europe, and eastern Asia live. Temperature increases in these latitudes will be at least 3°–3.5°C for the annual mean and 4°–4.5°C or more during the coldest winter months (Figure 19–13).

These temperature changes can best be understood by comparing them against modern seasonal changes. Today, on landmasses at 40°–60°N latitude, mean daily temperatures of ~0°C (32°F) in January rise to ~25°C (77°F) by July and fall back again the following winter. This 25°C shift over 6 months amounts to an average change of 4.25°C (7.5°F) per month, approximately the same amount as the anticipated cold season temperature increase at high latitudes caused by a CO_2 doubling. As a result, the temperature response of northern mid-latitudes to a CO_2 doubling should feel like a one-month shift of the seasons. Future Aprils will be like modern Mays and future Novembers like modern Octobers (Figure 19–14). Summer will last for an extra two months, while winters will be shorter and less harsh.

These changes will come on slowly enough and be sufficiently masked by typical year-to-year variability that the trend should not be readily apparent to a person whose life spans the rest of this century. If that same person fell into a Rip Van Winkle sleep of 90 years, however, the changes on awakening would be striking, although no doubt overshadowed by far more bewildering societal and cultural transformations.

Outside this somewhat narrow mid-latitude perspective, Earth's responses to climate changes will vary

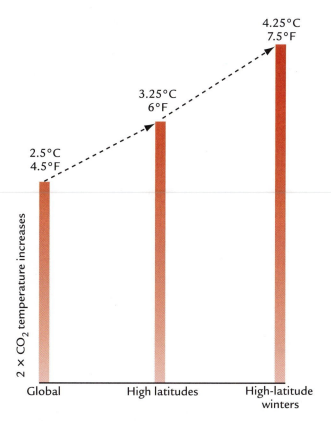

FIGURE 19-13 Larger temperature increases at high latitudes Temperature changes caused by a doubling of atmospheric CO_2 concentrations will be larger than the global average at high mid-latitudes, especially in winter.

In the tropics and subtropics, where 80% of humans live, the temperature increase will be smaller, but it will have a significant effect on moisture balances. The greater warmth will increase the global mean rate of evaporation and in dry regions put greater stress on other sources of water. In regions where evaporation increases but precipitation does not, agriculture will depend more on irrigation. Areas that depend on runoff from snowmelt and glacial meltwater for irrigation late in the growing season will suffer as both sources of water decrease. Regions that have for decades been pumping deep subsurface water for irrigation may run out of usable groundwater. In areas where rainfall increases, more precipitation will fall as localized summer cloudbursts than as the longer-lasting precipitation typically associated with the passage of frontal systems during cooler seasons.

The future sea level rise caused by melting of land ice and expansion of ocean water will have an adverse effect on coastal populations. The same factors that caused the 17-cm rise in sea level during the last 100 years will continue to operate: thermal expansion of ocean water, melting of mountain glaciers, and partial melting of the Greenland and perhaps the West Antarctic ice sheets.

Although the IPCC mid-range projection of a 30-cm sea level rise during this century sounds trivial, it will translate into a 300-m (~1000-ft) advance of the ocean across flat coastal regions where many people live. In heavily populated Bangladesh, a sea level rise of 30 cm would displace millions of people currently living within 1 m in elevation above the present sea level.

from region to region, and the range of effects on humans and other life will also vary. A few examples from the vast range of future changes are noted here. Some of the most striking changes will occur in regions where snow and ice retreat northward. For today's ski resorts at relatively low latitudes, a loss of two months of subfreezing winter cold may so truncate the snow-making season that business will be impossible. On the other hand, in far-northern regions of Hudson Bay and the Arctic coast of Siberia, the sea ice retreat will open up new year-round shipping lanes. The length of the shipping season in existing lanes is already lengthening, and nations are actively considering new (shorter) Arctic routes. In addition, less energy will be required for winter heating at middle and higher latitudes, but more energy will be needed for summer cooling at lower and middle latitudes.

Warmer winters at high latitudes should extend the northern limit for growing many crops, especially wheat across broad regions of Canada and Russia. Also, higher CO_2 levels in the atmosphere should allow some plants to obtain the CO_2 necessary for photosynthesis more quickly without exposure to the drying effects of evaporation.

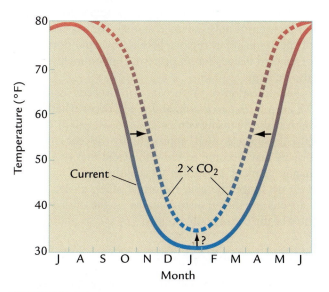

FIGURE 19-14 Changes in length of seasons The warming caused by $2 \times CO_2$ levels will shorten winters and lengthen summers by about one month at middle latitudes.

In built-up coastal regions such as the East Coast and coast of the Gulf of Mexico in the United States, ever-greater engineering efforts (construction of sea walls and pumping of sand to replenish beaches) will become necessary unless communities decide to retreat from vulnerable regions. Insurance companies have already begun to increase premiums sharply for homes and buildings in such areas. Some scientists have recently suggested that the intensity of the largest hurricanes will increase with global warming, but this prediction has been challenged.

Perhaps the greatest "wild card" in assessing the climatic future is the possibility that warming will push the Greenland ice sheet into a regime of irreversible melting. Although this possibility is still a minority opinion among climate scientists, it is no longer considered impossible. If the entire Greenland ice sheet were to melt over future centuries (or millennia?), the 6-m (20-ft) rise in sea level would be a true catastrophe for billions of people living along the world's coasts. People in flat rural areas would have to retreat inland many miles as the sea invaded. The fate of most of the great cities of the world is harder to predict. Each of them would have to decide whether or not it would make sense to build massive walls as armor against such a large rise in sea level.

Some of the largest changes caused by future global warming may have relatively little broad economic impact on humans and yet still be crucial from an ecosystem perspective. Large-scale melting of sea ice and permafrost and loss of tundra around the Arctic margins will hardly be a central issue in the lives of people who have never traveled to that barren land. But the shrinking of those habitats could devastate caribou, polar bears, and the rest of the polar ecosystem, including the cultures of the few native people who still rely on hunting of polar-adapted animals for survival. Loss of species in other regions because of rapid northward or upward dislocation of preferred environments could also be large.

Acidification of the oceans will proceed as long as we keep burning carbon, and it will occur whether we burn it slowly (using conservation measures) or quickly (the "business as usual" approach). By the year 2300, the oceans may become so close to acidic that one of the major planktic organisms (coccoliths; see Chapter 2) may no longer be able to form $CaCO_3$ shells. If this happens, further transformations of the entire chemistry of the ocean could occur.

Climate Modification?

Until recently, the possibility of altering Earth's future climate by engineering was rarely discussed among mainstream scientists, many of whom feel that we should minimize future CO_2 increases by conservation efforts. This subject is no longer entirely taboo, however, because the scope of what we have already done to past climate and will do to future climate has become clearer.

Humans face an interesting dilemma suggested by the imaginary sequence in Figure 19–15. If all our emissions of sulfate and carbon aerosols were suddenly eliminated, precipitation would wash the excess aerosols out of the lower atmosphere within a few weeks, and their cooling effect would quickly disappear. As a result of an attempt to clean up our aerosol emissions, the climate system would register a large short-term warming.

In contrast, a simultaneously abrupt and total end to global CO_2 emissions by humans would lead to a much slower reduction in atmospheric CO_2 levels. The ocean would take up one-half of the pulse of excess CO_2 produced by previous human activities within 50 years, but 10% of the total would remain in the atmosphere for

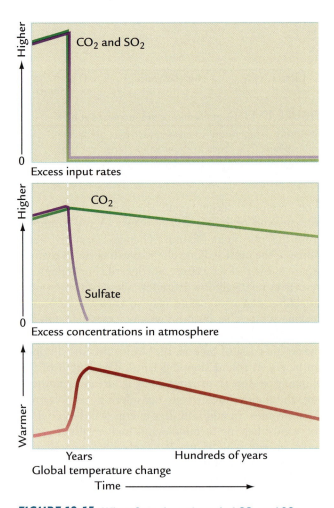

FIGURE 19-15 What if we abruptly ended CO_2 and SO_2 emissions? If humans instantly eliminated all industrial and other emissions of greenhouse gases and SO_2, the sulfates and their cooling effect would soon disappear, but the CO_2 and its warming effect would linger for centuries.

many millennia. Part of the anthropogenic warming will persist as long as this tail end of the excess CO_2 pulse lingers in the atmosphere.

One technological idea for avoiding adding huge amounts of CO_2 to the atmosphere is to pipe the excess industrial CO_2 into the ocean or into old oil fields. As with many technological solutions, cost is a major issue. Many CO_2-emitting sites lie far from the ocean or from oil reservoirs, and some of the CO_2 would have to be piped hundreds to thousands of miles at considerable cost. In addition, CO_2 sent directly into the ocean will increase the acidity of the ocean even faster than present rates.

The atmospheric chemist Paul Crutzen has suggested that sulfate aerosols could be used to offset some or all of the CO_2 warming. He proposed that very large amounts of sulfate aerosols be added to the stratosphere to simulate the effects of volcanically emitted aerosols in blocking solar radiation and cooling climate. By adding enormous amounts of aerosols, we might be able to cool Earth's climate even in the absence of effective action in reducing emission of CO_2 and other greenhouse gases.

Because gravity pulls stratospheric aerosols down into the troposphere within a few years (where they are soon removed by rainfall), sustaining a high enough concentration to offset global warming would require a constant supply of aerosols, by one estimate half a million tons of aerosols per year. Several means of delivering the aerosols have been proposed: balloons, artillery shells fired into space, a fleet of high-flying planes larger than any modern commercial airliners. By one estimate, the cost would be hundreds of millions of dollars per year, but even this price would be far lower than the expense in sequestering CO_2 after burning. One of several possible drawbacks is the fact that the additional sulfates would add to the acidification of the ocean and of terrestrial environments.

Another recent suggestion is to use genetic manipulation to develop microbes that feed on CO_2 and turn it into some form of carbon other than greenhouse gases. If such a transformation turns out to be possible, we could get rid of one greenhouse gas and turn it into another one that is a source of energy.

In the distant future, most of the pulse of excess CO_2 will have disappeared into the ocean, and climate will have cooled back toward its natural level (see Figures 19–3 and 19–5) but probably not to a glaciation. Part of the excess CO_2 pulse will linger in the atmosphere for tens of thousands of years and keep climate too warm for new ice to accumulate on North America or Eurasia. Ongoing methane emissions from irrigated areas and landfills will also prevent cooling to the point of glaciation.

In a world where climate can be manipulated, humankind will face the question of what "natural" is. By consensus, "natural" might be the climate that existed just prior to the industrial era. An equally viable view is that the true natural state would be one in which the now-overdue glaciation is underway (Chapter 15). In the future, humankind will probably be weighing the merits of engineering a warmer or a colder Earth.

Epilogue

Humans have been responsible for three major alterations of Earth's natural state. The first was the mass extinction of large mammals by late Stone Age peoples near 50,000 and again near 12,500 years ago. The second was the clearing of the forests of southern Eurasia during the last several millennia to open up land for agriculture. The third and largest has been the wide range of industrial era changes that have further altered Earth's surface during the past two centuries. Humans now move more debris (rock and sediment) than the combined action of all Earth's rivers, glaciers, and winds.

Estimated future changes in climate will be on a scale comparable to the largest natural changes of the past. The projected 5°C warming from the $4 \times CO_2$ emissions scenario would be equivalent to the amount of cooling at the most recent glacial maximum 20,000 years ago. The projected 2.5°C warming from the $2 \times CO_2$ emissions scenario would amount to half of the glacial maximum cooling. We are now ~0.7°C of the way into this huge new experiment in transforming our planet. Unless technology or extreme conservation efforts intervene, Earth is headed toward a warmer future at rates that are unprecedented in its 4.5-billion-year history.

Key Terms

methane clathrate (p. 347)   ocean acidification (p. 351)

Review Questions

1. What factors will determine how much CO_2 humans add to the atmosphere in the future?

2. Where will all the excess carbon eventually go?

3. In what way will the future CO_2 warming be like and unlike past CO_2 warmings?

4. In a $2 \times CO_2$ world, how many trends under way in recent decades are likely to continue?

5. In a $4 \times CO_2$ world, where will ice of any kind still be found on Earth?

6. Will future temperature changes be readily apparent to the average person? Why or why not?

7. What are the disadvantages in drastically reducing our industrial emissions of sulfur and carbon?

8. Yes or no: Do you think that future global warming is a major concern?

Additional Resources

Basic Reading

Henson, R. 2006. *The Rough Guide to Climate Change*. London: Rough Guides.

Schneider, S. H. 1997. *Laboratory Earth*. New York: Basic Books.

http://wcrp.wmo.int/ (World Climate Research Program)

www.ncdc.noaa.gov/oa/climate/globalwarming.html (National Climate Data Center)

www.ipcc.ch (Intergovernmental Panel on Climate Change)

Advanced Reading

Graedel, T. E., and P. J. Crutzen. 1997. *Atmosphere, Climate, and Change*. New York: Scientific American Library.

Intergovernmental Panel on Climate Change. 2007. "Climate Change 2007: The Physical Science Basis." Geneva: World Meteorological Organization.

National Research Council. 2003. *Understanding Climate Change Feedbacks*. Washington, DC: National Academy of Sciences Press.

Appendix 1

Isotopes of Oxygen

Oxygen is present in abundance in several key parts of Earth's climate system: as the second most abundant gas (O_2) in the atmosphere, as water vapor (H_2O_v) in the atmosphere, as a component of water molecules (H_2O) in the ocean and in lakes, and as frozen water in ice sheets. These reservoirs interact with each other and exchange oxygen.

Oxygen occurs in nature mainly as two isotopes. The lighter ^{16}O isotope accounts for almost 99.8% of the total amount, and the heavier ^{18}O isotope for most of the rest. The ratio of ^{18}O to ^{16}O is approximately 1/400, equivalent to a value of about 0.0025.

Individual measurements of the $^{18}O/^{16}O$ ratio in natural materials are reported as departures in parts per thousand (‰) from a laboratory standard:

$$\delta^{18}O \atop (\text{in ‰}) = \frac{(^{18}O/^{16}O)_{\text{sample}} - (^{18}O/^{16}O)_{\text{standard}}}{(^{18}O/^{16}O)_{\text{standard}}} \times 1000$$

All measurements of samples are referenced to a standard in order to establish a common reference point for analyses in all laboratories. The reason that the measured ratio is multiplied by 1000 is convenience: the multiplication converts the extremely small variations in an already small ratio to a more workable numerical form (values that range between about +3‰ and –55‰).

Samples with relatively large amounts of ^{18}O (compared with ^{16}O) are said to have more positive $\delta^{18}O$ values and are referred to as ^{18}O-enriched (or ^{16}O-depleted). Samples with relatively small amounts of ^{18}O have more negative $\delta^{18}O$ values and are referred to as ^{18}O-depleted (or ^{16}O-enriched).

Modern $\delta^{18}O$ Values in Water, Ice, and Water Vapor

Scientists are mainly interested in the ocean and the ice sheets because these reservoirs contain large amounts of oxygen and have exchanged it over time (Figure 1). The average $\delta^{18}O$ value of ocean water is set by definition near 0‰ (more specifically, –0.1‰), but the ratio actually varies between about –2‰ and +3‰ in different ocean regions and depths.

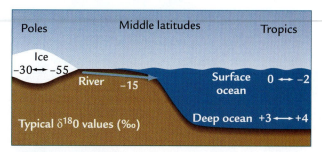

FIGURE 1 $\delta^{18}O$ **values in the modern world** In the modern ocean, $\delta^{18}O$ values vary from 0 to –2‰ in warm, tropical surface waters to as much as +3 to +4‰ in cold, deep ocean waters. In today's ice sheets, typical $\delta^{18}O$ values reach –30‰ in Greenland and –55‰ in Antarctica.

Most of this range results from changes in the temperature of ocean water. For each 4.2°C increase in temperature, the $\delta^{18}O$ ratio decreases by 1‰ (that is, ^{18}O becomes less abundant in relation to ^{16}O). As a result, warm waters in the surface ocean have more negative $\delta^{18}O$ values that range from 0‰ to –2‰, while colder waters deeper in the ocean have more positive values that range from +3‰ to +4‰.

The ice sheets are the other major oxygen reservoir of interest. The water vapor that supplies snow to the ice sheets comes from the ocean. The tropical atmosphere is rich in water vapor (H_2O_v) evaporated from the warm tropical ocean (companion Web site, pp. 12–13). The natural circulation of the atmosphere transports water vapor to higher latitudes and higher altitudes, where it condenses and then falls to Earth's surface as precipitation.

The transport of water vapor occurs through repeated cycles of evaporation and precipitation (Figure 2). Because the lighter ^{16}O isotope evaporates more readily, it tends to be preferentially extracted from the low-latitude ocean and sent toward higher latitudes. This transfer leaves the tropical ocean enriched in ^{18}O. In addition, the heavier ^{18}O isotope is more easily removed from the atmosphere when condensation and precipitation occur, leaving the water vapor that remains in the atmosphere even more enriched in ^{16}O and the low-latitude ocean still more enriched in ^{18}O. These enrichment processes are called **fractionation.**

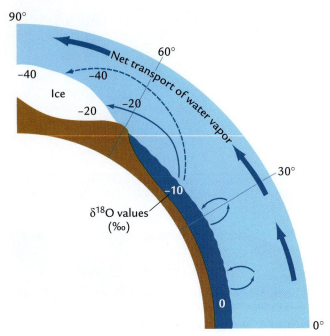

FIGURE 2 Isotope fractionation As water vapor moves from the tropics toward the poles, it is enriched in the ${}^{16}O$ isotope during each step of evaporation and condensation. This fractionation process makes the $\delta^{18}O$ values of snow falling on (and stored in) ice sheets more negative (${}^{16}O$-rich).

Each step in the fractionation process decreases the $\delta^{18}O$ value of the water vapor by ~10‰ in relation to that of the ocean water left behind. Because the water vapor that reaches the ice sheets is highly enriched in ${}^{16}O$, the $\delta^{18}O$ composition of the ice sheets is very negative: −30‰ over much of Greenland and −55‰ over central Antarctica. Fractionation also occurs at high altitudes in lower and middle latitudes because air that reaches high elevations has been through the same processes and has become similarly enriched in ${}^{16}O$. For this reason, glacial ice high on tropical mountains has relatively negative $\delta^{18}O$ values.

The process of isotopic fractionation is also accompanied by the progressive removal of water vapor from the air because cooler air holds much less water vapor than warmer air. As a result, air masses that are the most enriched in ${}^{16}O$ contain the smallest amount of water vapor.

Local Complications

As a result of the fractionation effect, $\delta^{18}O$ values in today's surface ocean do not follow the trend that would be expected if temperature were the only controlling factor. Modern tropical surface waters at 25°C have $\delta^{18}O$ values near 0‰. Using the temperature/$\delta^{18}O$ relationship defined earlier, high-latitude surface waters at temperatures of 0°C (just above the −1.8°C freezing point of seawater) should have $\delta^{18}O$ values of about 15‰. Instead, these waters have $\delta^{18}O$ values not much different from tropical surface waters.

The reason for these unexpectedly negative values is that high-latitude rivers carry water fed by precipitation with $\delta^{18}O$ values averaging near −15‰. The $\delta^{18}O$ value of each river depends on the degree of fractionation that has occurred in the precipitation that reaches its watershed. This annual delivery of ${}^{16}O$-rich river water amounts to just a small fraction of the total volume of the high-latitude surface ocean, but it drives the oceanic $\delta^{18}O$ composition toward far more negative values than those expected from the temperature relationship. Coastal surface waters heavily affected by such rivers are more negative in $\delta^{18}O$ than the tropical ocean, and even high-latitude surface waters in regions well away from rivers have values comparable to those of the tropical ocean. This dilution effect by river water is also closely related to similar effects on ocean salinity: each 1.0‰ decrease in the $\delta^{18}O$ value of ocean water is accompanied by a 0.5‰ decrease in salinity due to delivery of fresh (nonsaline) water.

Climatic Application 1: Changes in Seawater $\delta^{18}O$

Through time, the $\delta^{18}O$ composition of ocean water is affected by changes in ocean temperature and in the amount of ${}^{16}O$-rich water extracted and stored in the ice sheets. The effect of past temperature changes on the $\delta^{18}O$ values of ocean water is the same as that in the modern ocean: a 1‰ decrease for each 4.2°C warming (and conversely). Because the size of past temperature variations has varied from region to region, the effect on $\delta^{18}O$ has also varied on a local basis.

During colder climates of the past such as the last glacial maximum 20,000 years ago, the large ice sheets on North America and Europe held enough ${}^{16}O$ to leave the ocean enriched in ${}^{18}O$ by an average of just over 1‰. In contrast, when no ice sheets were present on Earth, the ${}^{16}O$-rich ice that is now trapped in the Greenland and Antarctic ice sheets was instead water in the ocean. At that time, the mean oceanic $\delta^{18}O$ value was about 1‰ lower than it is now.

These changes in past $\delta^{18}O$ values are recorded in the $CaCO_3$ shells of two kinds of foraminifera. Planktic foraminifera live mainly in the upper 100 meters, and their shells contain oxygen taken from bicarbonate ions (HCO_3^-) in the near-surface ocean. When these floating organisms die, their shells fall to the sea floor and accumulate as a permanent record of past values of seawater $\delta^{18}O$ at the surface. In comparison, benthic foraminifera live on the seafloor and within the uppermost layers of ocean sediment, and their shells contain oxygen taken from bicarbonate ions in deep water.

Sediment samples taken from the ocean are sieved to remove the mud and silt in order to isolate the sand-sized fraction from which foraminifera are individually picked. Typical $\delta^{18}O$ analyses require a few milligrams of sample, usually less than a dozen foraminifera. The $CaCO_3$ shells are dissolved in acid to produce CO_2 gas, which is then analyzed in a **mass spectrometer,** an instrument capable of detecting the small difference in atomic mass between the ^{16}O and ^{18}O isotopes.

Climatic Application 2: Changes in Ice Sheet $\delta^{18}O$

Changes in $\delta^{18}O$ values of layers in the ice sheets are also important to studies of past climate. Samples are drilled out of ice cores and melted, and the water is vaporized to form the gas H_2O_v for analysis on mass spectrometers. The $\delta^{18}O$ values within ice sheets can vary by 5‰ or more as climate changes. In the ice, more negative values are typical of colder climates and less negative values indicate warmer climates. In Greenland, interglacial values typically vary between –30‰ and –35‰, while glacial values fall between –35‰ and –40‰. In Antarctica, interglacial values are typically between –50‰ and –55‰, while glacial values fall between –55‰ and –60‰. Note that the $\delta^{18}O$ responses of the ocean and the ice sheets are opposite in direction: as marine $\delta^{18}O$ values become more positive during glacial climates, the $\delta^{18}O$ values in ice cores become more negative (and conversely).

These $\delta^{18}O$ changes in the ice sheet layers reflect several influences (Figure 3). One factor is the tempera-

TABLE 1   Causes of $\delta^{18}O$ Changes Recorded in Ice Cores

Negative ←	Change in $\delta^{18}O$ values	→ Positive
Colder	Air temperature over ice	Warmer
Distant	Proximity of source region	Close
Low $\delta^{18}O$	$\delta^{18}O$ composition of source	High $\delta^{18}O$
High	Elevation of ice	Low
Winter	Primary season of precipitation	Summer

ture of the snow that precipitates on the ice sheets. "Warm" (wet) snow can form at temperatures near freezing, but cold (dry, powdery) snow forms at much colder temperatures. The $\delta^{18}O$ value of snow falling on Greenland today trends 0.7‰ more negative with each 1°C drop in the temperature of the air in which it forms. This temperature/$\delta^{18}O$ relationship holds for today's seasonal changes in the temperature of the air masses that deliver the snow, and also for the cooling of air at higher elevations on the ice sheets. But this relationship did not always apply in the past: $\delta^{18}O$ values of glacial-age ice in Greenland suggest temperatures about 10°C colder than those today, but direct measurements of the temperature of the ice indicate a cooling of 15°C or more.

Several other factors can affect the $\delta^{18}O$ values recorded in ice cores, such as changes in the source of the water vapor, in the path of transport to the ice sheets, and in the season when the precipitation falls (Table 1). These complications arise because the water vapor that supplies snow to the ice sheets comes from several sources, each with a different initial $\delta^{18}O$ value, and it follows different paths of transport. The longer the distance the water vapor travels, the more negative is its $\delta^{18}O$ value, because it has evaporated and condensed repeatedly along the way (see Figure 2). Changes in the relative amounts of water vapor coming from different sources can alter the mean $\delta^{18}O$ value of the snow that falls on the ice through the span of a year. Changes in the seasonal balance of water vapor delivery can also affect mean annual $\delta^{18}O$ values in the ice: more snow in the colder winter season results in lower $\delta^{18}O$ values.

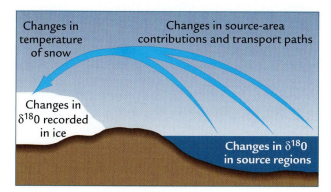

FIGURE 3 Controls on variations in ice core $\delta^{18}O$ The $\delta^{18}O$ values recorded in ice cores vary with changes in the local temperature of the snow that falls on the ice, changes in the relative contributions among the source areas of water vapor, and $\delta^{18}O$ changes within the source regions.

Appendix 2

Isotopes of Carbon

Both ^{13}C and ^{12}C are stable (nonradioactive) isotopes of carbon that occur naturally in Earth's vegetation, water, and air. The ^{12}C isotope accounts for more than 99% of all the carbon present on Earth, and ^{13}C accounts for most of the rest. A small amount exists as radioactive ^{14}C. Geochemists who analyze material for its carbon isotope composition measure small variations around the average $^{13}C/^{12}C$ ratio of less than 0.01.

Similar to the convention used for oxygen isotopes, measurements of $^{13}C/^{12}C$ ratios are reported as departures in parts per thousand (‰) from a laboratory standard:

$$\delta^{13}C \atop (‰) = \frac{(^{13}C/^{12}C)_{sample} - (^{13}C/^{12}C)_{standard}}{(^{13}C/^{12}C)_{standard}} \times 1000$$

All measurements are referenced to standards supplied by the National Bureau of Standards for use as a common reference point. Like the oxygen isotope ratios, carbon isotope ratios are multiplied by 1000 to convert the very small measured variations in an already small ratio to a more handy numerical form. As a result, $\delta^{13}C$ values for carbon that occurs in oxygen-rich conditions fall between −25‰ for some kinds of vegetation on land to +2‰ for carbon dissolved in ocean surface waters in some regions. For carbon that forms in the absence of oxygen (in "reducing conditions"), $\delta^{13}C$ values can be far more negative, around −50 to −60‰.

Carbon samples with relatively large amounts of ^{13}C compared with ^{12}C have more positive $\delta^{13}C$ values and are referred to as ^{13}C-enriched or ^{12}C-depleted. Samples with relatively small amounts of ^{13}C compared with ^{12}C have more negative $\delta^{13}C$ values and are referred to as ^{13}C-depleted or ^{12}C-enriched.

Fractionation during photosynthesis causes changes in $\delta^{13}C$ values (Figure 1). As the plants take inorganic carbon and turn it into organic carbon, they incorporate the ^{12}C isotope into their living tissue more easily than the ^{13}C isotope. This discrimination in favor of ^{12}C shifts the $\delta^{13}C$ of organic matter toward values that are more negative than the initial inorganic carbon source.

For example, plant plankton in the ocean take inorganic carbon from seawater with a $\delta^{13}C$ value near 0‰ and convert it to organic carbon with a $\delta^{13}C$ value near −22‰. Some of the organic carbon is sent to the deep ocean, but most is oxidized back to inorganic form and recycled within the ocean or sent back to the surface waters. The net export of a small fraction of ^{12}C-rich organic carbon to the deep ocean leaves the remaining

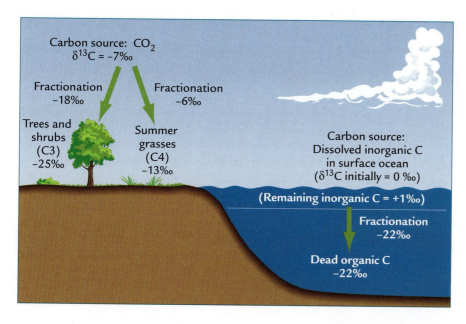

FIGURE 1 Photosynthesis and carbon isotope fractionation
Photosynthesis on land and in the surface ocean converts inorganic carbon to organic form and causes large negative shifts in $\delta^{13}C$ values of the organic carbon produced.

surface water enriched in ^{13}C, with δ^{13}C values of 1‰ or slightly higher.

Overall, organic carbon forms a small fraction of the total carbon reservoir in the ocean, where inorganic carbon predominates. The combined effect of the two reservoirs (a small amount of organic carbon with a δ^{13}C value near –22‰ and a large amount of inorganic carbon with a δ^{13}C value near +1‰) yields a mean δ^{13}C value of ~0‰ for all the carbon in the ocean.

Inorganic CO_2 in the atmosphere with a δ^{13}C value of –7‰ is the carbon source for photosynthesis by plants. All trees as well as most shrubs and cool-climate grasses use a type of photosynthesis called the C3 pathway, which produces organic tissue with δ^{13}C values in the range of –21‰ to –28‰ and an average value of –25‰. Some shrubs and most grasses that grow in hot climates during the summer season use a different kind of photosynthesis called the C4 pathway, which produces vegetation with less negative δ^{13}C values, ranging between –11‰ and –15‰ and averaging –13‰. By far the largest amount of organic biomass on Earth's land surfaces resides in C3 trees, and as a result the average δ^{13}C of vegetation on land (and of litter in soils) is –25‰.

For purposes of reconstructing past climates, the δ^{13}C composition of a wide range of fossilized materials contain can be analyzed on mass spectrometers. The same $CaCO_3$ shells of marine foraminifera that yield oxygen for δ^{18}O analyses also provide carbon for δ^{13}C analyses. Acid can be used to dissolve the shells and produce CO_2 for analysis. Living vegetation or its organic residue in soils as well as organic matter in plankton and ocean sediments can also be analyzed on mass spectrometers after the organic carbon is burned to produce CO_2 gas.

GLOSSARY

(Many definitions apply specifically to their use in climate studies.)

ablation The loss of snow or ice from a glacier by melting, calving, and other processes.

accumulation The addition of snow to a glacier.

adiabatic Having to do with an increase in pressure that raises the temperature of a parcel of air (adiabatic warming) or a decrease in pressure that lowers that temperature (adiabatic cooling).

aerosols Extremely small particles or droplets carried in suspension in the air.

albedo The decimal fraction or percentage of incoming solar radiation reflected from a surface.

albedo-temperature feedback A positive feedback that amplifies an initial temperature change by altering the amount of snow cover or sea ice and changing the amount of solar radiation absorbed by Earth's surface.

aliasing The misleading (unrepresentative) signals that result from sampling a record of climate change at too low a resolution.

alkenones Complex organic molecules found in fossil shells of plant plankton and used to reconstruct past temperature.

amplitude Half of the height between peaks and troughs in a regular wave form.

Antarctic bottom water A dense, cold water mass that forms near the Antarctic continent by extreme chilling of surface waters, sinks, and flows along the seafloor below a depth of 4 km.

Antarctic intermediate water A water mass that forms in the Southern Ocean by chilling of seawater exposed in or near sea ice, sinks, and flows northward at depths of 1 to 2 km.

anthropogenic CO_2 increase The steadily increasing concentration of CO_2 in the atmosphere over the last 200 years due to human activities.

anthropogenic forcing All human-related factors that cause climate change.

aphelion The point in Earth's slightly eccentric orbit at which it is farthest from the Sun.

asthenosphere A partially molten layer of rock in the upper mantle that is weak enough to flow and cause movement of the overlying lithospheric plate.

axial precession The wobbling movement of Earth's axis of rotation, which causes it to point in different directions over a cycle of 26,000 years.

back radiation Electromagnetic energy at long (infrared) wavelengths emitted from any material with a temperature above absolute zero (0 K).

Barents ice sheet An ice sheet that covered the present-day Barents Sea, north of Scandinavia, during orbital-scale glaciation cycles.

basal slip Rapid sliding of an ice sheet across its water-lubricated bed, especially in a region where ice streams lie above water-saturated sediment.

bedrock pinning point A high-standing protrusion of bedrock that lies beneath the margin of an ice sheet and slows the flow of ice into the ocean by frictional resistance.

benthic foraminifera Sand-sized organisms (protozoans) that live on and in the seafloor and form shells of $CaCO_3$.

biomass The amount of living matter in a region; also, organic matter used as a source of energy.

biome A region on Earth with a distinctive community of plants.

biome model A vegetation model that simulates the major vegetation type (for example, grassland or desert scrub) that can exist in a region under a given set of climatic conditions.

biosphere The part of the Earth system that supports life, including the oceans, land surfaces, soils, and atmosphere.

biotic proxy An index of past climate change based on measurable variations in the type or abundance of climate-sensitive organisms.

Black Sea flood hypothesis The hypothesis that melting ice sheets caused rising ocean waters to flood into an ancient glacial lake, displacing humans and forming the Black Sea.

BLAG (spreading rate) hypothesis The hypothesis that tectonic-scale climate changes are driven by variations in the global average rate of seafloor spreading, which alter the amount of CO_2 introduced into the atmosphere.

boundary conditions The initial configuration of Earth's properties chosen for a model simulation (such as land-sea distribution, mountain elevation, and atmospheric CO_2 concentration).

Brown Cloud Haze of carbon aerosols emitted by human activities, especially from cities in Asia.

burial flux The rate of deposition of a substance in a sedimentary reservoir measured in units of mass per unit of area per unit of time.

C3 pathway The means by which trees and most shrubs (about 95% of all land plants) obtain CO_2 from the air during the initial step of photosynthesis.

C4 pathway The means by which grasses that grow during the warm season (about 5% of all land plants) obtain CO_2 from the air during the initial step of photosynthesis.

calibration interval The interval of time (usually 50–100 years) over which the width or density of tree rings can be correlated with historical observations of temperature and precipitation change.

calorie The amount of energy required to raise the temperature of 1 gram of water by 1°C.

calving The process by which a large block of ice breaks off from the margin of a glacier and forms an iceberg that floats in the ocean or in a lake.

carbon isotopes Isotopes of the element carbon with different atomic masses, used to trace the movement of different kinds of carbon through Earth's climate system (12C and 13C) or to measure elapsed time indicated by radioactive decay (14C and 12C).

cardinal points The two equinoxes (spring and autumn) and the two solstices (summer and winter) in Earth's annual revolution around the Sun.

Celsius A temperature scale on which water freezes at 0° and boils at 100°.

channeled scablands A region in Idaho and eastern Washington State in which water impounded in glacial lakes suddenly rushed out and reshaped the landscape, perhaps repeatedly.

chemical weathering Dissolving or other alteration of minerals to a different form by chemical reactions in the presence of water.

chlorofluorocarbons (CFCs) Synthetic chemical compounds generated by human activity and containing chlorine or fluorine that can destroy ozone in the stratosphere.

CLIMAP (Climatic Mapping and Prediction Project) A large cooperative research group during the 1970s and 1980s that first mapped the surface of the ice-age Earth.

climate Fluctuations in Earth's air, water, ice, vegetation, and other properties on time scales longer than one year.

climate data output The climatic properties (such as temperature, precipitation, and winds) that are produced by simulations with climate models.

climate point The point where the equilibrium line that separates net ice melting from net ice accumulation intercepts sea level.

climate proxy A quantifiable indicator of climate change contained in a climate archive and covering an interval that precedes direct instrument measurements of climate.

climate science The study of climate changes and their causes.

climate simulation The use of a numerical model to reproduce climate for a specified set of boundary conditions.

climate system The components of Earth (air, water, ice, vegetation, and land surfaces) that participate in climate change.

climatic hypothesis of human evolution The hypothesis that a trend toward drier climates in Africa changed forest to savanna or grasslands and accelerated human evolution.

clipped responses Climatic responses that are truncated (cut off) in one direction.

closed system A system that does not exchange matter across its boundaries.

coccoliths Tiny disklike plates of $CaCO_3$ produced by algae living in ocean surface waters.

COHMAP (Cooperative Holocene Mapping Project) A cooperative research effort in the 1980s and 1990s that evaluated the causes of climate change during the most recent deglaciation by comparing geologic data with climate model simulations.

conifer forest Forest comprising evergreen, needle-bearing trees.

continental collision The occasional result of plate tectonic processes in which two continents are carried into each other by plate movements, creating high plateaus.

continental crust A layer of rock averaging 30 km thick, having the composition of granite, and comprising the continents.

continental ice sheet A mass of ice kilometers thick covering a continent or a large portion of a continent and moving independently of the underlying bedrock topography.

continental shelf A shallowly submerged extension of a continent beneath the ocean.

continental slope A ramplike structural edge of a continent that slopes into the deep ocean.

control case A model simulation run to reproduce Earth's present-day climate from its present-day boundary conditions.

convection The rising motion of a fluid (air or water) produced when its bottom layer is heated, accompanied by sinking of cooler, denser fluid elsewhere.

convergent margin A boundary between two lithospheric plates that move toward each other and cause a collision of continents or subduction of one plate beneath the other.

coral bands Annual banding in the structure of corals caused by seasonal changes in sunlight, water temperature, and nutrient content.

Cordilleran ice sheet A small ice sheet that covered western Canada and the far northwestern United States during orbital-scale glaciation cycles.

Coriolis effect The apparent deflection of a fluid (air or water) from a straight-line path because of Earth's rotation. The deflection is to the right in the northern hemisphere and to the left in the southern hemisphere.

CO_2 fertilization effect The increased growth rate of plants caused by adding CO_2 to the atmosphere.

CO_2 saturation The point at which the concentration of CO_2 in the atmosphere reaches so high a level that additional amounts do not increase the greenhouse effect.

Cretaceous An interval of warmer climates and higher sea level between 135 and 65 Myr ago.

Dansgaard-Oeschger cycles Oscillations in various properties (including dust and isotopes of oxygen) recorded in Greenland ice at intervals of 2000 to 7000 years during glacial intervals.

daughter isotope An isotope produced by radiometric decay of another isotope.

deforestation Cutting of forests by humans to clear land for agriculture and other processes.

deglacial two-step The irregular melting of ice sheets during the most recent deglaciation: fast-slow-fast.

δ¹³C aging The gradual shift toward more negative $\delta^{13}C$ values in slow-moving deep water caused by the downward rain of ^{12}C-rich organic matter from overlying surface water.

dendroclimatology The methods used to extract climate signals from changes in the width or density of tree rings, both of which are sensitive to extremes of temperature and precipitation.

dew point The temperature at which cooling air becomes fully saturated with water vapor and permits condensation.

diatoms Silt-sized algae that live in surface waters of lakes, rivers, and oceans and form shells of opal ($SiO_2 \cdot H_2O$).

diffusion The transfer of a property such as heat by random, small-scale movements from a region of higher to lower concentration.

diluvial hypothesis The hypothesis that unsorted sediments found on northern continents resulted from a great flood; these deposits are now recognized as deposits from glaciers.

dissolution A form of chemical weathering in which rocks such as limestone ($CaCO_3$) or rock salt (NaCl) are dissolved by water and produce ions that are removed by rivers.

divergent margin A boundary between two lithospheric plates that are moving apart, usually at the crest of an ocean ridge.

Earth system The complex system of Earth's atmosphere, hydrosphere, biosphere, and lithosphere, through which energy and matter circulate.

eccentricity The extent to which Earth's orbit around the Sun departs from a perfect circle.

elastic Capable of deforming rapidly under pressure (as is bedrock under the pressure of a glacier) and rebounding when the pressure is removed.

electromagnetic radiation Self-propogating electric-magnetic waves, which include visible light as well as infrared and ultraviolet waves.

electromagnetic spectrum The complete range of electromagnetic radiation at differing wavelengths.

El Niño A climatic pattern that recurs at intervals of 2 to 7 years and is marked by warm sea surface temperatures in the eastern tropical Pacific, off the west coast of South America.

enhanced greenhouse effect Trapping of Earth's back radiation by greenhouse gases produced by humans, in addition to the warming caused by natural greenhouse gases.

ENSO (El Niño Southern Oscillation) The combined oscillations of El Niño (temperature changes in the eastern Pacific) and the southern oscillation (atmospheric pressure changes in the western and south-central Pacific).

eolian sediments Fine sediments deposited by the action of wind.

equilibrium A state of climatic stability toward which the climate system is moving and at which it will eventually remain, unless disturbed.

equilibrium line The level in the atmosphere separating the zones of net addition and loss of ice.

equinoxes The two times during each year (spring and autumn) when the lengths of days and nights are equal.

equivalent CO₂ Changes in all greenhouse gases expressed in terms of an equivalent change in atmospheric CO_2 concentrations.

eustatic Characterized by changes in sea level that are global in scale, rather than the result of local factors such as tectonic uplift or subsidence of the land.

evaporites Minerals or rocks formed by precipitation of crystals from water evaporating in restricted basins in arid climates.

evolution The process by which particular forms of life give rise to other similar forms by gradual genetic changes.

faculae Bright rings that surround sunspots and emit large amounts of solar radiation.

Fahrenheit A temperature scale on which water freezes at 32° and boils at 212°.

faint young Sun paradox The paradox in which astronomical models indicate a much weaker Sun through Earth's early history but geologic evidence shows that Earth never froze.

feedback A process internal to Earth's climate system that acts either to amplify changes in climate (positive feedback) or to moderate them (negative feedback).

filtering The technique of extracting and isolating the shape of cycles at specific wavelengths or periods from complex signals.

fluvial sediments Sediments deposited by the action of water.

forcing Any process or disturbance that drives changes in climate.

fractionation A process favoring the transfer of one isotope of an element more than another.

frequency The number of full wave forms (each with one peak and one trough) that occur within a defined interval of time (usually one year). Also, the inverse of the period.

Gaia hypothesis A hypothesis that life regulates climate on Earth.

general circulation model (GCM) A three-dimensional computer model of the global atmosphere (or ocean) that simulates temperature, precipitation, winds, and atmospheric pressure.

geochemical model A model that quantifies the movement of geochemical tracers (minerals, elements, or isotopes) among reservoirs in the climate system.

geochemical tracer A chemical element or isotope whose movement between reservoirs in the climate system can be quantitatively tracked.

geological–geochemical proxy An index of past climate change based on measurable variations in physical or chemical properties of sediments, ice, or other archives.

Gondwana The large continent that existed in the southern hemisphere before the creation of the giant continent Pangaea.

greenhouse debate The controversy over the extent to which rising levels of atmospheric greenhouse gases have warmed Earth's climate during the last 200 years.

greenhouse effect The warming of Earth's surface and lower atmosphere that occurs when its own emitted infrared heat is trapped and reradiated downward by greenhouse gases.

greenhouse era An interval of warm climate on Earth, such as the Cretaceous, with ice sheets absent even in polar regions.

greenhouse gases Gases such as water vapor (H_2Ov), carbon dioxide (CO_2), and methane (CH_4), which trap outgoing infrared radiation emitted by Earth's surface and warm the atmosphere.

greenhouse surprise A climate change in the greenhouse world of the future that cannot be predicted.

grid boxes Geometric units within climate models that have uniform climatic characteristics and exchange heat, energy, and other properties with adjoining grid boxes.

Gulf Stream A narrow current of warm water that emerges from the Gulf of Mexico through the Florida Straits and flows northward along the southeastern coast of the United States.

gyre A spinning cell of water in an ocean basin, particularly at a subtropical latitude.

Hadley cell An atmospheric circulation cell in which air rises in the tropics, flows to the subtropics, sinks near 30° latitude, and flows back toward the tropics as surface trade winds.

half-life The time required for half the number of atoms of a radioactive isotope to decay.

harmonics Secondary cycles related to a wavelike climatic response with a period N, occurring at periods of $N/2$, $N/3$, $N/4$, and so on.

hardwood forest A forest comprising leaf-bearing (deciduous) trees.

heat capacity The amount of heat energy required to raise the temperature of 1 gram of a substance by 1°C.

Heinrich event An interval of rapid flow of icebergs from the margins of ice sheets into the North Atlantic Ocean, causing deposition of sediment layers rich in debris eroded from the land.

historical archives Sources of information on climate based on human observations of natural phenomena made before the era of instrument measurements.

hot spot A point on the surface of a lithospheric plate where magma rising from below causes frequent volcanic activity.

hydrologic cycle The movement of water and water vapor among the atmosphere, land, and ocean through evaporation, precipitation, runoff, and subsurface groundwater flow.

hydrolysis A form of chemical weathering in which water reacts with silicate minerals rich in silicon and oxygen to produce dissolved ions removed in rivers and clays left on the landscape.

hydrothermal Characterized by the circulation of hot fluids through rocks in Earth's outer crust, as at the Mid-Ocean Ridge system.

hypothesis An explanation of observations based on physical principles.

hypsometric curve A graph that summarizes the proportions of Earth's surface that lie at various altitudes above and depths below sea level.

ice dome A high, gently sloping central region of an ice sheet in which snow accumulates and away from which ice flows slowly.

ice-driven response A climate change produced by fluctuations in the size of an ice sheet.

ice-elevation feedback The positive feedback that results when an ice sheet grows to a higher elevation at which accumulation exceeds ablation.

ice flow model A model that simulates ice sheet processes of snow accumulation, internal ice flow, and ablation.

icehouse era An interval of cold climate on Earth, such as the present, with ice sheets present in polar regions.

ice lobe A rounded or arc-shaped outward protrusion of the margin of an ice sheet.

ice-rafted debris Sediments of widely ranging sizes eroded from the land by ice, carried to the ocean, and deposited on the seafloor.

ice saddle A ridge that connects multiple domes of an ice sheet at a slightly lower elevation.

ice shelf A wide body of ice usually hundreds of meters thick that is fed by ice flowing off a continent, partially floats on seawater, and produces icebergs as blocks of ice break off.

ice stream A region of an ice sheet in which the motion of ice is unusually rapid, generally because of water-saturated sediments at the base of the ice.

igneous rock Rock formed by the cooling and solidification of molten magma.

insolation The amount of solar radiation arriving at the top of Earth's atmosphere by latitude and by season.

instrument records Records of climate change measured by devices made by humans, from early thermometers through modern satellite-mounted instruments.

Intergovernmental Panel on Climate Change (IPCC) A large international group of scientists who reflect the current scientific consensus on the impact of greenhouse gases.

intertropical convergence zone (ITCZ) A narrow region within the tropics where warm moist air rises, cools, and loses its water vapor in heavy tropical rainfall.

iron fertilization hypothesis The hypothesis that iron-rich dust blown from the continents during glaciations enhances the productivity of the surface ocean, sends CO_2 into the deep ocean, and reduces the concentration of CO_2 in the atmosphere.

jet stream A narrow meandering stream of air moving rapidly (generally from west to east) at a high latitude and at an altitude averaging 10 km.

Kelvin A scale on which temperature is measured in Celsius degree intervals and on which water freezes at 273 K and boils at 373 K, and all motion ceases at 0 K.

La Niña A pattern opposite from that of El Niño, in which sea surface temperatures in the tropical eastern Pacific, off the west coast of South America, become unusually cold.

lapse rate The rate at which temperature falls with elevation in the atmosphere, averaging 6.5°C of cooling for each kilometer of altitude, but more for dry air and less for moist air.

latent heat The quantity of heat gained or lost as a substance changes state (liquid, solid, or gas) at a given temperature and pressure.

latent heat of melting The amount of heat gained as ice melts or released as water freezes.

latent heat of vaporization The amount of heat gained when water turns to water vapor or released when water vapor condenses back to water.

Laurentide ice sheet The largest of the northern hemisphere ice sheets that grow and shrink at orbital cycles, covering east-central Canada and the northern United States east of the Rockies.

lichen Primitive mosslike vegetation that lives on bare rock surfaces, uses sunlight to secrete acids and weather the rock, and slowly grows in a nearly circular shape.

lithosphere The outer rigid shell of Earth (including the upper mantle and oceanic and continental crust), characterized by strong, rocklike properties and divided into plates that move as rigid units during plate tectonic processes.

Little Ice Age An interval between approximately 1400 and 1900 A.D. when temperatures in the northern hemisphere were generally colder than today's, especially in Europe.

loess Silt-sized windblown glacial sediment.

longwave radiation Energy emitted in the infrared part of the electromagnetic spectrum by materials having a temperature above absolute zero (0 K).

macrofossils Larger fragments of vegetation (such as needles, twigs, or cones) that prove the local presence of vegetation on past landscapes, used as a supplement to pollen analysis.

magnetic field The lines of force that are generated by motion in Earth's outer core and cause iron-bearing materials to align in specific directions in relation to the magnetic north pole.

magnetic lineations Long, stripelike regions of ocean crust marked by stronger or weaker magnetism acquired when the crust formed by cooling from molten magma at a ridge crest.

mantle The middle (and largest) layer of the solid Earth, lying beneath the crust and above the core and composed of silicate minerals.

marine ice sheet An ice sheet whose base lies below sea level, as in western Antarctica.

mass balance The method of tracking the movement of materials within Earth's climate system by applying the law of conservation of mass.

mass balance model A model that tracks the movement of materials within Earth's climate system using conservation of mass to balance inputs and outputs among different reservoirs.

mass spectrometer An instrument that measures different isotopes of the same element (such as ^{16}O vs. ^{18}O and ^{12}C vs. ^{13}C) by separating them by mass.

mass wasting A downhill movement of rock or soil under the force of gravity.

Maunder sunspot minimum An interval between 1645 and 1715 A.D. when astronomers observed very few sunspots on the Sun's surface.

medieval climatic optimum An interval between 1100 and 1300 A.D. in which some northern hemisphere regions were warmer than in the Little Ice Age that followed.

Mediterranean overflow water A water mass that forms in the northern Mediterranean Sea by winter chilling of salty surface waters and flows into the North Atlantic at a depth of 1 km.

methane clathrate A partly frozen slushy mix of methane gas (CH_4) and ice.

Mg/Ca ratios The ratio of the elements Mg and Ca in shells of foraminifera, an index of past temperature changes.

Milankovitch theory The theory that orbitally controlled fluctuations in high-latitude solar radiation (insolation) during summer control the size of ice sheets through their effect on melting.

millennial oscillations Fluctuations in climate lasting thousands of years and generally larger during glacial than interglacial intervals.

modulation The tendency for peaks and troughs in a wave form to vary in size in a regular way, such that clusters of large peaks and troughs alternate with clusters of smaller ones.

moraine A pile of unsorted rubble (till) deposited by a glacier at its margin.

monsoon Winds that reverse direction seasonally, blowing onshore in summer and offshore in winter because of different rates of heating and cooling of land and water.

Monterey hypothesis The hypothesis that increased rates of burial of organic carbon on the margins of the Pacific Ocean 17 Myr ago reduced CO_2 levels in the surface ocean and atmosphere.

mountain glacier A body of ice tens to hundred of meters thick and kilometers in length confined to a valley at a high elevation. A mountain ice cap is a similar-sized body of ice lying on the rounded summit of a mountain.

no-analog vegetation A combination of types of vegetation in the past for which no similar (analogous) combination exists today.

nonlinear response A climatic response that occurs on other than a simple one-for-one basis in relation to the forcing.

North Atlantic deep water A water mass that forms in the high-latitude North Atlantic Ocean by winter chilling of salty surface water, sinks, and flows southward at depths of 2 to 4 km.

North Atlantic drift A warm, multipart current flowing northeastward into the high latitudes of the North Atlantic as a continuation of the Gulf Stream.

North Atlantic Oscillation A fluctuation in subtropical and North Atlantic temperatures that persists for more than a year.

ocean carbon pump hypothesis The hypothesis that changes in the amount of organic carbon taken up by ocean plankton during photosynthesis and exported to the deep ocean after they die control CO_2 levels in the surface ocean and atmosphere.

ocean crust A layer of rock averaging 7 km thick, having the average composition of basalt, and comprising the ocean floor.

ocean heat transport hypothesis The hypothesis that changes in the amount of heat transported toward polar regions by the ocean cause changes in polar climate.

oceanic gateway A narrow passage between continents that opens or closes and thereby alters ocean circulation.

orbital monsoon hypothesis The hypothesis that orbitally controlled changes in summer insolation at low latitudes drive the strength of the tropical summer monsoon.

orbital tuning The process of constructing a time scale by using the link between astronomically dated changes in solar radiation and the rhythmic climatic responses they cause on Earth.

orographic precipitation Precipitation on the upwind side of a mountain or plateau caused by the forced ascent of warm air to cooler elevations,where the entrained water vapor condenses.

outwash Layered sediments deposited by meltwater streams emerging from a glacier.

overkill hypothesis The hypothesis that the sudden extinction of many mammals 12,500 years ago resulted from human hunting rather than climatic stress.

oxidation A chemical reaction in which electrons are lost from an atom and its charge becomes more positive; also, the addition of oxygen to an element.

ozone A triple molecule of oxygen (O_3) formed by the collision of cosmic particles with normal (O_2) oxygen. Ozone in the stratosphere blocks harmful ultraviolet radiation from the Sun.

ozone hole A region centered over the Antarctic continent in which ozone (O_3) levels in the upper atmosphere (stratosphere) drop to very low values in the spring.

Pacific Decadal Oscillation A fluctuation in tropical and North Pacific temperatures that persists for more than a year.

paleomagnetism The study of patterns of ancient magnetism recorded in rocks or sediment.

pandemics Diseases that kill many millions of people on several continents.

Pangaea The giant supercontinent that existed between 300 and 175 Myr ago and consisted of all landmasses present on Earth.

parent isotope A radioactive isotope that naturally decays to a daughter isotope.

peat A deposit of decayed carbon-rich plant remains in a wetland environment with little oxygen.

peatlands hypothesis The hypothesis that the increase in methane levels between 5000 and 250 years ago was caused by expanded areas of methane-producing bogs in north polar regions.

perihelion The point in Earth's slightly eccentric orbit at which it is closest to the Sun.

period The time interval between successive peaks or troughs in a series of regular wave forms.

peripheral forebulge A region in which the weight of glacial ice sheets caused bedrock to flow out to the ice margins at great depths and produced a broad upward bulge of the land.

permafrost A permanently frozen mixture of rocks and soil occurring in very cold regions.

phase lag The amount by which one cyclic signal lags behind another signal of the same wavelength (or period).

photosynthesis The process by which plants use nutrients and solar energy to convert water and CO_2 to plant tissue (carbohydrates) and thereby produce oxygen.

physical climate model A numerical model that simulates Earth's climate on the basis of physical principles of fluid motion and transfers of radiative heat energy and momentum.

physical weathering Any mechanical process by which rocks are broken into smaller fragments of the same material.

phytoplankton Small floating organisms (usually algae) that use energy from the Sun and nutrients from the water for the process of photosynthesis.

plane of the ecliptic The plane within which Earth revolves around the Sun.

planktic foraminifera Sand-sized organisms (protozoans) that live in ocean surface waters and form shells made of $CaCO_3$.

plankton Organisms that float in the upper layers of oceans or lakes.

plate tectonics Tectonic interactions resulting from the movement of lithospheric plates.

polar front A sharp boundary zone in a polar ocean between cold, low-salinity waters and warmer, saltier waters; similarly, in the atmosphere, a sharp temperature boundary.

polar position hypothesis The hypothesis that ice sheets exist during intervals in Earth's history when landmasses are moved into polar regions by plate tectonic processes.

power spectrum A graphic display of the distribution of power (the square of wave amplitude) against the period (or frequency) of each cycle present in a signal.

precessional index The mathematical product (esinv) of Earth's sine-wave motion (sinv) around the Sun and the eccentricity of its orbit (e)

precession of the ellipse The slow turning of Earth's elliptical orbit in space.

precession of the equinoxes The movement of the solstices and equinoxes around Earth's elliptical orbit over cycles of 23,000 and 19,000 years.

preindustrial CO_2 level The concentration of CO_2 in the atmosphere (280 parts per million) that existed for several thousand years before the Industrial Revolution.

productivity The amount of organic matter synthesized by organisms from inorganic substances per unit of area per unit of time.

proglacial lake A short-lived lake that develops after the retreat of an ice sheet in the bedrock depression left by the weight of the ice.

radiation Electromagnetic energy that drives Earth's climate system. Ultraviolet and visible radiation emitted by the Sun affect climate on Earth, which emits infrared back radiation to space.

radiative forcing The effect of greenhouse gases in trapping (or blocking) solar radiation, expressed in the same units as those of solar energy: watts per square meter (W/m2).

radiocarbon dating Dating of relatively young carbon-bearing geologic materials by means of ^{14}C, a radioactive isotope that decays with a half-life of 5700 years.

radiolaria Sand-sized organisms that live in surface waters and form shells of opal ($SiO_2 \cdot H_2O$).

radiometric dating Determining the ages of rocks or sediments by measuring the amount of naturally occurring radioactive parent isotopes and their nonradioactive daughter products.

reconstruction A simulation run with a climate model by altering of several boundary conditions in an effort to reproduce a climate that existed at some time in the past.

red beds Sediments or rocks with a red color caused by the oxidation of iron (similar to rust).

regressions Relative motions of the ocean down and off the margins of the land.

reservoir A place of residence for an element or isotope that moves in a cycle.

residence time The average amount of time a tracer of any substance spends in a reservoir.

resolution The degree of detail detected in a climate signal by sampling at a particular interval.

resonant response A strong cyclic response of the climate system to perturbations occurring at the same or other cycles.

response Any change in the climate system caused by a change in climate forcing.

response time The time required for a climatic response to move a defined fraction of the way from its existing value to the value it would hold at its full equilibrium response.

salinity A measure of the salt content of seawater in parts per thousand (‰).

salt rejection The salt left in seawater when sea ice forms on the ocean.

sapropels Black organic-rich muds deposited on the Mediterranean seafloor as a result of strong inflow from the Nile River, which stifles delivery of oxygen to the deep parts of the basin.

saturation vapor density The maximum amount of water vapor that air can hold at a given temperature.

savanna A semi-arid region of grasses and scattered trees.

Scandinavian ice sheet An ice sheet that covered most of Norway and Sweden as well as the northern part of Germany and France during orbital-scale glaciation cycles.

seafloor spreading The mechanism by which new seafloor is created at an ocean ridge as the adjacent plates move away from the ridge crest at a rate of centimeters per year.

sediment drift A lens-shaped pile of fine sediment (clay and silt) picked up in a region where bottom currents move swiftly and then deposited in a region of the seafloor where currents slow.

sensible heat Heat energy carried by water or air in a form that can be easily felt or sensed, rather than hidden in latent form.

sensitivity test A simulation run with a climate model in which one boundary condition is altered from the (modern) control case to test its effect on climate.

shortwave radiation Electromagnetic energy emitted by the Sun in visible and ultraviolet wavelengths that deliver heat to Earth's climate system.

silicate minerals Minerals rich in silicon and oxygen and accounting for most of the rocks in Earth's crust and mantle.

sine wave A perfectly regular wave form in which successive peaks and troughs are evenly spaced, with each peak reaching a value of +1 and each trough a value of -1.

sintering The process by which bubbles of air become sealed off and preserved as snow turns to ice at depths of 50 to 100 m within an ice sheet.

snowball Earth hypothesis The hypothesis that Earth was frozen even in the tropics sometime in the interval between 850 and 550 Myr ago.

solstices The times during Earth's yearly revolution around the Sun when the days are longest (summer solstice) and shortest (winter solstice).

southern oscillation Naturally occurring fluctuations in which changes in lower atmospheric surface pressure in the far western Pacific, near northern Australia, are opposite in sense to those in the south-central Pacific, near Tahiti.

specific heat The amount of heat required to raise the temperature of 1 gram of a substance by 1°C.

spectral analysis A numerical technique for detecting and quantifying the distribution of regular (periodic) behavior in a complex signal.

speleothems Cave deposits made of $CaCO_3$ containing climatic signals.

Sporer sunspot minimum An interval between 1460 and 1550 A.D. when very few sunspots were observed on the Sun's surface.

stratosphere The stable layer of the atmosphere lying between 10 and 50 km above Earth's surface and containing most of Earth's ozone.

subduction The slow downward sinking of an ocean plate beneath a continent or an island arc as a result of plate tectonic processes.

sulfate aerosols Fine particles produced in the atmosphere from SO_2 gas emitted by volcanoes or by industrial smokestacks. These particles can block incoming solar radiation.

sunspots Dark areas of temperatures lower than average on the surface of the Sun.

sunspot cycle A natural 11-year cycle in the number of dark spots visible on the face of the Sun, reliably recorded by astronomers for more than four centuries.

surge A sudden and rapid forward movement of the margin of a glacier.

tabular iceberg A large flat-topped slab of ice produced by calving from the seaward margin of an ice shelf.

tectonic plates Divisions of the upper solid Earth (lithosphere) that are 100 km thick and thousands of kilometers in lateral extent and move as rigid units in plate tectonic processes.

termination A 10,000-year interval of rapid melting of ice sheets that brings to an end a longer (90,000-year) interval of slower ice growth.

theory A hypothesis that has survived repeated testing.

thermal expansion coefficient The volumetric expansion of seawater when it warms, and contraction when it cools, by 1 part in 7000 per degree C (for temperatures above 4°C).

thermal inertia The resistance of a component of the climate system to temperature change.

thermocline A layer of water in which temperature changes rapidly in a vertical direction.

thermohaline flow The vertical and lateral movement of subsurface waters in the ocean as a result of contrasts in density caused by differences in temperature and salinity.

thermostat A mechanism that senses changes in temperature and acts to moderate them.

threshold A level at which a sudden change in the basic nature of a climatic response occurs.

tilt The angle between Earth's equatorial plane and the plane of its orbit around the Sun, also equivalent to the angle between Earth's axis of rotation and a line perpendicular to its axis of rotation around the Sun. Also referred to as *obliquity*.

time-dependent model A geochemical model that tracks changes in the rate of movement of tracers among reservoirs within the climate system over time.

time-series analysis A group of techniques for extracting periodic signals (cycles) from complex signals and quantifying their strength, relative timing, and correlation.

transform fault margin A boundary between lithospheric plates in which the plates slide past each other.

transgression Relative movement of the ocean up and across the margins of the land.

transpiration The release of water vapor by plants into the atmosphere.

tree rings Annual bands formed by trees in regions of seasonal climate, with lighter layers formed during rapid growth in the spring and darker layers at the end of growth in the autumn.

troposphere The layer of the atmosphere just above Earth's surface (10 km or more in thickness) in which weather occurs.

tundra A high-latitude or high-altitude environment in which the ground freezes deeply in winter but thaws at the surface in summer, permitting low-growing plants to flourish.

2 x CO$_2$ sensitivity The amount of warming produced by an increase in atmospheric CO$_2$ levels from the preindustrial level (280 parts per million) to a level of 560 parts per million.

uplift weathering hypothesis The hypothesis that tectonic-scale climate changes are caused when uplift of plateaus and mountains alters the amount of CO$_2$ removed from the atmosphere by chemical weathering of fragmented rock.

upwelling The rise of cool, nutrient-rich subsurface water to the ocean surface to replace warm nutrient-poor surface water.

urban heat island An urban area where asphalt and other heat-absorbing surfaces absorb solar radiation during the day and radiate it back at night, keeping the area unusually warm.

varves Alternating layers of dark and light (or coarse and fine) sediment that accumulate in annual couplets in lakes or in ocean margin basins with no water turbulence near the bottom.

vegetation-albedo feedback A positive feedback that amplifies an initial temperature change through vegetation changes that alter Earth's surface albedo and the absorption of solar radiation.

vegetation-precipitation feedback A positive feedback that amplifies an initial precipitation change through vegetation changes that alter the transpiration of water vapor and the availability of moisture.

viscous Characterized by a consistency such that the material in question deforms slowly under pressure (as is bedrock under the pressure of a glacier) and only slowly regains its initial shape when the pressure is removed.

warm saline deep water Deep water proposed to have formed in warm salty tropical seas in the past, in contrast to modern sources of deep water in cold polar regions.

water vapor feedback A positive feedback that amplifies an initial change in temperature by altering the amount of water vapor held by the atmosphere and changing the amount of Earth's back radiation trapped by the atmosphere.

wavelength The distance between successive peaks or troughs in a series of regular wave forms.

weather Fluctuations in temperature, precipitation, and winds on time scales of less than a year.

Younger Dryas An interval during the middle of the last deglaciation (near 12,000 years ago) marked by slower melting of ice sheets and a major cooling in the North Atlantic region.

INDEX

Page numbers in *italic* indicate figures; those followed by b indicate boxed material; those followed by t indicate tables

373